Lecture Notes in Electrical Engineering

Volume 831

The book series *Lecture Notes in Electrical Engineering* (LNEE) publishes the latest developments in Electrical Engineering - quickly, informally and in high quality. While original research reported in proceedings and monographs has traditionally formed the core of LNEE, we also encourage authors to submit books devoted to supporting student education and professional training in the various fields and applications areas of electrical engineering. The series cover classical and emerging topics concerning:

- Communication Engineering, Information Theory and Networks
- Electronics Engineering and Microelectronics
- Signal, Image and Speech Processing
- Wireless and Mobile Communication
- Circuits and Systems
- Energy Systems, Power Electronics and Electrical Machines
- Electro-optical Engineering
- Instrumentation Engineering
- Avionics Engineering
- Control Systems
- Internet-of-Things and Cybersecurity
- Biomedical Devices, MEMS and NEMS

For general information about this book series, comments or suggestions, please contact leontina.dicecco@springer.com.

To submit a proposal or request further information, please contact the Publishing Editor in your country:

China

Jasmine Dou, Editor (jasmine.dou@springer.com)

India, Japan, Rest of Asia

Swati Meherishi, Editorial Director (Swati.Meherishi@springer.com)

Southeast Asia, Australia, New Zealand

Ramesh Nath Premnath, Editor (ramesh.premnath@springernature.com)

USA, Canada:

Michael Luby, Senior Editor (michael.luby@springer.com)

All other Countries:

Leontina Di Cecco, Senior Editor (leontina.dicecco@springer.com)

**** This series is indexed by EI Compendex and Scopus databases. ****

More information about this series at https://link.springer.com/bookseries/7818

Mikhailo Klymash · Mykola Beshley ·
Andriy Luntovskyy
Editors

Future Intent-Based Networking

On the QoS Robust and Energy Efficient Heterogeneous Software Defined Networks

Springer

Editors
Mikhailo Klymash
Lviv Polytechnic National University
Lviv, Ukraine

Mykola Beshley
Lviv Polytechnic National University
Lviv, Ukraine

Andriy Luntovskyy
Saxon Study Academy
BA Dresden University
of Cooperative Education
Dresden, Germany

ISSN 1876-1100 ISSN 1876-1119 (electronic)
Lecture Notes in Electrical Engineering
ISBN 978-3-030-92433-1 ISBN 978-3-030-92435-5 (eBook)
https://doi.org/10.1007/978-3-030-92435-5

This Springer imprint is published by the registered company Springer Nature Switzerland AG
The registered company address is: Gewerbestrasse 11, 6330 Cham, Switzerland

Preface

This volume is a collection of the most important research for the future intent-based networking deployment provided by different groups of researchers from Ukraine, Germany, Slovak Republic, Switzerland, South Korea, China, Czech Republic, Poland, Brazil, Belarus and Israel. The authors of the chapters from this collection present in depth extended research results in their scientific fields.

The volume consists of 28 chapters.

Chapter 1 presented by **M. Beshley, M. Klymash, H. Beshley, O. Urikova and Y. Bobalo** *Future Intent-Based Networking for QoE-Driven Business Models*, describes that, until now, business process management had been the driving force behind optimization and operational efficiency for companies. But the digital age that we are experiencing requires companies to be agile and responsive. To be part of this digital transformation, a new level of network automation is needed. These authors proposed a conceptual model for the construction of a heterogeneous software-defined intent-based network. The basic principle of the proposed IBN is to translate the information business intentions of users into appropriate network configurations for all devices based on network analytics and machine learning. One of the main types of intent refers to the specification of services. The service-level agreement (SLA) model is an example of service-specific intent used at different operation stack levels. The authors proposed a gradual transition from traditional SLAs to new experience-level agreements (ELAs) necessary to implement the concept of IBN for digital business models. So, the quality of experience (QoE) business aspect will be one of the media value chains of future networking.

Chapter 2 *Designing HDS under Considering of QoS Robustness and Security for Heterogeneous IBN* by **A. Luntovskyy and M. Beshley** focuses on designing paradigms for intent-based highly distributed system (HDS). A suitable framework for building HDS is often service-oriented architecture (SOA) and micro-services (MS), which combine agile development processes and flexible interaction models. HDS is systematically involved in the creation of artificial intelligence (AI). The neural networks and machine learning (ML) technologies are extensively integrated into the internal structures of HDS. A key goal of ML for HDS is the ad hoc deployment of certain workflow steps and learning algorithms that are performed without human intervention. The time factor and the amount of data required for learning are among the most important performance and quality of service (QoS) metrics for applications and HDS. The authors analyze how secure IBN solutions are in realistic scenarios using techniques such as VPN, PKI/ TLS, blockchain in desktop and mobile environments.

In Chapter 3 *Intent-Based Placement of Microservices in Computing Continuums*, **J. Spillner, J. Freitag Borin and L. Fernan-do Bittencourt** explore the programmable computing infrastructure that is becoming increasingly available in heterogeneous devices and data centers. This greater choice leads to the ability to run applications and network services on top of them with an improved match to required or desired performance. The remaining challenge is to account for computing resources without forcing software engineers to reflect them directly into the software design. In this chapter, the authors present continuum computing scenarios and describe the current state of the technology. In addition, the authors demonstrate how application placement can be managed declaratively based on meaningful human and business vocabulary within this intent-based abstraction. This method promotes application portability and resilience, important characteristics for the effective digital transformation of entire industries. Once knowledge of applications and resources is gained, it becomes possible to create seamless software that adapts to existing infrastructure. This is especially relevant for the initial deployment of software units, in particular microservices, as well as for any migration due to changing conditions. The authors proposed various strategies for deploying microservices in the future intent-based computing continuums.

In Chapter 4 *Infrastructure as Code and Microservices for Intent-Based Cloud Networking*, **M. Kyryk, N. Pleskanka, M. Pleskanka and V. Kyryk** present a new mechanism for deploying, managing infrastructure and creating and delivering microservices for future intent-based cloud networks. The process of building infrastructure is similar to the software programming process, where some scripts, modules, providers and version control systems are used together. The processes of building serverless microservices and how to create new content, reduce maintenance, scale easily and deliver new features to users faster have been explored. The main advantage of serverless platforms is that they allow you to focus on writing code without worrying about infrastructure management, auto-scaling or paying for more than you use. With Cloud Functions and Cloud Run, you can create high-quality microservices that improve the performance of your application or site. Cloud Run and Cloud Functions are serverless platforms offered by Google Cloud, but they have nuances that can make one preferable to the other in certain situations. The unique advantages and disadvantages of each platform are explored. To ensure continuous integration and deployment of applications and infrastructure, enterprises will need DevOps tools in the future, which include intent-based networks (IBNs). Numerous manual tools and automated IT operations platforms are being replaced by artificial intelligence (AI), machine learning (ML) and network orchestration. User intentions are defined in human language, so cloud service providers must translate them into IT policy using natural language processing (NLP) to ensure the quality of service (QoS)/quality of experience (QoE).

Chapter 5 *Intent-based Adaptation Coordination of Highly Decentralized Networked Self-Adaptive Systems* by **I. Shmelkin, D. Matusek, T. Kluge, T. Springer and A. Schill** proposes the concepts and approaches of self-adaptive systems as a promising solution to serve as a basis for designing and implementing intent-based networked systems. Particular attention is paid to role-based concepts that allow continuous design and implementation of system variability at runtime. An example scenario from the IoT domain is used to continuously illustrate the concepts and demonstrate how networked self-adaptive systems can benefit from the role-based concepts introduced.

In Chapter 6 *Intent-Based Routing in Delay- and Disruption-Tolerant Network*, **F. Walter, J. Irigon de Irigon, O.de Jonckère and T. Springer** propose the architecture of delay and disruption-tolerant networks (DTNs), which provides communication between nodes in networks with no continuous end-to-end connectivity. This introduces the bundle protocol, which encapsulates application data and allows it to be transmitted over heterogeneous channels with forwarding and saving. Although a myriad of routing algorithms has been proposed for DTNs, in current deployment scenarios, they are applied in a non-adaptive manner and are often configured statically for the entire network. With the emergence of intent-based networking technologies, it becomes plausible that the DTN domain can benefit tremendously from the portability of these concepts. In this context, the authors note the close connection between intention-based networks and existing work on self-adaptive systems. Based on this, methods for providing adaptive routing in DTNs are described and future predictions for improving node configuration and routing in DTNs using intent-based networking concepts are provided.

In Chapter 7 *QoE-Oriented Routing Model for the Future Intent-Based Networking*, **A. Pryslupskyi, M. Beshley, H.Beshley, Y. Pyrlh and A. Branytskyy** propose an IBN concept model using the northbound interface (NBI) of the SDN architecture to declare customer intentions. Using the proposed IBN architecture, authors offer the ability to accept incoming data from end users, configure networks according to customer intentions, verify the correct design, implement the necessary network configurations and then continuously monitor the execution of system intentions and make changes as needed. The quality of experience (QoE) intents in this approach are set as a score from 1 to 5, where the highest score means the best quality of service. The intents are then passed to the SDN controller and automatically translated by developed IBN manager into pre-assembled network policies in the form of QoE routing rules. The proposed QoE-aware routing implemented in the IBN system was compared and evaluated against the default routing system in traditional SDN, and the new system resulted in much less packet loss than the default routing and hence much higher video quality.

Chapter 8 *Complex Investigation of the Compromise Probability Behavior in Traffic Engineering Oriented Secure Routing Model in Software-Defined Networks* by **O. Lemeshko, O. Yeremenko, M. Yevdokymenko, A. Shapovalova and**

O. Baranovskyi is devoted to the study and analysis of the results of the behavior of the trade-off probability in a secure routing model in software-defined networks oriented to traffic engineering (TE). In the context of the study, the classical flow model, based on load balancing in accordance with the principles of the traffic engineering concept, has been enhanced and augmented with conditions that allow for the consideration of network security parameters in the process of obtaining a routing solution. Thus, the secure routing model TE was obtained, the novelty of which lies in the modified conditions of load balancing taking into account network characteristics such as topology, features of transmitted traffic, as well as bandwidth and probability of its compromise. This model reduces congestion on channels in a network with a high probability of compromise, while allowing more traffic to pass through secure channels without causing congestion. For comparison, the power and exponential forms of the functional dependence of the weight coefficients on the channel compromise probability were used to derive secure routing solutions. The investigated flow routing model based on secure traffic engineering is proposed for use in the data plane of a software-defined network.

Chapter 9 *Intelligent Traffic Engineering for Future Intent-Based Software-Defined Transport Network* by **V. Andrushchak, M. Beshley, L. Dutko, T. Maksymuk and T. Andrukhiv** addresses traffic engineering (TE) in the future software-defined infrastructures using machine learning (ML) and neural network techniques. The software-defined networking (SDN) architecture can be used to implement intent-based networks (IBNs), allowing the automation of network management tasks using elements of artificial intelligence (AI) and ML. An intent-based optical transport network infrastructure adapted to the use of SDN-based intelligent TE algorithms and optical label switching (OLS) technology is proposed. An algorithm for determining the states of the intention-based software-defined transport network (IBSDTN) based on ML k-means and c-means algorithms is proposed. An intelligent TE method using graph neural networks is proposed to provide the required quality of service (QoS) parameters based on user intentions during peak hours. Using a vector of network parameters, which also takes into account the energy consumption parameter, a given algorithm manages network resources to provide the necessary QoS parameters.

Chapter 10 *The Approach to Flow Management in Virtual Computational Environment for Up-to-Day Telecom Networks* by **L. Globa, M. Skulysh, D. Parhomenko and K. Yakubovska** proposes an "endless train" method to reduce the decision time for load balancing for the future SDN-based network architecture. This method, instead of analyzing the state of the input stream and simultaneously the state of the resources, analyzes only the state of the computational resources in order to decide on the necessary resources based on the current task requirements. This reduces the decision time to select the server serving the input stream. The chapter proposes a solution to the problem of organizing the application flow management system and its redistribution among the available computing resources. To verify the effectiveness of the proposed method, an implementation scheme using MS Azure was developed. The process of dynamic deployment of additional virtual servers (virtual machines) to handle flows in case of congestion was tested. The test results show the effectiveness of overload prevention, but it increases the consumption of computing resources.

Chapter 11 *Calculation of Quality Indicators of the Future Multiservice Network* by **B. Zhurakovskyi, S. Toliupa, V. Druzhynin, A. Bondarchuk and M. Stepanov** considers the main quality indicators of a future multiservice network, such as: delay; delay variation (jitter); number of packets with errors; and number of lost packets. Not all types of traffic are sensitive to packet delays, at least not to those delays that are typical of multiservice networks. The purpose of this study is to assess the quality of the multiservice network through simulation. The analysis of the behavior of the network when it receives different types of traffic, real-time traffic, data traffic and mixed traffic, was carried out. The obtained results showed that the most critical to changes in the parameters of the next-generation multiservice network is such a quality indicator as the delay.

In Chapter 12 *Intelligent Detection of DDoS Attacks in SDN Networks*, **N. Peleh, O. Shpur and M. Klymash** propose intelligent DDoS attack detection in SDN networks based on log analysis. Using SDN management and introducing a self-learning element, it is proposed to teach the SDN controller to detect attacks using information about the flow state, session duration and its source, using information from logs and flow tables. For this purpose, it is necessary to divide the total traffic flow into abnormal and normal. By identifying repeated client requests that are the result of a DDoS attack, it is possible to create appropriate rules to block them. To do this, the authors propose to define traffic behavior metrics using the Kullback–Leibler approach to identify flow anomalies over the course of a session. By using machine learning, the SDN controller will block the IP domains from which DDoS attacks are just initiated.

Chapter 13 *Mathematical Methods of Reliability Analysis of the Network Structures: Securing QoS on Hyperconverged Networks for Traffic Anomalies* by **N. Kuchuk, A. Kovalenko, H. Kuchuk, V. Levashenko and E. Zaitseva** considers the approach to QoS provisioning in hyperconverged networks with traffic anomalies. Analytical dependencies for calculating the statistical characteristics of traffic by its samples are proposed. The authors prove that all considered statistical characteristics are uniquely defined by means of only three parameters: fractal exponent; traffic process intensity; fractal tuning. The mathematical model of the anomalous traffic is offered. The model is adequate to real traffic and takes into account the fractal nature of the anomaly. The model uses the properties of scale invariance. Packet losses are compensated by an increase in message transmission time, which leads to the formation of long statistical time dependencies. In the resulting model, the effect of losses and the cause of extended dependencies are formally accounted for by introducing a fractional integration operation. The anomalous traffic model of the hyperconverged system was used to construct a short-term forecast. The prediction is fed into the system hypervisor and used to quickly reallocate resources. Experimental demonstration of QoS provisioning in a Cisco HyperFlex HX220c M4 Node hyperconverged system network during traffic anomalies uses the proposed approach.

Chapter 14 *Parametric Analysis of Statistical and Correlation Characteristics of Discrete Processes in Dynamic Systems with Non-Stationary Nonlinearities in Time for the Secure Intent-Based Networks* by **V. Dubrouski, A. Semenko, M. Kushnir and M. Steita** presents the results of numerical modeling of dynamic systems with nonlinear feedbacks based on first-degree polynomials with constraints on the dynamic range of possible values and nonstationary in time nonlinearities with two and three degrees of freedom. Requirements for generating sequences with acceptable auto- and intercorrelation functions for use in information transmission and security systems have been determined. Practically realizable structural and functional schemes of signal generation devices have been proposed. The regions of nonlinear system parameters, in which discrete processes with given spectral time and statistical characteristics are formed, are defined.

Chapter 15 *Methodology of ISMS Establishment against Modern Cybersecurity Threats* by **V. Susukailo, I. Opirsky and O. Yaremko** discusses an approach to creating an information security management system (ISMS) that provides the necessary controls to prevent currently widespread cybersecurity threats, including in the future intent-based networks. Authors analyze the most common attack vectors and techniques over the past three years to identify a set of information security practices that can minimize the risks associated with today's cybersecurity threats. An analysis of cybersecurity systems such as ISO 27001/2, CIS Top 18, NIST 800-53 and their distinguishing features was conducted. This chapter proposes an algorithm for creating an ISMS with a detailed explanation of each step and the controls needed to implement the system. The document defines cybersecurity technologies for management systems, which are defined according to the type of infrastructure. A document management structure and risk management methodology are proposed, current awareness strategies are analyzed, and a roadmap for training roles in the ISMS is defined.

Chapter 16 *QoE Estimation Methodology for 5G Use Cases* by **R. Odarchenko and T. Dyka** considers the problems of functioning of quality assurance systems in the fifth-generation networks as well as the importance of quality of service in wireless and mobile networks and analyzed the main features of 5G technologies. Standard definitions and the most important developed measurement methods are given. The chapter demonstrates significant improvements and approaches to service quality control to meet user experience expectations.

In Chapter 17 *Software Implementation Research of Self-Similar Traffic Characteristics of Mobile Communication Networks* by **I. Strelkovskaya, I. Solovskaya, J. Strelkovska and A. Makoganiuk**, a self-similar traffic characteristic study was carried out for a queueing system (QS) of the form WB/M/1/∞, which simulates the servicing of self-similar traffic using a two-parameter Weibull distribution. Using the Laplace-style transform, an analytical expression for finding the qualitative characteristics of self-similar traffic is obtained, for which software solutions based on Python are proposed. The obtained results will allow at the stage of planning, design and further operation of mobile communication networks to choose the configuration of connections between base stations by the criterion of average waiting time and in the real processes of network operation to take into account its construction.

In Chapter 18 *Universal Method of Multidimensional Signal Formation for any Multiplicity of Modulation in 5G Mobile Networks*, **Berkman, L. Kriuchkova, V. Zhebka and S. Strelnikova** propose a new method of forming a multidimensional signal with amplitude phase difference modulation (MAPDM signal) OFDM technology for 5G mobile networks. This method allows to increase the noise immunity of the reception in 2 times compared to the two-dimensional OFDM signals. Information parameters of the MAPDM signal are the amplitude, phase and temporal distance between the boundaries of the parcels, as well as the interval of signal integration. Improved noise immunity is achieved by increasing the equivalent energy of the signal determined by the distance between the two nearest points of the signal, thus increasing the resolution of the receiver. The use of MAPDM signal allows to approach the rate of information transfer to the channel capacity, which is necessary for the implementation of mobile 5G networks.

In Chapter 19 *AI-Enabled Blockchain Framework for Dynamic Spectrum Management in Multi-Operator 6G Networks*, **T. Maksymyuk, J. Gazda, M. Liyanage, L. Han, B. Shubyn, B. Strykhaliuk, O. Yaremko, M. Jo and M. Dohler** propose the fusion of blockchain and AI technologies to address the issue of multi-operator spectrum management in decentralized 6G deployment. The authors provide a framework for applying the blockchain and AI to the overall 6G network management by the unlimited number of mobile network operators via the distributed ledger infrastructure. A particular case of the spectrum management using deep learning is proposed, and some simulation results are outlined for determining the efficiency of the proposed framework in terms of resource allocation among multiple operators with different bandwidth demands.

In Chapter 20 *Estimation of Energy Efficiency and Quality of Service in Cloud Realizations of Parallel Computing Algorithms for IBN*, **I. Melnyk and A. Luntovskyy** propose and discuss the main approach for increasing the efficiency of paralleling algorithms based on the use of arithmetic logic relations and the theory of recurrence matrices, a method for computing computer cooling systems based on solving the Boltzmann thermodynamic balance equation and a comparative analysis of RS codes and convolutional error correction codes. Computational examples are also given to illustrate the discussed methods. Thus, the chapter defines and justifies a combined integrated approach for intent-based networking (IBN). Quality of service (QoS) parameters, such as better performance, security, data rate and latency, are fully guaranteed here through the implementation of parallel computing in cloud environments.

Chapter 21 *Modeling of 5G Energy Efficiency on Example of Germany as Technological Basis for Intent-Based Networking* by **D. Wasiutinski and V. Vasyutynskyy** examines whether the new 5G network, which must be not only fast but also energy efficient, would be a reasonable solution for this. In this regard, the authors compare the existing 3G and 4G mobile networks with the new 5G network in terms of workload and energy efficiency based on various options for the development of data volumes in German mobile networks. The authors estimate the energy consumption, future data consumption and especially the energy efficiency of the current networks and the fifth generation of mobile networks for the coming years in Germany by statistical analysis using various models and extrapolations. The result shows that 5G is indeed more energy efficient than 4G, but only for a certain amount of data per mobile cell. Nevertheless, with the growth of mobile traffic in the coming years, 5G will clearly lead to improved energy efficiency compared to previous technologies.

Chapter 22 *Methods of Signal Detection and Recognition to Perform Frequency Resource Sharing in Cognitive Radio Networks* by **V. Bezruk, S. Ivanenko, O. Fedorov and Z. Nemec** considers the problem of spectrum sensing in cognitive radio networks and poses the problem of improving the efficiency of signal detection procedures by using non-traditional methods of signal detection and recognition. Such methods make it possible to assign unknown signals to a special class of signals for which no prior information is provided. The chapter is devoted to the study of algorithms based on methods for detecting changes in the probabilistic properties of signals. Studies are conducted on samples of real signals, typical for both VHF/UHF and IEEE 802.22 frequency bands.

Chapter 23 *Model of Increase of Spectral Efficiency of Use of Frequency Resource of Low-Orbit System with Architecture of the Distributed Satellite* by **V. Saiko, S. Toliupa, V. Nakonechnyi, M. Brailovskyi and V. Domrachev** is devoted to a review of the known ways to improve the efficiency of OFDMA mobile communications systems. To improve the spectral efficiency of using the frequency resource of a low-orbit satellite network with a distributed satellite architecture a model proposed for cognitive multiuser access with OFDMA.

This model includes a block to determine the required number of frequency subchannels, as well as a new algorithm for determining distorted carriers and an algorithm for choosing the OFDMA (RU) working frequency section for the relevant frequency channels. To assess the effectiveness of the developed model, simulation of channels with a bandwidth of 20 MHz under the influence of interference with a bandwidth of 1 MHz. It follows that compared to the original ITU algorithm, the proposed RU suppression method gives a significant simulation gain, namely the efficiency of subcarriers has increased significantly compared to the original algorithm.

In Chapter 24 *Optical Signal Decay and Information Data Loss in Wireless Atmospheric Communication Links with Fading*, **I. Bronfman, I. Juwiler, N. Blaunstein and A. Semenko** investigate optical signal attenuation based on the effects of attenuation and scattering of gaseous structures and hydrometeors (rain, snow and clouds), as well as turbulent structures, which have a predominant influence on the rapid attenuation of optical signals propagating through atmospheric channels with attenuation. The signal data parameters including bandwidth, spectral efficiency and bit error rate (BER) have been analyzed to predict and improve QoS taking into account all the features occurring in atmospheric communication links. An optimal algorithm was found to predict the total signal attenuation considering the different meteorological conditions occurring in the real atmosphere at different altitudes and different frequencies of radiated signals.

Finally, a method was proposed to estimate the data flow parameters: throughput, spectral efficiency and BER, taking into account the effects of rapid attenuation of atmospheric turbulence, which distorts the information signals transmitted via such channels.

Chapter 25 *Control Methods Research of Indicators for Intelligent Adaptive Flying Information-Telecommunication Platforms in Mobile Wireless Sensor Networks* by **L. Uryvsky, O. Lysenko, V. Novikov and S. Osypchuk** is devoted to the formulation of the problem of controlling an intelligent adaptive network of flying information and telecommunication platforms (FITPs). The chapter considers features of construction and functioning of mobile wireless sensor networks (MWSNs) with FITP. An approach to creating new architectural, algorithmic and technical solutions for building intelligent control systems based on MWSN with FITP capabilities is proposed. The paper analyzes the construction methods and protocols of intelligent adaptive FITP. The concept of intelligent adaptive FITP control for application in emergency protection zones or critical infrastructure is developed. Methods to increase the throughput capacity of MWSN with FITP are investigated, and a mathematical model is presented. The methods for increasing channel interference immunity for MWSNs with FITP are analyzed. A general statement of the problem of investigating the interference immunity of MWSN channels with FITP based on the creation of a mathematical model is presented. Quality indicators of interference immunity of MWSN channels with FITP are presented. A mathematical model for interference immunity research and comparison for wireless communication systems in Gaussian and Rayleigh channels is proposed.

Chapter 26 *Technologies for Building Intelligent Video Surveillance Systems and Methods for Background Subtraction in Video Sequences* by **A. Babaryka, I. Katerynchuk and O. Komarnytska** is devoted to the analysis and improvement of video analytics functions in intent-based video surveillance systems in order to improve the detection efficiency of dynamic objects in video surveillance sectors. It is found that video analytics methods using background subtraction and object recognition methods have significant drawbacks, namely: Algorithms cannot distinguish an object from the background when contrast is low; some moving objects can be recognized as background; algorithms are critically dependent on lighting conditions, etc. Thus, the aim of the research is to improve the method of detecting dynamic objects in video sequences using background subtraction methods based on pixel analysis of frames using elements of expert systems theory. An intelligent video surveillance system (IVSS) based on intention is a video surveillance system with the ability to automatically analyze data from the video cameras and perform necessary tasks, such as generating alarms or warnings. Users can create the tasks themselves in such a system in the form of their intentions and transmit them to the main center of the intelligent system.

Chapter 27 *An Ontological Approach to Detecting Non-Relevant Information on Web-Resources and Social Networks* by **M. Dyvak, A. Melnyk and S. Mazepa,** addresses the important scientific and applied problem of identifying irrelevant and unreliable information on Web resources, which is an important direction for the development and implementation of methods of data mining in the future intent-based social networks. The analysis of modern methods and tools for the assessment of irrelevant and unreliable information is provided in terms of assessing sources of information. This analysis highlights the main problem areas, which arise in the process of their operation. The system of indicators for filtering unreliable and irrelevant information, which is derived from several sources, has been proposed. On the basis of this system, the method of checking the information from Web resources for relevance and reliability is implemented. This approach is based on the possibility of using a predetermined resource, the data from which is only reliable. The method of revealing unreliable and irrelevant information, taking into account the peculiarities of its distribution through the relevant pages in social networks and the use of multitask classification of information obtained from different data sources, was developed. The use of the proposed intelligent methods of data processing together with other methods of intellectual analysis used to evaluate the information obtained from the Internet will significantly improve the efficiency of the process of establishing the irrelevance and unreliability of information, as well as build an assessment of a particular Web resource for publication and distribution of such information.

Chapter 28 *Application Peculiarities of Deep Learning Methods in the Problem of Big Datasets Classification* by **B. Rusyn, O. Lutsyk, R. Kosarevych and Y. Obukh** is very important for the development of future intent-based network, because the learning systems (popularized by deep learning and neural networks) in the future intent-based networks are able to learn without depending on user programming or articulation of rules. However, they need a learning or training phase under required big datasets. In machine learning tasks, there is a correlation between the amount of data used to train a model and its subsequent accuracy with both test and control data. Most often, it arises in the problem of lack of data to create a quality sample of training. A representative sample of training is largely responsible for the correct training of the model in the classification. There is no universal approach that gives an unambiguous answer to the question of how much data and what size is needed to train a particular model with predictable accuracy. The authors optimize the total time spent on the creation of the training sample, its building and evaluation, as well as make it possible to automate the process of creating a quality training sample for arbitrary classification models. The authors proposed a new method for estimating the quality of a training sample of big datasets has been developed. The method is based on the assumption that the quality of a training sample can be represented by a set of a finite number of characteristics, each of which describes certain properties of data. The correlation between the characteristics of a training sample and the accuracy of a classifier trained on the basis of this sample is established using a linear regression model.

To reduce the computational complexity, it was proposed to use a method of data dimensionality reduction without loss of data structure, based on minimization of the Kullback–Leibler distance that is very vital for intent-based network performance. This allowed us to move to the construction of the characteristics of the training sample with much less computational cost and a compact representation of the feature space.

The experiments obtained on various test training samples showed that this method gives results comparable to those obtained by training a neural network. At the same time, the estimation time of the training sample by the proposed method increases the speed of obtaining the result by order of magnitude. This allows us to use it effectively for the preliminary estimation of the training sample, which makes it possible to adjust its size before training the network on big datasets.

July 2021

Mykhailo Klymash
Mykola Beshley
Andriy Luntovskyy

Introduction

So-called intent-based networking (IBN) is founded on well-known software-defined networking (SDN) and represents one of the most important emerging network infrastructure opportunities. The IBN is the beginning of a new era in the history of networking, where the network itself translates business intentions into appropriate network configurations for all devices. This minimizes manual effort, provides an additional layer of network monitoring and provides the ability to perform network analytics and take full advantage of machine learning. The centralized, software-defined solution provides process automation and proactive problem solving as well as centralized management of the network infrastructure. With software-based network management, many operations can be performed automatically using intelligent control algorithms (artificial intelligence and machine learning). As a result, network operation costs, application response times and energy consumption are reduced, network reliability and performance are improved, and network security and flexibility are enhanced. This will be a benefit for existing networks as well as evolved LTE-based mobile networks, emerging Internet of things (IoT), cloud systems and soon for the future 5G/6G networks. The future networks will reach a whole new level of self-awareness, self-configuration, self-optimization, self-recovery and self-protection.

Systems integration will require an interdisciplinary approach as many technologies come together for future network deployment, including machine learning, SDN, blockchain, artificial intelligence, network functions virtualization (NFV), network slicing, quality of service management, cloud computing and advanced security methodologies (Fig.1). With the emergence of new services and applications, 5G/6G intent-based networks are investigated to satisfy a wide range of users and meet their needs in terms of end-to-end latency, reliability and scalability, to promote the use of mobile devices and provide flexible and efficient network connectivity. SDN and NFV, on the other hand, can be considered key technologies for transforming current networks into software-defined networks where 5G/6G goals are achieved. SDN provides a programmability feature that is essential to facilitate 5G network operation while reducing operational costs. NFV allows network resources to be virtualized and decoupled from hardware platforms,

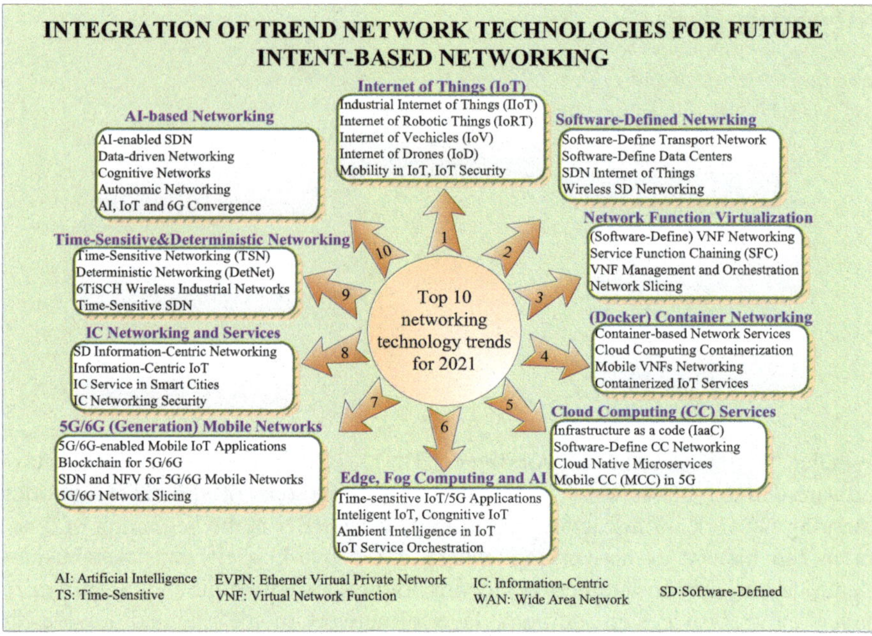

Fig. 1. Top 10 networking technology trends used in series for "future intent-based networking"

facilitating programmatic enhancement of network functionality, thereby making the network easier to deploy and more adaptable to possible changes.

Future intent-based 6G networks will require precise spectrum management, due to many available spectral bands with different characteristics and much higher number of mobile network operators with different business models and types of offered services.

With the development of network systems, customer needs and behavior have changed. The focus shifts from improving network performance to improving the perception of quality of experience (QoE). Providing according to the intentions of users of a given level of QoE for services and applications becomes a fundamental task in the implementation of end-to-end resource management in the concept of IBN. The main idea of using IBN is to change the paradigm of the network infrastructure: Now it is not the user with his application that adapts to the capabilities of the network, but the network changes its settings according to user requirements. Thus, for information systems based on IBN technology, the task of system administrators goes from manual configuration to programming and defining an intelligent network development strategy.

In this series, we provide a more detailed view of our own concept of IBN, which is built on SDN with subsequent network virtualization and software implementation of network functions with a significant improvement in service quality, which feels like the usual quality of experience (QoE), which can be every time monetized. The protection against usual risks for networked data from intrusions is guaranteed. Such networks already create convenient conditions for reducing their cost and cost of their operation (the expenditures, namely CAPEX and OPEX). In the future, IBN guarantees for energy efficiency, optimal power usage effectiveness (PUE) and minimizing CO_2 emissions (CO_2 footprint). Intent-based networking raises new challenges for researchers to investigate, redesign and develop SDN-based future networks in the 5G/6G era.

Mykhailo Klymash
Mykola Beshley
Andriy Luntovskyy

Acknowledgment

The Editorial Team thanks many colleagues and our good friends for the assistance and inspiration by preparing the manuscript, especially to Prof. Dr. habil. Igor Melnyk with NTUU "KPI I.Sikorsky" (Kyiv), PhD student Yuliia Pyrih and Dr. Halyna Beshley, both with LNPU "Lvivska Politechnika" (Lviv).

Contents

Future Intent-Based Networking for QoE-Driven Business Models

Mykola Beshley(✉), Mykhailo Klymash, Halyna Beshley, Oksana Urikova, and Yuriy Bobalo

Lviv Polytechnic National University, Lviv 79013, Ukraine
{mykola.i.beshlei,mykhailo.m.klymash,halyna.v.beshlei, oksana.m.urikova,yurii.y.bobalo}@lpnu.ua

Abstract. The Quality of Experience (QoE) business aspect will be one of the media value chains of the future networking. Therefore, new Service Level Agreements (SLAs) such as Experience Level Agreements (ELAs), which are based solely on QoE, should be the main pinnacle of future intent-based networking (IBN) projects to improve business and customers quality. Moreover, we present intent-based network management using Software-Defined Network (SDN) perspective in relation to QoE-business aspects. In this chapter we proposed a conceptual model for the construction of a heterogeneous software-defined intent-based network. This model allows providing effective distribution and redistribution of common resources adapting to the changing requirements of business customers regarding the Quality of Service (QoS) provision. It is proposed to use a comprehensive indicator of QoS for users, formed in the form of QoE assessment. This is the main criterion for adaptive management of resource reallocation in the context of changes in the importance of business processes in the IBN concept implementation. The proposed model of IBN allows to guarantee the ordered level of service by analyzing QoE estimates of users according to the new ELA contract. The model also makes use machine-learning capabilities to manage the network in response to changing business requirements. They are used to regulate and perform routine tasks, adjust policies, respond to system events, and verify that necessary goals are met and actions are taken. The system not only configures changes to the network, but also enforces them and makes the necessary adjustments. In addition, such a system is considered the next stage in the development of SDN and is based on the principles of intelligence and software infrastructure to provide a higher level of analysis and determine which tasks need to be automatized.

Keywords: Quality of experience (QoE) · Intent-based network (IBN) · Service level agreements (SLAs) · Experience level agreements (ELAs) · Software-defined network (SDN) · Quality of service (QoS)

1 Introduction

New information and communication technologies are increasingly penetrating all spheres of life. Their active use in business is caused primarily by the improvement

M. Klymash et al. (Eds.): *Future Intent-Based Networking*, LNEE 831, pp. 1–18, 2022.
https://doi.org/10.1007/978-3-030-92435-5_1

of business management and control systems [1]. First of all, this involves automation of business processes, such as introducing electronic document management, organization of video and audio content, provision of online services to the public and customers [2]. These in turn led to the emergence and spread of the concept of "digitalization" as a process of transferring information into digital form. Unfortunately, existing information systems and business processes in the digital transformation are no longer effective, old methods of communication are experiencing transformations, models and consumer behavior are changing [3]. Customers are becoming more and more important for adaptive service delivery, constant communication, individualization of offers from companies [4]. Business organizations, in turn, are interested in finding new ways to optimize their business processes and improve efficiency and competitiveness [5]. Currently, there is no unified concept of creating a network infrastructure management system. And also, there are no clear standards for the management of the redistribution of network resources. Modern resource management models and such traditional functions as monitoring and analysis network performance must also solve such vital problems as adaptive management of resource allocation and redistribution in info-communication systems under conditions of their constraints [6–8].

The relevance of this problem is due not only to the fact that always in the process of network functioning may arise a conflict when several users turn to the same service resource. But also, the fact that in any info-communication network, sooner or later, there is a situation when network resources become limited, and one of the services has to be preferred [9]. Limited resources can arise with the emergence and introduction of new services without installing additional physical resources (productive server equipment, router, switch), functional breakdown or maintenance of servers and network equipment, increased cost of maintenance, and reduced consumption of resources provided, for example, by telecommunications service providers [10]. Also temporary, but excessive consumption of resources by one of the services can lead to increased congestion on communication channels, network devices, and servers, etc. In the process of adaptive resource management, info-communication systems should be guided by the importance of all business processes in the infrastructure of corporate enterprises, respectively, and the types of services responsible for the efficiency of running the business processes themselves [11]. The system should also be working on analyzing the priority of service requests and assessing their needs in the overall resources of the info-communications network [12]. Naturally, in the course of the everyday activities of any corporate enterprise, the significance of business processes can change. These changes manifest themselves at the organizational level, caused by changes in the business objectives of the enterprise and force majeure contingencies. In particular, the emergence of COVID-19 pandemic has led to a significant increase in information traffic and a lack of resources for its quality maintenance in all classes of network infrastructure and the like.

For this reason, an essential task for both telecommunications market players and scientists aimed at developing future networks is the development of an autonomous system of adaptive management of resource reallocation of info-communication systems. In particular, it is essential under conditions of limited available network resources when new service requests appear [13]. For its execution, services require part of the resources already used by other services. To do this, the future network model must take into account the importance of business services, both those that are running and new ones, as well as the possibility of changing over time the priorities concerning the allocation of network resources due to changes in the importance of the business processes they support [14]. Accordingly, the new resource management model should be able to diagnose and evaluate the functions and procedures of business processes and their relationship to rationally manage network resources adapted in the context of the development goals of the infrastructure itself [15].

2 Intent-Based Networking for Adaptive Business Process Management

Thus, with the development of business, diversity of services and user requirements to the quality of service (QoS), the concept of Intent-Based Networks (IBN) comes to the fore as a tool for intelligent management of info-communication networks. It allows to abstract away from the details of configuration and functioning of separate network elements and focus on the behavior of the whole network, as a system for providing service in accordance with the requirements and ensuring quality of service on the basis of users' intents [16–18]. The basic principle of IBN is to translate the information business intentions of users into appropriate network configurations for all devices based on network analytics and machine learning. Functional architecture of the IBN is depicted in Fig. 1. There are four essential components of the intent layer: the Knowledge Base (KB), the data storage, the reasoning mechanism, and the agent architecture [19]. The KB contains an ontology of intent and specific knowledge, such as the current state of the system. The Data monitors network objects and is used for efficient storage. The Data contains the topology of the network and inventory information. The Data is primarily responsible for forwarding updates to KB dynamically whenever new objects are deployed to the network, an object is removed, or a topology change occurs. The Data ensures that the topology of the network is modeled and communicated to the Agent. Updating the model with more accurate estimates or when the business intent changes is the most vital state of the Data. The Reasoning Mechanism uses the knowledge graph and provides a central coordinating function for finding actions, assessing their impact, and sorting their execution. The Agent architecture finally allows the use of any number of models and services. The Agents can include machine-learning-based models or rule-based policies or realize the services needed in the cognitive reasoning process.

For an Agent to be usable, it must be registered and described in a knowledge base. Its description can be added and changed at any time, allowing the lifecycles of models, policies, and additional services to be separated from the overall lifecycle of the cognitive layer.

The metadata of an agent consists of a description of the agent's interface and its functions, roles, and features. For example, we implemented a machine learning model that can suggest base station configurations that optimize the quality of service. This model is registered as an agent in the role of "suggesting" actions on configurations. A separate lifecycle allows the model to be replaced with an improved version as it becomes available, regardless of cognitive release cycles.

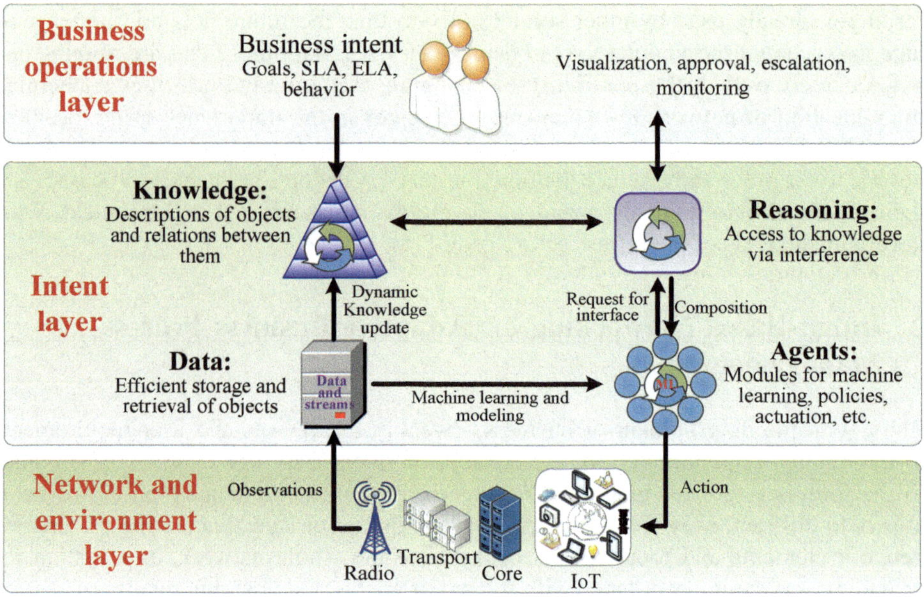

Fig. 1. Functional architecture of the IBN

We've also demonstrated agents in the "predictor" role with the capability to assess the impact of actions on Key Performance Indicators (KPIs). The agent in the "observer" role would monitor sources of data, maintaining up-to-date state knowledge. An agent in the "executor" role can perform actions on the network using, for example, the network management functions installed.

One of the main types of intent refers to the specification of services. Service-specific intentions define the expected functional and operational characteristics. The Service Level Agreements (SLAs) model is an example of service-specific intentions that are used at different levels in the operations stack [20].

SLAs are the objects of Business Support Systems (BSSs). SLA-based intentions define the promised service and include details of expected performance and business consequences, such as delivery charges and penalties in case of failure [21–23].

This scheme of making decisions based on a given intent and executing actions by sending intent to lower-level subsystems is the key mechanism for the interaction of intent-based operations, under which the entire operational stack of autonomous networks is built [24].

We propose an autonomous network resource management system that takes into account the importance of business services, both those that are running and new ones. This system provides the ability to re-prioritize the allocation of intermediate resources due to changes in the importance of the business processes they support. Accordingly, the new model of resource management has the ability to diagnose and evaluate business process functions and procedures and their interrelationships for the purpose of rational management of network resources adapted in the context of the development goals of the infrastructure itself.

To formalize the new model of adaptive redistribution of network resources, we introduce the following changes:

$E = \{e_i\},\ i = \overline{1,n}$ it is a space, the elements of e_i of which are the resources and services that are required in the IBN. The convergence of resources and services into a common space is done because information services may require a network resource, the allocation of which creates new network tasks for management. Space E includes all available resources of the IBN and all possible services that are transmitted in the system to ensure the operation of existing and planned business processes. In the process of adding network resources due to the expansion and modification of infrastructure or installation of new business services that were not provided, as well as in the case of irreversible removal of resources or services, the dimension of space E increases or decreases. Accordingly, to do this, we formalize the space in the form of two subspaces, namely:

$S = \{s_i\},\ i = \overline{1,n}$ it is a subspace of space E, the elements S_i of which are services that ensure the functioning of business processes of infrastructure (corporate enterprises);

$R = \{r_i\},\ i = \overline{1,n}$ is a subspace of space E, the elements r_i of which are responsible for the resources used by the services S_i.

To select resources or services from the space E, we introduce n-dimensional vectors $\overline{V}_r$ and $\overline{V}_s$, the first of which defines the elements of network resources, and the second - the elements of services. The elements of the vectors take the value 0 or 1, and 1 is set at those positions, for example, the vector $\overline{V}_r$ that correspond to the resource itself, and 0 - services.

We will set $w_1, \ldots, w_n$ a coefficients of importance of resources and services. These coefficients are determined by network management policies, and when policies change, they are subject to some adjustment to adapt to changing user requirements.

We set the matrix of the use of $C = \|c\|,\ i,j = \overline{1,n}$ resources, the c_{ij} element of which determines the normalized utilization factor of the e_i element of the e_j element.

Set the $\overline{X} = (x_1, \ldots, x_n)$ vector, which defines the services from the subspace S that are currently running:

$$x_j = \begin{cases} 1, & \text{if the service } S_j \text{ is under execution (transmission);} \\ 0, & \text{otherwise.} \end{cases} \quad (1)$$

Let's set the vector, which defines those services generated from the matrix P, execution (transfer), which can be canceled to free resources, or their use of resources can be reduced to ensure the management of the transmission of important new services that appear in the IBN:

$$t_j = \begin{cases} 1, & \text{if the service } S_j \text{ can be canceled;} \\ 0, & \text{otherwise.} \end{cases} \quad (2)$$

The maximum efficiency of the IBN will be ensured by a kind of centralized intelligent control system when the most important business process services required by users at a certain point in time of its operation will be performed, it is necessary to maximize the next target F function:

$$F = \max \sum_{i=1}^{n} w_i x_i. \quad (3)$$

This objective function (3) implies the achievement of the maximum total importance of services that are performed and planned to be performed taking into account the new values of importance factors, adjusted when changing network management policies.

Since the C_{ij} element in the C matrix is the normalized coefficient of use of the S_j service by the r_j resource, the total use of any resource cannot exceed one, the condition will always be fulfilled:

$$\sum_{i=1}^{n} p_{ij} \cdot x_i \leq 1 \tag{4}$$

to all $j = \overline{1, n}$.

Let a new service appear for a certain business process, $S_k^{'}$, $k = \overline{1, n}$, for which the $x_k = 0$ condition was previously fulfilled. If there are not enough resources to implement it, it is necessary to either free up resources to implement $S_k^{'}$, or decide that $S_k^{'}$ cannot be implemented because it is not important enough, given the resource constraint policy that is currently used in the centralized management system.

Determine the number of r_j free resources in the IBN:

$$r_j = 1 - \sum_{i=1}^{n} c_{ij} \cdot x_i \tag{5}$$

to all $j = \overline{1, n}$.

For each resource r_j required for a new service $S_k^{'}$, check the fulfillment of the condition:

$$r_j \geq c_{kj} \tag{6}$$

to all $j = \overline{1, n}$, and if this condition is met, it means that the IBN has a sufficient number of resources required to transfer the service $S_k^{'}$ additional management for the redistribution of resources is not required.

If the system does not have enough unused resources required for the $S_k^{'}$ service, the network management system automatically frees up some of the network resources allocated to less important services. Define $\Delta r_j, j = \overline{1, n}$ as the amount of r_j resource that is missing to pass the $S_k^{'}$ service:

$$\Delta r_j = r_j - c_{kj} \tag{7}$$

to all $j = \overline{1, n}$.

Then for adaptive management of redistribution of resources it is necessary to find all such S_j services which cancellation will release the resources necessary for $S_k^{'}$ service.

Let's define the main criteria which are considered at the decision of this problem.

1. Withdraw part of the resources from those services whose total importance is minimal:

$$\min \sum_{i=1}^{n} w_i t_i. \tag{8}$$

2. Stop transmitting the least number of less important services that are in the transfer stage:

$$\min \sum_{i=1}^{n} t_i x_i . \tag{9}$$

3. The release of a r_j resource may require the termination of some background services, which, given the interdependence of resources and services, may lead to the release of some r_{j^*} resources more than necessary, which will ultimately reduce network performance. The following criterion is taken into account to prevent excessive resource release:

$$\begin{cases} b_j = \sum_{i=1}^{n} c_{ij} t_i - \Delta r_j; \\ b_j = \rightarrow \min; \\ b_j \geq 0. \end{cases} \tag{10}$$

to all $j = \overline{1, n}$, and the value of b_j is the amount of resource that is released, will always be positive, because you need to release a resource not less than the required amount.

To solve management problems taking into account criteria 1–3, are typical tasks of combinatorics - to find a certain set of elements that meet the given criteria, genetic algorithms can be used. The difficulty of quickly finding a set of tasks that meet these criteria is to go through a large number of combinations for each of the resources. To simplify the search, we introduce the c_i^* parameter, which characterizes the value of the average use of resources of the IBN by the i-th service:

$$c_i^* = \frac{\sum_{j=1}^{n} c_{ij} U_j}{n^*} \tag{11}$$

for $j = \overline{1, n}$, where n^* is the number of non-zero components $c_{ij} U_j$, and U_j determines the use of the resource by the s_j service, and

$$U_j = \begin{cases} 1, & \text{if the service } s_j \text{ uses the resource } r_j; \\ 0, & \text{otherwise.} \end{cases} \tag{12}$$

The introduction of the c_i^* parameter is made under the assumption of the existence of a proportional relationship between the utilization factors of different resources by one service, ie it is assumed that if the S_j service. uses one of the resources with a large coefficient value, then other resources will also be used with a large coefficient. For example, if a service requires a large amount of bandwidth, it will use more network channel resources. In the process of adaptive management to resolve the conflict between the new $S_k^{'}$ services, and services that already use the same network resources, depending on the value of c_i^*, there are three options:

1. At $c_i^* = 0$ the S_i service uses resources that are not claimed by the new service $S_k^{'}$.
2. If $c_i^* \ll c_k^*$, then the S_i service uses small amounts of resources claimed by the $S_k^{'}$ service, in particular, the denial of S_i service will not affect the solution. For example, a slight reduction in the bandwidth of the current service channels will not affect the quality of the service itself, and the reduced resource will be allocated by the network to service the new service.
3. In other cases, the S_i task uses significant amounts of the same resources that $S_k^{'}$ claims.

Thus, the management system, allocating resources for a new $S_k^{'}$ task leads to the task of flexible redistribution of resources between existing S_i tasks when changing service policies, should search for services to be deleted among those executable services that can free up sufficient network space.

Given that the problem of management in real conditions requires large computational costs, we propose an algorithm for its solution, which can significantly reduce the list of options. Moreover, this algorithm gives an exact solution except in cases where one of the coefficients c_{ij} significantly deviates from the average value of c_i^*, which is very rare.

The essence of the algorithm, which is launched every time a new service appears or there is a redistribution of service priorities in network nodes, is as follows.

Initially, all elements are arranged in order of importance:

$$w_i + 1 > w_i. \tag{13}$$

Among the elements of space E stand out services: $S = \overline{V_z} \times E$.

These services are not considered, the importance of w_i which is more important than the w_k task of $S_k^{'}$. The i_{max} is determined - the upper limit for the considered tasks.

For each service, the average use of c_i^*, $i = 1, 2, ..., i$ max resources is calculated.

To free up resources, services that meet one of the following criteria are deleted:

1. The least important services should be deleted:

$$T_i = 1, \quad if \sum_{j=1}^{i} c_j^* < c_k^*, \quad for\ i = 1, 2, \ldots, i^{\max}. \tag{14}$$

2. The least number of services must be deleted:

$$T_i = 1, \ if \sum_{j=1\max}^{i} c_j^* < c_k^*, \quad for\ 1 = 1^{\max}, \ldots, 2, 1. \tag{15}$$

3. The most resource-intensive services with minimal importance should be removed. To do this, the services are arranged according to the values of $q_i = c_i^*/w_i$, so that $q_{i+1} < q_i$, and then the choice of services is made according to the second criterion.

After determining the services that can be removed in principle, the amount of released resources is compared and the resource with the maximum max Δr_j deficit is determined. The set of services which are blocked for release of resources on one of the above criteria is defined and is replaced by c_j^* value of the j-th resource.

The use of the proposed model in the management systems of information and communication systems allows for effective allocation and redistribution of common resources in the process of new services and change the importance of their business processes. The proposed formalized approach to resource allocation, taking into account the changing importance of services, can be used in intentionally oriented information and communication networks of the new generation.

3 Future Insights of QoE Based Network Management

However, it should be borne in mind that with the development of infocommunication systems, customer needs and behavior have changed. The focus shifts from improving network performance to obtaining the necessary additional resources (particularly for services responsible for important business processes) to improving the quality of experience of users (Quality of Experience, QoE) based on available network resources through rational managing them [25]. Thus, the paper proposes a gradual transition from traditional Service Level Agreements (SLAs) to new Experience Level Agreements (ELAs) necessary for the implementation of the concept of IBN [26–28].

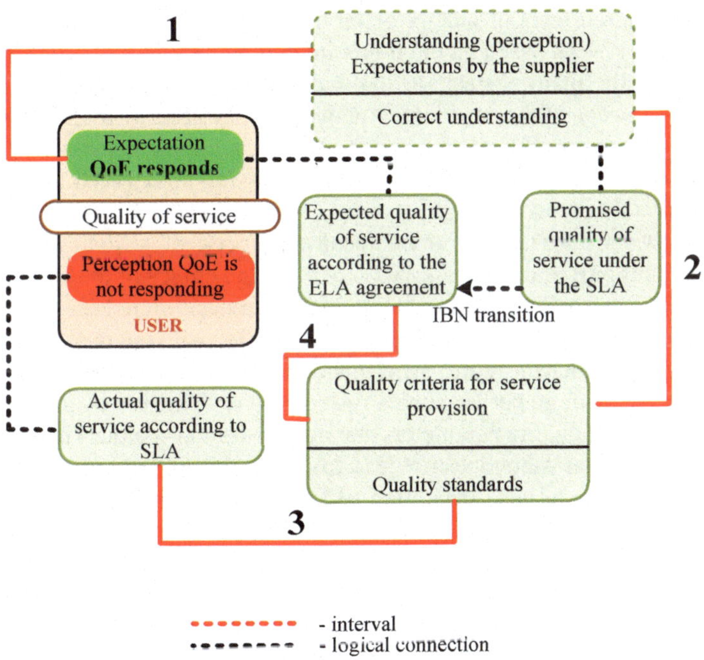

Fig. 2. Transition from SLA to new ELA intents for future IBN

In assessing the quality of service perception, there are four stages (intervals) that affect the assessment of the quality of its provision, which can be defined as the intervals between the expected and actual quality of service. In this paper, a scheme is proposed that takes into account the differences between the expected and actual quality of service delivery (Fig. 2).

The first interval is between the consumer's expectation of the required quality of service and the perception of these expectations by the service provider. If the service provider does not understand the wishes and expectations of the consumer, it is unlikely that the service will be provided to the consumer according to his expectations. Thus, it is necessary that the service provider understands the desire (intentions of the user) that the service was provided to the consumer according to his expectations.

The second interval is between a correct understanding of the consumer's expectations and the quality criteria of the service provided by the service provider in order to meet the expectations and expectations of the consumer. In this case, there is a transition from Service Level Agreements (SLAs) to Experience Level Agreements (ELAs).

The third interval is between quality standards and the actual quality of services, i.e the ability of the supplier to provide the required level of service quality. In fulfilling the requirements for the provision of services, the network provider must support this process with appropriate resources. Where, according to existing solutions, standard SLAs often do not justify the expected level of quality of service perception. Thus, using the IBN network there is a transition to agreements on the level of expected quality of service and, accordingly, the fourth interval is formed.

The fourth interval is between the expected ordered quality of the user in order to obtain the required level of quality of service perception and the quality of service provided by the network based on the user's intentions.

Accordingly, the main obstacle to the widespread implementation of ELA agreements for the concept of intent-oriented network is the lack of a unified view on the formation of input mathematical description of QoE evaluation systems based on known QoS criteria for various infocommunication services and unexplored cost optimization and according to the received profit.

Accordingly, most researchers argue that in the process of designing next-generation networks, including IBN, it is necessary to focus on the choice of the number of service quality indicators that are taken into account in the synthesis of the network. The number of partial parameters that characterize the quality of the real system can be very diverse and large. This means that the more partial quality parameters are taken into account when optimizing IBNs, the more perfect such a system will be. That is why in practice there is an optimal number of quality parameters that need to be considered. The introduction of additional quality parameters does not lead to improvement, but to deterioration of the results of optimization of new generation IBNs. However, most modern networks take into account the standard parameters of service quality, for optimization, each of which has its own values according to the established recommendations of telecommunications. The paradigm shift in the concept of service provision, which was associated with a general change in the functioning of standard networks in the direction of developing the concept of IBN, is expressed primarily in the fact that the roles of operator and user have changed significantly. Now the user and the operator act as allies in a single process of informatization, and such interaction can be considered the evolution of modern methods of providing services.

Thus, when developing a new quality management system for the provision of information services, it is logical to use a systems approach: the problem of quality assurance should be addressed not in isolation, but within the framework of interaction with the user. Satisfaction of user requirements includes both technical (parameters of the quality of network operation) and non-technical (subjective perception by end users) aspects [6]. In the process of adaptive service management, it is necessary to control both the compliance of service characteristics with regulatory indicators and, if necessary, to make adaptive adjustments to the standards.

Thus, based on the above, the formation of service quality includes both an objective assessment of network characteristics and a subjective expert assessment of the user. And if the parameters of the network can be determined using the appropriate equipment, then taking into account the perception of users of the quality of services received, this is done using the ratio of QoS offered by the operator and QoS perceived by the customer, or QoE.

At this time, the network must have a device that compares the difference between the required level of quality and the actual provided network provider, and if the comparison process deviates from the allowable value, then the control signals are reported to the required network transformation. The network must remember and analyze the state of the network and the appropriate assessment of the quality of customer service, as well as be able to configure the network based on experience. This approach can be implemented by introducing machine learning algorithms of the artificial intelligence subclass into the service management system, which is the main idea of IBN. Thus, the network configuration and functionality of the network equipment automatically change depending on the changing requirements of the user. The network not only responds to the user's current requests, but also analyzes its benefits and current environment, providing relevant information to the network controller, who is responsible for centralized management of the entire network.

Also important is the technological transformation of architectural models for building infocommunication systems using trend network technologies, in particular the main of which is the technology of software-configured networks, allowing you to literally program and reprogram networks in real time to meet specific business needs. and user requirements as they arise.

In this chapter we proposed a conceptual model of a heterogeneous software-defined intent-based network. This model, unlike the existing ones, allows efficient allocation and redistribution of shared resources, adapting to the changing requirements of business users regarding the quality of service provision. The main idea of the proposed IBN concept is in the change of paradigm of network infrastructure: now it is not the user and his application that adapts to the capabilities of the network, but the network changes its configuration to the requirements of the user. Provision of the specified level of QoE services by the user becomes a fundamental problem for the implementation of end-to-end resource management in the IBN concept [29–31]. Thus, for the development of a new system of adaptive quality management of information services, a system approach is used in this work. In particular, the quality problem is not solved in isolation by operators, but in close interaction with the users of services. To do this, the IBN controller analyzes the state of the network and the corresponding custom QoE scores of users, characterizing a certain level of service quality. This controller also automatically adjusts the network configuration based on the accumulated experience and developed new methods of resource allocation and traffic engineering at each level of the conceptual network. This approach is implemented by introducing machine learning algorithms of the artificial intelligence subclass into the service management system. In this way, the network configuration and functionality of the network equipment automatically changes according to changing user requirements. For this purpose, the conceptual model of the heterogeneous IBN network is based on the principles of centrality, programmability, abstraction and openness, using SDN, NFV, SDR, Big Data, IoT and Cloud computing technologies. The conceptual model of intent-based IBN with adaptive quality management of service provision for mobile operator is shown in Fig. 3.

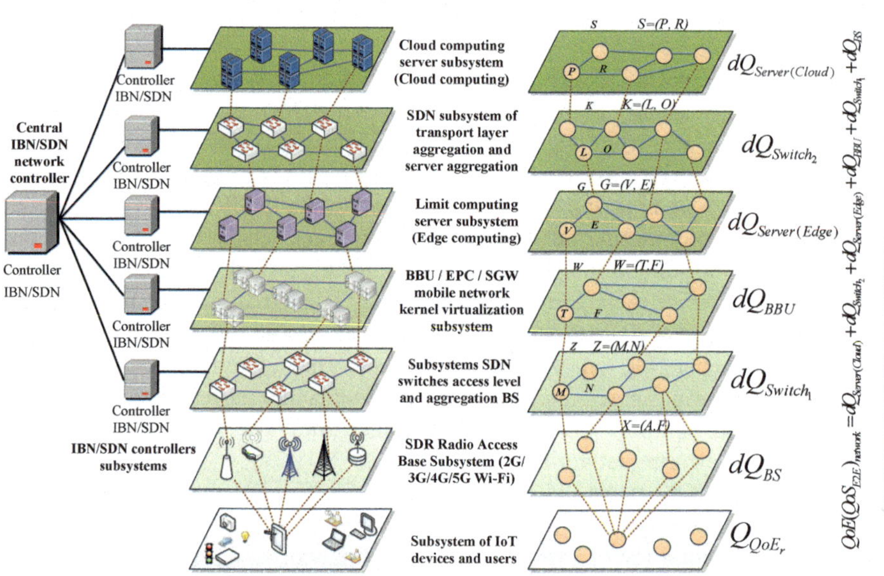

Fig. 3. A conceptual model of intent-based network for future mobile communication system

User service scenarios in a proposed future IBN mobile communication system can be formally described as follows. There is a set of N corporate services $\phi : \phi = \{S_1, S_2, ..., S_N\}$, a set Θ, there is a set of business user services, which are provided by providers $\Theta : \Theta = \{P_1, P_2, ..., P_w\}$, and a set F consisting of a set i and users $F : F = \{User_1, User_2, ..., User_i\}$. Each user can use N different networks belonging to the set $X : X = \{D_1, D_2, ..., D_N\}$ to access one or more information services, with a certain ordered level of quality of service Q_{QoE_r}. A user $User_i$ during a session in the system can be represented by a tuple:

$$c_{\langle m,i,QoE_r,w,N\rangle} = \langle m, User_i, S_N, \{P_1, P_2, ..., P_w\}, D_N, Q_{QoE_r}\rangle. \quad (16)$$

The value $c_{\langle m,i,QoE_r,w,N\rangle}$ is understood as follows: during the session number m the user $User_i$ has access to services through the network D_N, receiving services according to the set S_N, with a ordered quality Q_{QoE_r}. This formalism (16) is proposed to be used for two fundamental purposes necessary for the knowledge block of the IBN controller. Namely, to describe, represent, and control user behavior in a heterogeneous network and to obtain data on the change in user profile according to the criterion of ordered quality of service by investigating aspects of their activity in the network and intuitively explaining their behavior. In addition, the given user service model (Fig. 1) in IBN network allows intuitive formalization of clusters of users with the same type of quality of service Q_i requirements based on machine learning algorithms, in particular using k-means clustering method. Thus, for the users, which are included in a particular cluster $UC_{i\ Cluster_{QoEr}}$ is formed by IBN/SDN controller its own network configuration policy in the form of program code for adaptive resource management and to ensure the necessary quality of service provision. In the general case, the integral indicator of the quality of service provision Q is associated with a certain dependence with the partial indicators q_i, which can also be in functional dependence with each other:

$$F = Q(q_1, q_2, \dots, q_i, \dots, q_n). \tag{17}$$

Let in the given dependence (17) all partial indicators be independent changes. Accordingly, the influence of partial indicators on the complex quality indicator is formalized in the form of a complete differential of the function Q:

$$dQ = \frac{\partial Q}{\partial q_1}dq_1 + \frac{\partial Q}{\partial q_2}dq_2 + \dots + \frac{\partial Q}{\partial q_i}dq_i + \dots + \frac{\partial Q}{\partial q_n}dq_n. \tag{18}$$

Partial derivatives before dq_i values are considered as weights of partial q_1,q_2,...,q_i,...,q_n. quality indicators related to the functional dependence of the complex indicator Q. The $\partial Q_i/\partial q_i$ expression shows how the quality of Q services changes when the partial q_i quality indicator changes (at fixed values of other indicators). Based on the above, the expression is formalized:

$$w_i = \frac{\partial Q}{\partial q_i}\bigg|_{q_i = q_{i0}}, \quad i = \overline{(1, n)}, \tag{19}$$

where w_i is weighting factor of the i-th partial quality indicator.

Accordingly, Eq. (18) is written as a comprehensive indicator of the quality of service provision:

$$dQ = w_1 dq_1 + w_2 dq_2 + \dots + w_i dq_i + \dots + w_n dq_n. \tag{20}$$

Equation (20) is a consequence of the linearization of the function Q at a point whose coordinates $q_i = q_{i0}$, $i = (1,n)$.

$$w_i = f_i(q_1, q_2, \dots, q_i, \dots q_n). \tag{21}$$

From expression (21) it is seen that the w_i weighting factors expressed in (4) are functions of many variable partial q_i quality indicators. In cases where q_i values are specified, numerical w_i values are determined by substituting in Eq. (20) specific values of partial quality indicators.

Accordingly, the system of differential equations for adaptive intention-oriented quality management of services in a conceptually heterogeneous IBN network (Fig. 3) is formalized in the form (22):

$$\begin{cases} dQ_{Server(Cloud)} = \frac{\partial Q_{Server(Cloud)}}{\partial q_1}dq_1 + \frac{\partial Q_{Server(Cloud)}}{\partial q_2}dq_2 + ... + \frac{\partial Q_{Server(Cloud)}}{\partial q_i}dq_i + ... + \frac{\partial Q_{Server(Cloud)}}{\partial q_x}dq_x; \\ dQ_{Switch2} = \frac{\partial Q_{Switch_2}}{\partial q_1}dq_1 + \frac{\partial Q_{Switch_2}}{\partial q_2}dq_2 + ... + \frac{\partial Q_{Switch_2}}{\partial q_i}dq_i + ... + \frac{\partial Q_{Switch_2}}{\partial q_y}dq_y; \\ dQ_{Server(Edge)} = \frac{Q_{Server(Edge)}}{\partial q_1}dq_1 + \frac{Q_{Server(Edge)}}{\partial q_2}dq_2 + ... + \frac{Q_{Server(Edge)}}{\partial q_i}dq_i + ... + \frac{Q_{Server(Edge)}}{\partial q_n}dq_n; \\ dQ_{BBU} = \frac{Q_{BBU}}{\partial q_1}dq_1 + \frac{Q_{BBU}}{\partial q_2}dq_2 + ... + \frac{Q_{BBU}}{\partial q_i}dq_i + ... + \frac{Q_{BBU}}{\partial q_z}dq_z; \\ dQ_{Switch1} = \frac{\partial Q_{Switch_1}}{\partial q_1}dq_1 + \frac{\partial Q_{Switch_1}}{\partial q_2}dq_2 + ... + \frac{\partial Q_{Switch_1}}{\partial q_i}dq_i + ... + \frac{\partial Q_{Switch_1}}{\partial q_m}dq_m; \\ dQ_{BS} = \frac{\partial Q_{BS}}{\partial q_1}dq_1 + \frac{\partial Q_{BS}}{\partial q_2}dq_2 + ... + \frac{\partial Q_{BS}}{\partial q_i}dq_i + ... + \frac{\partial Q_{BS}}{\partial q_v}dq_v; \\ QoE(QoS_{E2E})_{network} = dQ_{Server(Cloud)} + dQ_{Switch2} + dQ_{Server(Edge)} + dQ_{BBU} + dQ_{Switch1} + dQ_{BS}; \\ QoE(QoS_{E2E})_{network} \approx Q_{QoE_r}. \end{cases} \quad (22)$$

From Eq. (22) it follows that to ensure the user-ordered level of Q_{QoE_r} service quality (in IBN ideology is understood as the user's intention), it is necessary, centrally, flexibly, adaptively and consistently manage $dQ_{Server(Cloud)}$, $dQ_{Switch2}$, $dQ_{Server(Edge)}$, dQ_{BBU}, $dQ_{Switch1}$, dQ_{BS} service quality at each level of the conceptual IBN network (Fig. 2), providing the ordered end-to-end q of $QoE(QoS_{E2E})_{network}$ service. Where QoE is the subjective evaluation of a service at the application level by the user who uses the service. QoS is a set of network and channel layer technologies, the use of which allows more efficient use of network resources, especially during streaming traffic to provide the required level of QoE.

Based on the above conceptual model of network construction, it is similarly possible to provide adaptive quality management services for wired operators, taking away the subsystems: SDR radio access base stations (2G/3G/4G/5G/Wi-Fi), SDN of access level aggregation switches and BS aggregation, BBU/EPC/SGW mobile network kernel virtualization subsystem. The model of building an IBN with adaptive quality management for the leading communication operator is shown in Fig. 4.

Accordingly, the system of differential equations for adaptive intentionally oriented quality management of service provision in the conceptual IBN network for the leading communication operator (Fig. 4) is formalized in the form (23):

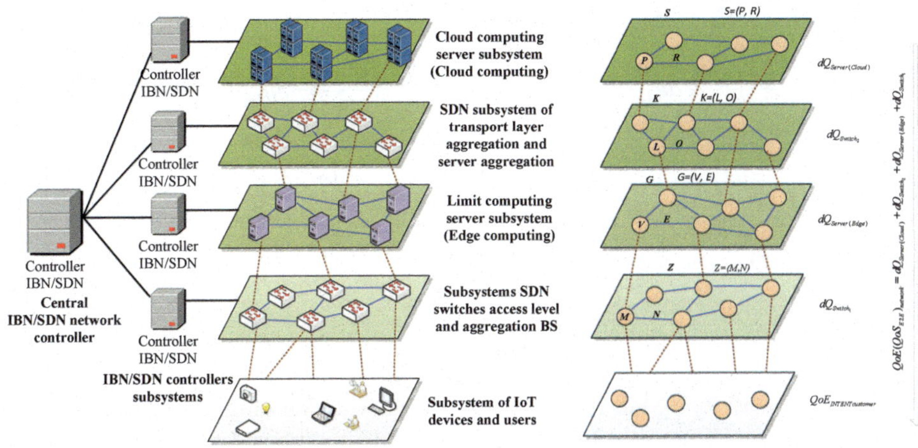

Fig. 4. A conceptual model of intent-based network for providers

$$
\begin{cases}
dQ_{Server(Cloud)} = \dfrac{\partial Q_{Server(Cloud)}}{\partial q_1} dq_1 + \dfrac{\partial Q_{Server(Cloud)}}{\partial q_2} dq_2 + \ldots + \dfrac{\partial Q_{Server(Cloud)}}{\partial q_i} dq_i + \ldots + \dfrac{\partial Q_{Server(Cloud)}}{\partial q_x} dq_x; \\
dQ_{Switch2} = \dfrac{\partial Q_{Switch_2}}{\partial q_1} dq_1 + \dfrac{\partial Q_{Switch_2}}{\partial q_2} dq_2 + \ldots + \dfrac{\partial Q_{Switch_2}}{\partial q_i} dq_i + \ldots + \dfrac{\partial Q_{Switch_2}}{\partial q_y} dq_y; \\
dQ_{Server(Edge)} = \dfrac{Q_{Server(Edge)}}{\partial q_1} dq_1 + \dfrac{Q_{Server(Edge)}}{\partial q_2} dq_2 + \ldots + \dfrac{Q_{Server(Edge)}}{\partial q_i} dq_i + \ldots + \dfrac{Q_{Server(Edge)}}{\partial q_n} dq_n; \\
dQ_{Switch1} = \dfrac{\partial Q_{Switch_1}}{\partial q_1} dq_1 + \dfrac{\partial Q_{Switch_1}}{\partial q_2} dq_2 + \ldots + \dfrac{\partial Q_{Switch_1}}{\partial q_i} dq_i + \ldots + \dfrac{\partial Q_{Switch_1}}{\partial q_m} dq_m; \\
QoE(QoS_{E2E})_{network} = dQ_{Server(Cloud)} + dQ_{Switch2} + dQ_{Server(Edge)} + dQ_{Switch1}; \\
QoE(QoS_{E2E})_{network} \approx Q_{QoE_r}.
\end{cases} \tag{23}
$$

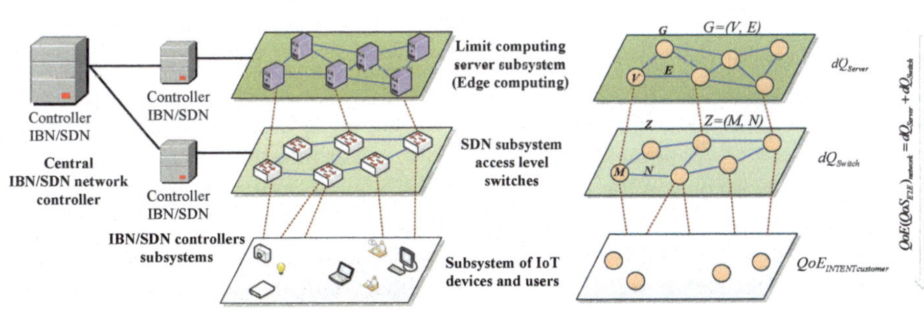

Fig. 5. A conceptual model of intent-based network for corporate class

Accordingly, the system of differential equations for adaptive intentionally-oriented quality management of services in the intentionally oriented network of the corporate class (Fig. 5) is formalized in the form (24):

$$\begin{cases} dQ_{Server} = \frac{\partial Q_{Server}}{\partial q_1} dq_1 + \frac{\partial Q_{Server}}{\partial q_2} dq_2 + ... + \frac{\partial Q_{Server}}{\partial q_i} dq_i + ... + \frac{\partial Q_{Server}}{\partial q_n} dq_n; \\ dQ_{Switch} = \frac{\partial Q_{Switch}}{\partial q_1} dq_1 + \frac{\partial Q_{Switch}}{\partial q_2} dq_2 + \ldots + \frac{\partial Q_{Switch}}{\partial q_i} dq_i + \ldots + \frac{\partial Q_{Switch}}{\partial q_m} dq_m; \\ QoE(QoS_{E2E})_{network} = dQ_{Switch} + dQ_{Server}; \\ QoE(QoS_{E2E})_{network} \approx Q_{QoE_r}. \end{cases} \tag{24}$$

We considered business perspectives by using SDN, SDR and NFV to the existing infrastructure, and it became clear that this approach proposed more opportunities for revenue generation in the value chain of services. Integrating SDN combined with NFV into IBN can bring exciting opportunities. NFV combined with SDN will result in lower CAPEX/OPEX, streamlined operations and shorter time-to-market for new entrants. Aligning QoE factors with service characteristics, business elements and system performance parameters is challenging. Because if one considers QoS only as a representative aspect of QoE, other non-technical aspects such as business factors (e.g., cost, promotions, advertising) are likely to be overestimated. Furthermore, in the future, new applications and services with high demands will be driven primarily by vertical industries. The billing schemes are further expected to evolve through the use of virtualization, revealing more dynamic characteristics.

4 Conclusion

We argue that new methods for measuring QoE at various connection points are needed to ensure QoE-centric business in the future networks. Also, the definition of new classes of QoE to ensure a common understanding among all stakeholders in the service delivery chain. Essentially, ELAs require the formulation of a new QoE-centric business framework and the development of effective marketing strategies when considering differentiated or personalized levels of QoE in the future IBN. We have proposed a mathematical model for adaptive management of resource reallocation of info-communication networks to implement the IBN concept. For this purpose, the model takes into account the importance of business services, both those that are running and new ones, as well as the possibility of changing priorities over time regarding the allocation of network resources due to changes in the importance of the business processes that they support.

We considered QoE-driven business perspectives by using SDN, SDR and NFV to the future IBN infrastructure, and it became clear that this approach proposed more opportunities for revenue generation in the value chain of services and customers satisfaction.

Acknowledgement. This research was supported by the Ukrainian government project №0120U102201 “Development of the methods and unified software-hardware means for the deployment of the energy efficient intent-based multi-purpose information and communication networks.”

References

1. Kir, H., Erdogan, N.: A knowledge-intensive adaptive business process management framework. Inf. Syst. **95**, 101639 (2021)
2. Torkhani, R., Laval, J., Malek, H., Moalla, N.: Intelligent framework for business process automation and re-engineering. Int. Conf. Intell. Syst. **2018**, 624–629 (2018). https://doi.org/10.1109/IS.2018.8710523
3. Klymash, M., Beshley, M., Koval, V.: The model of prioritization of services for efficient usage of multiservice network resources. In: Proceedings of International Conference on Modern Problem of Radio Engineering, Telecommunications and Computer Science, pp. 320–321 (2012)
4. EL-ezzi, Z.Q., Al-Dulaimi, A.M., Ibrahim, A.A.: Personalized quality of experience (QOE) management using data driven architecture in 5G wireless networks. In: 2020 4th International Symposium on Multidisciplinary Studies and Innovative Technologies (ISMSIT), pp. 1–10 (2020). https://doi.org/10.1109/ISMSIT50672.2020.9254863
5. Marchão, J., Reis, L., Martins, P.V.: Business areas and processes alignment in ICT framework. In: 2020 15th Iberian Conference on Information Systems and Technologies (CISTI), pp. 1–4 (2020). https://doi.org/10.23919/CISTI49556.2020.9141067
6. Romanchuk, V., Beshley, M., Polishuk, A., Seliuchenko, M.: Method for processing multi-service traffic in network node based on adaptive management of buffer resource. In: 2018 14th International Conference on Advanced Trends in Radioelecrtronics, Telecommunications and Computer Engineering (TCSET), pp. 1118–1122 (2018). https://doi.org/10.1109/TCSET.2018.8336390
7. Beshley, M., Kryvinska, N., Seliuchenko, M., Beshley, H., Shakshuki, E., Yasar, A.: End-to-end QoS "smart queue" management algorithms and traffic prioritization mechanisms for narrow-band internet of things services in 4G/5G networks. Sensors **20**(8), 2324-1–2324-30 (2020)
8. Kryvinska, N.: An analytical approach for the modeling of real-time services over IP network. In: Elsevier Transactions of IMACS, Journal Mathematics and Computers in Simulation (MATCOM), vol. 79, pp. 980–990 (2008). ISSN: 0378-4754. https://doi.org/10.1016/j.matcom.2008.02.016
9. Seliuchenko, M., Beshley, M., Kyryk, M., Zhovtonoh, M.: Automated recovery of server applications for SDN-based internet of things. In: 2019 3rd International Conference on Advanced Information and Communications Technologies (AICT), pp. 149–152 (2019). https://doi.org/10.1109/AIACT.2019.8847743
10. Jun, S., et al.: A cost-efficient software based router and traffic generator for simulation and testing of IP network. Electronics **9**(1), 40-1–40-24 (2020)
11. Kryvinska, N.: Intelligent network analysis by closed queuing models. Telecommun. Syst. **27**, 85–98 (2004). https://doi.org/10.1023/B:TELS.0000032945.92937.8f
12. Panchenko, O., et al.: Method for adaptive client-oriented management of quality of service in integrated SDN/CLOUD networks. In: 2017 4th International Scientific-Practical Conference Problems of Infocommunications, Science and Technology (PIC S&T), Kharkov, pp. 452–455 (2017)
13. Mandal, S.K., Ogras, U.Y., Rao Doppa, J., Ayoub, R.Z., Kishinevsky, M., Pande, P.P.: Online adaptive learning for runtime resource management of heterogeneous SoCs. In: 2020 57th ACM/IEEE Design Automation Conference (DAC), pp. 1–6 (2020). https://doi.org/10.1109/DAC18072.2020.9218604
14. Schulz, D.: Intent-based automation networks: toward a common reference model for the self-orchestration of industrial intranets. In: IECON 2016 - 42nd Annual Conference of the IEEE Industrial Electronics Society, pp. 4657–4664 (2016). https://doi.org/10.1109/IECON.2016.7792959

15. Farahnakian, F., Bahsoon, R., Liljeberg, P., Pahikkala, T.: Self-adaptive resource management system in IaaS clouds. In: 2016 IEEE 9th International Conference on Cloud Computing (CLOUD), pp. 553–560 (2016). https://doi.org/10.1109/CLOUD.2016.0079
16. Rafiq, A., Mehmood, A., Song, W.-C.: Intent-Based slicing between containers in SDN overlay network. J. Commun. **15**(3), 237–244 (2020). https://doi.org/10.12720/jcm.15.3.237-244
17. Singh, A., Aujla, G.S., Bali, R.S.: Intent-based network for data dissemination in software-defined vehicular edge computing. IEEE Trans. Intell. Transport. Syst. **22**(8), 5310−5318. https://doi.org/10.1109/TITS.2020.3002349
18. Rafiq, A., Afaq, M., Song, W.-C.: Intent-based networking with proactive load distribution in data center using IBN manager and smart path manager. J. Ambient. Intell. Humaniz. Comput. **11**(11), 4855–4872 (2020). https://doi.org/10.1007/s12652-020-01753-1
19. Hyun, J., Hong, J.W.: Knowledge-defined networking using in-band network telemetry. In: 2017 19th Asia-Pacific Network Operations and Management Symposium (APNOMS), pp. 54–57 (2017). https://doi.org/10.1109/APNOMS.2017.8094178
20. Wu, C., Horiuchi, S., Tayama, K.: A resource design framework to realize intent-based cloud management. IEEE Int. Conf. Cloud Comput. Technol. Sci. **2019**, 37–44 (2019). https://doi.org/10.1109/CloudCom.2019.00018
21. Ujcich, B.E., Sanders, W.H.: Data protection intents for software-defined networking. IEEE Conf. Netw. Softwarization **2019**, 271–275 (2019). https://doi.org/10.1109/NETSOFT.2019.8806684
22. Beshley, M., Vesely, P., Prislupskiy, A., Beshley, H., Kyryk, M., Romanchuk, V., Kahalo, I.: Customer-oriented quality of service management method for the future intent-based networking. Appl. Sci. **10**(22), 8223-1–8223-38 (2020)
23. Wang, L., Delaney, D.T.: QoE oriented cognitive network based on machine learning and SDN. In: 2019 IEEE 11th International Conference on Communication Software and Networks (ICCSN), Chongqing, China, pp. 678–681 (2019)
24. Beshley, M., Pryslupskyi, A., Panchenko, O., Seliuchenko, M.: Dynamic switch migration method based on QoE- aware priority marking for intent-based networking. In: 2020 IEEE 15th International Conference on Advanced Trends in Radioelectronics, Telecommunications and Computer Engineering (TCSET), Lviv-Slavske, Ukraine, pp. 864–868 (2020)
25. Barakabitze, A.A., et al.: QoE management of multimedia streaming services in future networks: a tutorial and survey. IEEE Commun. Surv. Tutorials **22**(1), 526–565 (2020)
26. Lewis, B., Fawcett, L., Broadbent, M., Race, N.: Using P4 to enable scalable intents in software defined networks. In: 2018 IEEE 26th International Conference on Network Protocols (ICNP), Cambridge, pp. 442–443 (2018)
27. Beshley, M., Pryslupskyi, A., Panchenko, O., Beshley, H.: SDN/cloud solutions for intent-based networking. In: 2019 3rd International Conference on Advanced Information and Communications Technologies (AICT), Lviv, Ukraine, pp. 22–25 (2019)
28. Abbas, K., Khan, T.A., Afaq, M., Song, W.-C.: Network slice lifecycle management for 5G mobile networks: an intent-based networking approach. IEEE Access **9**, 80128–80146 (2021). https://doi.org/10.1109/ACCESS.2021.3084834
29. Medvetskyi, M., Beshley, M., Klymash, M.: A quality of experience management method for intent-based software-defined networks. In: 2021 IEEE 16th International Conference on the Experience of Designing and Application of CAD Systems (CADSM), pp. 59–62 (2021). https://doi.org/10.1109/CADSM52681.2021.9385250
30. Beshley, M., Kryvinska, N., Beshley, H., Yaremko, O., Pyrih, J.: Virtual router design and modeling for future networks with QoS guarantees. Electronics **10**(10), 1139 (2021)
31. Flores Moyano, R., Fernández, D., Merayo, N., Lentisco, C.M., Cárdenas, A.: NFV and SDN-based differentiated traffic treatment for residential networks. IEEE Access **8**, 34038–34055 (2020). https://doi.org/10.1109/ACCESS.2020.2974504

Designing HDS Under Considering of QoS Robustness and Security for Heterogeneous IBN

Andriy Luntovskyy[1(✉)] and Mykola Beshley[2]

[1] BA Dresden Univ. of Coop. Education, Saxon Study Academy, Hans-Grundig-Str. 25, 01307 Dresden, Germany
Andriy.Luntovskyy@ba-dresden.de

[2] Lviv Polytechnic National University/ITRE, Stepan Bandera Street 12, Lviv 79000, Ukraine
mykola.i.beshlei@lpnu.ua

Abstract. So-called Intent-Based Networking (IBN) is founded on well-known SDN and represents one of the most important emerging network infrastructure opportunities. Their potential economic and social benefits cannot be overestimated, as networks will also reach a whole new level of self-awareness, self-configuration, self-optimization, self-healing and self-protection. This will be a benefit for existing networks as well as evolved LTE, emerging Internet of Things (IoT) and Cloud systems, and soon for future 5G/6G networks. The running software, data and content within IBN is loose-coupled and highly-distributed. HDS means "Highly-Distributed Systems".

Keywords: IBN · SDN · QoS and QoE · Energy efficiency · IDS · IPS · Blockchain · Sensor networks · 5G/6G · ML · AI · CAPEX · OPEX

1 Motivation and Introduction: HDS for IBN

In recent years, IBN has increasingly found its way into heterogeneous Software-Defined Networks (SDN).

The distinguishing features of the IBN that we feel sets them apart from others in the networking field [1–5]:

- Multiple consideration of the IBN aspects based on SDN
- IBN provide robust QoS and usual QoE as well as attractive up-to-date content
- IBN are deployed under wide function softwarization and virtualization (NFV)
- Any modern (fixed, radio, SAT-based, wireless or mobile) network technology can be configured and dynamically used as well as networking heterogeneity can be rapidly consolidated
- IBN are mostly energy-efficient and economizing of expenditures (CAPEX – Capital Expenditures, OPEX – Operational Expenditures), as well as in mid-term also CO_2 footprint minimizing
- IBN content is per definition intrusion protected as well as their workflow and the transactions are secured via usual PKI and up-to-date Blockchain technology

M. Klymash et al. (Eds.): *Future Intent-Based Networking*, LNEE 831, pp. 19–37, 2022.
https://doi.org/10.1007/978-3-030-92435-5_2

- And the last but least: IBN content, data and software, use service-oriented concept, is highly-distributed and loose-coupled; IBN content is integrated, composed and processed via SOA and Micro-Services.

We will provide for you a clearer demarcation of our own concept of IBN, i.e. "intent-based networking", which is built on SDN with subsequent network virtualization and software implementation of network functions with a significant improvement in service quality, which feels like the usual QoE (Quality of Experience), which can be every time monetized. The protection against usual risks for networked data from intrusions is guaranteed. Such networks already create convenient conditions for reducing their cost and cost of their operation (the expenditures, namely, CAPEX and OPEX). In the future, IBN will guarantee energy efficiency, optimal PUE (Power Usage Effectiveness) and minimizing CO_2 emission (CO_2 footprint).

The analysis of modern literature highlights this IBN trend in the development of heterogeneous IP networks, which include sensor networks and the routes of IoT (e.g. NB-IoT, LTE-CatM, LoRa, 6LoWPAN, EnOcean), DSL, ATM, MPLS, WiFi-6, as well as hierarchical 4G, 5G cells and radio relay areas [14, 15, 22, 24].

When distinguishing the categories of OoS and QoE by the method of expert assessments, we would like to introduce a specific quality parameter Q, which consider also the network security, for example, as the intrusion probability (IDS), the percentage of Intrusion Detection & Intrusion Prevention [6–12].

The rest of the chapter is organized as follows:

- Section 2 shows the used Design Paradigms for so-called HDS (Highly-Distributed Systems for IBN).
- Section 3 depicts Service Composition via Micro-Services for IBN.
- Section 4 of this chapter considers QoS Robustness and IBN Security.
- Conclusions and outlook finalize the work and show why we mean this work as a WIP: Work-in-Process.

2 Designing HDS

2.1 Distributed Systems

The term "Distributed Systems" has been used for many years for applications, which operate in modern combined wired-wireless-mobile networks under clear cooperation goals, as well as have no centralization in memory access or synchronization in the clocking. The distributed applications are constructed on the sample n-tier and often possess redundancy in form of server and database replications. They follow established SOA (service-oriented architecture) concept and can be often organized as cloud-centric structures. Significant architectural transformations in network services and distributed systems characterize an ongoing trend nowadays [3, 6, 12, 19, 21, 23, 32]. The clouds, clusters with explicit cooperation goal (e.g. parallelized computing) as well as grids belong to the above-mentioned systems.

2.2 Highly-Distributed Systems

The term "Highly-distributed systems" include all modern combined, wireless, and mobile networks and have a complex internal structure. This technology should be well protected, maintaining the maximum possible value of the QoS parameter (higher DR and availability, low latency).

HDS allows us to deploy flexible structures based on SOA and micro-services, as well as deploys efficient communication models like P2P, cloud-fog or M2M, which will allow us to resolve distribution conflicts in a short time and support fast access to data analytics.

Since 2005, P2P systems, the C-S communication model, as well as server-less structures (SLMA, robotics) have gained popularity. In 2011, cloud solutions became popular too. With the predominant use of load-balanced "thin clients" with the delegation of functionality to the cloud [1–3, 6]. So what do we mean by the term "highly distributed systems"? The distinguishing features of HDS are as follows (Fig. 1):

1. Advanced communication models (C-S with Clouds, Fog, P2P, M2M).
2. Advanced methods for performance management and optimization as well as for QoE (Quality of Experience (Fig. 2) increasing.
3. Advanced SWT (Fig. 3) (agile approaches like XP, DevOps, Kanban, Scrum, Micro-Services [18, 22, 24]).
4. Advanced Data Analytics regarding solving of "Big Data" shortcomings [17] via Machine Learning (ML), refer Fig. 4.

HDS systems can be efficiently used and provide powerful modern apps only if they achieve needed QoS requirements and parameters (refer Fig. 1), such as: Performance; Reliability; Scalability as well as support necessary security and privacy. HDS cooperate with basic modern technologies and/or offer the apps for such technologies on existing digital platforms, e.g. such subject like: Blockchain (Decentralized Transactions, Data Mining, Smart Contracting) [13, 26–31]; Machine Learning and Neural Networks; IoT, 5G networks and Robotics. For rapid development of efficient HDS the novel SWT approaches and process models are widely used, like Scrum and DevOps (refer Fig. 3).

The appropriate construction basis for HDS are often SOA and Micro-Services (refer Sect. 2), which combine development process agility with flexible communication models [7–11]. Systematically, HDS are involved in the creation of AI (Artificial Intelligence). The neural networks and ML (Machine Learning) techniques are widely integrated into internal structures of HDS. The main purpose of ML for HDS is a special deployment of some workflow steps and learning algorithms, which are performed without human intervention. The time factor and amount of data, which are required for training, belong to the most important indicators of performance and QoS for apps and HDS.

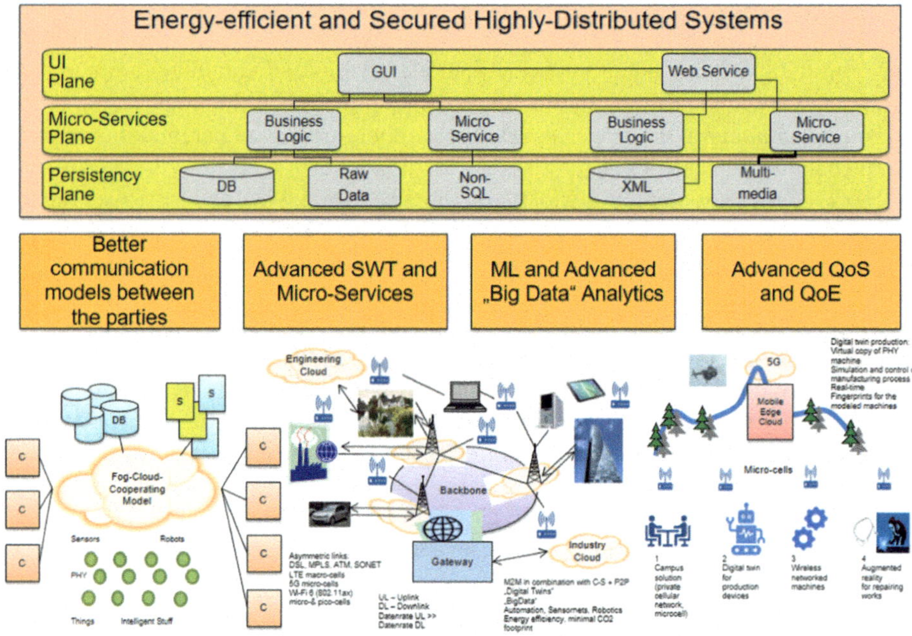

Fig. 1. To the motivation on HDS [18, 22]

2.3 HDS Construction Paradigms π1-π4

The following HDS construction paradigms π1-π4 can be formulated taking into account the above-discussed positions:

- Paradigm π1. The first paradigm is the use of advanced architecture models (M3-M5) for efficient communication, providing advanced QoE and energy autarky within IoT Apps.
- Paradigm π2. The next one is the deployment of modern SWT process models (like Scrum) and use of flexible Micro-Services, leading to the efficient decentralized highly-distributed systems (so called HDS), which provide better scalability, reliability and reconfiguration.
- Paradigm π3. The further paradigm regards security, privacy, authentication and compulsoriness of HDS workflow steps, modules and services under use of Blockchain technology. However, as the main disadvantage of the Blockchained IoT, the performance reduction by real-time services as well as energy consumption become a critical position.
- Paradigm π4. In addition to the classic algorithms, ML must be also integrated within HDS applications imperatively to overcome Big Data problems.

The SOA (Service-Oriented Architecture) via Web Services and Micro-Services are frequently the regular construction materials for apps in the world of IBN, i.e. Distributed and Highly-Distributed Systems (HDS) like e.g. so-called “Lego pieces”. The providers of such basic elements are often called SOA providers, which follow non-monolithic software modules concepts. Due to Service Composition rules the modern distributed apps can be loosely coupled or constructed as well as EAI (Enterprise Application Integration based on Services) can be provided, inter alia for the clouds [22].

2.4 Demarcation of Web Services and Micro-services

The demarcation of Web Services and Micro-Services is given below. Firstly, let us to illustrate and explain the evolution for app implementation in Distributed Systems and HDS, please use the one depicted below in Fig. 2:

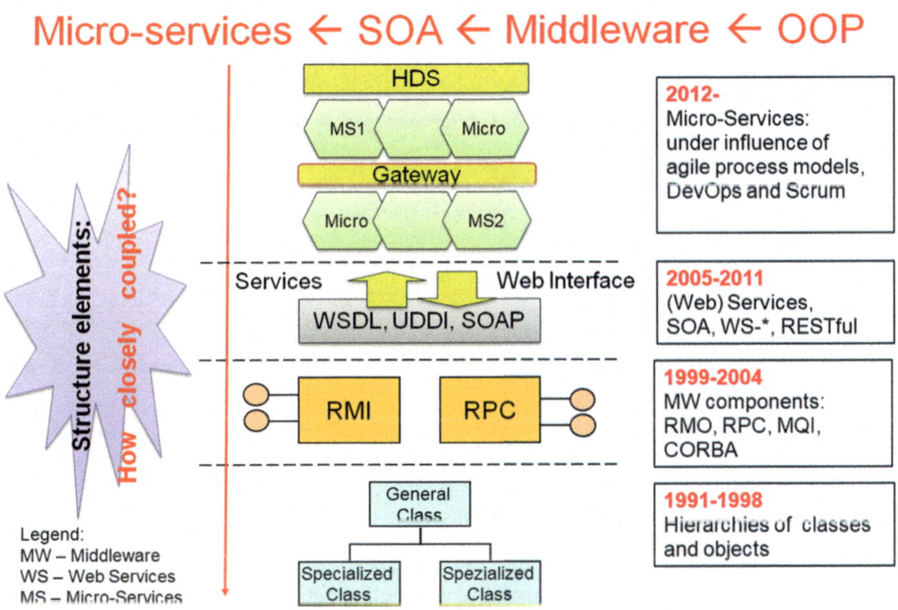

Fig. 2. Evolution of app implementation [6, 12]

The given diagram presents the short periodization in the app implementation in distributed systems or in HDS. Since the 1990s, the implementation of apps in distributed systems has been striving for ever looser coupling of the components with each other or convenient adaptation or EAI under heterogeneity conditions. The cost factor (CAPEX/OPEX) is also reduced through source code unification and increased reusability. The depicted evolution goes from initial OOP thru middleware (MW) components under use of RPC, RMI to SOA via WS and Micro-Services [3, 6, 12, 14, 15, 18, 22]. The approach follows consequently to increasingly looser coupling (refer Fig. 2).

2.5 Three Main Questions

The following three questions arise in front of us:

1. Which of known SWT (software technology) process models offer an efficient basis for the expansion and deployment of Micro-Services?
2. Which architecture components use Micro-Services?
3. Which main platforms for Micro-Services Deployments can be used?

The first question can be answered based on Fig. 3:

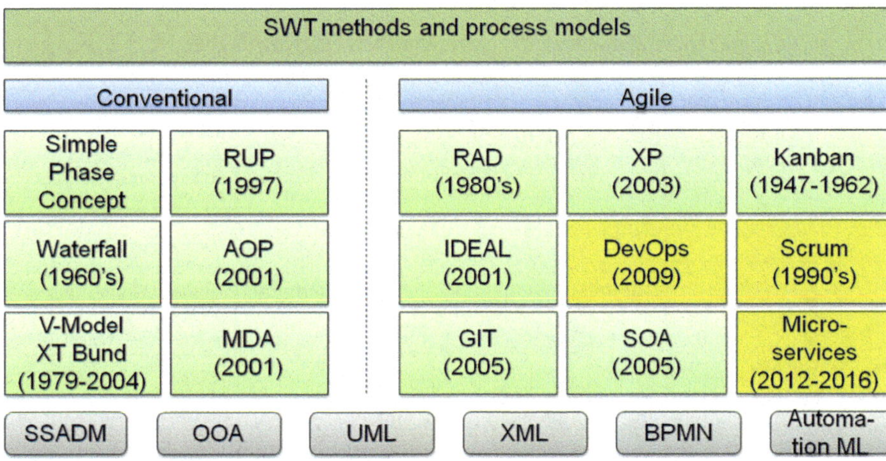

Fig. 3. SWT process models in context [18, 23]

A process model organizes a workflow for creative production (SW development) into various, structured sections and phases, which are assigned to corresponding technologies, tools, languages, protocols and methods of the organization (refer Fig. 3). The mostly known of them are listed below:

1. AOP – Aspect-Oriented Programming (Xerox PARK);
2. BPMN – Business Process Model and Notation (OMG);
3. IDEAL – SW process of Carnegie Mellon University;
4. MDA – Model-Driven Architecture (OMG);
5. OOA – Object-Oriented Analysis (P.Coad, E.Yourdon);
6. RAD – Rapid Application Development (J.Martin, B.Boehm);
7. RUP – Rational Unified Process (P.Kruchten, Rational);
8. SOA – Service-Oriented Architecture (Web Services);
9. SSADM – Structured systems analysis and design method (T.de Marco, E.Yourdon);
10. UML – Unified Modeling Language (G.Booch);
11. V-Model – SW process federative model (Germany);

12. XP – eXtreme Programming (M.Fowler).

For instance, Scrum as an efficient process model (refer Fig. 4) can be considered [3, 6, 12, 18, 23].

Scrum is a virtual analogy to rugby sports and describes the successful teams, which are working on the development of a project or SW product together and involved in close human communication.

Fig. 4. On scrum method explanation [7–11]

As an iterative SWT process model, Scrum does not describe classic project phases, but instead attaches the values of so-called continuously usable results (artifacts) right from the beginning of the project.

The Scrum is oriented on three basic principles for three people (Roles), four meetings (Sprint) and six results (Artifacts).

Scrum is nowadays a "de facto" agile SWT standard and is interoperable with advanced SOA as well as Micro-Service architectures.

Based on Conway's Law, the micro-services are involved in flexible SW development [7–11].

For unified definition of a software project, roles, events and artifacts (i.e. obtained handmade results) must be assigned. The defined roles belong to the Scrum workflow: Development Team, Scrum Master, Product Owner. Furthermore, a so-called Sprint Process with regular meetings with validation and review of project results must be processed.

The typical slogans by software development under use of Micro-Services via Scrum process model (refer Fig. 3 and 4) are as follows:

- Slogan 1: "One for all and all for one" (a musketeer slogan);
- Slogan 2: "Do twice in half time" (s. XP);
- Slogan 3: Unix-Philosophy („Do One Thing and Do It Well!");
- Slogan 4: Two pizzas teams (6–8 people can be satisfied with two big pizzas) – refer Fig. 5.

Fig. 5. Two pizzas concept [7–11]

IBN content uses normally the Micro-Services as building material for the running distributed apps. The use of Micro-Services: what does it mean indeed?

1. Concept for modularization, specific organization and SWT approach
2. Component basis: a single Micro-Service, independent and technically oriented
3. Flexible architecture: Micro-service architecture is represented as a large entity across all individual micro-services.

The mostly popular citations about [7–12]:

- James Lewis: "Micro-Service is a small application that can be deployed, scaled and tested independently. This application and its code should be easy to understand and do exactly one job."
- Sam Newman: "Micro-Service is a self-executable software component that collaborates with other software components within an integrated application. Micro-Services should take on a role within a specialized business domain."

The next question is as follows: which architecture components use Micro-Services?

Using, for example, Scrum approach, a much more flexible MS-based structure can be derived from WS-based 3-tier structure (Macro-Services). Using an application scenario for a supermarket, one can design simpler and more casual coupling of the services, whereby the access times to the layers involved are saved, especially on the data layer (refer Fig. 6) with individual database (DB) control.

Which main platforms for Micro-Services Deployments can be used? The main platforms and frameworks for IBN software implementation with Micro-Services are as follows: Netflix, Spring, Kubernetes. The comparison of their main properties is given in Fig. 7.

Let's give an example. A gateway with MS acts as a service provider and can deliver the required codes from the content specific MS1, MS2, MS3 under use of widespread protocols and interfaces like HTTP/REST, AMQP/JSON-RPC, WebSockets/ WAMP to several apps like Browser App, Native App, Server App. An appropriate example of MS use is given below (Fig. 8):

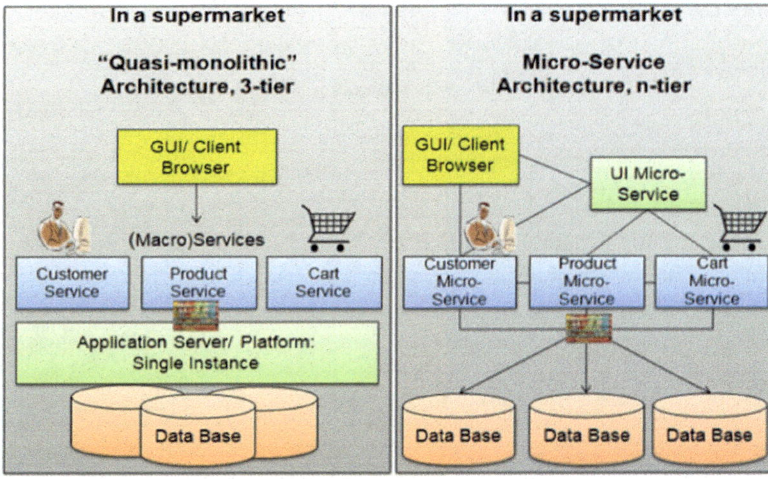

Fig. 6. Quasi-monolithic approach vs. micro-services [18, 23]

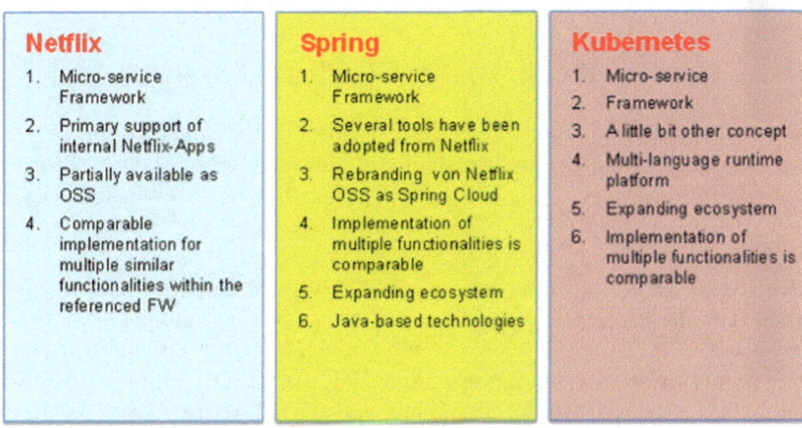

Fig. 7. Micro-services platforms at glance [7–11]

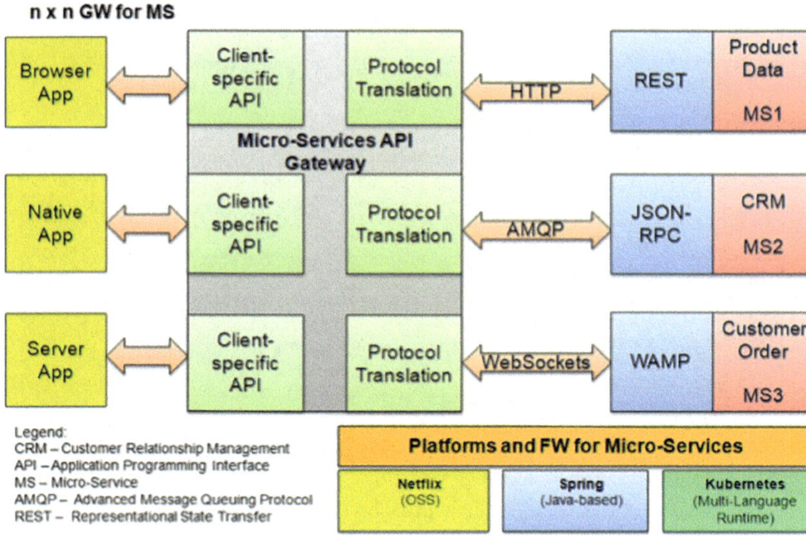

Fig. 8. Micro-services as architecture components [18, 23]

3 Service Composition in Highly-Distributed Systems

And, furthermore, the next question for use of HDS in IBN must be answered. What does "service composition" mean? Are there differences in service composition for Web Services and Micro-Services? The following distinguishing features are to be determined [12–25, 32]:

- Service composition is as a rule process- and service-oriented;
- Web services, Micro-Services and their platforms are very suitable as an implementation basis;
- Service composition offers component and source code reusability (Fig. 9);
- The approach provides solution embedding and offers fine granularity for the process business logic (Fig. 10);
- Flexible design paradigms are available (refer previous Sections);
- Much looser coupling, in contrast to so-called quasi-monolithic architectures of applications in convenient distributed systems and networks (refer previous Sections).

The difference in use of Web Services and Micro-Services can be explained with Fig. 11. Web Services are usually oriented for EAI (external coupling) in the style "Business-to-Business" or "Business-to-Consumer" and heterogeneity breakdowns. In contrast, by the use of Micro-Services is acting about an internal flexible structure. Some known authors (Sam Newman) call the convenient approach under use of Web Services and SOA as "quasi-monolithic" or, even, "monolithic", which is completely incorrect (refer Fig. 2). They would like only polemicize and, therefore, underline the importance to the Micro-Service Architecture transition. Let's compare the both (refer Fig. 11). As you can see, there is n-tier-structure with more flexible composition and intrinsic interaction between the services as a rule.

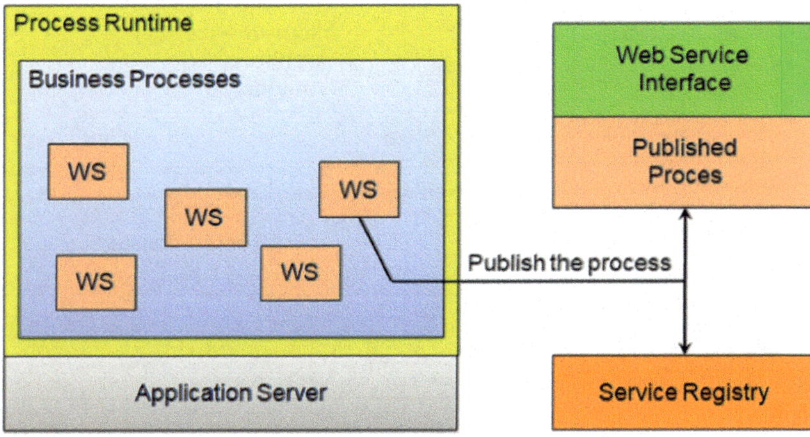

Fig. 9. Micro-services: reusability [6, 12]

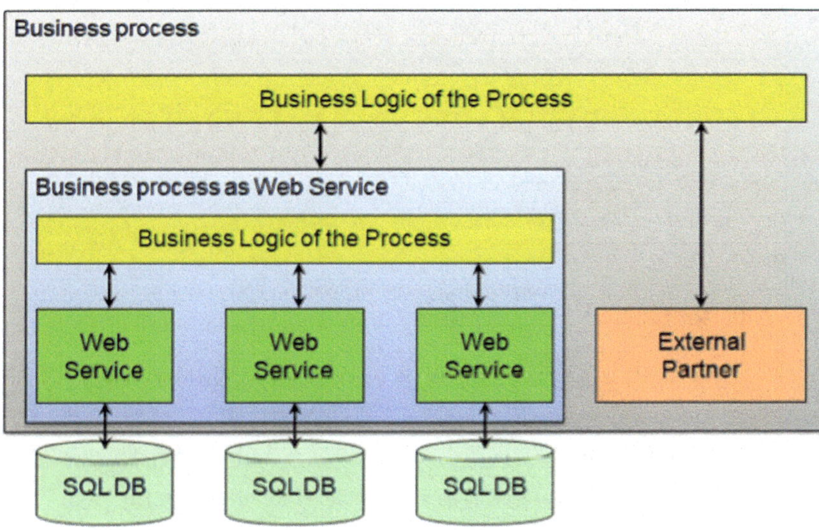

Fig. 10. Micro-services: embedding and granularity [6, 12]

The following questions can be asked on this point:

1. What are the advantages of centralized or decentralized composition of services (Web Services or Micro-Services)?
2. Clarify the differences between the both concepts: orchestration and choreography (Fig. 12).

The service composition is generally possible via two general approaches (refer Table 1): a) Service Orchestration (Centralized Approach); and b) Service Choreography (Decentralized Approach).

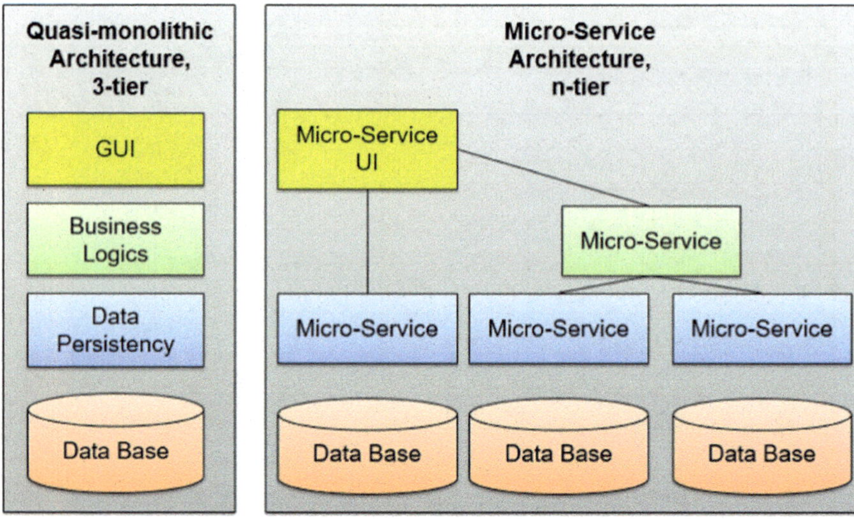

Fig. 11. Micro-services: intrinsic view [18, 23]

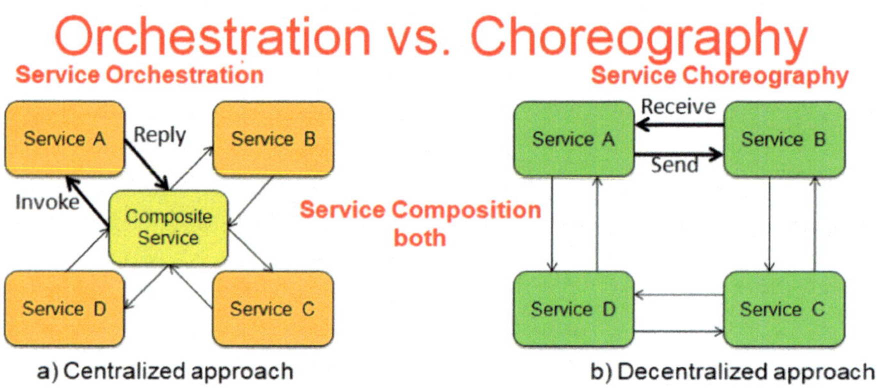

Fig. 12. Micro-services as architecture components [6, 12]

The given comparison (refer Table 1) depicts that choreography differs from an orchestration with respect to where the logic that controls the interactions between the services involved should reside.

Table 1. Comparison orchestration via choreography

Centralized approach	Decentralized approach
Service orchestration represents a single centralized executable business process (the orchestrator) that coordinates the interaction among different services. The orchestrator is responsible for invoking and combining the services	Service choreography is a global description of the participating services, which is defined by exchange of messages, rules of interaction and agreements between two or more endpoints
The relationship between all the participating services are described by a single endpoint (i.e., the composite service)	Choreography employs a decentralized approach for service composition
The orchestration includes the management of transactions between individual services	The choreography describes the interactions between multiple services, where as orchestration represents control from one party's perspective
Orchestration employs a centralized approach for service composition	Choreography employs decentralized approach for service composition

The Micro-Services' demarcation to SOA is given in Table 2 below:

Table 2. Comparison SOA and micro-services

SOA	Micro-Services
Both SOA and Micro-Services use services as architectural elements	
SOA uses the services for integration of different apps (EAI) Frequently external integration, as internal also available	Micro-Services bring a structure to an App under use of the services Better internal integration
So-called quasi-monolithic structure	n-tier structure
The combination of the services (Service Composition) can be "orchestrated" or "choreographed" (so-called Orchestration or Choreography), and the portals can provide a common GUI for all services (central or decentralized composition)	Each Micro-Service can include a GUI and implement the business processes in similarity to a SOA with Orchestration (centralization)
Conclusion for the comparison: loosely coupled, more flexibility by Micro-Services	

4 Considering of QoS Robustness and Security

4.1 QoS Robustness and QoE

With the development of IBN and HDS, the user needs and behaviour have significantly changed. The focus shifts from improving network performance and QoS parameters to improving the perception of Quality of Experience (QoE).

Providing according to the intentions of users of a given level of QoE for services and applications becomes a fundamental task in the implementation of end-to-end resource management in the concept of IBN.

The main idea of using IBN is to change the paradigm of the network infrastructure: now it is not the user with his application that adapts to the capabilities of the network, but the network changes its settings according to user requirements. Thus, for information systems based on IBN technology, the task of system administrators goes from manual configuration to programming and defining an intelligent network development strategy. In the future, the developed information technology IBN will automate the management of all domains of the network, including campuses, branches, WAN, IoT, 5G, 6G and Big Data, providing a significantly new level of automation, improving service efficiency, innovation and network infrastructure.

Intent-based networking raises new challenges for researchers to investigate, redesign, and develop SDN-based networks in the 5G/6G era.

4.2 CIDN for IBN

The widespread modern Intrusion Detection Systems (IDS) evaluate and prohibit the potential hackers' attacks that are directed against a computer system or a network. IDS increase data security significantly in opportune to the classical firewalls which lonely deployment is not satisfying. Intrusion Prevention Systems (IPS) are the enhanced IDS which provide the additional functionality aimed at discovering and avoiding of the potential attacks [1, 2]. Nevertheless, as a rule the classical IDS/IPS are operated autonomously. They are not able to detect temporary unknown hackers' threats, which become more sophisticated and complex year by year. Those dangerous threats can serve to disrupt the operation of IBN: data centres, IoT and robotic clusters round-the-clock in 24/7-mode. Therefore, the cooperation and collaboration of the IDS within a network is of great meaning [1, 2]. A CIDN is a further concept for a collaborative IDS/IPS network intended to bridge over the disadvantage of the standalone defence against the unknown dangerous attacks (Fig. 13). The CIDNs allow the participating IDS as the network peers to share the detected knowledge, experiences and best practices oriented against the hackers' threats [1–3, 15, 25].

The main requirements to the construction of a CIDN and the support of such functionality are as follows: efficient communication at short up to middle distance, robustness of the peers (IDS) and links, scalability and mutual compatibility of individual participating peers (single IDS). The typical interoperable networks are as follows: LAN, 3-5G mobile, Wi-Fi, BT and NFC. Collaborative intrusion detection networks (CIDN) consist from multiple IDS-solutions under use of multiple "things", robots, gadgets, PC,

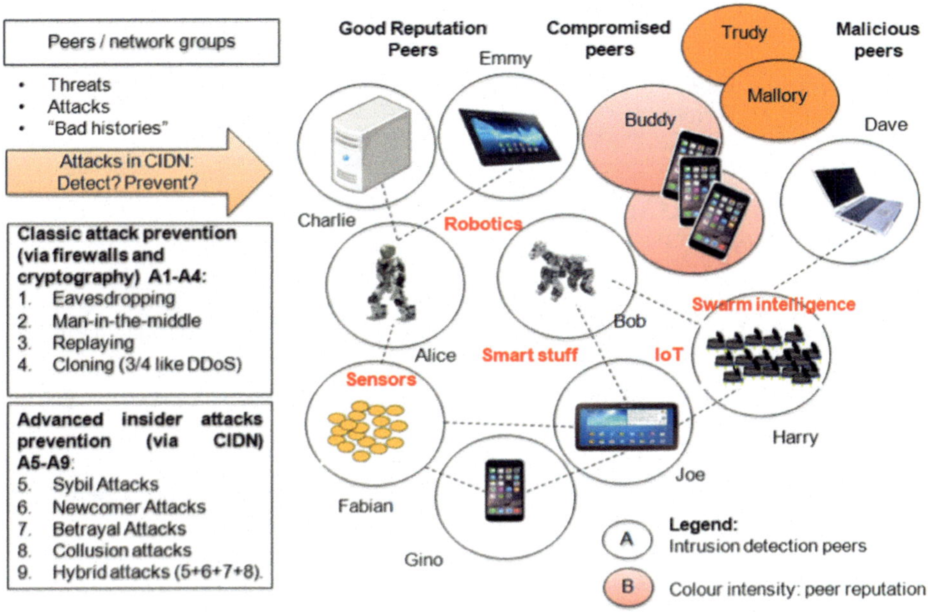

Fig. 13. Example of peer cooperation within a CIDN [6, 15]

end radio-devices and installed firewalls as well of the groups of users which are divided into the clusters – peers (titled as users Alice, Bob, Charlie, Dave etc.).

The coupling between the groups is loosely or tightly. However, the reputation of the users is quietly different (cp. Fig. 13): good, compromised (refer Buddy), malicious (refer Trudy and Mallory). Additionally, the insider attacks to CIDNs are possible (e.g. by Emmy with temporary "good" reputation).The CIDN can efficiently prevent such multiple attacks A1–9 (cp. Fig. 13) by peer-to-peer cooperation. This type of networking improves the overall accuracy on the threats assessment. The cooperation among the participating single peers (IDS-collaborators) became more efficient within a CIDN.

Unfortunately, the CIDN can become itself a target of attacks and malicious software. Some malicious insiders within the CIDN may compromise the interoperability and efficiency of the intrusion detection networks internally (Table 3). Therefore, the following tasks must be solved: selection of the peers (collaborators), resource and trust management, collaborative decision making [1–3, 15, 25].

Table 3. CIDN functionality

Certain CIDN examples	Detection and prevention of the attacks A1–A9	Topology type	Focus	Specialization on the further threats
Indra	+	Distributed	Local	SPAM
Domino	+	Decentralized	Global	Worms
Abdias	+	Centralized	Hybrid	Trojans
Crim	+	Centralized	Hybrid	Social Engineering, web appl. firewalls

5 Summarizing

The final question can be formulated here: what role do Micro-Services play in the development of HDS in IBN?

Distributed Systems

- The term "Distributed Systems" has been used for many years for the applications, which operate in modern combined wired-wireless-mobile networks under clear cooperation goals, as well as have no centralization in memory access or synchronization in the clocking.
- The distributed applications are constructed on the sample n-tier and often possess redundancy in form of server and database replications. They follow established SOA (service-oriented architecture) concepts and can be often organized as cloud-centric structures. Significant architectural transformations in network services and distributed systems characterize an ongoing trend nowadays.
- The clouds, clusters with explicit cooperation goals (e.g. parallelized computing) as well as grids belong to the above-mentioned systems.

Highly-Distributed Systems

- Since 2005 the P2P systems (Internet of Things, fog) in combination with convenient C-S communication model as well as server-less structures (SLMA, robotics) have gained in popularity. Then the Cloud-based solutions became a trend (2011) under predominant use of the load-balanced "thin clients" with functionality delegation to the clouds under use of macro- and micro-services.
- The access to the resources and data follows under use of service concept. The more flexible Micro-services bring a structure to the HDS instead of convenient external app integration (EAI).
- Under use of fog computing the IoT solutions are constructed.
- The workload is shifted on the edge to the energy autarky and resource economizing small nodes.
- The service execution is guaranteed and secured, inter alia, via CIDN and Blockchain.

6 Conclusions and Outlook

This work can be positioned as WIP: Work-in-Process. What else do we have?

In any case, in addition to the convenient n-tier and cloud-based structures, we need further comparison and analysis of further models: P2P, M2M etc. Which implementation basis do you choose for this target Web Services, Micro-Services or else) and why?

When comparing both convenient solutions with new solutions (HDS in IBN), we need an evaluation of decreasing development complexity (KLOC, manly hours) and costs (CAPEX/ OPEX), the reusability of the source code (components, SOA, patterns), the use of a possible current process model (such as Scrum, or alternatively, XP, DevOps), scalability and so-called utility analysis.

In addition, the assessment of the performance and QoS parameters in realistic scenarios is very important (data volumes, throughput, and latencies).

In the case of HDS software development, up to 25% of the total costs are attributable to security-related routines. For this reason, the analysis of how secure the IBN solutions are in realistic scenarios using methods such as VPN, PKI/ TLS, Blockchain in the desktop area and in the mobile environment.

References

1. Luntovskyy, A., Gütter, D.: Moderne Rechnernetze: Protokolle, Standards und Apps in kombinierten drahtgebundenen, mobilen und drahtlosen Netzwerken, 481 Seiten, 263 Abb., Springer Nature (2020). ISBN: 9783658256166, https://www.springer.com/gp/book/9783658256166
2. Luntovskyy, A., Gütter, D.: Moderne Rechnernetze - Übungsbuch: Aufgaben und Musterlösungen zu Protokollen, Standards und Apps in kombinierten Netzwerken, 145 Seiten, 44 Abb., Springer Nature (2020). ISBN: 9783658256180, https://www.springer.com/gp/book/9783658256180
3. Schill, A., Springer, T.: Verteilte Systeme: Grundlagen und Basistechnologien: Kompakte Darstellung der Grundlagen und Techniken Verteilter Systeme. Springer-Verlag, Heidelberg (2012). ISBN: 9783642257957, https://www.springer.com/gp/book/9783642257957#otherversion=9783642257964
4. Luntovskyy, A., Spillner, J.: Architectural transformations in network services and distributed systems: current technologies, standards and research results in advanced (mobile) networks, p. 344. Springer Vieweg, Wiesbaden (2017). ISBN 9783658148409, https://www.springer.com/gp/book/9783658148409#otherversion=9783658148423
5. Luntovskyy, A., Gütter, D., Melnik, I.: Planung und Optimierung von Rechnernetzen. Methoden, Modelle, Tools für Entwurf, Diagnose und Management im Lebenszyklus von drahtgebundenen und drahtlosen Rechnernetzen: Planung von Rechnernetzwerke theoretisch anspruchsvoll und praxisnah, Springer/Vieweg+Teubner, Wiesbaden (2012). 415 Seiten, 245 Abb, ISBN 9783834814586, https://www.springer.com/gp/book/9783834814586#otherversion=9783834882424
6. Luntovskyy, A., Klymash, M.: Software technologies for mobile apps, apps for fog computing. Robotics and Cryptoapps, Lviv, p. 247 (2019). Monograph, Ukrainian, ISBN 978–617–642–399–7
7. Newman, S.: Building Microservices: Designing Fine-Grained Systems (2014)
8. Newman, S.: Building Micro-Services, p. 473. O'Reilly Media, USA (2015). ISBN 978-1-491-95035-7

9. Newman, S.: Monolith to Microservices: Evolutionary Patterns to Transform Your Monolith (2019)
10. Fowler, M., Sadalage, P.J.: NoSQL Distilled (2012)
11. Fowler, M., Beck, K.: Refactoring (2018)
12. Luntovskyy, A.: Programming Technologies of Distributed Applications, Monograph, P. 474. Ukrainian, Kiev DUIKT (2010).
13. Luntovskyy, A., Guetter, D.: Cryptographic technology blockchain and its applications. In: Ilchenko, M., Globa, L. (eds.) Advances in Information and Communication Technologies, Springer, LNCS "Processing and Control in Information and Communication Systems (Int. Conf. UkrMiCo-2019)", pp. 14–33. Springer, Heidelberg (2019). ISBN: 978-3-030-16770-7, https://doi.org/10.1007/978-3-030-16770-7
14. Luntovskyy, A., Globa, L.: Performance, Reliability and Scalability for IoT. In: International IEEE Conference IDT-2019, Zilina, Slovakia, 25–27 June 2019, p. 6. IEEE Xplore (2019). https://ieeexplore.ieee.org/document/8813679, https://doi.org/10.1109/DT.2019.8813679
15. Luntovskyy, A., Klymash, M.: Robotic apps and platforms: mobility, localization, management and security aspects. In: International IEEE Conference AICT-2019, Lviv, Ukraine, 2–6 July 2019, p. 6. IEEE Xplore (2019). https://ieeexplore.ieee.org/document/8847741, https://doi.org/10.1109/AIACT.2019.8847741
16. Maksymyuk, T., Han, L., Larionov, L., Shubyn, B., Luntovskyy, A., Klymash, M.: Intelligent spectrum management in 5G mobile networks based on recurrent neural network. In: 15th IEEE International Conference on the Experience of Designing and Application of CAD Systems (CADSM), in Polyana, Ukraine, 26 Feburvary–2 March 2019, p. 6. IEEE Xplore (2019). https://ieeexplore.ieee.org/document/8779301, https://doi.org/10.1109/CADSM.2019.8779301
17. Luntovskyy, A., Globa, L., Shubyn, B.: From big data to smart data: the most effective approaches for data analytics. In: Advances in Information and Communication Technology and Systems, Springer Nature, Part of the "Lecture Notes in Networks and Systems Series", MCT 2019, vol. LNNS 152, Chapter 2, pp. 23–40 (2020). ISBN 978-3-030-58359-0, https://doi.org/10.1007/978-3-030-58359-0_2
18. Luntovskyy, A., Shubyn, B.: Highly-distributed systems based on micro-services and their construction paradigms. In: IEEE Conference TCSET 2020, Lviv-Slavske, Ukraine, 25–29 February 2020, p. 8. IEEE Xplore (2020). https://ieeexplore.ieee.org/document/9088603, https://doi.org/10.1109/TCSET49122.2020.235378
19. Hara, T.: Analyses on tech-enhanced and anonymous Peer Discussion as well as anonymous Control Facilities for tech-enhanced Learning, PhD Dissertation (TU Dresden) (2016)
20. Hara, T., Luntovskyy, A., Braun, I., Kubica, T.: Designing IT curriculum to foster self-regulated learning through peer instruction and audience response systems. In: IEEE Conference TCSET 2020, Lviv-Slavske, Ukraine, 25–29 February 2020, p. 8. IEEE Xplore (2020). https://ieeexplore.ieee.org/document/9088585, https://doi.org/10.1109/TCSET49122.2020.240939
21. Kubica, T., Hara, T., Braun, I., Kapp, F., Schill, A.: Choosing the appropriate audience response system in different use cases. In: Proceedings of the 10th International Conference on Education, Training and Informatics (ICETI 2019) (2019)
22. Luntovskyy, A., Shubyn, B.: Advanced architectures for IoT scenarios. In: 5th International IEEE Conference on Smart and Sustainable Technologies (SpliTech), Split and Bol, Croatia, 23–26 September 2021, p. 6. IEEE Xplore (2021). https://ieeexplore.ieee.org/document/9243784, https://doi.org/10.23919/SpliTech49282.2020.9243784
23. Luntovskyy, A., Shubyn, B., Maksymyuk, T., Klymash, M.: Highly-distributed systems: what is inside? In: 2020 Int. IEEE Scientific-Practical Conference on Problems of Infocommunications. Science and Technology PICS&T-2020, Kharkiv, Ukraine, 6–9 October 2020, p. 6 (2020)

24. Luntovskyy, A., Shubyn, B.: Energy efficiency for IoT. In: van Gulijk, C., Zaitseva, E. (Eds.) IEEE Workshop RECI-2020 on "Reliability Engineering and Computational Intelligence", Zilina, Slovakia, 27–29 October 2020 (online), in Springer LNCS series "Reliability Engineering and Computational Intelligence - Studies Comp. Intelligence", vol. 976, p. 16 (2021). ISBN: 978-3-030-74555-4
25. Luntovskyy, A., Shubyn, B., Maksymyuk, T.,.Klymash, M:5G slicing and handover scenarios: compulsoriness and machine learning. In: Vorobiyenko, P., Ilchenko, M., Strelkovska, I. (eds.) "Current Trends in Communication and Information Technologies", in Springer Lecture Notes in Networks and Systems series, Based on Conference IPF-2020 "Infocommunications - Present and Future", Odessa, 16–19 November 2020, vol. 212, p. 32 (2021). ISBN 978-3-030-76342-8
26. Luntovskyyy, A., Shubyn, B., Scherm, I.: Blockchaining for modern HDS, BA Magazine "Wissen im Markt", pp. 38–44 (2019). ISSN 2512-4366
27. Zobjack, T., Luntovskyy, A.: Blockchained IoT: Verbindlichkeit in der dezentralisierten Welt smarter Dinge, BA Magazine "Wissen im Markt", Berufsakademie Sachsen, pp. 75–81 (2020). (in German), ISSN 2512-4366, https://www.ba-sachsen.de/.
28. MIT Blockchain Course (2020). http://executive-education.mit.edu/MIT-Blockchain/Online-Course/
29. Blockchain as a Service: Teamwork ohne Daten-Grenzen (2020). https://www.t-systems.com/blockchain/
30. Survey on Crypto-platforms (2002). https://hackernoon.com/top-blockchain-platforms-to-watch-out-in-2019-aa80e336a426/
31. Wuest, K., Gervais, A.: Do you need a Blockchain? ETH Zurich & Imperial College London (2020). https://eprint.iacr.org/2017/375.pdf
32. Schneider, U., (Hrsg.): Kapitel Software Engineering, in Taschenbuch der Informatik, Hanser Verlag, 736 S (2012). ISBN 978-3-446-42638-2. (in German)

Intent-Based Placement of Microservices in Computing Continuums

Josef Spillner(✉), Juliana Freitag Borin, and Luiz Fernando Bittencourt

ZHAW Zurich School of Engineering, Obere Kirchgasse 2, 8400 Winterthur, Switzerland
Josef.Spillner@zhaw.ch

Abstract. Programmable computing infrastructure is increasingly available at heterogeneous locations across devices and data centres. This greater choice leads to opportunities to run applications and network services on top with improved matching of required or desired characteristics. A remaining challenge is to address the computing resources without forcing software engineers to reflect them directly into the software design. An emerging helpful notion on the infrastructure level is that of computing continuums of various types, such as fog-cloud or sensor-edge-cloud continuum. The emerging continuum computing paradigm is hence seen as evolution of cloud computing. It exploits recent execution technologies to guarantee loose coupling between applications and infrastructure as well as late and dynamic placement optimisation. Thus, it leads to fluid and osmotic distributed applications with inherent knowledge about how well specified constraints are fulfilled at any point in time. We suggest that continuums are therefore a suitable abstraction to learn in networked software engineering with benefits for application design, implementation and deployment. In this chapter, we present continuum computing scenarios and outline the current state of technology. Furthermore, we demonstrate how application placement can be controlled declaratively based on meaningful human and business vocabulary within this abstraction. This method contributes to application portability and resilience, important characteristics in the efficient digital transformation of whole industries.

Keywords: High-performance computing · Microservices · Cloud · Intents · Fog

1 Introduction

Computing capacity has become available everywhere in ubiquitous form, through visible and invisible devices. Some of that capacity is typically idle and not contributing meaningfully to complex data processing, while the remainder may be exhausted in burst scenarios. A key concern is therefore for any planned computing activity: Where do we compute?

Given that the computing activity can be not only a monolithic process, but can be modular for instance in the form of composite microservices, that concern can then be refined to: How do we decompose and position our computation in order to maximise the capacity utilisation or other, user- defined utility goals?

M. Klymash et al. (Eds.): *Future Intent-Based Networking*, LNEE 831, pp. 38–50, 2022.
https://doi.org/10.1007/978-3-030-92435-5_3

Due the heterogeneity of hardware and networking topologies, addressing this concern in a general way requires a powerful abstraction and notation.

Continuums have been proposed as a novel abstraction layer to express a continuous range of capacities [1]. Continuums range from networked sensor nodes referred to as the Internet of Things (IoT), handheld devices, robots, industrial production equipment and other far edge hardware, over edge and fog intermediate infrastructure potentially consisting of multiple layers, to more centralised cloud, multi-cloud and even high-performance computing (HPC) facilities. Inspired by the mathematical concept of continuously (albeit not necessarily linearly) progressing curves, such computing continuums provide similar curves, but for a variety of non-functional characteristics [2]. For instance, the latency to detect an anomaly in a data stream increases the more far away the detection code is executed from the point of data acquisition. It is therefore, from a latency perspective, desirable to detect anomalies locally at a sensor rather than in the cloud. In contrast, the available memory is usually growing with this distance and is maximised in the cloud. This means that continuums are multi-spectral targets for any compute activity, and mathematical optimisation can be applied, constrained by technological aspects such as network protocols and software artefacts, to address the concern of decomposing and placing the compute activity to maximise the utility.

Figure 1 shows typical computing continuums and their use cases along with the typically associated characteristics for one of the continuums. It is clear that given enough investment, many of the characteristics can be harmonised, for instance, by bringing more compute power to the edge by installing more processor cores. However, this comes at a cost whereas the concept of continuums is able to exploit existing hardware with its resource differences. Moreover, several sensor nodes and other IoT equipment are not capable of running user-provided code at all or only in specific forms, limited to certain programming languages and frameworks such as MicroPython [3].

The intelligent computing continuum has been specifically proposed to automate some of the placement decisions. This paves the way for merely expressing the intent as utility function, raising the abstraction level further and thus allowing for more dynamic continuums whose topology and capacity changes with the computing activities being adjusted to still converge against the goal. Intelligent continuums promise resource and data management at different levels with largely heterogeneous computing and networking capabilities. They fulfil the promise by proper resource management, placement and scheduling with explicit knowledge, including a consideration of partial lack of knowledge, about the continuum and application characteristics within the ecosystem. Eventually, intelligent continuums may turn into intelligent data fabrics for elastic cloud-to-edge intelligence [4].

Intelligent continuums are therefore useful to translate intent from the soft- ware design level to operational practice. But this translation is subject to many unsolved challenges, including insufficient coverage with wireless connectivity capable of network slicing, unknown energy consumption trade-offs, too complex federation concepts and overload through massive mobility [5]. The main challenges from a software engineering and placement perspective, related to orchestration, are:

1. Understanding and modelling application requirements, in terms of non-functional characteristics. An example would be a licence plate detection application based on a

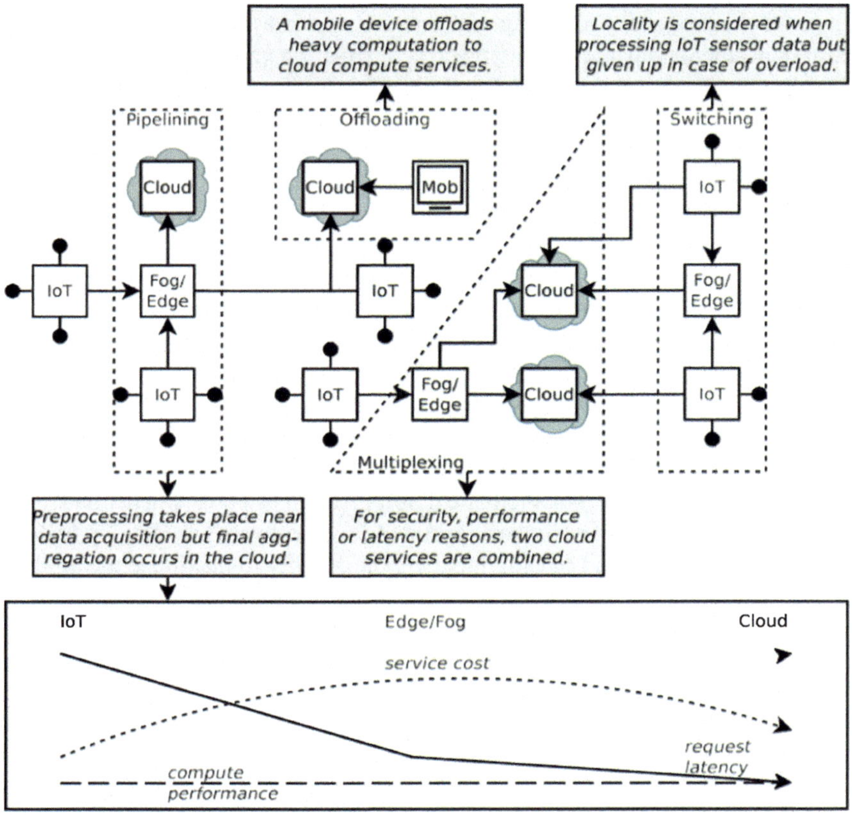

Fig. 1. Typical computing continuums and resulting computing characteristics

video camera that would require low latency (to alert police before the car leaves the area), high throughput (to raise the likelihood of detection with better quality video streams), high reliability (ability to process a secondary video stream), and high privacy (early blurring of any faces recognised in the streams, as only the licence plates are of interest).

2. Modelling and monitoring of infrastructure capacity. This primarily means tracking the available resources in terms of CPU/GPU cores, frequencies and instruction sets, available main memory, persistent storage, network links, hardware security tokens and secure computation enclaves, and other special features that form the basis of matching the application requirements. Apart from the basic hardware description, in dynamic contexts this information would change at high frequency by taking also the current load for all shared resources and the current reservation status for all exclusive resources into account.

3. Decision making on where the processing should occur. This is a match- making process optimising the combination of application requirements with resource availability at any point in time, including the initial deployment of applications, redeployment in dynamic scenarios as well as per-request planning for already deployed applications

even in static scenarios. At this stage, the user intent on what constitutes optimality must be expressed and evaluated. The decision making is furthermore subject to external constraints in privacy, security [6] and, primarily setting limits to scalability and usage density, the utility capacity for electricity and connectivity.
4. Application orchestration. When a decision has been made, all application components must be managed, including seamless service offloading preferably without downtime and data transfers. This challenge also involves technical decisions on caching, cross-layer replication and migration, as well as on execution choices involving virtual machines and containers, both as long-running services and as per-request instantiations. The necessity to mi- grate can in particular be triggered either by changing conditions in the infrastructure and the application workload, or by device mobility, causing the intent utility functions to yield different optima that will be described in the next section.

The concrete realisation of algorithms and systems to meet all challenges depends largely on the far edge devices. For instance, mobile and non-mobile devices can share edge and fog resources, but a proper resource allocation in the layered hierarchy is then needed. An appropriate notion is that of a cloudlet, a specific part of the computation not executed in the cloud but in a fog environment close to the devices and itself communicating with the cloud as needed. Moreover, the realisation depends on the permitted cost/performance ratio driven by overhead costs for the modelling and monitoring processes as well as bottlenecks for the dynamic migrations. Eventually, the optimality of the modelled utility functions may be relaxed with near-perfect results that can be achieved in a fraction of the time. This is especially crucial in massively parallel scenarios where thousands of sensors, people and devices are coupled and 90% optimality for all is better than the optimum for 90% with unforeseen consequences for the remaining 10%. It is also important for integrated microservice engineering and quality assurance processes, in particular in DevOps contexts, that require sub-second checking of the feasibility of placement during code commits.

To convey the advantages of a continuum for engineering better software, in this chapter we describe two concrete processes: An initial placement matchmaking between applications and infrastructure with multiple implementations (RBMM [7] and Continuum Deployer [8]), and a simulated migration and re-placement with wireless handoff and application migration implemented as MobFogSim [9]. A general overview about software engineering concerns for continuums is given beforehand.

2 Placement, Migration and Other Software Engineering Concerns

Initial placement and migration are two closely related steps within a wider software service lifecycle. Software optimised for continuums shall be adaptive and liquid, allowing for optimal execution in any constellation of underlying hardware infrastructure. 'Coding the Continuum' has been one of the first works that looked at continuums from a software engineering perspective and explored techniques to achieve such software [10]. While earlier abstractions such as network-procedure calls (RPC) allowed software to cross hardware boundaries, the inherently high latency led to the manifestation of fallacies and therefore to functionality and usability problems [11]. The fallacies still apply, but they manifest less often, so that continuums are now set as next- generation abstraction. Hence, coding the continuum departs from conventional fixed terms such as 'cloud' and 'edge' and rather puts the (sometimes dynamic) characteristics of these resources into the focus. These characteristics combine basic properties such as speed and location into holistic measures so that a function executing on a slower computer nearby may deliver its results faster than the same function executing on a far-away faster machine. A continuum-aware programming model thus combines function, data and trust fabrics along with the associated cost. Therefore, the user intent can be expressed as a set of utility functions, for instance to maximise performance as location-speed trade-off that is shifted depending on device proximity due to the distance-based networking cost.

Similar ideas are represented in 'Harnessing the Computing Continuum', with emphasis on abstract machine models so that programming is elevated from the device level to intellectual goals and intents [12]. A mapping of user intent to the resource level is encapsulated by the abstract machines.

However, exploiting resource characteristics requires high-quality knowledge about them. The quality is determined by valid, accurate, up-to-date specifications. Without the knowledge, blind computing ensues. Various approaches to capture cost, characteristics and constraints exist, including CloudPick [13], SCRIMP/ParaOpt [14] and FaaSCC [15]. Apart from static capturing, monitoring can be used to get dynamic information on resource utilisation. Monitoring can also be used to detect the geo-coordinates of mobile devices to support mobility-based placement and migration.

On a practical application level, descriptions of microservice resource constraints also need to be collected. There are de-facto standards from industry, for instance, resource limits in Kubernetes deployment objects such as Docker container images (and therefore in Helm charts) and AWS Lambda function configuration (and therefore in Serverless Application Model packages). Nevertheless, these facilities are not always used, with only around 25% of Docker images and 63% of Lambda functions declaring resource limits according to a public sample [8]. Dynamic approaches such as the Microservice Artefact Observatory can be used to determine unspecified resource constraints [16].

Once the knowledge about both applications and resources is available, liquid software that adapts to the available infrastructure becomes feasible. This concerns especially the initial placement of software units, in particular microservices, as well as any migration due to changing conditions. In the next sections, we assume the existence of the knowledge and refer to the previously mentioned works for strategies on how to retrieve missing information. Table 1 shows exemplary knowledge on deployment decision factors based on the categorisation proposed in [7], scoped to apply to either the application A and/or the resources R. Additionally, external context information including jurisdiction, business strategy, date/time and system load as well as user intent must be available in machine-processable format.

3 Matchmaking as Planning Phase for Initial Placement

The aim of achieving an initial placement by matching application requirements, broken down to individual microservice requirements, and infrastructure capacity is the

Table 1. Subset of deployment decision factors

Scope	Name	Values (examples)
A, R	Memory	128 MiB, 2 GiB
A, R	Runtime	Python:3, java
A, R	Latency	5 ms
A, R	Duration	900 s
A, R	Zone	Intranet, dmz, internet
A	Vulnerability	Backdoor, CVE-477
A	Consistency	True, false
A	Complexity	High, medium, low
A	Port	9233
R	Country	gb, cn
R	Geolocation	WGS84 47°29′N 8°43′E
R	Trust	High, low
R	Billing	Monthly, pay-per-use, free
R	Gpu	True, false

rapid transition from development to operation. In this section, a sample application is described, formalised preliminaries are presented and four matchmaking approaches – greedy, rule-based, SAT and advanced intent-based matching – are described. The rule-based approach is based on RBMM [7], whereas the greedy and SAT-based approach is based on the Continuum Deployer [8].

3.1 Example Description

To better convey the methods for initial placement, a running scenario is introduced that references the characteristics from Table 1. A machine manufacturer intends to deliver real-time insights into the machine performance, including in potential faults in the context of predictive maintenance. The machine ships with an on-board computer C1 equipped with 2 GB RAM and a 3 GHz CPU. Moreover, the machine operator can attach the machine to a cloud service C2 with the equivalent of 2 GHz CPU performance and 1024 MB RAM instances. The insights application consists of two microservices M1 and M2 implemented as containers, each requiring at least 800 MB RAM (or more depending on the workload), and fully exhausting the CPU capacity during heavy analytics, represented by a nominal 1.5 GHz requirement per microservice. Moreover, a third microservice M3 implements the predictive maintenance forecast. It is implemented as a workflow consisting of three consecutively invoked cloud functions F1–F3, each requiring 100 MB RAM and a small share of the CPU.

3.2 Preliminaries: Definitions and Models

We define a composite application A to consist of a number of software artefacts a_i which are loosely coupled and instantiated as application execution units, or parts, with certain scaling factors, i.e. $A = \{s_0 \times a_0, s_1 \times a_1, \ldots\}$. A continuum resource collection R consists of independent resources r_i whose owners or operators can differ, leading to further differences in location as well as technical characteristics including the resource level (infrastructure, platform, middleware).

Both artefacts and resources have certain properties, although not all of them are guaranteed to be explicitly expressed in machine-readable descriptions. Sometimes, they are also loosely expressed, for example a runtime environment *java* without corresponding version number. Therefore, we assume those *deployment factors* to contain a measure of uncertainty, i.e. $F = \{u_0 \times f_0, u_1 \times f_1, \ldots\}$. In component notation, resource factors are *offered* whereas artefact factors are *required*. Some factors only exist for either artefacts or resources, while others exist for both; this is expressed by the *factor scope* (*A*, *R* or both).

A deployment plan (assignment) is the projection $A_{\rightarrow C+P} R_{used} R$ under a set of conditions C and a set of preferences P. Conditions must be fulfilled (e.g. sufficient memory to run the application, monthly cost not more than a certain limit) whereas preferences are used to determine the winning resource combination out of several that fulfil the conditions (e.g. smallest possible latency). Multiple preferences can be combined by weights, i.e. $P = \{w_0 \times p_0, w_1 \times p_1, \ldots\}$. All conditions and preferences are expressed with application-specific rules (Ψ) referencing arbitrary deployment factors F and applying to any pair (a_i, r_i).

Our model is limited by not taking data dependencies or workflows into account. Specifically, we assume for simplicity that any application part can generate data and transmit it freely to any other part. We acknowledge this limitation while claiming that even the simplified model advances systematic deployment methodologies beyond current deployment tool designs for clouds and continuums. On the other hand, the model is flexible by allowing to skip certain factors so that a subsequent matchmaker or deployment tool can perform further micro-optimisation. These skipping rules, together with the deployment rules as well as propagation and accumulation rules, lead to a highly flexible approach that fits multiple deployment patterns and topologies.

3.3 Greedy Matching

A greedy strategy returns the first matching association of microservices to the available resources for each service. This is a very fast strategy, but also one that may get stuck, not finding or a combination that works although one exists, or deliver a suboptimal result instead of finding the best combination. Consider the following case documented in Table 2. Although a match is found after four steps, the allocation is not optimal. Placing the maintenance service M3 first would require only one cloud instance instead of three. Of course, this example is simplified and does not take data affinity, pricing or runtime compatibility into account. Nevertheless, it shows that greedy matching is fast at the cost of a risky outcome.

Table 2. Greedy matching protocol

Round	A	R	Result
1	M1	C1	Fits; adjust C1: 1.5 GHz, 1.2 GB RAM
2	M2	C1	Fits; adjust C1: 0 GHz; 0.4 GB RAM
3	M3	C1	Does not fit (will starve)
4	M4	C2	Fits; requires 3 instances

3.4 Rule-Based Matching

Rules (Ψ) applying to factors are composed of propagation rules (Ψ_π), skipping rules (Ψ_σ) deployment rules (Ψ_δ) and accumulation rules (Ψ_α). They are applied in this order: First, propagation rules use invariants to complement missing factors or change existing factors in application compositions and resource sets. On this basis, skipping rules temporarily hide factors – marking them to be skipped during processing – so that they remain intact in the output mapping and serve as input for further post-processing. Then, matchmaking is performed with deployment rules, and all successful assignments imply the use of accumulation rules to adjust post-deployment resource characteristics. In case an assignment is reverted, for instance through back- tracking, the accumulation rules are executed in inverse order to roll back forecasted resource modifications.

3.4.1 Rules

Propagation Rules
By considering a hierarchical application and resource model, it is possible to specify which factors at lower levels invariably influence those at higher levels, and vice-versa, as well as lower level siblings (up-, down- and side- propagation). The hierarchy can have multiple levels, although on the application side many composition formats adopted in industry only support two levels. Resources can however have sub-resources, for instance a VM instance offering both CPU and GPU computing access. Through propagation, further efficiency gains can be achieved when only modelling some of the factors that have to be modelled manually or whose automated acquisition consumes a lot of time. For instance, a software composition affected by a security vulnerability in one constituent unit is considered itself tainted, representing an up-propagation, whereas another component in the same composition remains unaffected by itself. Specifically, the following propagation rules, including two trivial ones, are useful in continuums and need to be expressible as pre-processing step in matchmaking.

1. Replication. All factors in *a* trivially apply to all other instances of the same artefact. This applies to all static factors as well as, assuming they share the same resource, to dynamic runtime-related ones such as maximum task processing rate. However, in our work we apply these rules before deployment and thus do not consider dynamic factors.
2. Subsumption. The resource needs of *A* are trivially defined as the con-junction of all resource needs of the constituent *a* including scaling factors (up).

3. Bounding. The upper bound of latency in a is mirrored in A (up).
4. Tainting. Any quality deficiency or security vulnerability in a is mirrored in A (up), and any trust level in r is mirrored in the subset of R that shares the same operator (side).

Skipping Rules
A potential use case of matchmaking is generating a subset of valid deployment mappings, and further post-processing it with another matchmaking or optimisation tool that might use different algorithms for achieving increased output precision. Skipping rules mark attributes such as CPU or memory needs by the application, and the corresponding offered capacity by the re- sources, effectively leading to them being skipped during the matchmaking. Afterwards, they get reinstated on the resulting mapping. Possible skipping rules are:

1. Context. Skip CPU and memory factors, deferring these technical details to later, and instead perform matchmaking primarily on context factors.
2. Feasibility. Pre-check whether a deployment is technically feasible at all by skipping non-technical factors such as trust, country or geolocation.

Deployment Rules
These rules set constraints on where each application part a can be deployed. We require the deployment rules to express the following scenarios:

1. An application processing sensitive personal information shall not be deployed in hosting locations whose jurisdiction does not support certain minimum guarantees on privacy.
2. An application subject to a vulnerability shall only be deployed into the demilitarised zone (DMZ), not in the internal network behind the firewall.
3. Any application part a needs to be deployed into a resource with sufficient memory. For latency-sensitive applications, the entire application A needs to fit within one resource.

Accumulation Rules
Due to resource sharing and resource utilisation in general, each deployment leads to some changes in factors. We differentiate between constant factors unimpeded by any deployment (e.g. location of a datacentre) and those changing their values according to accumulation rules, for instance, available memory being reduced by any running application. More accurately, we assume that resource access is either unlimited, shared, or exclusive.

A concrete set of rules might look as follows:

1. The amount of free memory in r is reduced by the memory claimed by a (shared).
2. The range of free port numbers in r is reduced by any allocated port in a (shared).
3. A GPU available as sub-resource in r is occupied by any a claiming to perform GPU computing (exclusive).

The distinction into resource access models influences the permissible algorithms. Under the assumption that resources are infinite ($|R| = \infty$) or largely available beyond what can possibly be consumed by A, as in most clouds, a simple combinatorial matchmaking can be performed. Otherwise, a complex assignment and satisfiability problem needs to be solved. For constrained devices, we propose a depth-first recursive tree search where after each candidate assignment its validity is determined by successful matchmaking of the remaining subtree, otherwise rolled back.

3.4.2 Rule-Based and Weighted Matchmaking Concept

The matchmaking process guarantees that if a deployment of A including its constituent parts, e.g. microservices, to R is possible, a valid deployment plan is returned.

Design and Architecture
RBMM's matchmaking component operates as a service to be queried by deployment tools after scanning the composition description and artefacts to be deployed. Figure 2 outlines the process of acquiring the factors through automated scanning (acquirer tools) and manual curation, creating an instance of the models of A and R, and submitting them to the matchmaker to yield a resource-aware deployment plan.

The goal of the matchmaking process is to achieve an optimal deployment by creating, rating and ranking all possible assignment combinations of $A \rightarrow R$.

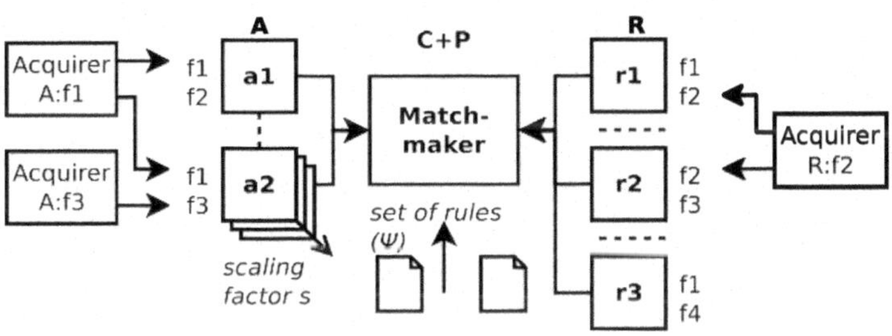

Fig. 2. Matchmaking based on collected factors, rules and constraints

We briefly describe an exemplary *fast combinatorial algorithm* assuming infinite resource availability and a more thorough recursive tree search assuming finite resources. These algorithms are suitable for simple scenarios but would be replaced with more capable ones in actual deployment systems.

Combinatorial Algorithm
The iterative combinatorial algorithm attempts to perform a mapping of all resources on all application parts. For any part a mapping has been found and validated according to deployment rules, the accumulation rules are applied. As further resources are then skipped, the algorithm complexity is approximately $O(n \times \frac{n}{2})$ for $|A| = |R| = n$.
Tree search Algorithm
In the recursive tree search, again for each application part a mapping is attempted. Any successful mapping of a_i applies the accumulation rules, followed by a recursive

invocation of $A \backslash a_i$, i.e. the set of application parts without the one already mapped. In case the invocation returns a valid result, it is proven that all application parts have been mapped successfully. Otherwise, the accumulation is reversed and the next resource is mapped for a_i. The complexity is approximately $O(n \times \frac{(n-1)^2}{2})$.

Matchmaker Library Implementation
First, we implemented the matchmaking algorithms as a Python library, and complemented it with a test tool to synthetically generate applications, resources and rules.

In an experiment with 10'000 application parts and the same number of resources, i.e. 100 million possible combinations, around 488 million factor comparisons were generated. On a single-core Intel i7 processor with 2.60 GHz, using only deployment and accumulation rules, the iterative combinatorial matchmaking took 3.3 s.

For a more modest scenario with 200 application parts and resources, i.e. 40'000 possible combinations and 175'275 factor comparisons, the combinatorial matchmaking finished in less than 0.05 s. For this scenario, the recursive algorithm implementation became feasible and finished in 80.1 s.

3.5 SAT Solver Matching

The constraint satisfaction problem (CP-SAT) solver [17] offers multiple options with regard to the optimisation target. Evaluated are six single and multiple targets, including the maximisation of idle CPU, the minimisation of idle memory, or the maximisation of idle combined resources. This solver uses constrained programming to define rules and constrains that describe the resource matching problem in mathematical terms. Afterwards this optimisation is solved as optimal as possible. The results of this solver differ from the greedy ones: if this solver cannot come up with an optimal solution the run will fail and all resources are displayed as unschedulable. This feasibility constraint is enforced on each label group (if labels are defined).

3.6 Towards Advanced Intent-Based Matching and Placement

The previously explained matching approaches (greedy, rule-based and SAT) are concerned with system-level metrics that are understood by software engineers and operators, but not on the business level. Intent-based placement becomes possible if the intent can be expressed in business terminology. This requires replacing static resource constraints with utility functions as well as, to ensure their understanding, typical domain-specific performance profiles. For instance, in the domain of artificial intelligence, a user wants to train 50 AI models in one hour and needs appropriate resources (mostly involving GPUs) to perform that task. The models themselves are rather small, therefore resource locality is less of an issue and remote cloud resources can be considered. In the domain of e-commerce and web hosting, a user wants to ensure 1000 Wordpress page views per minute are possible to reach a certain audience. In the factory automation domain, a user wants to ensure that a million observation transactions can be stored in an in-memory database per day.

The proposed approach foresees a rich application description on the level of microservice metadata as shown in the following example that is based on Kubernetes manifests:

```
resources:
  requests:
    cpu: 300m + x*30m
    memory: 512Mi + x*20Mi
    storage: 10Gi + x*1Gi
profiles:
  poc:
    x: 0
  production:
    x: 100
```

The application thus self-declares its ability to accommodate certain intents in using this application just in a test scenario or in a large-scale production environment, and adapts its resource utilisation to the chosen performance profile that most closely represents the desired intent.

4 Conclusion

The chapter contains theoretical methods and approaches to the placement of microservices. The simulation and other practical results aimed to Application Migration for Changing an Existing Placement will be published in another way. These experimental routines will show the readers how to simulate mobility and migration in cloud-fog continuums, particularly on the fog level as well as proof the examined theoretical methods and approaches for the placement of microservices. It will be based on a technical realisation as MobFogSim, an extension of iFogSim with mobility concerns.

References

1. AbdelBaky, M., Zou, M., Zamani, A.R., Renart, E., Diaz-Montes, J., Parashar, M.: 2017 IEEE 37th International Conference on Distributed Computing Systems (ICDCS), pp. 1815–1824 (2017). https://doi.org/10.1109/ICDCS.2017.323
2. Rosendo, D., Silva, P., Simonin, M., Costan, A., Antoniu, G.: 2020 IEEE International Conference on Cluster Computing (CLUSTER), pp. 176–186 (2020). https://doi.org/10.1109/CLUSTER49012.2020.00028
3. Gaspar, G., Fabo, P., Kuba, M., Dudak, J., Nemlaha, E.: Intelligent algorithms in software engineering. In: Proceedings of the 9th Computer Science On-line Conference 2020, vol. 1, Advances in Intelligent Systems and Computing, vol. 1224 (2020)
4. Silhavy, R.: Advances in Intelligent Systems and Computing, vol. 1224, pp. 388–394. Springer (2020). https://doi.org/10.1007/978-3-030-51965-0_34

5. Theodorou, V., Gerostathopoulos, I., Alshabani, I., Abelló, A., Breitgand, D.: The 35th International Conference on Advanced Information Networking and Applications (AINA)/M2EC: The 3rd International Workshop on Multi-Clouds and Mobile Edge Computing (M2EC) (2021)
6. Bittencourt, L., et al.: Internet of Things vol. 134, pp. 3–4 (2018). https://doi.org/10.1016/j.iot.2018.09.005. https://www.sciencedirect.com/science/article/pii/S2542660518300635
7. Mäkitalo, N., Ometov, A., Kannisto, J., Andreev, S., Koucheryavy, Y., Mikkonen, T.: IEEE Softw. **35**(1), 30 (2018). https://doi.org/10.1109/MS.2017.4541037
8. Spillner, J., Gkikopoulos, P., Buzachis, A., Villari, M.: 2nd International Workshop on Cloud, IoT and Fog Systems (CIFS)/13th IEEE/ACM UCC (2020)
9. Hass, D., Spillner, J.: The 35th International Conference on Advanced Information Networking and Applications (AINA)/M2EC: The 3rd International Workshop on Multi-Clouds and Mobile Edge Computing (M2EC) (2021)
10. Puliafito, C., et al.: Modeling and simulation of fog computing. Simul. Modell. Pract. Theory **101**, 102062 (2020). https://doi.org/10.1016/j.simpat.2019.102062. https://www.sciencedirect.com/science/article/pii/S1569190X19301935
11. Foster, I.: 2019 IEEE International Parallel and Distributed Processing Symposium (IPDPS), p. 1 (2019). https://doi.org/10.1109/IPDPS.2019.00011
12. Rotem-Gal-Oz, A.: Fallacies of Distributed Computing Explained (2005). https://arnon.me/wp-content/uploads/Files/fallacies.pdf
13. Beckman, P., Dongarra, J., Ferrier, N., Fox, G., Moore, T., Reed, D., Beck, M.: Harnessing the Computing Continuum for Programming Our World, chap. 7, pp. 215–230. Wiley (2020). https://doi.org/10.1002/9781119551713.ch7
14. Dastjerdi, A.V., Garg, S.K., Rana, O.F., Buyya, R.: CloudPick: a framework for QoS-aware and ontology-based service deployment across clouds. Softw. Pract. Exp. **45**(2), 197 (2015). https://doi.org/10.1002/spe.2288
15. Wu, C., et al.: 2019 IEEE International Conference on Cloud Computing Technology and Science (CloudCom), Sydney, Australia, 11–13 Dec 2019, pp. 255–262. IEEE (2019). https://doi.org/10.1109/CloudCom.2019.00045
16. Spillner, J., Boruta, D.: FaaS Characteristics and Constraints Knowledge Base (2020). https://github.com/serviceprototypinglab/faascc
17. Gkikopoulos, P.: 2019 IEEE World Congress on Services, SERVICES 2019, Milan, Italy, 8–13 July 2019, pp. 319–322. IEEE (2019). https://doi.org/10.1109/SERVICES.2019.00089
18. Col, G.D., Teppan, E.: Proceedings 35th International Conference on Logic Programming (Technical Communications). In: Bogaerts, B., et al. (eds.) ICLP 2019 Technical Communications, Las Cruces, NM, USA, 20–25 Sept 2019, EPTCS, vol. 306, pp. 259–265 (2019). https://doi.org/10.4204/EPTCS.306.30

Infrastructure as Code and Microservices for Intent-Based Cloud Networking

Marian Kyryk[1(✉)], Nazar Pleskanka[1], Mariana Pleskanka[1], and Vladyslav Kyryk[2]

[1] Lviv Polytechnic National University, Stepan Bandera street 12, Lviv 79013, Ukraine
marian.i.kyryk@lpnu.ua, mariana__p.m.v.9@ukr.net

[2] Warsaw University of Technology, plac Politechniki 1, 00-661 Warsaw, Poland

Abstract. Infrastructure as Code (IaC) has been recently receiving increasing attention in the research community, mainly due to the new approach in software design, development, deployment, and operations management. DevOps engineers efficiently maintain and continuously improve Infrastructure as Code. In this chapter, we present a new mechanism of infrastructure deployment, management and microservices building and delivery for future intent-based cloud networking. The process of setting up an infrastructure is similar to the process of programming software, when some scripts, modules, providers, and Version Control System are used together. The processes of how to build serverless microservices and how to create new content, reduce maintenance, scale easily, and deliver new features to users faster have been investigated. The main benefit of serverless platforms is that they let you focus on writing code without worrying about managing infrastructure, auto-scaling, or paying for more than what you use. With Cloud Function and Cloud Run, you can create high-quality microservices that will enhance the experience of your app or website. Cloud Run and Cloud Functions are serverless platforms offered by Google Cloud, but they have nuances that can make one preferable to the other in certain situations. The unique advantages and disadvantages of each of the platforms have been investigated.

Keywords: Automated network deployment · Infrastructure as Code · DevOps · AIOps · Deployment · Intent-based cloud networking · Cloud run · Cloud function · Google Kubernetes cluster · Version control

1 Introduction

Up until recently, IT infrastructure management was a laborious task. All hardware and software for the applications had to be manually configured and controlled by system administrators. However, new trends, including cloud computing, have improved approaches to organizing, developing, and maintaining IT infrastructure. The most important component of this trend is «Infrastructure as code - IaC».

The IaC is a model in which the process of setting up an infrastructure is similar to the process of programming software. Essentially, applications can contain scripts that create and manage their own virtual machines. Infrastructure as Code uses a high-level coding language for description of the entire existing infrastructure in the form

M. Klymash et al. (Eds.): *Future Intent-Based Networking*, LNEE 831, pp. 51–68, 2022.
https://doi.org/10.1007/978-3-030-92435-5_4

of code as well as the tools, which automate the provisioning of IT infrastructure and implementation of a real infrastructure from it [1–6].

Infrastructure as code is the process of managing data centers, not by physically configuring equipment or using interactive configuration tools, but by utilizing configuration files. System administrators and technicians should focus on service development, rather than IT infrastructure maintenance. The manual process often causes several problems that have to be addressed [7–9]:

- *High cost.* You need to hire many professionals to manually set up the necessary IT environment. In addition to paying all these people, they also need to be managed. This, in turn, leads to higher overhead costs and adds more complexity to communication.
- *Scalability and availability* (slow installs). To manually set up an infrastructure, engineers need to rack the servers, and then configure the hardware and network to the proper settings. Those processes are time-consuming and often errors arise.
- *Monitoring and performance visibility.* When you have a problem with your infrastructure, how do you determine exactly where the problem comes from? Is this a network, server, or application problem? You need tools that give you a comprehensive overview of how your entire IT infrastructure works.
- *Environment inconsistencies.* As the infrastructure grows, more and more people are joining the manual deploying configurations, which causes inconsistencies and over time it becomes difficult to reproduce the same environments. These inconsistencies lead to critical differences between development, quality assurance, and the production environments and cause deployment problems.

With the implementation of the IaC tools, productivity increases, quality of any software installation improves, and the software development life cycle becomes more efficient. Developers will be able to complete more projects in less time. Infrastructure as Code is made possible by the Infrastructure as a service (IaaS) platforms, which enable on-demand provisioning and requesting of cloud resources through remote APIs [10, 11].

2 AIOps and Benefits of Infrastructure as Code

There are two categories of tools for IaC:

- Configuration management tools, which are designed to install and manage software on existing servers (like Ansible, Chef, Puppet);
- Provisioning tools (like CloudFormation and Terraform), which are designed to provision the servers on their own, leaving the job of configuring those servers (and rest of your infrastructure, like load balancers, databases, networking configuration, etc.) to other tools.

In fact, most configuration management tools can perform some degree of provisioning, and most provisioning tools perform some level of configuration management, the only difference being that one type or another is better suited to certain tasks [12, 13].

IaC means to manage your IT infrastructure using configuration files. So, infrastructure configuration takes the form of a code file.

Since it is just text, it can be easily edited, copied, and distributed. You can put it under source control (git, cvs, bitbucket etc.), like any other source code file. In turn, it allows you to quickly create as many instances of your entire infrastructure as you need, in a different location from your source code file. You will be able to create resources consistently without error while reducing management and manual setup time. The schema for Infrastructure deployment is below (Fig. 1):

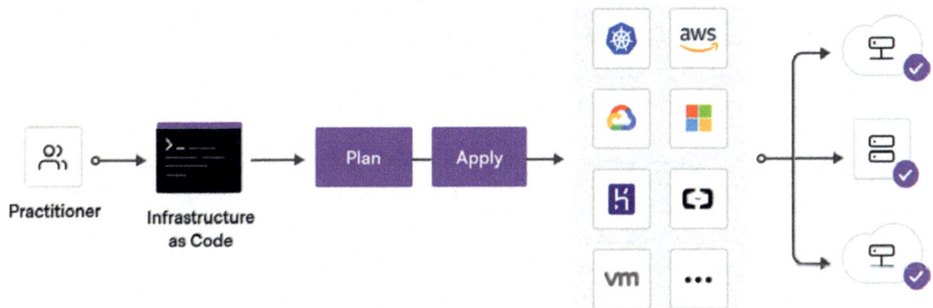

Fig. 1. Infrastructure deployment in the cloud platform

The key components of this schema:

- Developers/Authors – the persons, who will create infrastructure design and develop the code;
- Code in VCS – all the components in your infrastructure (buckets, IAM, network, VMs, K8s, etc.);
- Infrastructure provider – Cloud provider, that provides platform, where all infrastructure components will be stored (GCP, AWS, Azure, etc.);
- Up and running environments;
- Services – list of microservices running on prepared environments.

Within an IaC workflow, you can deploy the infrastructure multiple times in a standardized way, which means that software development and testing is accelerated because development, intermediate stages, quality assurance testing, and production environment are separated. Once the infrastructure you describe meets all the requirements, you will be able to deploy it in the cloud environment the right way. When new requirements appear, you can consider them and repeat the process as many times as needed.

However, even when using IaC we have to control our infrastructure, use some monitoring and health status mechanism to collect metrics and logs. Those metrics and logs allow us to control the status of our Infrastructure and Services and apply some scaling mechanism on demand. To ensure continuous integration and deployment of applications and infrastructure, enterprises will need DevOps tools that incorporate Intent-based networking (IBN) [14]. Multiple manual tools and automated IT operations platforms are substituted by artificial intelligence (AI), machine learning (ML), and network orchestration. Users' intents are defined in a human language, therefore cloud providers have to translate them into IT infrastructure policy using natural language processing (NLP) to guarantee quality of service (QoS)/quality of experience (QoE).

According to Gartner, by 2023, 30% of large enterprises will be using artificial intelligence for IT operations (AIOps) platforms and digital experience monitoring technology. Moreover, AIOps platforms will become the prime tool for analysis of monitoring data [15].

AI is used for management and maintenance of IT infrastructure and operations. AIOps Platform enables DevOps teams to respond more quickly (even proactively) to slowdowns and outages, with a lot less effort and more efficiency.

The Infrastructure as Code Workflow diagram below (Fig. 2) is a very simple model that shows the logical flow of IaC operation.

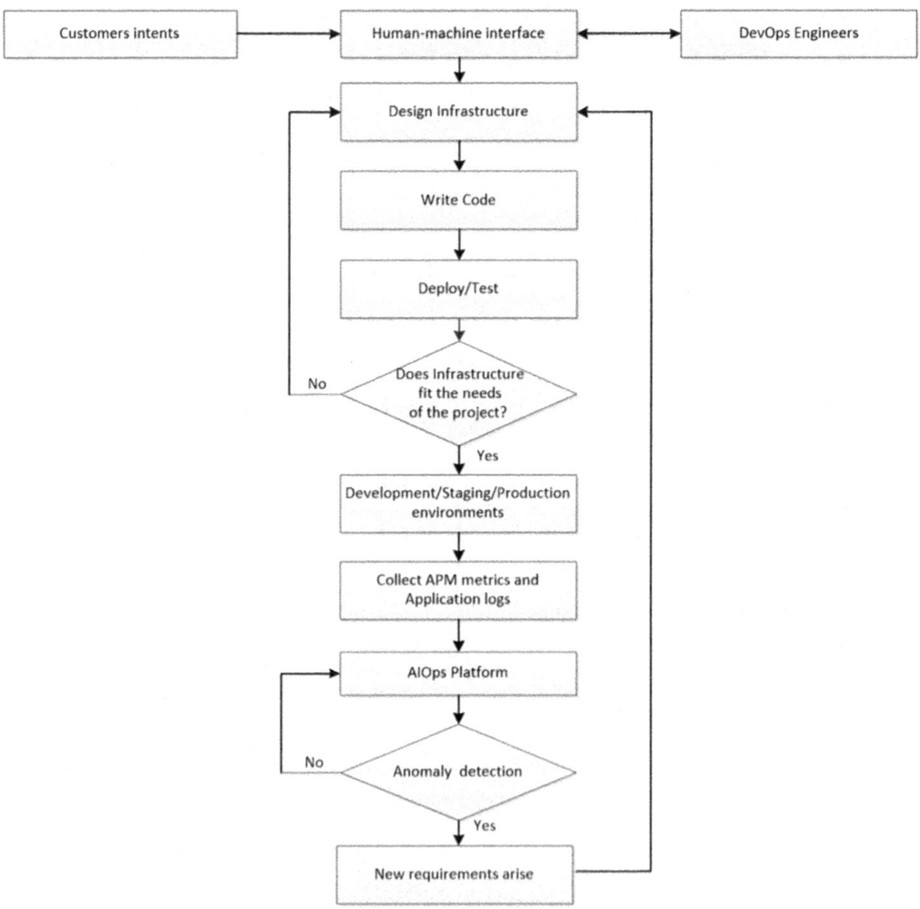

Fig. 2. Infrastructure as code workflow for intent-based cloud networking

Storing the infrastructure description in the version control system makes it accessible, and changes are reflected for all members of your team. In addition, you can always revert to the previous version if the new version shows errors. In addition, the version control software can be configured to automatically launch the tool to update the infrastructure in the cloud when approved modification is added. The automation IaC workflows helped reduce the complexity of cloud systems administration.

Infrastructure as Code brings a lot of benefits and is getting more popular among the DevOps Engineers. The main benefits of IaC are (Fig. 3):

- *Speed.* You can deploy different environments, from development to production including QA, and more; launch virtual servers and preconfigured databases, load balancers, storage systems, network infrastructure, or any other cloud service according to your requirements. By writing and running code, you can organize backup and disaster recovery by deploying your infrastructure environments in other locations where your cloud provider operates.
- *Consistency.* Despite efforts to maintain some consistency in the infrastructure deployment process by using standard operating procedures, manual infrastructure management will result in misconfiguration because people are not perfect. And no matter how hard they try, human error always appears. IaC resolves that problem and reduces the possibility of errors by using standardizing the setup of infrastructure from the configuration files.
- *Accountability.* Since IaC uses source control (or version control), we can track who and when made a specific change to the infrastructure component, and we can try to understand for what purpose.

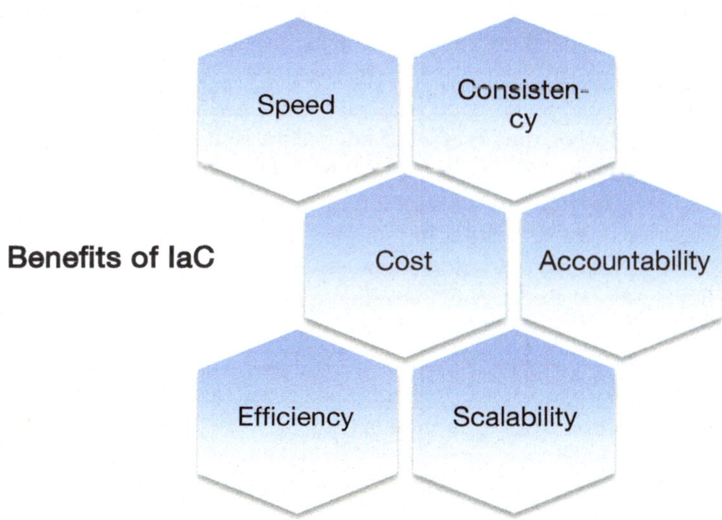

Fig. 3. The main benefits of IaC

- *Scalability.* Infrastructure deployments with IaC are repeatable and stable. You can write code once, then reuse it many times rapidly and reliably, avoiding manual setup and misconfiguration dependencies.
- *Efficiency.* You can increase software development efficiency by employing the infrastructure deployment from code. It makes the whole process safer and the issues or errors can be identified earlier, before running your infrastructure in production. Programmers and QA engineers can use sandbox environments for development and testing, and, in addition, developers can be more involved in the code review process. It's very important for Continuous Deployment (CD) process [16].
- *Lower Cost.* The automation of infrastructure configuration and lowering the costs of infrastructure management is one of the main benefits of IaC. By reducing the time for manual work, productivity is increased and staff costs are saved. The use of IaC deployment automation tools makes the process of building and configuring the infrastructure more efficient, reducing the costs and effort involved, and allows IT specialists to spend more time looking at innovative solutions.

3 GCP Cloud Functions for Microservices

The Google Cloud Platform (GCP) lets us build, deploy and run applications, websites and services on Google's infrastructure. Google Cloud Functions is a serverless execution environment for creating and connecting cloud services by writing simple single-purpose functions that are joined to events transmitted from your cloud infrastructure. It's one of the trending technologies nowadays.

Although we will cover only Google Cloud features, you can use similar services from other providers (AWS Lambda, Microsoft Azure, IBM Cloud Functions etc.). The choice of platform depends on various factors, such as functionality, performance, pricing, and more.

It is easy to run and scale code in a fully managed environment and the Cloud Function is triggered when the observed event occurs, so there is no need to provide any infrastructure or manage any servers. Cloud Function provides the ability to users to pay only for what they use (Function as a service, FaaS).

Cloud Functions can be written in multiple programming languages (Python, Go, Java, .NET, Ruby, Node.js, PHP, etc.) and are executed in language-specific runtimes.

Cloud Functions takes care of managing servers, configuring software, updating frameworks, and patching operating systems. The software and infrastructure are fully managed by Google, so that you just add code. Furthermore, the provisioning of resources happens automatically in response to events and automatically scales based on the load.

GCP Cloud Function Investigation plan (Fig. 4):

micro - just to check what is the minimum needed for setup and basic execution time of requests when function is dummy;
small - check if there is possibility to use Flask and native routing capability as well as emulate execution time and time outs;
basic - by usage of Flask, Connexion and other SDK provide ability to have API to get and update info in DB, log execution process, cache data.

Execution of Micro:

```
$ gcloud functions deploy cfpoc-1 --allow-unauthenticated --entry-point=test_v1 --memory=128MB --runtime=python37 --source=./ --trigger-http
Deploying function (may take a while - up to 2 minutes)...
$ #<<< Cold Function
$ time curl https://us-central1-sandbox05a33547.cloudfunctions.net/cfpoc-1
real    0m0.430s
$ #<<< Warm Function
$ time curl https://us-central1-sandbox05a33547.cloudfunctions.net/cfpoc-1
real    0m0.287s
```

Micro Summary:

- Deploying from local machine may take up to 2 min
- To be able to deploy Cloud Function, Cloud Build service must be enabled. (Is it true for any deployment? e.g. from repo)
- There is a difference between cold and warm execution.

Execution of Small:

```
#<<< Deploy v2 function with routing
$ time gcloud functions deploy cfpoc-2 --allow-unauthenticated --entry-point=test_v2 --memory=128MB --runtime=python37 --source=./ --trigger-http
Deploying function (may take a while - up to 2 minutes)…
$ #<<< Cold execution for V2 function
$ time curl https://us-central1-sandbox05a33547.cloudfunctions.net/cfpoc-2/hi/Nazar
Hi there, Nazar
real    0m1.498s
$ #<<< Warm execution
$ time curl https://us-central1-sandbox05a33547.cloudfunctions.net/cfpoc-2/hi/Test_V2
Hi there, Test_V2
real    0m0.334s
$ time curl -X POST https://us-central1-sandbox05a33547.cloudfunctions.net/cfpoc-2/hi/Test_V2/congrats -d '{"some": "data"}'
Hi there Test_V2, congrats on b'{"some": "data"}'!
real    0m0.285s
```

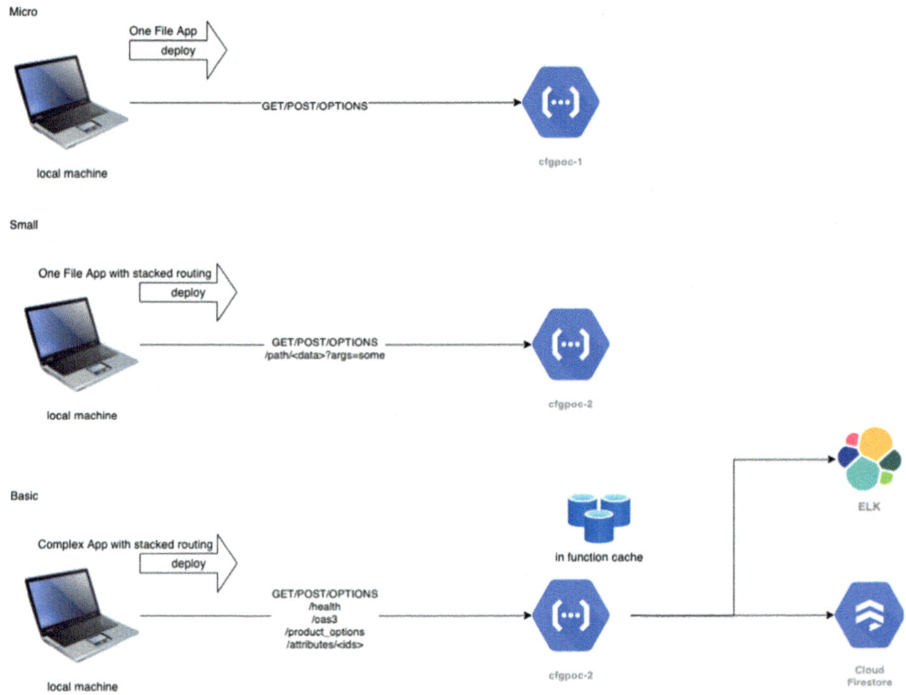

Fig. 4. GCP cloud function flow and execution plan

Small Summary:

- Deploying from local machine may take up to 2 min, but took 58 s;
- With a stacking of request processing in Flask, ability to do native flask routing was achieved;
- There is a difference between cold and warm execution and here it is more obvious;
- With a few tricks there is ability to transfer parameters and data to stacked process handler.

Execution of Basic:

```
$ #<<< Deploy v3 function with routing and OpenAPI validator between client and handler
$ time gcloud functions deploy cfpoc-3 --allow-unauthenticated --entry-point=test_v3 --memory=128MB --runtime=python37 --source=./ --trigger-http
Deploying function (may take a while - up to 2 minutes)…
$ #<<< Cold check for cfgpoc-3
$ time curl https://us-central1-sandbox05a33547.cloudfunctions.net/cfpoc-3/health
{
  "status": "OK"
}
real    0m2.628s
$ #<<< Warm check for first time of cfgpoc-3
$ time curl https://us-central1-sandbox05a33547.cloudfunctions.net/cfpoc-3/health
{
  "status": "OK"
}
real    0m1.298s
$ #<<< Warm check for second time of cfgpoc-3
$ time curl https://us-central1-sandbox05a33547.cloudfunctions.net/cfpoc-3/health
{
  "status": "OK"
}
real    0m1.205s
```

Basic Summary:

- Deploying from local machine took almost 2 min, but includes multiple files;
- There is a difference between cold and warm execution and here it is more obvious;
- There is a possibility to use cache as global variables (Table 1).

Performance Results

During the Cloud function deployment process it is easy to establish the requirements.txt file and an entry point has to be provided. By execution of one command it can be deployed to Cloud. It is also possible to deploy from a zip file that is located locally or on Google Storage as well as from Source Control or Google Registry.

After deployment the new function version will be active and the entire traffic will be directed to a new function without the ability to split it. Traffic management can be done by other services.

Cloud Function Application pushes some important metrics and logs to Cloud Log and NewRelic Cloud. All these metrics and logs will be used for AIOps and allow us to automate and accelerate many tasks more scalably, predictably, rapidly and efficiently

Table 1. Cloud function performance results

	Micro		Small		Basic	
	Client side	GCP logs	Client side	GCP logs	Client side	GCP logs
First call	476	90	1499	1150	2638	2268
Second call	373	103	327	25	568	119
40 calls avg	320	52	350	79	343	79

than manual methods alone. We can make real-time predictions and prevent potential errors by using AI's ability to perform continuous log analysis, anomaly detection, predictive maintenance, root cause diagnostics and other critical functions [17].

GCP Cloud Function Summary
Cloud Function can be used for a small size microservice with the following in mind:

- Cloud Function warming up issue:
  - warming can take more than 2 s and depends on initial implementation;
  - amount of dependencies influences the warming up process;
  - potentially Cloud Function may stay warm up to 15 min (but there is no guaranty by GCP);
  - potentially "ping" like activity may be used to warm function, but there is still 50/50 chance that it will be warm;
  - cold start may be minimized by trimming package size and dependencies and by usage of lazy-load technique.
- Inability to set custom FQDN for Cloud Function HTTP type [18];
- Anyway CNAME is available and can be configured on GCP;
- URL path always contains function name as a prefix;
- HTTP requests are always redirected to HTTPS;
- Background tasks are not recommended and should be avoided [19];
- Sending email from Cloud Function is forbidden. It is also a marker that in future list of restricted outbound ports can be extended;
- Cloud Function execution is limited to 60 s by default but can take up to 9 min;
- Out of the box there is no possibility to do traffic separation for Green/Blue deployment [20];
- Deployment:
  - Extremely easy to deploy function from local machine;
  - Local debugging is fast and easy with a help of functions-framework;
  - There is ability to deploy cloud function from repository and follow CI/CD process automatically.

4 GCP Cloud Run - Highly Scalable Containerized Applications

Cloud Run is a managed compute platform that enables you to run stateless containers that are invocable via HTTP requests. Cloud Run is serverless: it abstracts away all infrastructure management, so you can focus on what matters most — building great applications [21].

Since docker image is mandatory as input element for Cloud Run, developer is free to use any software to provide the functionality and handle requests, so we decided to implement it as python application with use of uWSGI as http server, Flask as web framework and Context for API handlers.

Requirements for Application

- Application can handle POST and GET requests,
- Application should handle different paths and arguments in path as well as in query,
- Possibility to use cache in functions,
- Possibility to talk to other services inside and outside of GCP (e.g. Firestore DB, on-prem ELK as Access log collector),
- Cold and warm starts and how they influence the execution,
- Measure request execution time,
- Compare the solution to Cloud Function option.

Cloud run Deployment and Execution Flow (Fig. 5):

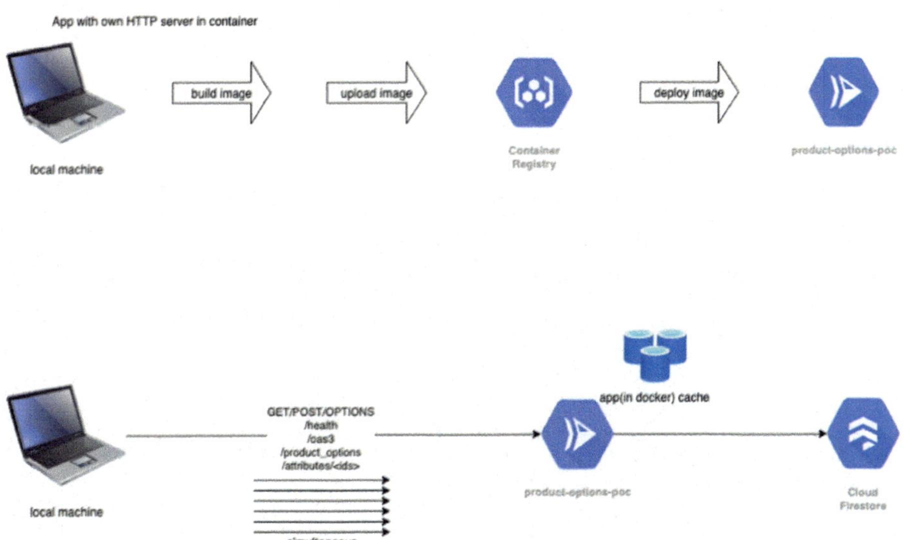

Fig. 5. GCP cloud run flow and execution plan

There are 2 steps to run GCP Cloud Run. The first one is to build docker image from source code and push it to Google registry. The next one is deployment of application image from GCR to Cloud Run. Cloud Run accepts container images built with any tool which follows the container contract [22].

The Summary of Requirements:

- The container must listen for requests on 0.0.0.0 on the port defined by PORT environment variable
- The container should not implement any transport layer security directly, it is terminated by Cloud Run for HTTPS and proxied as HTTP to the container without TLS;
- Responses must be sent within the time specified in the request timeout setting after it receives a request, including the container instance startup time;
- Environment variables K_SERVICE as name of the Cloud Run service and K_REVISION as revision being run;
- The filesystem is writable but it is in-memory filesystem and does not persist when the container instance is stopped;
- After startup, computation can be done only in scope of a request. Instance does not have CPU allocated out of request processing;
- An instance can be shut down at any time. SIGTERM signal is sent on termination with 10 s period before being shut down;
- Instance will not be kept idle for longer than 15 min.

There are several ways to build a container. For this one investigation, Dockerfile option has been selected [23].

Some calculations from performance testing are presented below (Fig. 6):

- Request rate is around 10 millions per 5 days:

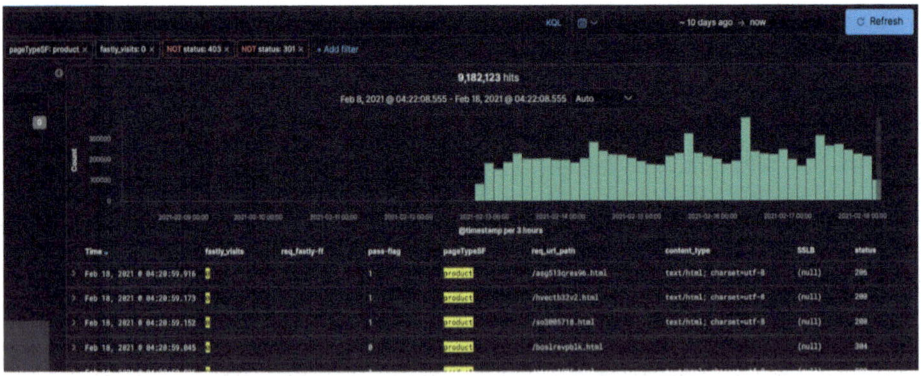

Fig. 6. Serverless microservices performance testing

```
per day = 10 000 000 / 5 days = 2 000 000
per hour = 2 000 000 / 24 hours = 83300
per second = 83300 / 60 minutes / 60 seconds = 23,14 = ~ 23.
To be able to utilize CPU and Memory resources properly from
time prospective, 10-20 concurrent users per instance were de-
fined.
$ # Cold start
$ time curl --location --request GET 'https://test-poc-
k5kabakx3a-uc.a.run.app/health'
{
  "run": "COLD",
  "status": "OK"
}
real    0m3,776s
$ # Warm start 1
$ time curl --location --request GET 'https://test-poc-
k5kabakx3a-uc.a.run.app/health'
{
  "run": "WARM",
  "status": "OK"
}
real    0m0,364s
$ # Warm start 2
$ time curl --location --request GET 'https://test-poc-k5ka-
bakx3a-uc.a.run.app/health'
{
  "run": "WARM",
  "status": "OK"
}
real    0m0,304s
```

As it can be seen the first call is cold and run variable value is "COLD" that shows the first application start that takes 3,7 s to execute. Next two calls are warm and take less than 400 ms for client to finish. Also the run variable value is "WARM" that is cached in the first call, so it shows that this is a good approach - to use cache ability to save CPU time (Table 2).

Performance Result

There is some period of time, when serverless microservice is warming up. In that time, response time is much longer than it can be in general. This is so-called First Call problem. We can see this situation on the Figs. 7, 8. Let's look at a visual representation:

As it is shown even if applications are warmed there is still a period of time when the service is warming up, but in general request latency is low. Cloud Run seems to be more beneficial in comparison with Cloud Function from a performance point of view.

Table 2. Cloud function/cloud run response time

	Cloud function		Cloud run	
	Client side	GCP logs	Client side	GCP logs
First call	5923	5399	5254	3574
Second call	1685	851	1376	290
40 calls avg.	1053	744	414	83
40 calls/sec	937*	859*	352	69

* - Cloud function and cloud run were warmed up before test.

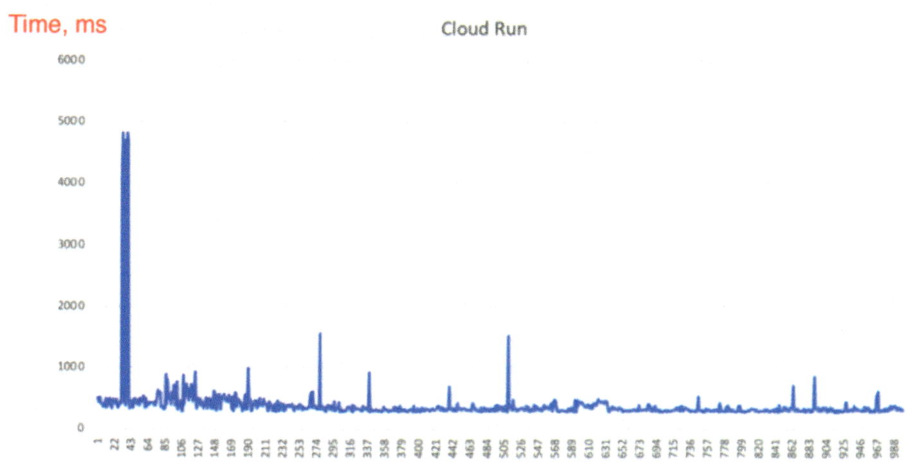

Fig. 7. Cloud run application response time

Fig. 8. GCP execution time (logs to firestore)

Applications running in Cloud Run have integration with Application Performance Management (APM) to provide better availability, efficiency and automation for deployment and management of cloud infrastructure.

Below are results of application performance monitoring for Internal Services and integrations with External Services (Figs. 9, 10 and 11):

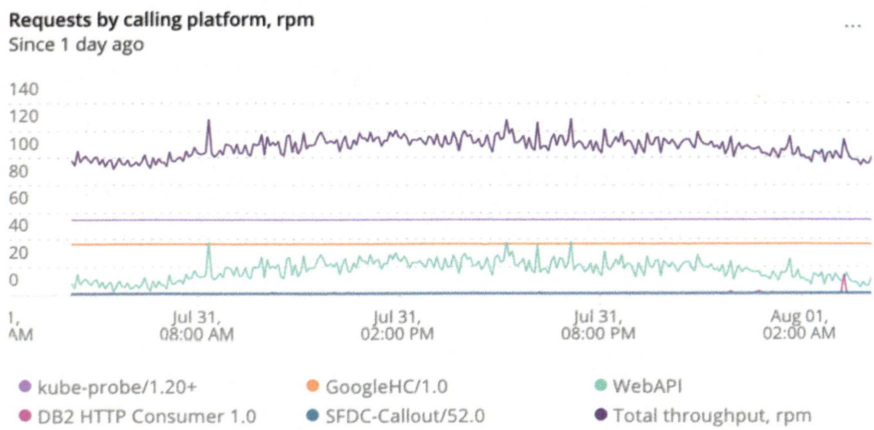

Fig. 9. GCP service, amount requests per minute

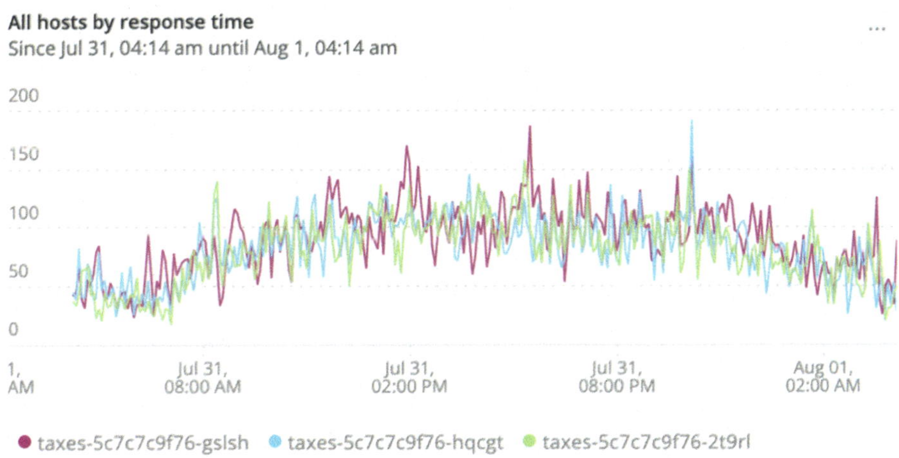

Fig. 10. GCP service, average response time

GCP Cloud Run Summary

Cloud Run warming up issue:

- warming can take more than in Cloud Function use case, but the whole startup flow is in container, so it can be tweaked;

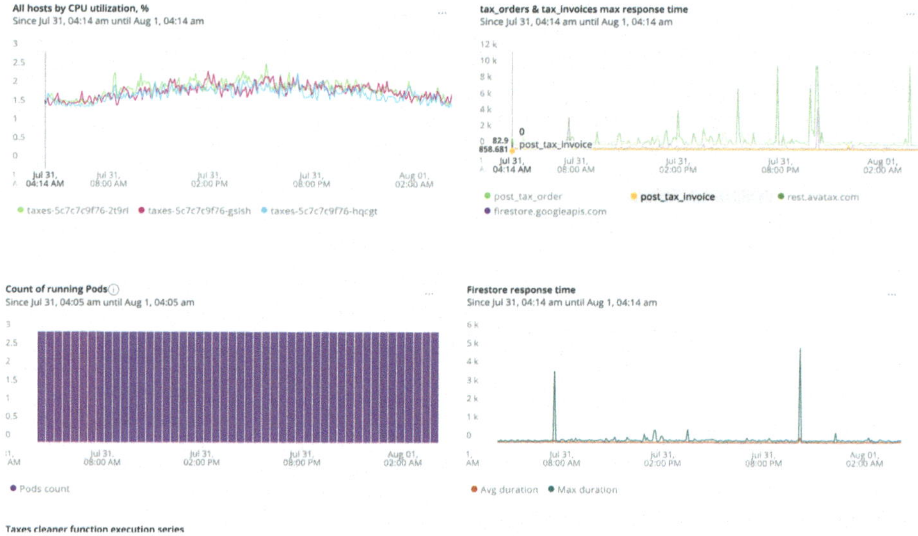

Fig. 11. GCP service, resources utilization

- amount of dependencies influence the warming up process the same way as for Cloud Functions;
- to speed up warming the recommendations are the same as for Cloud Function;
- start the process inside container quickly. GCP detects that instance is ready by checking port availability;
- use global variables to introduce object caching inside of application and concurrency features by periodical review of concurrency/CPU/Memory settings;
- build small images that do not influence the container start time, but influence the image deployment;
- setup a minimum amount of instances that need to be always warm or periodically poll instance;
- Cloud Run natively supports ability to set custom FQDN per service [24];
- HTTP requests are always redirected to HTTPS;
- it is not possible to define external port other that 443;
- background tasks are not recommended and should be avoided [25];
- Cloud Run execution is limited to 300 s by default, but can be changed;
- out of the box there is possibility to do traffic management during new version deployment (rollbacks, gradual rollouts, and traffic migration) [26].

5 Conclusion

In this chapter, authors explain a new approach to deploy and manage any infrastructure and microservices. Authors point out that correct and simple way to do that is by using Infrastructure as a Code - IaC and Version Control System - VCS. Those techniques enable you to manage infrastructure with configuration files rather than through a graphical user interface, so you can build, change, and administer your infrastructure in a safe, consistent, and repeatable way by defining resource configurations that can be versioned, reused, and shared.

Generally, there are a few ways, how to deploy microservices. In this article two of them were investigated: Cloud Function and Cloud Run. Each of these methods has some limitations and some key features.

Authors show performance results for both methods. Based on the results, the main problem is Cold Start, which takes too much time, which can be unacceptable for some applications. There are a few features that differentiate Cloud Run from Cloud Function: Cloud Run was developed to cover HTTP API use case; Cloud Run requires container to spin up instances, so there is a freedom from development point of view; Cloud Run natively supports ability to set custom FQDN per service; HTTP requests are always redirected to HTTPS; there is possibility to do traffic management during new version deployment; minimum number of instances can be configured, so there always exists a warm instance.

So, IaC process together with version control, enables us to quickly deploy, destroy, re-deploy or change our infrastructure. With Cloud Run microservice deployment, we can achieve higher performance for the application and reduce costs, because GCP resources will be only on demand in most cases.

Microservices, running as Cloud Function or Cloud Run all the time pushes key metrics and logs to Cloud Log and Cloud APM. These metrics and logs will be used for AIOps and will allow us to make online detection of issues and prevent potential errors using the AI's and ML's ability to learn from previous errors and prevent it happening again. Finally, all these tools will help with developing and deploying microservices on a continuous basis with CI/CD.

References

1. The Art of Service - Infrastructure As A Code Publishing. Infrastructure As A Code A Complete Guide - 2021 Edition, p. 319 (2020)
2. Fleming, S.: DevOps and Microservices Handbook: Non-Programmer's Guide to DevOps and Microservices (Continuous Delivery)/Fleming, p. 246 (2018)
3. Morris, K.: Infrastructure as Code: Managing Servers in the Cloud, 1st edn., p. 362. Kief Morris (2016)
4. Artac, M., Borovsak, T., Nitto, E.D., Guerriero, M., Tamburri, D.A.: Devops: Introducing Infrastructure-as-Code, ICSE (Companion Volume), pp. 497–498. ACM (2017)
5. Young, A.: Infrastructure as Code: A Comprehensive Guide to Managing Infrastructure as Code Kindle Edition, p. 118. Austin Young (2019)
6. Morris, K.: Infrastructure as Code: Dynamic Systems for the Cloud Age, 2nd edn., p. 430. Kief Morris, O'Reilly Media, Inc. ISBN: 9781098114671 (2020)
7. Romanchuk, V., Beshley, M., Polishuk, A., Seliuchenko, M.: Method for processing multi-service traffic in network node based on adaptive management of buffer resource. In: 2018 14th International Conference on Advanced Trends in Radioelecrtronics, Telecommunications and Computer Engineering (TCSET), pp. 1118–1122. IEEE (2018). https://doi.org/10.1109/TCSET.2018.8336390
8. Campbell, B.: The Definitive Guide to AWS Infrastructure Automation: Craft Infrastructure-as-Code Solutions, 1st edn., p. 367. Edition/Bradley Campbell (2019)

9. Joyner: DevOps For Beginners: DevOps Software Development Method Guide For Software Developers and IT Professionals
10. Davis, J., Daniels, R.: Effective DevOps: Building a Culture of Collaboration, Affinity, and Tooling at Scale
11. Lin, F.T., Shih, T.S.: Cloud computing: The emerging computing technology. ICIC Express Lett. Part B Appl. **1**(1), 33–38 (2010)
12. Brikman, Y.: Terraform: Up & Running: Writing Infrastructure as Code, 2nd edn., p. 368. Yevgeniy Brikman (2019)
13. Brikman, Y.: Why we use Terraform and not Chef, Puppet, Ansible, SaltStack, or CloudFormation/Yevgeniy Brikman (2016). https://blog.gruntwork.io/why-we-use-terraform-and-not-chef-puppet-ansible-saltstack-or-cloudformation-7989dad2865c
14. Beshley, M., Vesely, P., Prislupskiy, A., Beshley, H., Kyryk, M., Romanchuk, V., Kahalo, I.: Customer-oriented quality of service management method for the future intent-based networking. Appl. Sci. **10**(22), 8223-1–8223-38 (2020)
15. Gartner, Inc.: Deliver Cross-Domain Analysis and Visibility with AIOps and Digital Experience Monitoring. Charley Rich, Padraig Byrne (2018)
16. Farley, D., Humble, J.: Continuous Delivery: Reliable Software Releases Through Build, Test, and Deployment Automation, p. 464. Addison-Wesley Professional, Boston (2011)
17. Ahmed, M., Mahmood, A.N., Hu, J.: A survey of network anomaly detection techniques. J. Netw. Comput. Appl. **60**, 19–31 (2016)
18. Google Cloud Functions Documentation. https://cloud.google.com/functions
19. How to configure custom domain for Google Cloud Functions for rendering HTML. https://stackoverflow.com/questions/56469437/how-to-configure-custom-domain-for-google-cloud-functions-for-rendering-html
20. Blue green deployments with Google Cloud Functions. https://stackoverflow.com/questions/54224139/is-it-possible-to-do-blue-green-deployments-with-google-cloud-functions
21. Container runtime contract. https://cloud.google.com/run/docs/reference/container-contract
22. Google Cloud Run documentation. https://cloud.google.com/run/docs#training-and-tutorials
23. Building locally and pushing using Docker. https://cloud.google.com/run/docs/building/containers#docker
24. Mapping custom domains. https://cloud.google.com/run/docs/mapping-custom-domains
25. Avoiding background activities. https://cloud.google.com/run/docs/tips/general#avoiding_background_activitie
26. Rollbacks, gradual rollouts, and traffic migration. https://cloud.google.com/run/docs/rollouts-rollbacks-traffic-migration

Intent-Based Adaptation Coordination of Highly Decentralized Networked Self-adaptive Systems

Ilja Shmelkin, Daniel Matusek, Tim Kluge, Thomas Springer(✉), and Alexander Schill

Technische Universität Dresden, Dresden, Germany
{ilja.shmelkin,daniel.matusek,thomas.springer, alexander.schill}@tu-dresden.de
tim.kluge1@tu-dresden.de

Abstract. One challenge, that information technology faces, is how to cope with continuous contextual changes and the need for adjustments of the underlying networked system during runtime. In application domains like the Internet of Things (IoT) among others, system adjustments can't be performed in a device-by-device manner due to a large number of spatially distributed autonomous nodes. As a consequence, innovative approaches for autonomic management are required for such systems. A promising solution for network infrastructures are intent-based approaches. With the idea of declaratively specified goals for network functions, operators can holistically manage the network infrastructure at a high level of abstraction. Anyway, concrete design and implementation concepts are required to enforce intents within a networked system of autonomous nodes in a coordinated manner to ensure consistent behavior of the system according to the declared intent. In this chapter, concepts and approaches of self-adaptive systems are explored as a promising solution to serve as a basis for designing and implementing intent-based networked systems. The authors focus particularly on role concepts that allow a continuous design and implementation of system variability at run time. An example scenario from the IoT domain is used to continuously illustrate concepts and to demonstrate how networked self-adaptive systems can benefit from the introduced role-based concepts.

Keywords: Internet of Things · Intent-based Networks · Self-adaptive systems · Roles · MAPE-K · Distributed systems

1 Introduction

Application systems in many fields like smart cities, autonomous driving, robotics, or networking infrastructures rely on large sets of networked units that operate autonomously but need to collaborate to contribute to an overall objective. With their growing complexity and scale, such systems cannot be managed

M. Klymash et al. (Eds.): *Future Intent-Based Networking*, LNEE 831, pp. 69–100, 2022.
https://doi.org/10.1007/978-3-030-92435-5_5

in a device-by-device manner. Rather, autonomic management approaches are necessary that enforce operational goals and outcomes specified declaratively by operators allowing holistic management at a higher abstraction level. This declaratively specified operational guidance is called **intent** in the domain of Intent-based Networks (IBN).

This book chapter explores the use of concepts from the domain of self-adaptation for intent-based management of networked autonomous units. A scenario from the domain of smart farming is introduced in which a set of drones needs to continuously adjust their behavior to eventually fulfill the collaborative task intended for the swarm. After identifying major challenges for intent-based management of networked self-adaptive systems based on the scenario, the authors describe a conceptual framework with a strong focus on the concept of roles for modeling and implementing self-adaptive systems. This includes the notion of natural types and role types, the Compartment Role Object Model (CROM) as well as SCROLL, a method-call interception domain-specific language.

To enforce intents in a networked system of autonomous units, information flow and control structures need to be carefully designed. Thus, concepts for decentralized control based on the MAPE-K feedback loop are reviewed and concepts for each stage of the MAPE-K control loop are introduced to allow efficient management of changes of the networked system of autonomous nodes. The resulting overall system architecture comprises a monitoring component, a graph rewriting system for analyzing and planning, as well as a protocol for the coordinated execution of adaptation operations based on the notion of adaptation transactions. With a case study from the IoT domain, the authors demonstrate how networked self-adaptive systems can benefit from the introduced self-adaptation concepts based on roles.

2 Application Scenario

A scenario from the domain of smart farming will be used throughout the chapter to illustrate the challenges and concepts for decentralized coordination of autonomous units with an intent-based approach. In the scenario, a swarm of drones gets assigned the task to explore a field area, e.g., for soil analysis, crop monitoring, the need to apply fertilizers, or the impact of rain and storms. Therefore, the drone swarm explores a larger area of fields to capture the current state. The authors assume that the drones are equipped with communication devices for cellular networks and ad-hoc communication based on WiFi as well as a camera that can operate in low and high-resolution modes.

Individual drones operate either in roaming or inspecting mode. The **roaming mode** is used to examine an area according to potential anomalies. It is characterized by high flight speed, altitude, and distance to neighbor drones as well as a low camera resolution that limits required processing power, memory consumption, and energy for operation. Once a potential anomaly is detected, a subset of drones will examine the territory in more detail by entering the inspecting mode. In the **inspecting mode**, flight speed, altitude, and distance to neighbor drones are decreased to low, the camera resolution, however, is set to high. The image detection processes the high-resolution camera image to detect and record areas with anomalies.

Following an intent-based approach, the intent for the swarm can be formulated as follows: *The swarm should explore the field area completely and with minimal redundancy while discovering and recording all anomalies.* All drones in the swarm need to operate in a coordinated manner to cooperatively fulfill the intent.

The initial formation of the drones in the example is a line with a set of drones, that have the **roaming mode** activated. All drones belong to the swarm and are coordinated by one master drone, that is determined by leader election. When a drone detects a potential anomaly, the formation needs to be changed, since a subset of drones should now change the operation mode to **inspecting** to further examine the anomaly, while the majority of the drones continue to explore the field. To allow the drones in the **inspecting mode** to operate independently from the main swarm, the swarm is split by the swarm master. The drones in the sub-cluster elect a leader among themselves while the swarm master continues to manage the rest of the drones from the main cluster. After the split, the drones in the main cluster rearrange to form again a line with constant spacing and continue to explore the field area. The sub-cluster starts to inspect the field area with the potential anomaly at low altitude, speed, and distance to the neighbor. When done, the master of the sub-cluster initiates a reintegration with the main cluster which is coordinated by the master drone of the main cluster.

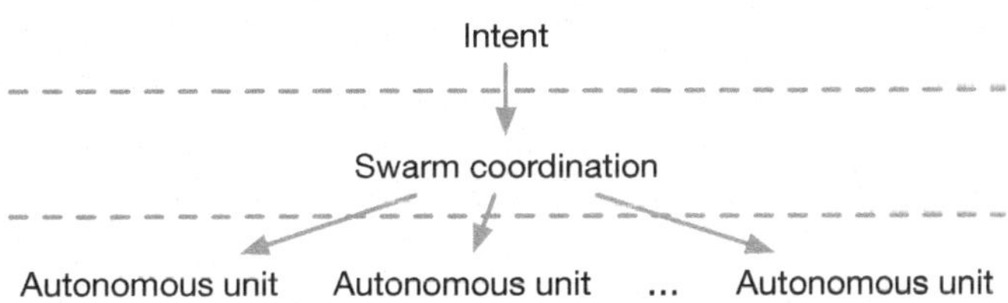

Fig. 1. Control layers for intent-based adaptation coordination

As shown in Fig. 1, intent-based adaptation coordination usually involves multiple layers of control. An **intent**, as specified for the drone swarm above, is, e.g., defined by a drone operator. It declares a goal that the swarm should achieve. The intent needs to be decomposed into a set of sub-intents that can be achieved step-wise by the swarm and that eventually lead to the fulfillment of the overall intent. These sub-intents need to be mapped to specific operations that can be performed by the swarm and finally by single drones in the swarm. At the layer for **swarm coordination**, the operations that have to be performed by particular drones are identified. In addition, swarm coordination triggers the changes which are necessary for the swarm (e.g., splitting or joining the swarm, changing the operation mode of a subset of drones) to fulfill a particular sub-intent. Finally, the **autonomous units** adapt their operation mode according to the plan provided by the swarm coordination and collaboratively perform the necessary operations to fulfill particular sub-intents.

Driven by the intent, the drone swarm should adapt itself to enforce the intent declared by the drone operator. Thus, the drone swarm needs to be designed and implemented as a distributed self-adaptive system. Especially, fully centralized control structures are strong limitations as drone swarms and sub-clusters should operate autonomously. In the following, the authors describe an approach for designing self-adaptive systems of autonomous units that can be managed by intents. The approach is based on the concept of roles for continuously modeling and implementing self-adaptive systems and a decentralized control loop. The fundamental concepts are introduced next.

3 Intent-Based Self-adaptation with Roles

Autonomic computing systems are *"computing systems that can manage themselves given high-level objectives from administrators"* as defined by Kephart and Chess (2003). Self-management - *"that frees system administrators from the details of system operation and maintenance"* - is considered as key property. Thus, autonomic computing systems can be directly linked to intent-based systems. *Intents* can be considered as high-level objectives declared by an administrator that free administrators from the details of system operation and maintenance.

Salehie and Tahvildari (2009) organize properties as self-management and self-configuration, also known as self-* properties, in a hierarchy with self-adaptiveness at the top level. Accordingly, self-adaptiveness can be considered as generalized property subsuming all self-* properties. Similarly, self-adaptive systems are a general term for systems that implement self-* properties.

Self-adaptive software is defined by Oreizy et al. (1999) as software *"that modifies its own behavior in response to changes in its operating environment.*

The goal of self-adaptation is to let the system itself collect additional data about, amongst other things, the operational environment, dynamics in the availability of resources, and variation of user goals to manage itself based on its high-level system goals (Weyns 2020). This shifts away the maintenance effort from the system administrator to the system itself. In the literature, two strategies for how self-adaptation can be used in practice (i.e., internal and external self-adaptation) are described.

Internal Self-adaptation makes systems able to cope with changes in the operational environment, dynamics in the availability of resources, and variations of user goals by itself with minimal to no required human intervention (Weyns 2020). In the self-adaptive community, such changes are often referred to as *uncertainties*. The self-adaptive system handles those uncertainties by being hardwired to do so within its program code. Such systems use low-level mechanisms like exceptions and time-outs to detect a problem close to its error source

(Cheng et al. 2005). This approach, however, is limited as it makes it difficult to detect and correct overall system anomalies. Also, modifications and maintenance of the system are difficult as the self-adaptation logic may be scattered around the system code (Cheng et al. 2005).

External Self-adaptation overcomes the limitations of the internal approach by separating the self-adaptive capabilities of a system from its domain-specific purpose. This is done by splitting the system into a *managing* and a *managed* part. The managing part is a control layer that interacts with the domain-specific, the managed, part of the system. This interaction comprises monitoring of the system and its environment as well as the execution of adaptations within the system to cope with uncertainties. The managing layer today is often realized as a control loop, from which the MAPE-K control loop is a popular representative. The authors identified this approach to be a good candidate for creating the intent-based system that was introduced in Sect. 2. Therefore, this chapter is continued by introducing the principles of the MAPE-K control loop.

3.1 The MAPE-K Control Loop

The Monitoring, Analysis, Planning, Execution and Knowledge Components (MAPE-K) feedback loop (see Fig. 2) is implemented using the autonomic manager (Kephart and Chess 2003) and represents a decomposition of the steps needed to allow for autonomic computing (Brun et al. 2009). Sensors first are used to monitor information about the context and computational environment and effectors, secondly, perform changes on the underlying system (IBM 2005). In between, the monitored data is analyzed and a suitable plan for autonomic change is created. The feedback loop is working as follows:

1. **M**onitoring: The monitoring component is using sensors or probes in applications to gather information about the environment, a system, or the internal state of the managed application. The information that was collected this way is then written into the knowledge component for future reuse.
2. **A**nalysis: The gathered information is used in combination with historical data and further information from the knowledge to make a diagnosis about the current system state with respect to high-level system goals.
3. **P**lanning: The planning component generates adaptation plans based on the diagnosis from the analysis stage. These plans are afterwards propagated to the execution component. This could either be a single instruction or a complex sequence of actions.
4. **E**xecution: The execution component of the MAPE-K loop performs actions from the generated plans using effectors, i.e., interfaces to the managed application that allow to adapting it.
5. **K**nowledge: The control loop uses the knowledge base for information sharing about metrics, adaptation plans, historical adaptations performed, etc. The knowledge could either be created locally by the autonomic manager, obtained externally, or passed manually into the repository.

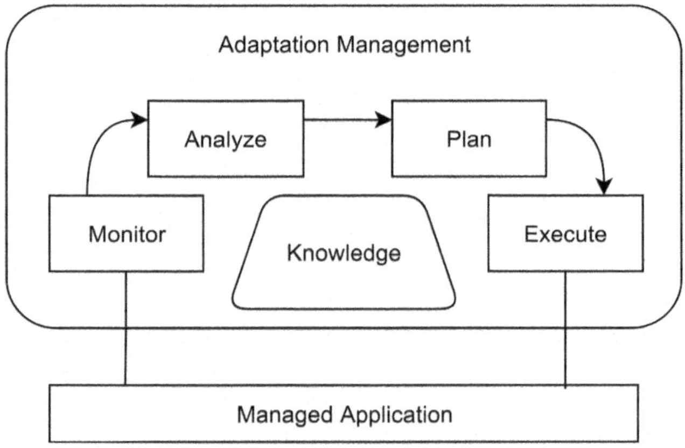

Fig. 2. MAPE-K feedback loop

3.2 Patterns for Decentralization in Control Loops

Although the overall intent (or phrased differently, the high-level system goals) is set in the application scenario, the organization of MAPE-K components that is used could be structured in various ways. Weyns et al. (2013) introduced different patterns for decentralizing and designing self-adaptive systems with each having its advantages and disadvantages. Generally speaking, the already presented abstract concept of the MAPE-K feedback loop in Fig. 2 does not make any assumptions about the concrete architecture and distribution of the MAPE-K components of a distributed self-adaptive system. By using an application scenario, the authors want to introduce possible architectural patterns and discuss their practicability for the scenario.

Decentralized Coordination Pattern. A good starting point for a decentralization of the MAPE-K components is decentralizing every aspect of the feedback loop. This is conceptualized by the "decentralized coordination" -pattern. Every drone would have its own MAPE-K feedback loop implemented and thus monitors its context and internal state. The monitored data is then analyzed, plans are generated and afterwards executed. In this approach, there is no central instance for either of the MAPE-K stages. This generally allows for good scalability, low communication overhead for local adaptations, and no single-point-of-failure. On the other side, finding a consensus for many participating drones could be difficult and scalability might be compromised in that case. Besides that, consistency could be lowered in a fully decentralized approach and sub-optimal adaptation decisions would be the consequence. Problems of maintaining consistent adaptations in an acceptable time for larger system sizes have been shown in experiments by Weißbach et al. (2017b).

Information Sharing Pattern. Better scaling with respect to the communication could be achieved by the "information sharing pattern". This pattern is a special case of the decentralized coordination pattern (Weyns et al. 2013). It provides better scalability and has less stringent interaction. Only the monitoring components of the drones are exchanging information and monitoring data. The A-P-E components do not need any coordination and allow for a more direct execution of the adaptation. With this pattern, local optimal objectives are more likely reached but global goals are approached worse (Weyns et al. 2013).

Master/Slave Pattern. The Master/Slave pattern could overcome the problem of worse accomplishment of global goals. It is suitable for scenarios in which slave components monitor themselves but let master nodes decide about the adaptation and decision making. Adaptations are executed locally afterwards. This has the advantage of efficient decisions for global objectives and provides guarantees, but generates overhead since data must be collected at the master and the adaptation actions must be distributed among all slaves afterwards. In large systems, this might lead to bottlenecks and could introduce a single point of failure. With an appropriate leader election mechanism, this problem could be overcome.

Regional Planning Pattern. A related, but more scalable approach, is the pattern of regional planning. A layered separation of concerns is introduced, which means that several MAPE-loops are delegating their planning to interconnected planning components. Each region, i.e., a group of collaborating drones performing a task together, has its own planner which is connected with all the other planners in the overall system and thus enables a layer for separation of concerns. The amount of data and frequency of interaction with the planner is reduced, but the run-time coordination of the execute phase among different regions is not possible. Additional overhead due to the aggregation of local analysis with the other planners might be introduced because coordination is needed.

Hierarchical Self-adaptation Pattern. The last pattern to be discussed is the hierarchical self-adaptation pattern. The adaptation logic is structured and the complexity of self-adaptation can be managed with that approach. By using this approach, separation of concerns can be achieved, since bottom layers focus on concrete adaptation for their own and higher levels of the system have a broader perspective on all devices. A downside of this method is the fact, that separation of all concerns is hard to accomplish, especially when the goals of different subsystems are interfering. Weyns et al. (2013) claim that there is no guarantee that an overall solution for an adaptation is optimal for all participants (Fig. 3).

All of the presented patterns have their advantages and disadvantages, depending on the use case. The authors' application scenario includes a collective monitoring unit with sensors and monitoring probes on each of the drones,

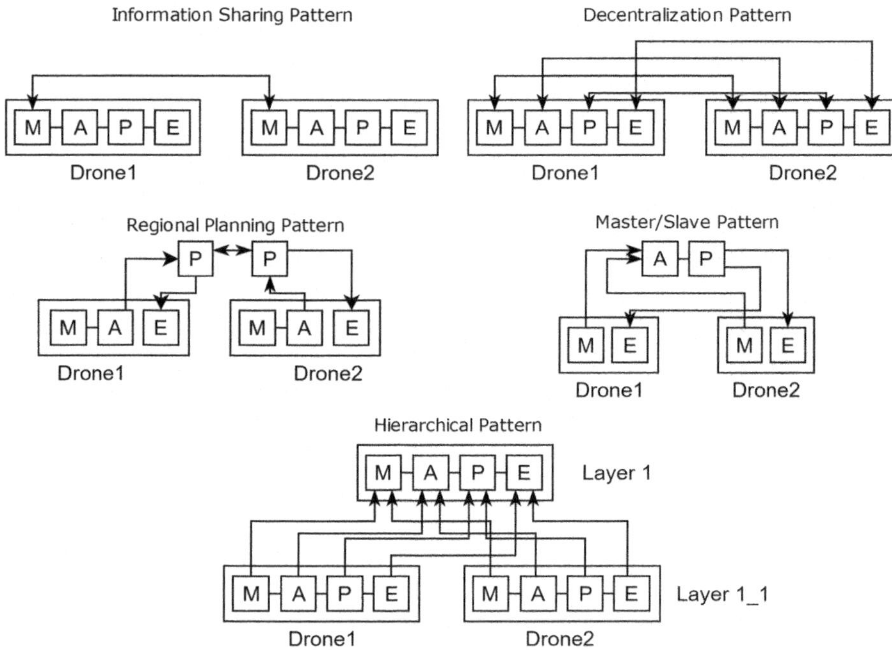

Fig. 3. Patterns for decentralize MAPE functions

which is not supported by any of the proposed patterns from Weyns et al. (2013). Besides that, the master drone should contain an analysis and planning component to generate adaptation plans for the whole drone swarm, but with the slave drones also having those two components integrated with a deactivated state to allow every drone to play the master role in case a new leader is elected. Therefore the master/slave pattern is not suitable, because the A-P components are available only once in for the whole system. Moreover, emerging clusters for specific tasks in the scenario should be able to have their own A-P components which would allow for fast local changes. The decentralization pattern has all of the needed components integrated, but the concept of decentralization proscribes a central unit which the authors' concept needs for the monitoring stage, as it functions as one central interface for other drones in the cluster to send monitoring data to. Besides that, the decentralization pattern allows every drone to plan adaptations for the whole cluster, whereas only the master should take on this task. The information sharing and regional planning patterns are not suitable for the scenario, since the execution of the foreseen adaptations must be coordinated consistently among all participants and these two patterns do not consider communication and coordination between the execution components.

Due to the downsides of the existing patterns for the chosen scenario, a hybrid pattern will be used for the drone swarm that was designed based on the following assumptions:

- Each drone contains a drone runtime capable to perform the roaming and inspecting operations. This runtime is the managed element according to ?.
- Each drone contains all MAPE components and is capable to act as a master drone for a (sub-)swarm. The MAPE components are part of the autonomic manager.
- Each drone can communicate using WiFi and cellular. Especially, it is assumed that each drone has a stable link to the network with a very low probability of disconnections.
- The number of drones in the swarm is moderate with a common size of 10 - 30 drones.

The hybrid pattern foresees a collective monitoring component that also contains a database with the system knowledge as depicted in Fig. 4. Independent of the swarm organization, each drone monitors its state and propagates state information to the collective monitoring component. Based on the assumption of a small swarm size the number of messages exchanged between for monitoring the swarm is low compared to the information sharing or decentralization pattern where all drones pairwise exchange monitoring messages. Due to the stable network connection, the collective monitoring component is also permanently available. Replication inside the monitoring component further ensures that the collective component does not become a bottleneck for the system.

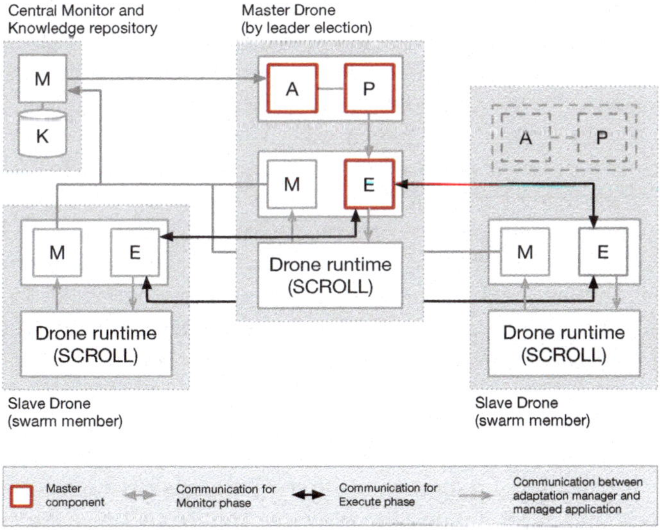

Fig. 4. Hybrid control loop for intent-based swarm coordination

For analysis and planning the Master/Slave pattern is used. To avoid a strong dependency on a single drone, potentially all drones can perform the analysis and planning based on the system knowledge provided by the centralized component. The master drone that performs analyzing and planning is determined on-demand by a leader election. This avoids a single point of failure in case the master drone becomes unavailable. At the same time, overhead for decentralized analysis and planning can be avoided. Simultaneously, master drones for managing sub-swarms can be dynamically determined.

The adaptation plan is propagated to the execution component residing at the master drone. The execution phase is performed according to the decentralization pattern, i.e., each drone performs the necessary adaptation operations locally while coordination messages are exchanged between all involved execution components to ensure coordinated execution of all adaptation operations.

In Sect. 2, the authors identified that role-based self-adaptive systems are one promising solution to manage an intent-based networked drone swarm. The basics of self-adaptivity together with the advanced concepts of control loops and various control loop patterns were already introduced in this section. The chapter is continued by introducing the notion of *roles* and technologies that are based on roles, i.e., the *Compartment Role Object Model* and SCROLL, a method-call interception domain-specific language that allows implementing role-based CROM-models.

3.3 Foundations of Roles

With each new domain that identifies the benefits of using computer systems, and therefore, software, to support their goals, also new challenges for modeling and implementation of the needed software arise. In contrast to currently predominantly used modeling techniques, which focus on modeling entities and relationships, as well as their object-oriented counterpart for implementing software, the notion of roles and role-based technology promises an intuitive and natural way of modeling and implementing software.

Bachmann's and Daya's notion of roles (Bachman and Daya 1977) is treated as the first occurrence of roles in the scientific literature. The authors identified that to this time, the *most conventional file records and relational file n-tuples are role oriented* (Bachman and Daya 1977), as they usually organize entities like employees, customers, patients, or students that indeed are roles that are played by individual persons (Steimann 2000). Based on this finding, Bachmann and Daya introduced a network model that amongst other characteristics, allowed to re-use the same piece of code to iterate through a set of syntactically distinct fields (Steimann 2000).

This initial idea on roles has then received increased attention as multiple researchers developed it further. Sowa first distinguished between *natural types* and *role types* and later made the addition that role types are a subtype of natural types to strengthen his new modeling concept (Sowa 1984), (Steimann 2000). Guarino separated those concepts even more by identifying that natural types are characterized by semantic rigidity and in contrast, role types are not (Guarino 1992).

For example, consider a cluster of drones in a networked system that operates towards identifying structural anomalies in a field, as described in the application scenario in Sect. 2. In its essence, one drone is a *natural type* as it requires no relationship to exist and has rigid, static attributes. When this drone starts to operate in a cluster to fulfill a specific mission (i.e., roaming the field or inspecting the crops), a relationship is forged between the drone, its cluster, and the current mission. The relationship between the drone and its cluster can be described by the non-rigid *role types ClusterMaster* and *ClusterSlave*, whereby only one ClusterMaster can be present within one individual cluster. Referring to the two modes for operation (i.e., roaming and inspecting) that were introduced in the application scenario, a drone on a mission can also be described by a *role type*, i.e., *RoamingDrone* or *InspectingDrone*.

Steinmann et al. have surveyed then-contemporary literature on roles and identified 15 role features which were later extended by Kühn et al. by an additional 11 features (Kühn et al. 2014). As Steinmann et al. already denoted, *some features conflict with others, and hence that there is no single definition of roles integrating all of them* (Steimann 2000). That also means, that each role-based modeling and programming approach is not able to express the full set of features but has to focus on a subset. Therefore, one can derive multiple classes of approaches on role-based languages from the literature (Leuthäuser 2017):

1. Languages like UML describe the **relational nature**[1] of roles. A role name explains how an object participates in a relationship. The participating object, however, can also be a role. For example, as mentioned above, a drone can participate in a cluster as the ClusterMaster role type. The ClusterMaster, however, can then play the role *Coarse* on a specific mission.
2. Next, there is a **behavioral nature**[2] of roles. In that, object behavior is controlled by the currently active context. Changing the context during execution also changes the behavior of the involved objects. ObjectTeams/Java is one example for such a language (Herrmann 2005). Referring to the application scenario, a drone can change its behavior depending on the mission that is currently active. Another example would be, that a ClusterMaster has superior capabilities compared to a ClusterSlave, however, both still are role types of the same natural type "Drone".
3. The **contextual nature**[3] of roles describes the ability of role types to be active within a defined context, e.g., there cannot be a cluster without a ClusterMaster. There are currently still no known languages that focus only on the contextual nature of roles (Leuthäuser 2017).

The Compartment Role Object Model (Kühn et al. 2015) is an approach which is **combining all three natures** of roles. As described above, the application

[1] Roles are usually related to other roles (Kühn 2017).
[2] Role's ability to change the objects' behavior (Kühn 2017).
[3] Roles are defined within a certain action, stage, or more generally context (Kühn 2017).

scenario relies on all of them, hence, CROM is a suitable technology to rely on for modeling. An overview of CROM's concepts and definitions is given in the next section.

3.4 The Compartment Role Object Model (CROM)

In the preceding section, the metamodel CROM was identified as one fitting technology for modeling the role-based view on the application scenario. CROM is defined as a tuple over the concepts M = (NT, RT, CT, RST, fills, parts, rel). Each of the tuple's elements is described below. The description is assisted by Fig. 5:

A **natural type** (NT) is a *rigid*[4] *non-founded*[5] type with a unique *identity,*[6] that contains a minimal set of necessary fields and methods that are statically bound to the very nature of the instance. In Fig. 5, the *Player* Drone is the only natural type.
With *anti-rigid, founded* and from their player types derived identity, a **role type** (RT) is the opposite of a natural type. Figure 5 features several role types, i.e., *ClusterMaster* and *ClusterSlave* (both existent within the compartments *Inspecting* and *Roaming*, as well as *Coarse*, which is only existent in the Roaming compartment and *Focused*, which is only existent in the Inspecting compartment.
A **compartment type** (CT) represents a *founded* type with *unique identity* that is *rigid.* It can only exist with at least one role type related to it. The compartment type describes the context in which natural types fill different role types. Therefore, when the compartment Roaming is active, a natural type Drone, that is a ClusterSlave or a ClusterMaster, can fill the Coarse role type in which its speed and altitude are specified to fixed values.
The **relationship type** (RST) brings at least two role-types into relation with each other. It is a *rigid, founded* type with an identity, that is *derived from the role-types* it relates to each other. Figure 5 features the relationship type *fills* between a role type and a role type, a natural type and a role type, or a compartment type and a role type.
A **fills relation** relates a *player type*[7] with a *role type.* The player can be a natural type or a compartment type. It is allowed for a role type to have multiple player types bound to it. In the example, the player Drone fills either a ClusterSlave or a ClusterMaster role type. Furthermore, the Coarse and Focused role types can be filled by the respective ClusterMaster and ClusterSlave role types.

---

[4] Here, rigidity denotes that an instance has to be a member of this type for its whole lifetime (Albuquerque and Guizzardi 2013).

[5] Foundedness means that a type cannot exist without the existence of other types (Albuquerque and Guizzardi 2013).

[6] Here, identity characterizes whether the instance of a type has a unique, derived, or composed identity (Albuquerque and Guizzardi 2013).

[7] A player is an object that is bound to a role during run time.

The **parts function** assigns each role type to a specific compartment type. A role type cannot be part of multiple compartment types. This also tells, that the ClusterMaster and ClusterSlave role types within the Roaming compartment type are indeed different roles than the ones with the same names but within the Inspecting compartment type.
Finally, the **rel function** assigns relationship types to specific role types that are part of the same compartment type.

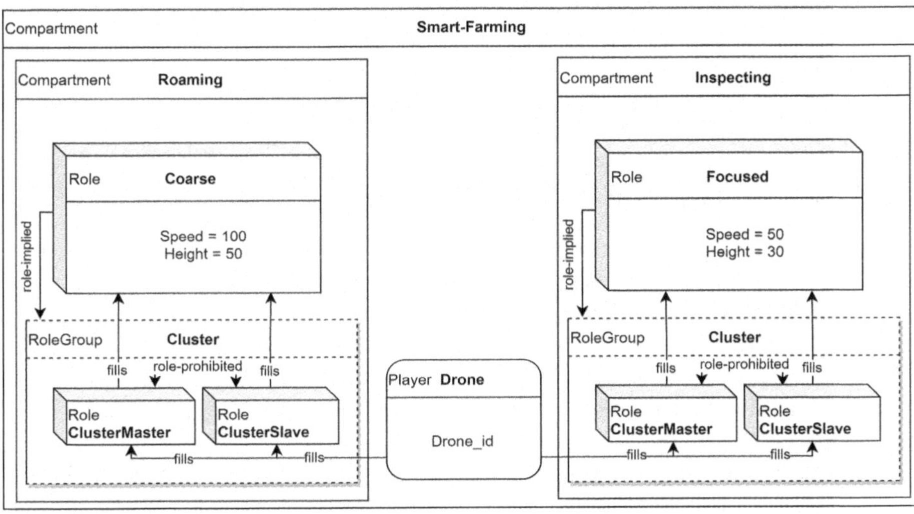

Fig. 5. CROM/CROI Model for a drone in the application scenario.

Additionally, to allow modeling an instantiated version of a CROM, *Compartment Role Object Instance* (CROI) was created. CROI is a tuple over the concepts I = (N, R, C, type, plays, links) and its elements are introduced below:

A **natural** (N) is the instantiated version of a natural type (NT).
A **role** (R) is the instantiated version of a role type (RT).
A **compartment** (C) is the instantiated version of a compartment type (CT).
The **type function** simply returns the type of the given instance, i.e., type(Coarse) = role type.
The **plays relation** assigns exactly one player to a role. Also, a compartment can be assigned to a role. The play relation is the instantiated equivalent to the fills relation.
The **links function** returns all related roles for the input relationship type.

CROI, additionally, is extended by a constraint model, that constrains roles and relationships. In the following, possible constraints are listed:

role-implied constraint: An object playing a role of role type A is required to also play a role of role type B (but not necessarily the other way round). For example, it is necessary that each drone that plays the role Coarse in the Roaming compartment also plays a role in the role group *Cluster*, i.e., role-implied[Coarse, Cluster] (see role groups below).
role-prohibited constraint: An object playing a role of role type A is never allowed to play a role of role type B, and vice-versa. In Fig. 5, a drone that plays the role of a *ClusterMaster* cannot play the role of a *ClusterSlave* and vice-versa, i.e., role-prohibited[Master, Slave].
role-equivalent constraint: An object playing a role of role type A is required to also play a role of role type B, and vice-versa.
role-dontcare constraint: No constraint is applied for an object playing a role defined in the pair (A,B).

Furthermore, the constraint model extends CROI by adding **role groups**, which allows defining constraints over a group of roles instead of only one specific role. Role groups are defined by the three definitions describing their syntax, an atoms function, and the semantics of role groups which are discussed in detail in the dissertation thesis of Kühn (Kühn et al. 2015), (Kühn 2017)). Also, Kühn introduces formal proofs and additional aspects of CROM, CROI, role groups, and constraints, however, the authors omitted details that are not required for the modeling process in Sect. 4. It is to be noted, that CROM is also the basis for multiple tools in the role-based community, including FRaMED,[8] FRaMED-io[9] and SCROLL.[10]

So far, the authors have introduced how self-adaptivity can be built into systems. Also, the authors have identified that the role-based approach provides an intuitive and natural way of modeling and implementing systems. Based on that, CROM was introduced as a meta-model for role-based systems. The chapter is continued by introducing SCROLL, which allows implementing the concepts of CROM into software.

3.5 Scala Roles Language (SCROLL)

The domain-specific language (DSL) SCROLL is an extension of its host language Scala. It is designed as an embedded method-call interception DSL (Lämmel 2002),(Mernik et al. 2005). SCROLL is fully implemented in its host language Scala and requires no additional language compiler or tooling. It is compiled to Java Virtual Machine (JVM) bytecode and, therefore, can be run by almost any machine. SCROLL implements the concepts of CROM, therefore, allowing users to implement programs that use roles and their surrounding compartments. It also supports the advanced features of CROM, allowing to define constraints, role groups, and role group constraints as described in the previous section. SCROLL is designed with a layered approach, that consists of

[8] https://github.com/Eden-06/FRaMED-2.0.
[9] https://github.com/Eden-06/FRaMED-io.
[10] https://github.com/max-leuthaeuser/SCROLL.

the four layers *usage layer*, *configuration layer*, *meta-object protocol layer*, and *specification layer* (only the usage layer is relevant for users of SCROLL). As identified earlier in this contribution, roles are a fitting candidate to support flexibility during modeling. SCROLL allows to translate those models into running code. Therefore it is very supportive to achieve the goal of designing and implementing a role-based self-adaptive system that has an intent. To understand how SCROLL is structured and more importantly, how it can be used by programmers, the usage layer will be described in the following.

The Usage Layer provides programmers with the basic functionality to embed roles in their code. With SCROLL, programmers can augment any class with new behavior during run time. For simplicity, the features provided by this layer will be explained with an example software that provides a drone with the ability to solve a task, that is determined by an active mission. For that, a class (the role) can be bound as an extension to another class (the player) (see Listing 1.1, line 44). This provides the player with potentially new attributes and functions that are provided by the role. In the example below, the drone can fly to a certain location while playing the Coarse-Role, however, when the Focused-Role is active, the drone will analyze an anomaly (see Listing 1.1, lines 51 and 24). CROM specifies that dynamic binding of classes (i.e., roles) is possible when the roles are enclosed within a *compartment*. SCROLL supports that by allowing a class to inherit from the compartment *trait* (see Listing 1.1, lines 1, 3, 19), which adds SCROLL's basic API functionality to the inheriting class, allowing to use the *play*-call together with the option to access functions and attributes of the played role by using the +-operator. Additionally, the usage layer of SCROLL supports the definition of constraints, role groups, role group constraints, and role restrictions, all of which are covered in more detail within the thesis of Leuthäuser (Leuthäuser 2017).

Listing 1.1. Example of a SCROLL application that implements the CROM-Model from Figure 5.

```
1  object SmartFarming extends Compartment {}
2
3  class Roaming extends Compartment {
4      class Coarse() {
5          val speed = 100
6          val altitude = 50
7          def fly(x,y) {
8              move_to_coordinate(x, y, speed, height) {...}
9          }
10     }
11     class RoamingClusterMaster() {
12     /* ...Master-Routines... */
13     }
14     class RoamingClusterSlave() {
15     /* ...Slave-Routines... */
16     }
17 }
18
19 class Inspecting extends Compartment {
20     class Focused() {
21         val speed = 50
22         val altitude = 30
23         def fly(x,y) {
24             analyze_anomaly(x, y, speed, height) {...}
```

```
        }
    }
    class InspectingClusterMaster() {
    /* ...Master-Routines... */
    }
    class InspectingClusterSlave() {
    /* ...Slave-Routines... */
    }
}
case class Drone(Drone_id: uuid)

main_loop(){
    val smart_farming = new SmartFarming {

        val drone = new Drone(1)


        if(!anomaly) {      /* No anomalies found - Roam! */
            val roaming_mission = new Roaming {
                drone play new RMaster() play new Coarse()
                drone.fly(x,y)
            }
        }
        else {              /* Anomaly found - analyze it!*/

            val inspecting_mission = new Inspecting {
                drone play new IClusterMaster() play new Focused()
                drone.fly(x,y)
            }
        }
    }
}
```

In the example, three compartments are defined. The *SmartFarming* compartment acts as the root compartment. In the main loop, a natural Drone scans the fields for anomalies. If no anomaly is present, a nested *Roaming* compartment instance is created, in which the natural Drone plays the role of a master and a role that implements the Coarse searching behavior. If an anomaly is found, an *Inspecting* compartment instance is created, where the Coarse search role is replaced by a Focused search role.

As depicted, the code implements a minimal loop that continuously adapts the system behavior by introducing and toggling role instances. However, the definition of adaptations is hidden in the code. Thus, with increasing complexity, it becomes difficult to grasp and reason about the specifications, which makes the system hard to develop and maintain. Following the philosophy of model-driven software engineering, declarative modeling of the adaptations is more feasible. The user-defined adaptation models can then be utilized in a feedback loop provided by the framework. In the following sections, the foundations for modeling such adaptations with rewriting rules and constraints instead of code are discussed.

3.6 Modelling Runtime Adaptations

As described above, the CROM meta-model can be used to model the architecture of role-based software architectures. However, only possible structural adaptations are modeled with roles, i.e., *what* to adapt. Adaptation scenarios that describe *when* to adapt have to be implemented as code. However, the declarative modeling of adaptations is more feasible as it enables more concise problem definitions and improves software maintainability.

Modeling adaptations for self-adaptive software has been investigated extensively in the Models@Runtime (M@RT) community, which aims to bring model-driven software engineering to self-adaptive systems and make use of software models at runtime (Giese et al. 2012). Models to adapt include structural and behavioral models of the system. Adaptations can then be modeled as rules and constraints defined on the model.

Reference Architecture. A reference architecture for Models@Runtime system as proposed by Aßmann et al. (2014) is depicted in Fig. 6.

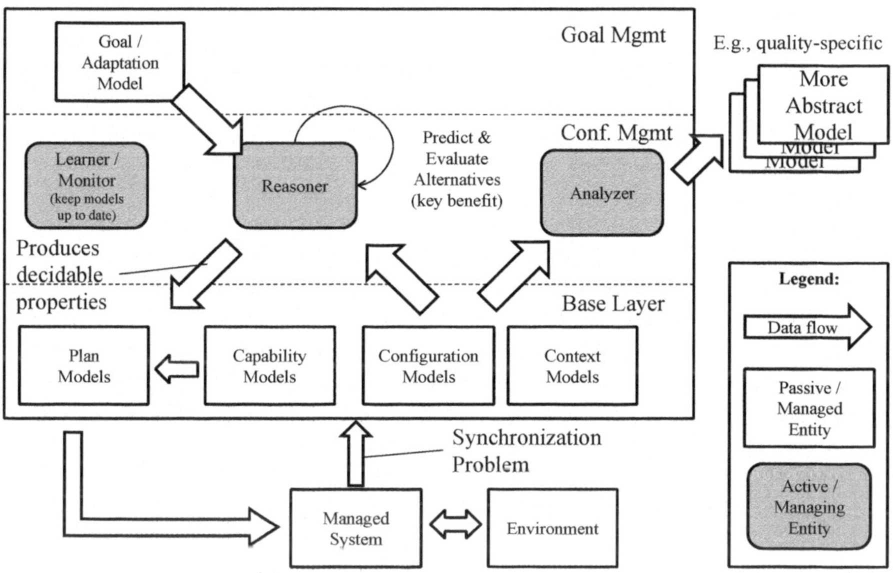

Fig. 6. M@RT reference architecture (Aßmann et al. 2014)

Here, the system is comprised of three individual layers:

- The **base layer** contains the models of the managed system. This includes context-models, configuration-models, capability-models, and plan-models of the system. Context-models describe the current state of the environment as measured by sensors, e.g., the current temperature. The configuration-models of the system describe the current configuration, usually focused on the architecture of the managed system, e.g., a component model depicting the current architecture of the system. On the other hand, capability-models describe how the system can be adapted. Finally, plan-models describe what actions should be executed to realize an adaptation.

- The **configuration management layer** contains a feedback loop that reconfigures and adapts the managed system. The central component is the reasoner, which derives possible variations of the current system configuration based on the context- and configuration-models from the base layer. A monitoring component keeps these models up to date for access by the reasoner. Possible adaptations are evaluated against the goal model contained in the goal management layer. The analyzer component then computes an adaptation plan, if necessary, to fulfill the system's goals. The feedback loop can be implemented following the MAPE-K pattern presented in Sect. 3.1; in this case, the reasoner fulfills the roles of analyzing and planning, while the analyzer component relates to the execution part in the MAPE loop.
- The **goal management layer** comprises models that describe the system's goals. Goal models can change over time, e.g., in response to changes in the context-models of the system.

As outlined in Sect. 3.6, the proposed framework is based on the reference architecture for systems using models at runtime. Thus, the core of the system consists of the same active components depicted in Fig. 6, which form the MAPE-K feedback loop as described in Subsect. 3.1:

- The **monitoring component** aggregates the current state of the system and directly corresponds to the monitoring step in the MAPE-K loop.
- The **reasoning component** consists of an analysis and planning component and derives a plan to adapt the system based on the information that was aggregated by the collective monitoring component. In the presented case, this results in a set of modifications of the role graph, i.e., role additions, removals, and state updates. The reasoning component implements the analyzing and planning steps in the MAPE-K feedback loop.
- The **execution component** equates to the same step in the MAPE-K loop and decides how to apply the derived plan and executes it.

As the proposed framework builds on roles, the authors use CROM models to describe the system architecture of the managed system. Therefore, capability-, configuration-, and context-models are modeled with CROM. Plan-models consist of graph rewriting rules and graph constraints which in turn are defined on the CROM architecture models.

As mentioned, all relevant system models are available to the central reasoning component in the configuration management layer, including the plan-models used to describe how to adapt the system. Rule-based models are commonly used for this purpose.

Graph Rewriting. Of these, graph rewriting rules are a prominent subclass. Graph rewriting is a technique to declaratively rewrite patterns in graph-based models. The following section provides a high-level overview based on Heckel and Taentzer (2020). A complete introduction to graph rewriting including formal foundations can be found in Rozenberg (1997).

Simple graphs are defined by a tuple (V, E), where V is a set defining the graph's vertices and E is a set of tuples (A, B) each defining an edge in the graph from vertex A to B. Graphs can be decorated with labels, which are defined as elements of an additional set included in the graph definition. Labels are associated with edges and vertices by (partial) functions $lab_v : V \rightarrow L_V$ and $lab_e : E \rightarrow L_E$. Furthermore, graphs can be typed by mapping each vertex and element to a respective element defined in a type graph. This results in a pattern shared with domains like object-oriented programming which defines classes and objects. Typed graphs can be attributed with additional named information, commonly data sets with algebraic operations which can be computed by expressions (e.g., integer, string, and boolean values).

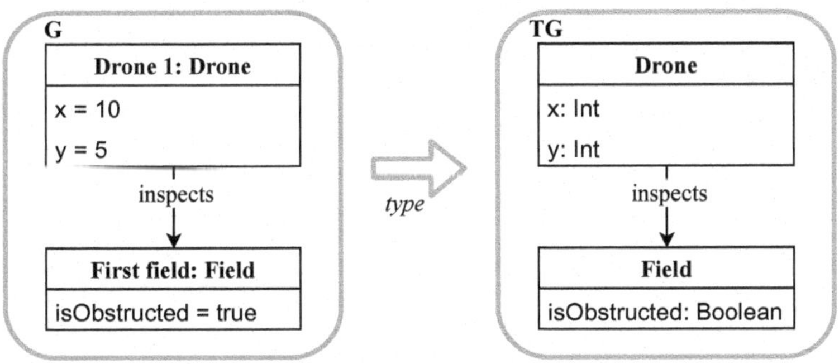

Fig. 7. Example typed graph for a minimal drone model

A minimal example for an attributed labeled and typed graph is depicted in Fig. 7. As shown, typed attributed graphs can represent data structures. Further features known from object-oriented design, such as inheritance, can be modeled by additional edges of a certain type, or by decorating other graph elements with new attributes.

Based on typed attributed graphs, rewriting rules can be modeled. Rules generally consist of a left-hand and right-hand side, where the left-hand side describes the pattern to be matched and the right-hand side describes the graph pattern it should be rewritten to. Upon application of the rule to a graph, one occurrence of the pattern is rewritten. Here, a rule is defined that matches a drone vertex in the graph that does not currently inspect a field. This pattern is called a Negative Application Condition (NAC). Upon application of the rule, a field is matched and a corresponding edge is added.

To restrict the application of a graph rewriting rule, constraints may be defined on the model. Constraints can enforce the presence or absence of a specific pattern in the graph. Consider the following example for the minimal drone example presented, graphically represented in Fig. 8: A drone has to be inspecting a field at any point in time. The constraint defines an invariant that may never occur in the graph, which in this example is that no drone may be present that *does not* inspect a field.

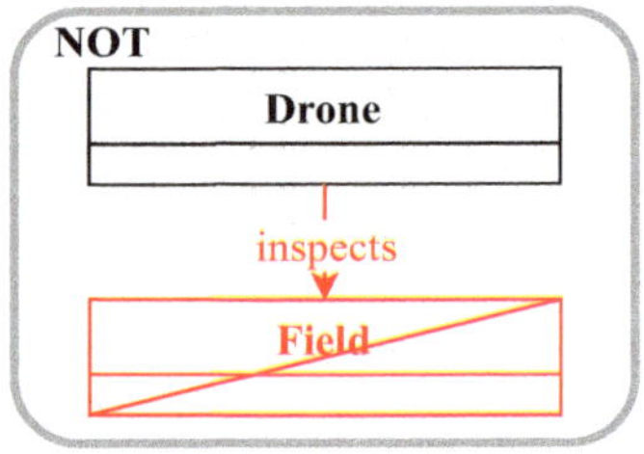

Fig. 8. Graphical representation of a graph constraint

As such, OCL constraints known from object-oriented modeling can be mapped to graph constraints.

To summarize, graph rewriting rules and constraints allow declarative modeling of architectural adaptations and invariants.

3.7 Conceptual Framework for Intent-Based Adaptation Coordination

The conceptual framework proposed by the authors assembles the described concepts to support Intent-based self-adaptiveness in distributed networked systems. Intents and sub-intents are modelled as CROM compartments. Each compartment contains a set of graph rewriting rules and graph constraints as introduced in Sect. 3.6, as well as nested compartments that represent sub-intents. Thus, rules and constraints implement the behaviour necessary to map intents to adaptation plans. Adaptation plans consist of the role change operations for all involved autonomous nodes that are necessary to fulfil the respective intents.

Each autonomous node in the network system contains an implementation of an autonomic manager according to the M@RT reference architecture. The reasoning component of the M@RT reference architecture is the key component for mapping intents. It calculates a set of architectural modifications required to service the system's goals. As described, graph rewriting rules and constraints are used as planning models. The rules reference the CROM models that describe the role-based architecture of the system and possible role-plays.

CROM models are directly related to typed graphs by mapping naturals, roles, and compartments to records. Role-plays, as well as relationships between roles, are mapped to directed edges. The authors differentiate between rules that *must* be executed when the selection part is matched and optional rules that *can* be executed. Constraints on the other hand represent invariants that must hold in any valid state of the system. Optional rules are executed as necessary to fulfil the defined constraints. Naturals can be created programmatically by SCROLL code, e.g., for the creation of the drone natural instances that represent the drones in the real world, as well as the creation of an anomaly object if detected. The creation of roles and thus adaptation of the system is limited to the graph rewriting system.

The main feedback loop as described in Subsect. 3.2 is implemented by the following steps according to the M@RT reference architecture:

1. Periodically, a request is sent to the collective monitoring component. If drones or anomalies are sensed that are currently not represented in the CROI world model, their representation as naturals is created.
2. The system checks for rewriting rules present in active compartments. Mandatory rules are executed for every match until no rule matches. Afterwards, optional rules are executed if necessary to fulfil the constraints defined in active compartments.
3. If no more rules are applicable, the difference between the original and modified architecture is calculated. The set of differences forms a transaction with an adaptation plan which brings the system from one valid state into another. The final component, the analyzer, then executes the transaction.

To conclude, the planning step in the feedback loop consists of calculating an adaptation plan out of declared intents that defines a set of roles to be played or revoked, as well as updates to their state. Details about the intent-based adaptation process are illustrated in the next section using the smart farming example introduced in section Sect. 2.

4 Case Study

In the following section, the authors present a case study to illustrate the implementation of the proposed conceptual framework. As a concrete use case, the following situation is selected. The drone swarm started to roam the field in a line formation. One drone is elected as master of the swarm. While roaming, a drone has detected a potential anomaly and reported the information. Following the implemented M@RT reference architecture, the actions performed by the three active components, i.e., Monitor, Reasoner, and Analyzer, are discussed in detail which are necessary to adapt the swarm. The intended reaction is a split of the swarm into a main and a sub-cluster with the sub-cluster of drones inspecting the anomaly.

4.1 Monitoring Component

The monitoring component represents the first stage of the managing subsystem's control loop, as depicted in Fig. 2. For a highly dynamic intent-based system of autonomous drones, numerous different patterns for the control loop were compared regarding their applicability for the presented application scenario in Subsect. 3.2. The authors concluded, that it is best to use a hybrid approach of multiple patterns, as it offers the most benefits regarding resilience, fault tolerance, system's state representation, and persistence (see Fig. 4).

Consequently, the monitoring stage for the intent-based adaptation of the drone swarm consists of two levels in the hierarchy, as depicted in Fig. 4, including a centralized monitoring instance that aggregates data about all components of the self-adaptive system and distributed monitoring sensors, which are located once on each drone.

The distributed monitoring sensors provide each drone with the feature to summarize all their properties, in-memory variables, and contextual information as a JSON object and push this object to a collective monitoring instance via HTTP/S in a fixed recurring interval. This allows the centralized monitoring component to have a picture of the state of the whole swarm, while each drone may stay agnostic of it. For that, in the case study, drones provide data about

- their orientation (IMU data, GPS coordinates),
- their cluster and if they are assigned with the leader role,
- information regarding their mission, i.e., inspected checkpoints, found anomalies, recorded pictures, and available storage capacity,
- the battery charge level,
- and the integrity of their system, e.g., if any part of the drone is damaged.

The centralized monitoring component in its essence is a programmable interface, that allows drones to send HTTP/S packets about their properties, in-memory variables, and contextual information to a database in a structured way. The interface is designed as a REST API, which accepts *JSON Objects* as input and takes care of organizing the data that is received by individual drones to picture the state of the whole swarm. For that, the drone's information is transformed and aggregated within a database. This database also acts as a knowledge repository, as the hybrid pattern suggests (Fig. 4). By design, the main monitoring component is highly scalable, as it acts only as an interface to a scalable database, hence, it does not impose any bottlenecks or single-point-of-failures on the system. This, however, limits the number of database systems from which can be chosen from, as the database has to be horizontally scalable as well, and needs to support time-series data. The support for time-series data is important, as monitoring systems in a control loop fashion essentially always produce time-series data.

The data that is provided by the drones is aggregated to extract additional knowledge about

- how many clusters exist,
- the state of each cluster, i.e., which drones participate in a cluster and which role do they play (ClusterMaster/ClusterSlave),
- the state of the field, i.e., how many anomalies exist and where they are located,
- the area, which was already covered by the swarm,
- the position of each drone in the area.

The entirety of this data can then be used to reason about a set of high-level system goals. This is the intent, that is declaratively specified for the swarm. This step is carried out by the reasoning component, which is active on the drone that plays the ClusterMaster. It can query the REST API of the centralized monitoring component to receive a JSON Object, that provides a detailed overview of the swarms' current state.

Consider the following state of the intent-based system as an illustrative example: One drone just reported an anomaly to the centralized monitoring component, while all drones are currently active in only one cluster. Consequently, all drones currently are in roaming mode. Once the ClusterMaster queries the centralized monitoring component for the aggregated system knowledge, the following set of data is transmitted:

- There is **one** cluster available.
- A list of all drones within this cluster and which role they play, i.e., one ClusterMaster and multiple ClusterSlaves.
- There is **one** anomaly that is **not yet investigated**.
- Data about the area, that was already covered by the swarm,
- and the position of each drone in this area.

4.2 Analysis and Planning

In the following section, an exemplary implementation of adaptation rules for the application scenario envisioned in Sect. 2 is given.

As outlined in the application scenario introduction, all drones start in a straight line with the intent of searching the field for anomalies. The system is initialized by selecting a master drone via leader election. In the beginning, all drones form a single cluster with a single elected master drone. This single root cluster compartment corresponds to the intent predefined in the application scenario. Sub-intents are represented as nested compartments.

Subsequently, the Reasoner component, which is responsible for analyzing and planning adaptations, is started on the master drone. Then, the main planning feedback loop of the system is run. This consists of the three steps lined out in Sect. 3.6. After pulling the current state of the system from the monitoring component, the set of graph rewriting rules and constraints contained in all active compartments is evaluated until no more mandatory rules are matched and a solution to satisfy all constraints is found.

In the application scenario, a subcluster of drones is split off as soon as an anomaly is found. This can be implemented as a constraint contained in the root intent compartment: If an anomaly object is matched, there has to be a compartment present in which three drones play the **inspecting mode** role.

Also, an optional rule is defined that matches a drone and an anomaly and implies the subcluster compartment as well as the role played by the drone. In the subcluster compartment, a constraint defines that drones have to hold a constant spacing. The reintegration with the main swarm is realized by a mandatory rule defined in the root compartment: If all anomalies that have been assigned to a subcluster have been investigated, the compartment and contained roles are removed. The drones are then reassigned to the main cluster. This behavior implements the fulfillment of the intent.

The rules and constraints necessary to implement the desired behavior for the intent defined by the scenario are described in Table 1. A concise pseudo-code specification is used, which corresponds to the graphical definition given in Sect. 3.6. The syntax used splits the matched pattern and the derived pattern by an arrow `->`. Individual objects are assigned to temporary variables. Role-plays are written as "player" **plays** "role" **in** "compartment". Attribute values of objects (e.g., the property *inspected* of an anomaly) are accessed using *object.property*. Negative Application Conditions (NAC) are depicted by a leading *!* followed by a subpattern to be matched in parenthesis. An exclamation mark is also used to remove matched patterns on the right side of the rule. For rules, the definition reads as "for each match of the left part, the right part has to be created". For constraints, the definition reads as "for each match of the left part, the right part must be matched anywhere in the model". Arrays are defined by square brackets, e.g., `[3]Drone()` defines an array of three drone objects. Consider the following iteration of the feedback loop as an illustrative example: After requesting the most recent snapshot of the system from the monitoring component, the planning component evaluates the ruleset. The monitoring component reports that there are drones in coarse flying mode and an anomaly that is not yet under investigation. Thus, the planning component has to find a configuration where the currently unfulfilled constraint number 1 is satisfied. This is achieved by matching optional rule number 2 on three separate drones and the anomaly, thereby creating an inspection role played by each drone in the compartment. Finally, the planning component constructs a transaction that defines the adaptations needed to transform the system architecture. It does so by calculating the difference between the new architectural model derived by evaluating the defined rules and the old architecture. In this case, the transaction contains the replacement of the coarse inspection roles played by the drones with focused inspection roles in a new cluster compartment.

In the following section, the execution of the derived transaction is discussed.

4.3 Decentralized Execution of Adaptation Plans

The execution component according to the hybrid pattern approach introduced in Fig. 4 is responsible for distributing and executing the derived adaptations by the planning component to fulfill a sub-intent. Furthermore, the success of the execution process must be monitored by the component. First, an overview of the adaptation execution process and its steps will be given.

The execution process as referenced by the MAPE-K feedback loop starts by receiving an adaptation plan from the analysis and planning component.

One of the participants takes the role of the execution master, which can be assigned dynamically. In the case study, the execution master role is assigned to the master drone of the (sub-)cluster. The master is responsible for the coordination of the transaction execution among the participants, which is crucial because a central unit for steering the execution does not exist. To ensure safe and consistent changes, coordination between the peers is mandatory. On each local node runtime, changes have to be performed in a way that no data gets lost

Table 1. List of rules and constraints defined for the case study

	Type	Description	Definition
1	Constraint in root	If an anomaly is found, there has to be a compartment with three drones playing the inspecting role	`a = Anomaly(inspected = false) -> c = Cluster(), ds = [3]Drone(), ds[0..2] play Inspecting in c, c.anomaly == a`
2	Optional rule in root	For every pair of drones and anomalies, there can be a subcluster with the drone inspecting the anomaly	`a = Anomaly(inspected = false), d = Drone() -> c = Cluster(), d play i = Inspecting in c, c.anomaly = a`
3	Constraint in cluster	An even spacing between drones is maintained at all times	`c = Cluster(), d = Drone(), d2 = Drone(), d play Inspecting in c, d2 play Inspecting in c, d != d2 -> |d.position - d2.position| == 10`
4	Mandatory rule in cluster	If an anomaly has been investigated, the cluster is removed	`c = Cluster(), d = Drone(), a = Anomaly(), c.anomaly == a, a.investigated -> !c`
5	Constraint in root	Every drone has to be part of a cluster at all times	`c = Cluster(), d = Drone() -> d play Coarse in c || d play Inspecting in c`
6	Mandatory rule in root	If there is a drone that is not currently member in any cluster, it defaults to the coarse roaming role in the root cluster	`c = Cluster(), d = Drone(), !(d play Inspecting in Cluster || d play Coarse in Cluster) -> d play Coarse in c`

and no inconsistencies occur. This state in which changes can be done safely is called a quiescent state. Those steps are discussed in more detail in the following.

The authors attempt to use the decentralized Weißbach (Weißbach et al. 2017a, 2017b) protocol for role-based coordinated self-adaptation execution. A big advantage of the Weißbach protocol is the support of decentralized adaptations. The decentralization allows the drones to organize themselves in swarms to fulfill subtasks and lets the emerging sub-cluster of drones adapt independently of each other. Decentral coordination is essential for the transaction execution since the drones flying in the field act autonomously without any central coordination unit and so must be able to control the adaptation execution on their own. This allows for better scalability according to Subsect. 3.2 while a single point of truth and bottleneck is avoided. The protocol allows for runtime execution of adaptation plans and so introduces a variability of the system, either parametric or structural adaptations. Drones can execute local adaptations autonomously but refer to the adaptation plan provided by the master to ensure consistent adaptation of the swarm. However, the small overhead due to the coordination of those adaptations is unavoidable to distribute the actions among the participants.

The execution protocol is based on a two-component architecture on the respective drones in the application scenario. Each drone contains a so-called Adaptation Manager (AM), which can be referred to as the execution component in the hybrid pattern approach as depicted in Fig. 4, and a Role Runtime, which is the SCROLL runtime in the application scenario. The AM of the master drone is responsible for the coordination and distribution of the execution and initiation of the plan on the respective drone runtime. The drone runtime contains the application logic based on the role concept already introduced in the previous chapter.

The aim of the Weißbach protocol is the adaptation of distributed nodes at runtime, which requires a methodology to safely change the internal structure in a running system. If the system is suddenly halted or performed the updates regardless of the current internal state and ongoing computations, it would risk data loss and errors since important tasks could be executed at the moment of the execution. Therefore the role runtime and drone runtime respectively have a lifecycle with different operational states, which determine whether the runtime can be adapted or updated or not. A simplified version of an exemplary role lifecycle is illustrated in Fig. 9. Generally, a running application has the current role in the bound state and the respective role is active. In this state, it would be disruptive to perform adaptations at run-time. To safely adapt the application, the application and its dependencies have to be passivated. In this state, the played role can be unbound and changed into the desired one. Note, that the *Active* state has two internal states, i.e., *Idle* and *Processing*.

Together with the lifecycle of the role runtime the Weißbach protocol introduces the usage of a quiescent or safe state. Kramer and Magee (1990) introduced the quiescent state of a node as a state where it is not engaged in a transaction that was initiated by the node and that the node will not initiate any new transactions. Furthermore, it is not engaged in servicing other nodes' transactions. The quiescent state is very disruptive and requires the affected nodes and the indirectly affected nodes to halt during the change process, hence the system is not allowed to process data and fulfill its tasks until the whole change has been performed and the adaptation transaction has been finished. Note, that the term transaction in the definition of quiescence differs from the authors' understanding of an adaptation transaction. The adaptation transaction in the understanding of Weißbach et al. (2017b) refers to a unit of different adaptations to be performed atomically, i.e., all changes should be performed or none at all. Transactions have a unique id and consist of one or multiple adaptation operations. The goal is to perform transactions consistently. In case of a failure, intermediate changes will be rolled back before activation of the changes.

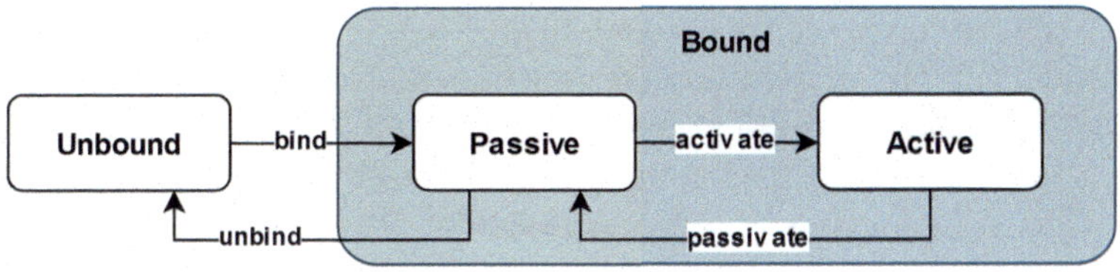

Fig. 9. Simplified role runtime lifecycle derived from Weißbach et al. (2017b)

Concerning the case study, the starting point of the execution is receiving the adaptation operations generated by the planner on the master drone as depicted in Fig. 4. The plan is passed to the execution component of the master drone, which is then responsible for distributing the plan among all affected participants using the Weißbach protocol (Weißbach et al. 2017b). An adaptation transaction which is initially distributed by the execution module on the master drone contains information about the drone to be adapted, the current roles, and the roles to be played after the adaptation, the current compartment, and the goal compartment. It is also possible to perform group adaptations, i.e., multiple adaptations on several drones belong to one transaction which is then executed atomically. In the case of an error, all temporary changes of the respecting transaction are reverted to recover a consistent state and the drones stay unaffected and remain in their current role, i.e., roaming or inspecting. The distributed adaptations are only executed and activated if all of the affected peers acknowledge the changes and successfully prepare the activation of the changes. Otherwise, no changes will be performed.

A part of an exemplary transaction structure is presented in Fig. 10. This transaction is the result of the plan by the planning component to switch three drones into the inspecting mode since an anomaly on the field was found. The figure shows the transaction with the adaptation operation for one drone. The other two drones' operations are structured as the shown one, respectively, and belong to the same transaction 'transaction1'. The three operations must finish successfully since they belong to one transaction, otherwise, all changes get reverted. An adaptation operation has a unique id and type of operation. In the figure, the current role 'Coarse' of 'drone1' will be exchanged by the role 'Focused' with the 'exchange' operation. Other operation types are the addition, removal, or cloning of the role to another entity, migration of a role, or the creation of a collaboration of certain roles, and disbanding the collaboration. Besides that, operations contain an order number for a successor/predecessor relationship, which is '1' in the example. In the case that the drone contains internal state information, the protocol can be used to transfer the internal state of the old to the new role to preserve important information. This can be required by the planning component. Most important, information about the target and source node to be adapted is provided in the adaptation operation. In the example, drone1 with the current roles Coarse and ClusterSlave in the compartment Roaming with the IP-address 192.168.0.10 will play the roles Focused and ClusterSlave in the compartment Inspecting. The source node information is used to identify the correct node to be changed. The three drones intended to be adapted form an independent swarm in the compartment 'inspecting' with an own master drone, which is responsible for the coordination in the new partial swarm.

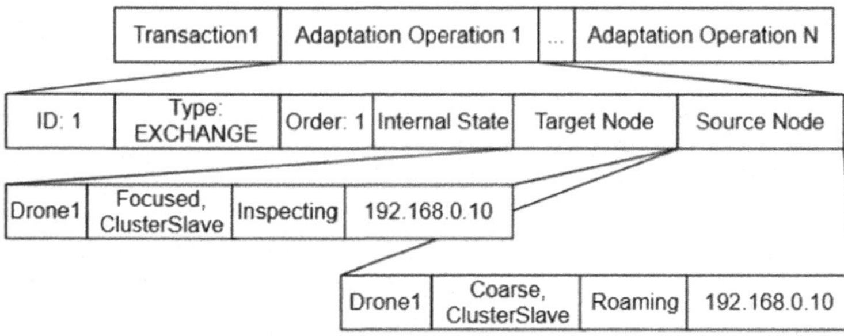

Fig. 10. Exemplary transaction in the application scenario

The drone in the application scenario which is currently playing the roles ClusterSlave and Coarse will be adapted to play the roles ClusterSlave and Focused afterward, as in the exemplary transaction in Fig. 10. To achieve that, the affected role Coarse needs to be passivated first as depicted in Fig. 9. If the role was passivated successfully, the new role 'focused' gets added to the system and is first in the unbound state. It will be bound then and remains in the passive state, too. If all drones agree on the respective transaction and are prepared for the execution, the passive new role can be activated and the role Focused is now played actively. The old role Coarse can be unbound and removed from the system.

To allow the drone to fly in a quiescent state, i.e., a safe state for adaptation is reached, a stronger separation of concerns regarding the played roles for the inspection of the field and the drone control needs to be defined since not everything should be halted during that phase. The roles which are designated to change are allowed to impact the flight height and speed, but should not contain the drone control since the passivation of the drone control would result in a crash when the drone remains passive or quiescent for too long. In that phase, basic processes as gliding and flying should work and only role-specific behavior as scanning behavior and specific communication in compartments should be affected by the passivation. That guarantees that the drone is still able to fly but the behavior which is meant to be changed is not performed during the adaptation transaction.

5 Conclusion and Future Work

Due to the highly decentralized and distributed nature of networked systems of autonomous units, their management is a challenging task. While keeping a human in the loop is often required to create trust in systems deployed in practice, it is almost impossible for an administrator to manage such systems in a device-by-device manner.

Using intents is a promising approach since it allows administrators to declare high-level goals for the system but relieves them from manual reconfiguration since the system manages itself in an autonomic manner driven by operational guidance from operators.

To bridge the gap between high-level intents and change operations performed at each autonomous node in a coordinated way, the authors propose the use of self-adaption concepts. Particularly, a conceptual framework for intent-based adaptation coordination has been introduced in this chapter, that builds upon roles, control loops, and decentralization patterns for them to structure the system and provide decentralized control structures where necessary.

For the implementation, each autonomous node contains an autonomic manager according to the Models@Runtime reference architecture. The reasoning component is the key component to map intents. It relies on the modeling of intents and sub-intents as compartments in CROM and calculates necessary architectural modifications based on the graph rewriting rules and constraints specified for each compartment. It uses the information about each autonomous unit and the overall state of the networked system gathered during the monitoring phase and produces an adaptation plan in form of an adaptation transaction. The execution of the adaptation transaction ensures that all autonomous nodes perform changes atomically, i.e., either all changes at all involved nodes are performed or non at all. This guarantees that the networked system is transformed from a valid configuration to another valid configuration for every adaptation transaction.

While the authors presented with the case study how an intent-based approach for adaptation coordination can be designed and implemented, there are still challenges that need to be addressed in the future to build robust, scalable, and secure solutions for practical use.

In the current version of the proposed framework, each autonomous unit comprises a local role runtime, i.e., SCROLL. Even though role-plays are triggered by the graph rewriting system, business logic inside of roles is still implemented with code. As such, a middleware abstracting the distribution of objects across the network is desirable in order to enable the developer to communicate between objects without reimplementing a complete network transport and concurrency layer. To address this, the authors plan to introduce a distributed role runtime that extends the actor model of concurrency as proposed by Hewitt and Baker Jr (1977) with roles.

Regarding the graph rewriting environment, the system is currently not able to adapt to situations and intents unforeseen by the developer, i.e., there is no learning aspect. In future iterations, a reinforcement learning approach based on probing optional graph rewriting rules can pose a solution to this aspect. Additionally, the augmentation of graph rewriting rules with control flow and optimization aspects is to be investigated to support the development of more complex scenarios.

The implementation of the monitoring stage, which was inspired by IBM's blueprint for autonomic computing IBM (2005), is specifically tailored to the presented application scenario. This has the consequence, that implemented artifacts in that stage cannot be re-used for different scenarios. Reasoning from that, a more general solution for the monitoring stage, which can function in different application scenarios, associated with different domains, is desirable. As introduced in the case study, roles may be used to dynamically adapt a system's functions. In the future, one could research the applicability of the role-based approach to the MAPE-K stages of the managing subsystem.

Currently, the execution of the adaptation transactions follows the concept of atomic changes, i.e., all changes are performed successfully or none at all. In addition, due to the presented quiescence criteria, the system is halted for a long time and no tasks can be started during that state. Therefore, further investigation into the direction of execution with the concept of eventual consistency for self-adaptive systems (Tomforde and Gruhl 2020) is a promising direction. If an error occurred during the adaptation process and all changes must be reset to the current configuration, the eventual consistency approach would allow the successfully adapted nodes to work in their new configuration and the erroneous node must be dealt with separately. This would allow already adapted parts of the system to return to work again and so the global downtime gets reduced.

Acknowledgements. This work is partially funded by the German Research Foundation (DFG) within the Research Training Group "Role-based Software Infrastructures for continuous-context-sensitive Systems" (GRK 1907).

References

Albuquerque, A., Guizzardi, G.: An ontological foundation for conceptual modeling datatypes based on semantic reference spaces. In: IEEE 7th International Conference on Research Challenges in Information Science (RCIS), pp. 1–12. IEEE (2013)

Aßmann, U., Götz, S., Jézéquel, J.-M., Morin, B., Trapp, M.: A reference architecture and roadmap for models@run.time systems. In: Bencomo, N., France, R., Cheng, B.H.C., Aßmann, U. (eds.) Models@run.time. LNCS, vol. 8378, pp. 1–18. Springer, Cham (2014). https://doi.org/10.1007/978-3-319-08915-7_1

Bachman, C.W., Daya, M.: The role concept in data models. In: Proceedings of the Third International Conference on Very Large Data Bases, vol. 3, pp. 464–476 (1977)

Brun, Y., et al.: Engineering self-adaptive systems through feedback loops. In: Cheng, B.H.C., de Lemos, R., Giese, H., Inverardi, P., Magee, J. (eds.) Software Engineering for Self-Adaptive Systems. LNCS, vol. 5525, pp. 48–70. Springer, Heidelberg (2009). https://doi.org/10.1007/978-3-642-02161-9_3

Cheng, S.-W., Garlan, D., Schmerl, B.: Making self-adaptation an engineering reality. In: Babaoglu, O., et al. (eds.) SELF-STAR 2004. LNCS, vol. 3460, pp. 158–173. Springer, Heidelberg (2005). https://doi.org/10.1007/11428589_11

Giese, H., et al.: Graph transformations for MDE, adaptation, and models at runtime. In: Bernardo, M., Cortellessa, V., Pierantonio, A. (eds.) SFM 2012. LNCS, vol. 7320, pp. 137–191. Springer, Heidelberg (2012). https://doi.org/10.1007/978-3-642-30982-3_5

Guarino, N.: Concepts, attributes and arbitrary relations: some linguistic and ontological criteria for structuring knowledge bases. Data Knowl. Eng. **8**(3), 249–261 (1992)

Heckel, R., Taentzer, G.: Graph Transformation for Software Engineers. Springer, Heidelberg (2020). https://doi.org/10.1007/978-3-030-43916-3

Herrmann, S.: Programming with roles in objectteams/java (2005)

Hewitt, C., Baker Jr., H.: Actors and continuous functionals. Technical report, Massachusetts Inst of Tech Cambridge Lab for Computer Science (1977)

IBM. An architectural blueprint for autonomic computing. IBM White Paper, pp. 1–34 (2005). https://www-03.ibm.com/autonomic/pdfs/AC%20Blueprint%20White%20Paper%20V7.pdf

Kephart, J.O., Chess, D.M.: The vision of autonomic computing. Computer **36**(1), 41–50 (2003). https://doi.org/10.1109/MC.2003.1160055

Kramer, J., Magee, J.: The evolving philosophers problem: dynamic change management. IEEE Trans. Softw. Eng. **16**, 1293–1306 (1990). https://doi.org/10.1109/32.60317

Kühn, T.: A family of role-based languages (2017). https://nbn-resolving.org/urn:nbn:de:bsz:14-qucosa-228027

Kühn, T., Leuthäuser, M., Götz, S., Seidl, C., Aßmann, U.: A metamodel family for role-based modeling and programming languages. In: Combemale, B., Pearce, D.J., Barais, O., Vinju, J.J. (eds.) SLE 2014. LNCS, vol. 8706, pp. 141–160. Springer, Cham (2014). https://doi.org/10.1007/978-3-319-11245-9_8

Kühn, T., Böhme, S., Götz S, Aßmann, U.: A combined formal model for relational context-dependent roles. In: Proceedings of the 2015 ACM SIGPLAN International Conference on Software Language Engineering, pp. 113–124. Association for Computing Machinery, New York (2015). https://doi.org/10.1145/2814251.2814255

Lämmel, R.: A semantical approach to method-call interception. In: Proceedings of the 1st International Conference on Aspect-Oriented Software Development, pp. 41–55. Association for Computing Machinery, New York (2002). https://doi.org/10.1145/508386.508392

Leuthäuser, M.: A pure embedding of roles. Technische Universität Dresden (2017). https://doi.org/10.1145/1118890.1118892

Mernik, M., Heering, J., Sloane, A.M.: When and how to develop domain-specific languages. ACM Comput. Surv. (CSUR) **37**(4), 316–344 (2005). https://dl.acm.org/doi/10.1145/1118890.1118892

Oreizy, P., et al.: An architecture-based approach to self-adaptive software. IEEE Intell. Syst. **14**, 54–62 (1999)

Rozenberg, G. (ed.): Handbook of Graph Grammars and Computing by Graph Transformation: Volume I. Foundations. World Scientific Publishing Co., Inc., USA (1997)

Salehie, M., Tahvildari, L.: Self-adaptive software: Landscape and research challenges. ACM Trans. Auton. Adapt. Syst. **4**, 14:1-14:42 (2009)

Sowa, J.F.: Conceptual Structures: Information Processing in Mind and Machine. Addison-Wesley Longman Publishing Co., Inc., Boston (1984)

Steimann, F.: On the representation of roles in object-oriented and conceptual modelling. Data Knowl. Eng. **35**(1), 83–106 (2000)

Tomforde, S., Gruhl, C.: Fairness, performance, and robustness: is there a CAP theorem for self-adaptive and self-organising systems? In: Proceedings - 2020 IEEE International Conference on Autonomic Computing and Self-Organizing Systems Companion, ACSOS-C 2020, pp. 54–59 (2020). https://doi.org/10.1109/ACSOS-C51401.2020.00029

Weißbach, M., Chrszon, P., Springer, T., Schill, A.: Decentrally coordinated execution of adaptations in distributed self-adaptive software systems. In: 11th IEEE International Conference on Self-Adaptive and Self-Organizing Systems, SASO 2017, Tucson, AZ, USA, 18–22 September 2017, pp. 111–120. IEEE Computer Society (2017a). https://doi.org/10.1109/SASO.2017.20, http://doi.ieeecomputersociety.org/10.1109/SASO.2017.20

Weißbach, M., et al.: Decentralized coordination of dynamic software updates in the Internet of Things. In: 2016 IEEE 3rd World Forum on Internet of Things. WF-IoT 2016, pp. 171–176 (2017b). https://doi.org/10.1109/WF-IoT.2016.7845450

Weyns, D.: An Introduction to Self-adaptive Systems: A Contemporary Software Engineering Perspective. John Wiley & Sons, Hoboken (2020)

Weyns, D., et al.: On patterns for decentralized control in self-adaptive systems. In: Lecture Notes in Computer Science (including subseries Lecture Notes in Artificial Intelligence and Lecture Notes in Bioinformatics), vol. 7475 LNCS, pp. 76–107. Springer, Heidelberg (2013). https://doi.org/10.1007/978-3-642-35813-5_4, https://link.springer.com/chapter/10.1007/978-3-642-35813-5_4

Intent-Based Routing in Delay- and Disruption-Tolerant Networks

Felix Walter(✉), José Irigon de Irigon, Olivier de Jonckère, and Thomas Springer

Technische Universität Dresden, Dresden, Germany
{felix.walter,jose.irigon,olivier.de_jonckere, thomas.springer}@tu-dresden.de

Abstract. The Delay- and Disruption-tolerant Networking (DTN) architecture enables communication between nodes in networks that lack continuous end-to-end connectivity. For this purpose, the Bundle Protocol is introduced, encapsulating application data and allowing their store-carry-forward transmission over heterogeneous links. Although a huge number of routing algorithms has been proposed for DTN, in current deployment scenarios, they are applied in a non-adaptive manner and are often configured statically for the whole network. With the advent of Intent-based Networking technologies, it becomes plausible that the DTN domain could vastly profit from a transfer of these concepts. The chapter investigates this general thesis. In this context, the authors point out a strong relationship between Intent-based Networking and existing work on self-adaptive systems. Based on that, techniques for enabling adaptivity in DTN routing are outlined and a future perspective on enhancing node configuration and routing in a DTN with the application of Intent-based Networking concepts is given.

Keywords: Delay- and disruption-tolerant networking · DTN · IB-DTN · DTN routing · Self-adaptive systems · Adaptive DTN routing · Runtime adaptation · MAPE-K loop · Multicast forwarding · CGR · SPSN · IMCEB

1 Introduction

Delay- and Disruption-tolerant Networking (DTN) technologies enable communication in environments where nodes cannot rely on permanent, bidirectional, and low-latency end-to-end connectivity. For this purpose, DTN protocols exploit a series of *contacts* between nodes to deliver messages based on a store-and-forward data transmission scheme. These contacts can even be unidirectional and, in between them, nodes may be moving and "carrying" the data physically. The DTN architecture can be used in manifold scenarios to support communication between humans or machines, e.g., using satellites, submarines, various sorts of vehicles, or even humans as "carriers".

With no stable end-to-end connectivity available, monitoring the network state and keeping configurations of devices consistent across a DTN deployment at scale is particularly challenging. A device-by-device management of nodes is only possible in an asynchronous manner (e.g., via the Asynchronous Management Protocol [6]), but becomes

M. Klymash et al. (Eds.): *Future Intent-Based Networking*, LNEE 831, pp. 101–127, 2022.
https://doi.org/10.1007/978-3-030-92435-5_6

a complex issue in large-scale deployments. Furthermore, due to inherent end-to-end delays, no timely reaction to changes in the network may be possible. An attractive solution to this problem is to enable the nodes to *manage themselves* in an autonomic manner driven by operational guidance from operators or outside systems as proposed for Intent-based Networking (IBN). Driving network configuration by **intents**, i.e., a *set of operational goals and outcomes defined in a declarative manner* [1], allows an operator to manage DTN deployments holistically at a higher abstraction level.

This book chapter discusses the use of IBN concepts for the management of DTN deployments. After analyzing the particular challenges of applying IBN in conjunction with DTN, a conceptual approach for enabling *Intent-based DTN (IB-DTN)* is introduced. Two key concepts implied by the overall approach are described in detail afterwards, namely, a concept for adaptive routing that enables nodes to autonomously configure themselves according to propagated intents and changed network state, and a multicast routing and forwarding technique to deliver such monitoring information and intents reliably to a set of nodes. The chapter especially highlights the parallels between IBN and autonomic computing and demonstrates the use of concepts for self-adaptation with a special focus on decentralization for enforcing intents at the nodes in a DTN deployment in an autonomous manner.

2 Foundations

2.1 Delay-/Disruption-Tolerant Networking

In extreme networks such as *mobile ad hoc networks* (MANET), *vehicular ad hoc networks* (VANET), underwater networks, or space networks, end-to-end communication is challenging because link disruptions and delays may respectively be frequent and high. For instance, in space networks, disruptions may occur due to the occultation of a node, and delays on a single link may be high due to the extreme ranges and the propagation of the signal being limited by the speed of light.

Such high delays and disruptions render the use of traditional (i.e., Internet) protocols and routing techniques in these environments complicated. Indeed, an end-to-end connection is in most cases not possible and topology changes as well as end-to-end delays often prevent acknowledgments through the same path. The mentioned topology changes also render most of the routing algorithms used in conventional networks inapplicable.

2.1.1 DTN Architecture

It was only in the late 1990s that challenging networks became areas of increased interest, with the democratization of the use of wireless protocols. In 2001, Vint Cerf and other scientists from the Jet Propulsion Laboratory (JPL) proposed the *Interplanetary Internet* (IPN) [2].

Soon afterwards, Kevin Fall proposed that the IPN principles could be applied to other kinds of challenging topologies and proposed the generalized term "Delay-Tolerant Networking" (DTN) for this domain [3].

In 2007, the *Delay-Tolerant Networking Architecture* [4] has been presented, allowing nodes to operate in a store-and-forward manner. In a Delay-tolerant Network, messages are buffered until the next transmission opportunities. In parallel, the *Bundle Protocol* [5] defines a format to allow the messages to contain enough data to enable an application to progress upon reception of a single message. Due to this particularity, the Bundle Protocol operates between the application and the underlying transport or network layer. Therefore, the *Bundle Protocol Agent* requires a *Convergence Layer Adapter* (CLA) to bridge with the underlying protocol it relies on. To this end, several CLAs exist, e.g., for the *Licklider Transmission Protocol* (LTP) [13] applicable to space links, as well as for TCP, UDP, and other transport protocols.

However, the whole Bundle Protocol stack operates independently to the underlying network protocols it uses. For identifying application endpoints, the DTN architecture relies on *Endpoint IDentifiers* (EID). Any Bundle Protocol Agent can *subscribe* to such an EID to deliver locally the bundles which have this EID as destination. EIDs use the syntax of URIs [4], composed of a scheme name and a scheme-specific part and, thus, offer a lot of flexibility.

2.1.2 DTN Routing

Routing and forwarding a *bundle* in a Delay-tolerant Network includes defining possible next hops for the bundle, storing it persistently until a transmission opportunity occurs, and, in some cases, enqueueing it for transmission via a specific contact to a specific next hop node. Determining the most-viable route (or even the most-viable next hop) in DTN is a challenging task, as the topology is rarely static and the intervals of connectivity between the nodes are not always predictable, therefore, finding an end-to-end path for a message is not a trivial task. Furthermore, bundle transmissions may be optimized for a variety of factors, such as end-to-end delay, overall delivery probability, or resulting network load, that cannot be precisely estimated. Thus, many routing algorithms for DTN have to rely on opportunistic and/or probabilistic mechanisms [7].

Routing algorithms like Epidemic Routing [8] or Spray an Wait [9] leverage a simple flooding-based approach: Bundle copies are transmitted to neighbors regardless of their properties to maximize the chance to have any of the copies reaching its destination. In the case of Spray and Wait, flooding is limited to a maximum number of copies. Routing algorithms like PRoPHET [10] leverage a *history of encounters* to infer probabilistic patterns from observed connectivity in an attempt to maximize the delivery probability.

Conversely to these opportunistic routing techniques, another class of networks exists in the DTN domain: Deep space missions, satellite networks, and public transportation networks (PTNs) are suitable candidates for *schedule-aware* or *deterministic* routing schemes (e.g., [11]), as upcoming contacts between the nodes are predictable. Indeed, in the deterministic case, each DTN node knows about the intervals of connectivity between all nodes in the network in advance thanks to the *schedule* or *contact plan.*

The contact plan has to be updated periodically on each node for the continuous functioning of the routing algorithms which rely on it, to, on the one hand, remove the expired contacts, and, on the other hand, push forward the *horizon* of the contact plan by adding further expected contacts to it. As long as a contact plan is available, nodes can leverage path finding algorithms to schedule bundles, rendering it possible to determine a next hop for a given message based on the minimal expected delay and only forwarding a single copy of the message via the associated contact.

2.2 Intent-Based Networking

The general motivation of this chapter is to provide a perspective on employing Intent-based Networking concepts in Delay- and Disruption-tolerant Networking. However, there is no unique, clear definition of IBN available in literature. Thus, the definition employed in this chapter is briefly sketched here.

Overall, the current Internet Research Task Force draft on Intent-based Networking concepts [1] is taken as a conceptual basis. The authors of this draft break up the functions of an Intent-based Network into two categories, *Intent Fulfillment* and *Intent Assurance*. While the former processes user[1] input in the form of *intent* specified to the system and attempts to realize it in the network, the latter provides validation and monitoring capabilities, to ensure that the network's operation is aligned with the specified intent.

The *Intent Life-cycle* proposed in this draft document is shown in Fig. 1. It becomes evident that additional processing steps are added that are not present in conventional network configuration concepts: In the course of *Intent Fulfillment*, the user intent is translated into a *course of action*, which is used to feed a process continuously monitoring, analyzing, and validating the network's compliance with the user intent. Based on that and the course of action derived from the user intent, the process plans and optimizes the network configuration (see Sect. 7 of [1]). *Intent Assurance*, however, leverages the results of analyzing network observations performed within the aforementioned process. Those are abstracted into information related to the user intent in order to enable reporting to the user space.

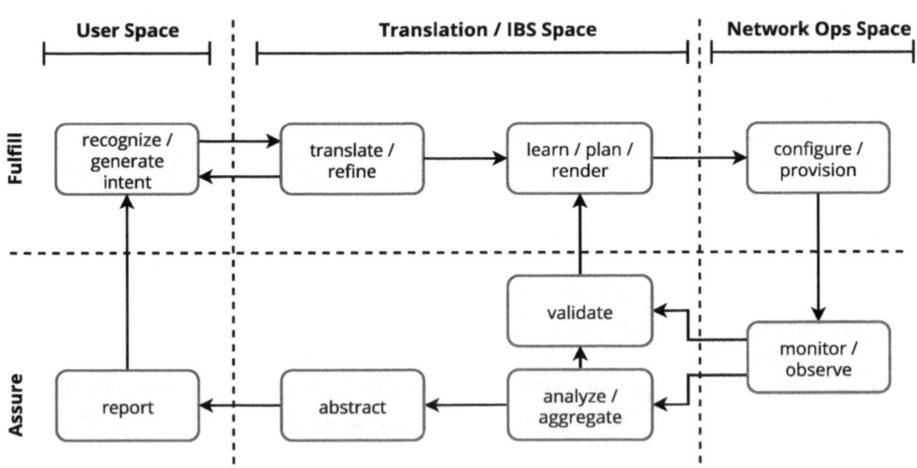

Fig. 1. The Intent Life-cycle proposed in [1], Fig. 1

[1] It shall be noted that a *user* is identified to be "generally, an administrator of the responsible organization" (Sect. 6.1, [1]). Thus, the terms *user* and *operator* are used interchangeably in this chapter.

Overall, the Intent Life-cycle represents two stacked, adaptive processes or *control loops*: The inner loop, i.e., the right half of Fig. 1, is executed by network devices autonomously and provides adaptivity without any user interaction. The outer loop involves the user, however, on a much more coarse-grained and abstract basis than necessary in traditional network management. This way, an Intent-based Network facilitates configuration and management with a largely reduced need for runtime user intervention.

2.3 Self-Adaptive Systems

Kephart and Chess [15] define *"computing systems that can manage themselves given high-level objectives from administrators"* as autonomic computing systems. They considered self-management as the key concept that frees system administrators from the details of system operation and maintenance with four aspects, namely *self-configuration* (automated configuration following high level policies), *self-optimization* (automated continuous improvement), *self-healing* (automated problem detection, diagnosis, and repair), and *self-protection* (automated defence against malicious attacks or cascading failures). These properties are considered as *self-CHOP* properties and are a subset of so called *self-* properties* introduced in the literature.

Salehie and Tahvildari [18] organize self-* properties in a unified hierarchical set with the four mentioned properties forming the middle level of three levels, called *major level*. The lowest level, called *primitive level*, consists of self-awareness (the system is aware of its own state and behaviors) and context-awareness (the system is aware of the operational environment). The top level, named *general level*, contains global properties of self-adaptive software summarized as properties of self-adaptiveness. Accordingly, self-adaptiveness can be considered as generalized property subsuming all self-* properties and self-adaptive systems as general term for systems that implement self-* properties.

According to Oreizy et al. [17], self-adaptive software modifies its own behavior in response to changes in its operating environment. The term *operating environment* refers to anything observable by the software system, such as end-user input, external hardware devices and sensors, or program instrumentation. As part of an conceptual approach for self-adaptive software systems they propose a life-cycle for adaptation management including evaluation and monitoring of observations, planning of changes and deploying change descriptions.

A parallel can be drawn between the proposed life-cycle and the *MAPE-K* control loop that is widely accepted for the implementation of autonomic and self-adaptive systems. Kephart and Chess [15] define autonomic systems as interactive collections of autonomic elements. They structure an autonomic element into an *managed element*, i.e., the software system to be adapted, and an *autonomic manager* that monitors and controls the managed element. The autonomic manager implements a MAPE-K loop to control the managed element as specified in [16]. It *monitors* the managed element for self-awareness and the operating environment for context-awareness, *analyzes* the monitored information to determine if changes are required, *plans* necessary changes and *executes* the plan to adapt the behavior of the managed element. *Knowledge* contains information such as monitoring history, behavioral constraints, and policies that is shared among the four functions.

To accomplish control in distributed self-adaptive systems, Weyns et al. [19] identified a set of decentralization patterns for the MAPE-K loop, namely the *Decentralized Coordination Pattern*, the *Information Sharing Pattern*, the *Master/Slave Pattern*, the *Regional Planning Pattern*, and the *Hierarchical Self-adaptation Pattern*. The selection of one of these patterns is highly use case dependent. As advantages and disadvantages are discussed based on an application scenario in [19], the paper can be used as a reference to guide the decision process.

3 Toward an Intent-Based DTN

In this section, an example DTN scenario for interplanetary communication is introduced that will be used throughout the chapter to illustrate the discussed concepts. The scenario description is followed by the proposal of a principle approach to apply Intent-based Networking concepts to manage DTN deployments.

3.1 Example Scenario

As foundation for describing the challenges and opportunities of applying Intent-based Networking in the DTN context, the concrete example scenario depicted in Fig. 2 will be used in the rest of this chapter. The example scenario represents an interplanetary DTN deployment, connecting a Mars inter-network with persistently-connected as well as intermittently-connected sub-networks via ground stations and relay satellites to the terrestrial Internet. It is assumed that the network is controlled by operators located on Earth and leverages different techniques for routing and forwarding in different parts of the network: While connectivity on Earth and within some sub-networks on Mars is persistent, the satellite links are only available intermittently (but can be scheduled). In both cases, a deterministic routing algorithm can be used as transmission opportunities

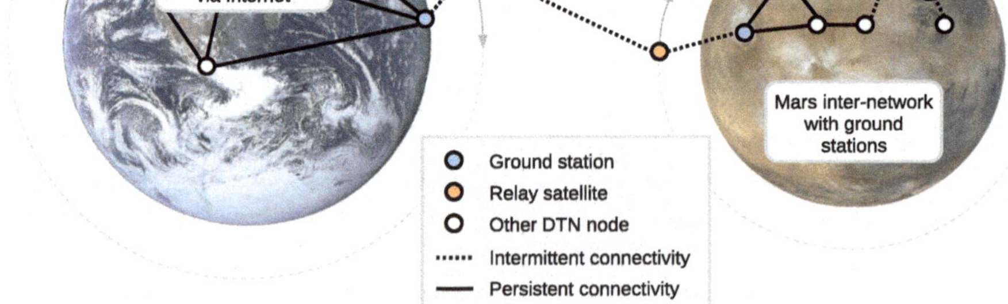

Fig. 2. Example scenario for the application of intent in a self-adaptive DTN

(contacts) can be scheduled. Within some opportunistic sub-networks on Mars, however, replication-based heuristics have to be applied for forwarding data. It is further assumed that, by default, network nodes shall optimize their forwarding decisions for a high delivery probability and low end-to-end delays. This policy is considered as the *intent* that is initially specified by the network operator and expected to be realized by all nodes. In consequence, the deterministic routing algorithm may select routes based on end-to-end delivery delay and the opportunistic algorithms may use a comparably high replication count to achieve high delivery probabilities with low delay.

At some point in the lifetime of the sketched network, a relay satellite fails, leading to a decrease in overall link capacity between Earth and Mars. The rate of data generation, however, does not decrease, leading to the same amount of network traffic that has to flow via the remaining links. In response to the outage, some of the applied routing techniques may even increase load on the network as alternative paths have to be selected or the replication count is increased to achieve an increased delivery probability. Overall, this behavior may result in a congestion situation, which may only be observed in a part of the network and cannot be quickly detected by all nodes. Especially nodes that are generating traffic and are, thus, responsible for the high load put on the remaining links, may not be aware of any issue with the overall network performance.

On the contrary, it can be expected that operators become aware of the link outage as well as the resulting degradation in network performance due to monitoring data provided by the Intent Assurance functions, which may indicate that the specified intent cannot be fulfilled anymore. On that basis, they may trigger processes to fix or re-deploy the failed relay satellite. Additionally, it is necessary to resolve the occurring degradation of service to allow continued operation of the network, e.g., to ensure the delivery of critical science data as well as communication related to repairing the infrastructure. At this point, in an IBN-enabled DTN deployment, an operator could specify a new intent to a) prioritize such data and b) optimize forwarding of all other data to reduce congestion. Nodes can react to this, e.g., by reducing the number of copies created or preferring rarely-overloaded paths over those with minimum latency. They may also choose a different routing and forwarding technique appropriate to the new intent, which mitigates the overall congestion situation.

In this example, Intent-based Networking would be an ideal tool to enable operators to facilitate smooth operation of the network while giving them appropriate time to address any technical issue.

3.2 An Intent-Based Networking Approach for DTN

A. Lerner [14] identifies that an Intent-based Networking system requires to "[ingest] real-time network status" and "continuously validates (in real time) that the original business intent of the system is being met". Obviously, these criteria cannot be met in a DTN scenario with a centralized system as, by design, no real-time operation is feasible.

To enable intent-based concepts in such a network nevertheless, modifications of the IBN control loop sketched in Sect. 2.2 are necessary to cope with the specific characteristics of DTN. Specifically, for facilitating intent-based DTN operation, a further subdivision of the intent processing life-cycle is proposed by the authors; a split into two distinct *segments*, based on the expected quality of network connectivity (i.e., expected delays and disruptions) relative to the operator (or intent-based network *user*).

Figure 3 depicts the proposed adaptation to the intent life-cycle (adapted from Sect. 7 of [1]). The *Operator Source Segment (OSS)* consists of systems for which good connectivity characteristics (i.e., low end-to-end delays, infrequent disruptions) toward the network operator are expected. One or multiple *Target Configuration Segments (TCS)*, on the other hand, are expected to ingest and act upon the intent specified by the operator, although they may not be well-connected to the respective OSS.

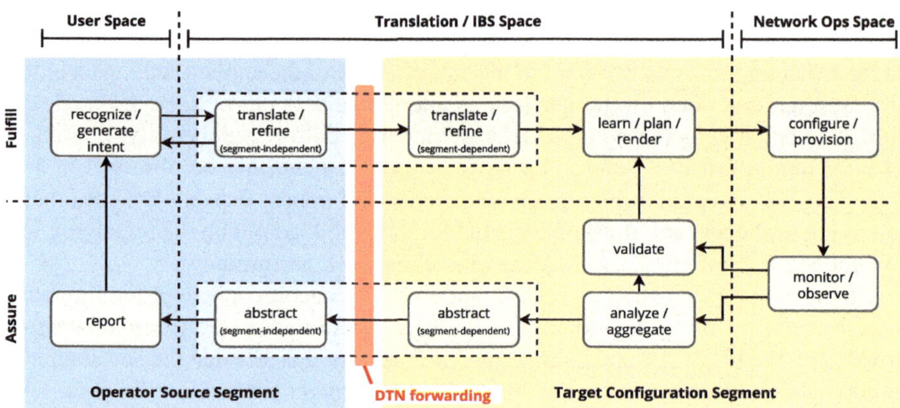

Fig. 3. Modified intent life-cycle for IB-DTN

The necessary *asynchronous* and *delay-tolerant* mode of operation is enabled by splitting the IBN functions to *translate* between intent and actions of the network into two steps each; a network-segment-independent and a network-segment-dependent processing step. Between those steps, the native DTN architecture is leveraged to forward necessary information via DTN bundles between the two defined network segments.

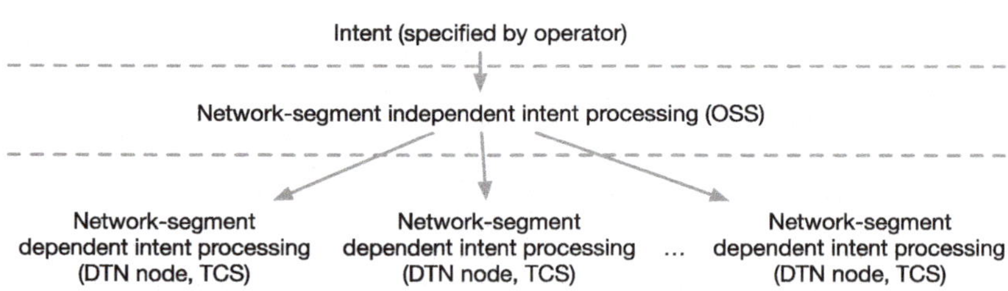

Fig. 4. Layered process of intent translation and enforcement

The resulting layered process of intent enforcement is depicted in Fig. 4. As described in the example scenario, the intent is declared and ingested by the network operator in the user space. A DTN node in the Operator Source Segment translates the initial intent to a more specific but still segment-independent intent specification. The resulting intent specification is propagated to the DTN nodes in the Target Configuration Segment where each DTN node autonomously translates and enforces the intent.

3.3 Self-Adaptation Concepts for Intent-Based DTN

As pointed out in Sect. 2, an analogy can be identified between the Intent Life-cycle depicted in the IRTF draft [1] and MAPE-K feedback loops used to control self-adaptive systems. According to the layered control structure introduced for intent enforcement, two nested control loops can be used to translate and enforce the intent declared by the operator at the Operator Source Segment at the one hand and the Target Configuration Segments, i.e., the DTN nodes at the other hand.

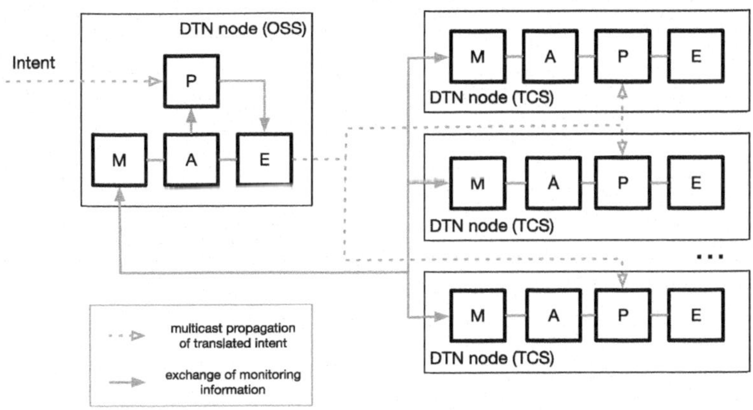

Fig. 5. Feedback loop structure for Intent-based DTN

The authors designed the feedback loop structure shown in Fig. 5 based on the patterns for decentralized control in self-adaptive systems proposed in [19]. To implement the intent life-cycle in a topology of distributed and autonomous DTN nodes, the Intent is first handled by the **planning component** of a DTN node in the Operator Source Segment. The planning component performs the segment-independent translation/refine operation of the intent as specified for the modified intent life-cycle in Fig. 3. The **execute component** is responsible for forwarding the refined intent to the DTN nodes in the Target Configuration Segment.

The DTN nodes in the Target Configuration Segment are structured based on the Information Sharing Pattern specified in [19]. According to that decentralization pattern, DTN nodes share state information captured in the *Monitoring* phase but perform *Analyze*, *Planning*, and *Execution* of intent-related adaptations autonomously.

Information sharing includes also the DTN node in the Operator Source Segment, i.e., monitoring information is forwarded within the DTN network between the two segments. The control cycle is closed when changes of the network state become apparent to the operator based on the monitoring information reported to the User Space. As described in the example scenario, the operator can react on such changes by declaring a new intent.

3.4 Open Issues for Intent-Based DTN

Based on the introduced overall concept, two major gaps can be identified, which need to be addressed in an IB-DTN approach:

1. The inner loop of the intent life-cycle needs to be specified in detail. Self-adaptivity is essential for realizing intent in a DTN deployment, because due to mobility, resulting disruptions, or long end-to-end delays, the network characteristics in the *Target Configuration Segment* may change faster than information about them changing can even reach the *Operator Source Segment*. Thus, intent has to be specified in a fashion independent of the network characteristics, and self-adaptivity is required in the *Target Configuration Segment* of the DTN.
2. Considering the *DTN forwarding* step, information (e.g., updates of the intent or monitoring information for *Intent Assurance*) has to be distributed throughout the network. However, it does not have to be distributed to all nodes in most cases; e.g., if the routing behavior should be controlled via intent, only nodes that may be intermediate DTN routers need to be addressed.

Section 4 focuses on technologies to address the first gap. While a principle control structure based on the MAPE-K control loop is specified for this purpose in Fig. 5, the following subsections focus on the technical details for adaptive DTN routing and present two concrete state-of-the-art approaches. Afterwards, in Sect. 5, an exemplary solution to the *DTN forwarding* issue for deterministic networks is outlined.

4 Adaptive DTN Routing

Self-adaptive concepts in DTN routing enable nodes to autonomously react to changes in network state and characteristics and facilitate optimization of their forwarding decisions. As such, self-adaptive techniques are important building blocks for intent-based operation of the network: The goals for which bundle forwarding shall be optimized can be specified by an operator via *intent* and the network will autonomously adapt to reach these goals.

As pointed out in Sect. 3.3, the adaptation occurs on each network node, in an iterative process that represents the inner loop of the modified Intent-based Networking life-cycle outlined in Sect. 3.2 and, thus, also maps to the MAPE-K loop from the theory of self-adaptive systems as described in Sect. 2.3.

Figure 6 depicts the resulting adaptive process and shows individual operations to be performed in the four core processing steps when applying it to DTN routing:

1. **Monitoring** of local network characteristics. The basis of any system reacting to changing characteristics of a network is the observation of these characteristics. This can include traffic and link state monitoring, the collection of contact and node properties reported by local subsystems such as the communication system, and the consolidation of the collected data into an easily storable and exchangeable format.

2. **Analysis** of observations. In the second step, the observation data are analyzed, aggregated, and validated to derive the information necessary for the next step. This can include the discovery and prediction of changes in the network, e.g., based on the comparison to historical data.
3. Derivation of a **plan** to address the changes. From the accumulated information about changes in network characteristics, an appropriate reaction, e.g., a new routing algorithm or a new set of parameters, is derived in this step.
4. **Execution** of the plan. Finally, the planned reaction is executed, possibly including the (re-)configuration of subsystems or devices.

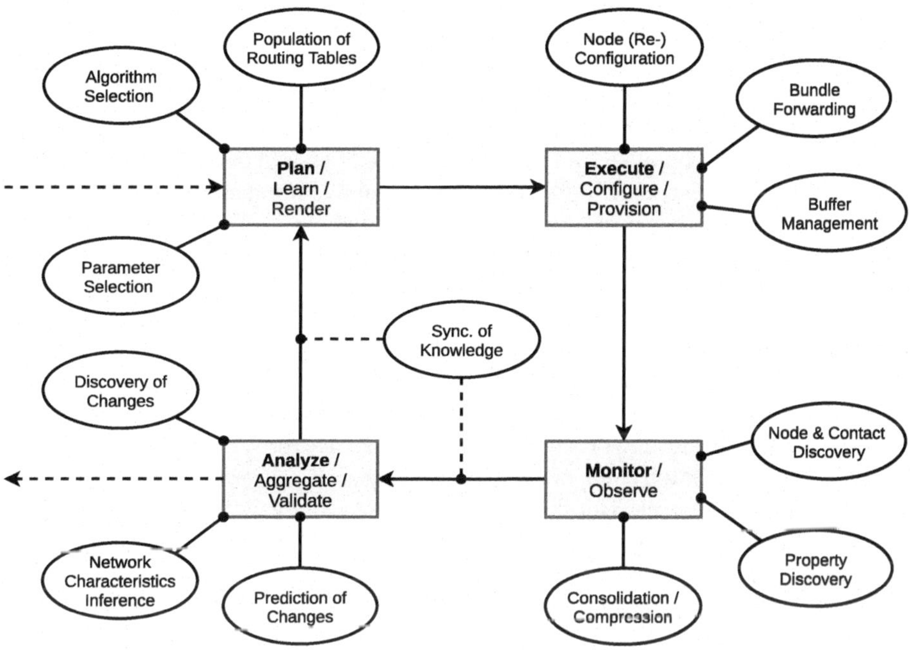

Fig. 6. Loop representation of an adaptive DTN routing technique

One further aspect in this cycle is especially important in DTN routing, the *synchronization of knowledge*. A central knowledge base readily-accessible to all nodes cannot exist, thus, to enable adaptive DTN routing, nodes have to be aware of network characteristics around them (e.g., current and expected future neighbors, contact probabilities, or even concrete contact intervals, expected data rates, and so on), to an extent which depends on the type of adaptations necessary. Hence, to make nodes aware of these network characteristics as well as the effects of previous re-configurations of the system, knowledge synchronization becomes an integral part of the loop. This synchronization can occur at two points: Either an "early" synchronization is performed based on the results of the *monitoring* step, or the output of the *analysis* step is distributed. While the former allows for maximum flexibility by enabling distributed, local analysis, the latter may provide for a more compact representation (i.e., summarized, analyzed data instead of raw observations) and less network load resulting from the information exchange. A combination of both is also imaginable.

It should be noted that the amount of data to be distributed also varies with the used consolidation and compression approaches as well as with the size of the network: While in small DTN deployments it might be viable to synchronize knowledge of all nodes and contacts, larger networks may be segmented into *regions* [20, 21] or using a similar concept with dedicated gateways between individual parts of the network, which reduces the amount of data that has to be synchronized to those relevant to the local segment or region. Additionally, due to the inherent delays and disruptions in a DTN, it cannot be expected that all DTN nodes share the same level of knowledge at all times. Rather, a principle of "eventual convergency" of state has to be applied.

With respect to the above-mentioned cycle, it can be discovered that some opportunistic DTN routing algorithms already perform these steps to some extent, e.g., the PRoPHET routing protocol [10] derives probabilities to reach every possible destination node via every encountered neighbor and then calculates transitive probabilities based on values distributed by other nodes. However, in any case known to the authors, such mechanisms are very specific to the routing protocol that implements them.

Overall, besides other existing models for classification (e.g., those presented in [7]), DTN routing techniques can be divided into different classes based on their degree of adaptivity:

1. **No adaptation**. The routing technique does not possess any form of reactivity to changing network characteristics.
2. **Adaptation of parameters**. The routing technique discovers and learns specific parameters from the network, which it uses to enhance forwarding decisions, e.g., by controlling replication count or choosing routes based on a specific metric depending on the network characteristics. In this case, the same base algorithm is used, but gets provided with parameters derived from the network characteristics.
3. **Adaptation of the algorithm**. The routing technique is able to select from a set of different routing algorithms depending on discovered or learned network characteristics. Such an approach can be also described as a "meta routing algorithm" which wraps a set of specific routing algorithms and chooses one that fits the current state of the network best. The selected algorithms may again be part of class 2, i.e., be able to adapt parameters.

Sections 4.1 and 4.2 outline two exemplary approaches to providing routing adaptivity of classes 2 and 3, respectively.

4.1 Contact Prediction-Based DTN Routing

The first exemplary approach adapts the data leveraged by a routing algorithm: In a network that allows for contacts being scheduled, it is possible to react to minor inaccuracies or ambiguities of the contact plan through the inference of contact properties. An example for a network allowing for this approach to be used is a small-satellite data-ferry network. So-called *Ring Road* networks [24], as shown in Fig. 7, facilitate Internet connectivity in remote areas. For that purpose, low earth orbit (LEO) satellites transport DTN bundles via their physical motion between decoupled ground stations (*cold spots*) and Internet-connected ground stations (*hot spots*).

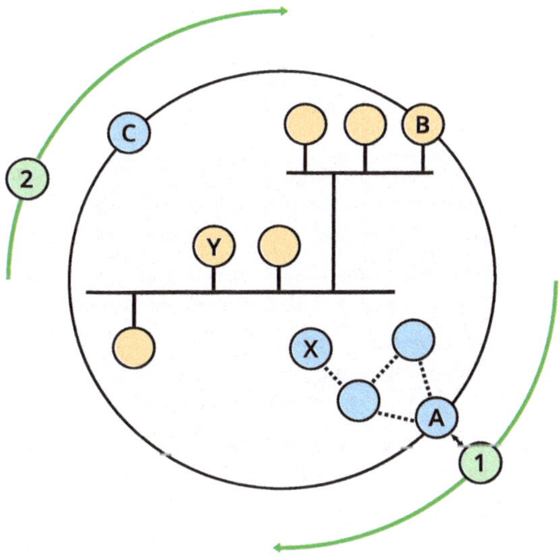

Fig. 7. *Ring Road*: A small-satellite data-ferry network (Fig. 3.1 from [25])

As satellite orbits are deterministic, satellite-to-ground-station contact intervals can be pre-computed. This way, it becomes possible to plan future transmissions using deterministic routing algorithms, e.g., Contact Graph Routing (CGR) [11]. However, some aspects cannot be computed as easily in advance: Due to low-cost hardware being leveraged in these networks and probabilistic effects (e.g., weather and interference) impairing transmissions, contacts may fail unexpectedly or offer less bandwidth than expected. Furthermore, some nodes may be more unreliable than others. In summary, this means that not all aspects associated with the contacts are fully known in advance. As a result, it may be necessary for a node to plan data transmissions differently, e.g., by increasing the amount of redundancy or planning less data for upcoming contacts. This requires a specific form of adaptation; the inference of contact characteristics.

A concept enhancing deterministic DTN routing by such a technique is proposed in [25] and consists of the following core processing steps:

1. **Contact Observation**. In a first step, encounters with other nodes are observed and recorded. Based on this information, the contact time intervals plus further factors such as the available data rate during each contact can be measured.
2. **Node Metric Inference**. From the available contact observations, so-called *node metrics* are calculated, which consist of a numerical representation of factors that determine the properties of contacts between two given nodes. For example, metrics related to the reliability of a specific node as well as the quality of the communication link between two specific nodes can be inferred.

3. **Contact Prediction**. Node metrics allow for improving the predictions of future contacts. These are still based on known initial data (e.g., location vectors and orbital parameters), but can be enhanced by the knowledge gained through the metric-based description of involved nodes.
4. **Routing**. The *contact plan* resulting from the previous step is used for calculating routes via an extended variant of Contact Graph Routing. A node can then schedule bundles for transmission via these routes and, again, observe the contact intervals during which the transmission does (or does not) occur.

To make other nodes aware of locally-discovered network characteristics, a **Metric Distribution** step is added before performing the **Contact Prediction**. In this step, available information about node characteristics is synchronized throughout the network. Compared to distributing contact observations, performing the data exchange after the **Node Metric Inference** step reduces the amount of data which has to be exchanged to the pre-processed form of node metrics. This way, even in large networks, only a minor amount of data has to be exchanged.

Overall, it becomes obvious that this concept represents a closed loop as well; the four processing steps indicated above are repeatedly executed in order and can be transferred to a loop representation as depicted in Fig. 8, representing a specific realization of the generic loop shown in Fig. 6.

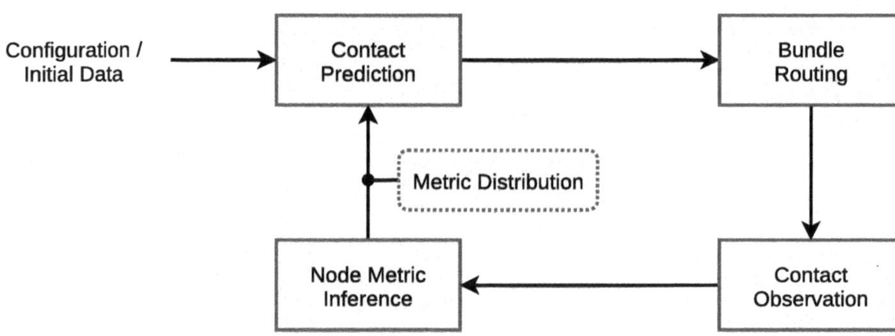

Fig. 8. Contact prediction-based routing modeled as an adaptive loop

4.2 Adaptation of the Routing Algorithm

In some use-cases, the mobility or connectivity of DTN devices may change more drastically than what can be compensated with changes in parameterization. Especially in terrestrial use-cases, accidents or disaster situations may impair a network. Additionally, the load may change radically in different parts of the network due to unforeseen traffic changes. Environmental characteristics (e.g., mobility and load behavior) define an operating environment recognized by nodes at run time, which can be taken as a basis to switch and parameterize routing algorithms autonomously. For that purpose, each node decides autonomously when to switch, selects and parameterizes the routing algorithm to achieve the network operator's intent based on the sensed operating environment.

In [23], Dabideen and Ramanathan adapt the routing algorithm depending on the connectivity available in the node's region. In a dense region, nodes switch to a MANET[2]-based routing algorithm assuming end-to-end connectivity; in sparse regions, a DTN alternative not making this assumption is chosen. CARTOON [22] proposes the adaptation of the DTN algorithm according to congestion likelihood. The author considers Epidemic routing as the best strategy to spread information within a DTN, as long as congestion does not happen. Therefore, nodes use Epidemic routing as long as the frequency of encounters is lower than a well-defined threshold; otherwise, a probabilistic alternative is used.

Deterministic algorithms, such as CGR [11], are expected to achieve a high end-to-end delivery probability with fewer replicas than their opportunistic counterparts, further minimizing congestion providing that contacts adhere to the schedule. However, for example in a terrestrial public transport network, in the event of a disaster or accident that blocks paths unexpectedly, the use of an outdated contact plan could preclude communication and partition a network applying such an approach exclusively. As soon as this mobility change is recognized, a node stops predicting future mobility behavior based on the history of contacts. Instead, a replication-based approach, e.g., Spray and Wait, could be used to improve the likelihood of bundle delivery, ignoring the inaccurate contact plan. Some use-cases such as public transport networks are characterized by repetitive behavior even if some paths are blocked, which allows for the use of more sophisticated algorithms: When, e.g., a tram line deviates, vehicles adopt a new trajectory that often remains the same until the cause of the change is resolved. Therefore, over time, nodes may recognize the new operating environment as "predictive" (i.e., having recurring patterns) and switch to a routing algorithm able to explore the history of contacts, avoiding the replication of messages to nodes that are unlikely to support their delivery.

In the following, an approach facilitating such an autonomous adaptation of the routing algorithm is outlined. A key concept to understand the presented approach is *context*: In Sect. 2.3, the operating environment has been defined as the information set that describes the state of the network (i.e., the reality from the node's viewpoint). This state combines multiple metrics, e.g., information about contacts, link utilization, buffer utilization, and energy. Frequently, it helps to define ranges for a metric to describe a situation. For example, a node may be considered prone to congestion if the buffer utilization is greater than 80%; If this metric decreases below 60%, or if a decrease over time is noticed, it can be said that this node is not prone to congestion anymore.

Sometimes, a combination of states within specific ranges defines a situation: For example, suppose a node notices that the observed contact opportunities match the configured contact plan considering an acceptable error margin, and contact opportunities are most of the time sufficient to exchange the stored bundles. In that case, a node may deem this situation as *normal*. If, over time, the buffer utilization increases and the number of bundle candidates that cannot be routed during a contact opportunity

[2] Although some DTN use-cases can be considered "mobile" and "ad-hoc", the term *MANET* was coined before the emergence of DTNs and generally refers to ad-hoc networks providing end-to-end connectivity.

increases, the combination of those metrics could indicate build-up of a *congestion* situation. Alternatively, if the number of observed contact opportunities not matching the contact plan rises, the situation may be labeled as *random* by the node. In summary, based on the exchanged information, a node senses its operating environment. Then, it uses this information to select a class of metrics able to describe roughly a situation that is referred to as **context** in this work. Discretizing the classes of metrics from the operating environment and their combinations allows for defining rules to determine the behavior a node should follow when it realizes it is in a given situation.

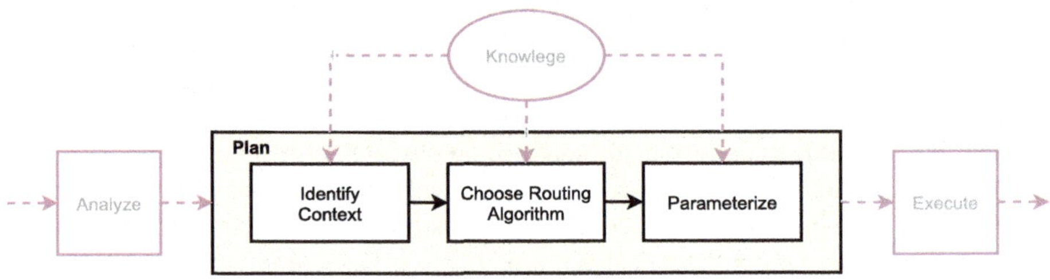

Fig. 9. Adapting the routing algorithm at runtime

To enable adaptive switching of the routing algorithm, the authors propose a subdivision of the planning phase of the self-adaptation process performed on each node, as shown in Fig. 9:

1. **Identify context**. A node maps the operating environment into a *context* based on a set of rules and the current characteristics of the local network acquired at runtime. Considering the examples mentioned earlier, different *contexts* could be defined according to the network density (e.g., *connected* and *disconnected*) in [23]; In [22], *contexts* were defined based on the frequency of encounters (e.g., *congestion-prone* and *congestion-free*); Finally, e.g., in public transport networks (PTNs), *context* could be defined according to the knowledge about future encounters: *scheduled*, if a schedule is given and respected; *probabilistic*, if a schedule is not provided or respected, but mobility is repetitive; *random*, if mobility seems to be unpredictable.
2. **Routing algorithm choice**. In this step, the node chooses one among the routing algorithms available for the present context category that is considered the most suitable according to the defined intent. A set of well-defined rules that map contexts and intents onto routing algorithms should be provided. In proposals such as [22, 23] the choice is trivial since there is only one routing algorithm per category. However, in the PTN example, there could be multiple routing algorithms in each category.
3. **Parametrization**. Similarly, a set of rules derived from extensive simulations or practical experience maps contexts and intentions into a set of parameters for the chosen routing algorithm. Such parameters can be values used in an objective function (e.g., parameters to PRoPHET's delivery predictability calculation) or affect the routing outcome to take into account aspects of congestion control (e.g., limit the maximum number of copies or hops for messages).

To translate the intent into a concrete algorithm and parameterize it, it is additionally necessary to relate each intent to a set of contexts of interest. In each context there might be a set of one or more suitable routing algorithms. All relevant contexts and the related routing algorithms have to be defined during the design of the individual sub-networks with respect to their characteristics and topology. For example, the most appropriate routing algorithm and its parametrization for a given operational environment can be defined through experimentation or simulations.

4.3 Application to Example Scenario

In the concrete example described in Sect. 3.1, nodes can be divided into zones (or delay- and disruption-tolerant sub-networks) with different characteristics. The effect caused by the change of intent can be distinct in different sub-networks. Assuming that the initial intent is to maximize delivery probability and minimize end-to-end delays, an adaptive routing technique (see Sect. 4.2) may define two possible contexts: *deterministic* or *opportunistic*. When two nodes meet, a node is in a *deterministic* context if it is possible to calculate a route to the destination node based on the contact plan; otherwise, it is in an *opportunistic* context. Thus, nodes on Mars that have persistent connectivity are expected to be most of the time in a *deterministic* context. In a *deterministic* context, a schedule-aware routing algorithm such as Contact Graph Routing (CGR) [11] can be considered the most suitable approach to forward bundles. As CGR uses an objective function to calculate the "distance" from source to destination and find the shortest path for a given metric, it can calculate paths with the least expected end-to-end delay. However, some DTN nodes in the presented example face intermittent connectivity, i.e., they could have opportunistic contacts. In this case, it is not possible to calculate a shortest path in advance, since such contacts cannot be precisely planned. A node in this situation finds itself in an *opportunistic* context, in which other routing algorithms are more appropriate. If the opportunistic contacts occur repeatedly, a history-based routing algorithm such as PRoPHET [10] could be chosen. Otherwise, a simpler algorithm like Epidemic [8] could be used to propagate information. The most appropriate algorithm is chosen autonomously by each node from the information about the network behavior collected and shared during the contact opportunities, as described in Sect. 4.2.

In the example scenario, after the failure of one of the links, the transmission capacity between Mars and Earth is reduced. Over time, this can lead to congestion and bundle drops. When operators become aware of the situation they react, changing the intent to *reduce traffic between Mars and Earth*. This change may modify the contexts to be recognized, and consequently, the routing algorithms suitable for each context. For the discussed use case, the new intent could result in three possible contexts: *deterministic* with congestion awareness, *opportunistic*, and *congestion-prone opportunistic*. The modification to the *deterministic* context could lead to a different parametrization of the schedule-aware routing algorithm to use an alternative objective function as distance, taking into account link utilization or least-used paths and avoiding paths via Mars-Earth links if possible. Furthermore, nodes that are in an *opportunistic* context can autonomously evaluate if they are in a *congestion-prone opportunistic* context according to the operational environment, e.g., based on the rate of bundles sent and received over time, the rate of dropping bundles, the local buffer utilization, or the rate of bundles forwarded that are expected to be sent via a Mars-Earth link at a later point in time. If the node recognizes its context as *congestion-prone opportunistic*, it could choose a routing algorithm that tries to avoid congestion or at least limit the number of replicas. Nodes in such a context can instruct the applied routing algorithm via parametrization to reduce and limit the number of copies created per bundle.

Overall, the change in intent in the example scenario leads to different input to the planning phase of the adaptive process performed on each network node, as outlined in Sect. 3. On that basis, a simple, intuitive specification of a new intent by a network operator can lead to adaptive re-configuration of nodes specific and optimized to the individual sub-networks.

5 A Multicast Approach for Intent-Based Node Configuration

For enabling intent-based DTN, beside facilitating self-adaptivity in the nodes, a central challenge to overcome is the communication of intent plus monitoring data to the responsible nodes in the network, which has to occur at the boundary of both segments identified in Sect. 3. On the "forward path" of the intent life-cycle depicted in Fig. 3, communication toward all network nodes responsible of realizing the intent needs to occur, while on the "backwards path" the forwarding of monitoring data back to the network operator is necessary.

As this boils down to a set of nodes in both cases, which, however, does not have to consist of all nodes in the network, a multicast forwarding technique is necessary. Such a technique for the case of scheduled networks is presented in the following subsections. In the other case of an opportunistic network in which the set of configurable nodes cannot be made available to the operator, a simple broadcast approach may be leveraged in a first step. However, reducing the overhead incurred by such a scheme and restricting the group of intent receivers is a problem requiring further research and is, thus, only discussed briefly in the outlook in Sect. 6 of this chapter.

5.1 Multicast in Deterministic DTN

This section focuses on the specifics of multicast approaches which leverage a contact plan, i.e., those applicable to deterministic networks. In many cases, the multicast algorithm can be considered independently to the underlying path-finding algorithm (for example, the DTBR routing approach presented below could use interchangeably one path finding algorithm or another). Thus, the authors cover these two topics in two different subsections. First, an overview of multicast algorithms for scheduled DTNs is given, followed by an overview of possible underlying path-finding and unicast routing algorithms.

5.1.1 Multicast Algorithms

Algorithms which leverage a schedule for path-finding can be called "tree-based routing" [29]. The authors present in this section several important variants from the state of the art.

In Static Tree-based Routing (STBR) [29], the source node computes a shortest-path tree to the intended receivers for the bundle, and forward the message to the next nodes.

These *forwarding nodes* will then compute a tree from the source to the destinations, and forward the message if they are part of the tree. This naive static approach seems to be meant to introduce shortest-path tree utilization for multicast purposes but has important issues: Firstly, all of the nodes need to share the same contact plan. Even if it is often the case in practice, we can imagine networks where the nodes do not share the same contact plans for optimization reasons, or networks where maintaining an accurate contact plan on each node is challenging, leading to periods when the contact plan for some nodes is inaccurate. Secondly, this approach does not distribute the computational load in an optimal manner: The forwarding nodes need both parts of the tree, i.e., the upstream and downstream branches, even though the bundles will only be sent through the downstream branches. Finally, the nodes are not allowed to leverage the knowledge of their close neighborhood to find better paths for the remaining destinations to serve, leading to a lack of flexibility.

To tackle those issues, Dynamic Tree-Based Routing (DTBR) [29] was presented. The approach relies on a special header extension which contains the list of remaining destinations to be served for the specific bundle. When a bundle has to be forwarded, a shortest-path tree to the remaining destinations is computed with the forwarding node as source. In contrast to STBR, this reduces necessary computations only to the downstream branches. If the tree has more than one first hop, the forwarding node splits the set of remaining destinations accordingly for each copy of the bundle. This header approach can be considered an adaptation of Xcast [30] from the IP world to the DTN world and is more flexible compared to STBR because the nodes along the path do not have to agree on the paths to take, allowing a node with better knowledge of its neighborhood to schedule bundles for better downstream paths.

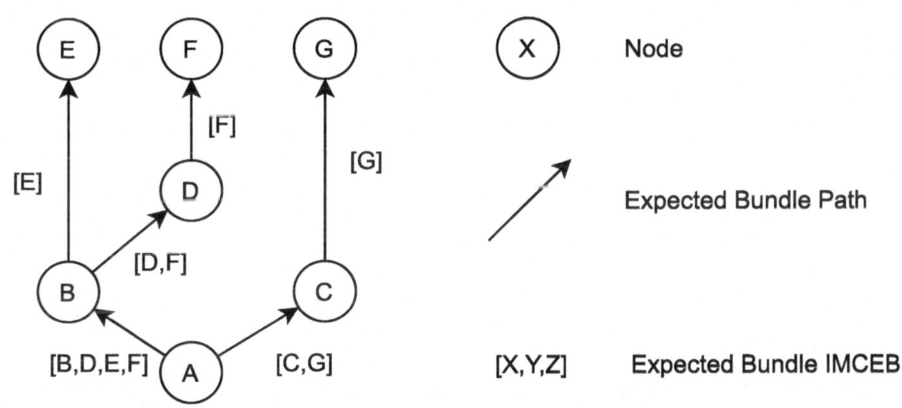

Fig. 10. Multicast bundle forwarding with IMCEB to a multicast group containing all nodes in the network, with node A as the source

It was later proposed to move the set of remaining destinations from the message header to a bundle extension block [31], and a prototype implementation of this concept has been developed within the Interplanetary Overlay Network (ION) [27] software using Contact Graph Routing as the underlying path-finding algorithm. As depicted in

Fig. 10, the extension block shall gather the remaining destinations to serve along a downstream branch. Of course, the downstream nodes do not have to agree on the paths that were expected by the upstream nodes: A node defines the IMCEB only for the transfer to the neighboring nodes and the responsibility to serve the remaining nodes of a branch is transferred to the according neighboring node with the IMCEB, which allow this next node to apply a different branching than the one an upstream node would have predicted. As a consequence, a node might receive two copies of the same bundle having a different IMCEB attached. The decision to leave the two versions of the bundle in the memory and route them separately, or to merge the two IMCEB to keep a single copy of the bundle in the memory (and possibly adapt the route volume consumption of the already scheduled copy) is the responsibility of the routing algorithm and does not conflict with the IMCEB's functioning.

Similar to DTBR, Space Minimizing Tree Based Routing (SMTBR) [28] is an algorithm which aims to compute *thin trees* (trees with fewer edges) to reduce the number of participating nodes and, thus, the number of bundle copies. The algorithm is built on top of a generic shortest-path tree search, and uses the concept of *importance* to select relay nodes. The simulations show that having a reduced number of relay nodes increases drastically the delivery latency, while on the other hand reducing the number of message copies. The two variants of the algorithm take an extra value as parameter, which impacts the number of relay nodes in the resulting tree. However, the DTBR approach has some issues: The first difficulty is to set this value in an optimal manner. Furthermore, the increased delays could cause issues if gateways to other network partitions are affected by the increased latency.

5.1.2 Path-Finding Algorithms

The reference unicast routing algorithm for scheduled networks is Contact Graph Routing (CGR) [11, 26], with its reference implementation being part of ION. CGR is a delay-tolerant single-copy routing technique leveraging a version of Dijkstra's shortest-path algorithm and was originally designed for deep space missions. CGR indeed supports huge delays on single links induced, e.g., by an interplanetary range between nodes. The routes, which in CGR represent the entire path from a local node to a destination, are stored in a *route list*, and CGR selects a suitable route (route selection) to schedule a bundle or computes new ones if needed (route search).

The very first particularity of CGR is the utilization of a time-varying graph of contacts. In analogy to airline scheduling, the vertices can be seen as flights and the edges as layovers. [12] The contact graph can be considered as a simple and intuitive way to manage the time variance in connectivity, much as flights are an easy way to manage time variance in air travel planning. The main problem of this design is that CGR's complexity increases with the number of contacts rather than the number of nodes in the network, which, depending on the horizon of the contact plan, can be arbitrarily high.

The second particularity of CGR is the tracking of the volumes of contacts and routes. Indeed, only a certain volume can be transmitted via a single contact. CGR assumes this volume simply being the contact bandwidth (average data rate) multiplied by the contact duration.

Moreover, the volume tracking of the routes increases the complexity, as scheduling a bundle for a route possibly affects the volume of the other routes sharing a contact with this route. Consequently, the volume of an arbitrary route has to be recalculated after each bundle scheduling, as CGR doesn't support tracking of the dependencies between the routes.

Finally, the proposed strategy to compute alternative routes in CGR relies on Yen's algorithm, which is known to have a high complexity. Moreover, the number of alternative routes that will be needed cannot be known in advance and depends on the load, rendering the configuration of Yen's algorithm non-trivial as well.

With such challenges, the optimization of CGR remains an active research topic. Indeed, to tackle those issues, the shortest-path tree approach for routing in space networks (SPSN) [32] has been proposed to reduce the algorithmic and software complexity inherent to CGR.

To reduce the complexity on the algorithmic level, the first modification done by SPSN was to switch back to a *node graph* approach, using a multigraph respecting the chronological ordering of the contacts. Secondly, the Dijkstra search leveraged by SPSN computes a shortest-path tree rather than a single path, to reduce the computational effort on a macro level, by pre-computing the routes for other destinations before they are needed for almost the same cost as computing a single route. Thirdly, Yen's algorithm is not used by SPSN. Rather, a capacity check is integrated into the Dijkstra search (capacity-oriented search). Indeed, CGR is not able to differentiate a suitable route from a route with an exhausted residual volume during the route search but only during the route selection, rendering Yen's algorithm (or another approach) needed to find the alternative routes. With SPSN, if a path sees its residual capacity exhausted for a given bundle size, the next spanning tree computation will omit this path during the route search.

To further reduce the software complexity, SPSN drops the concept of a route list, and introduces the so-called *spanning tree router*. In that context, SPSN maintains a single spanning tree, which permits the algorithm to take into account the dependencies between individual routes recursively. This way, the scheduling operations can be applied to the tree directly, which permits to avoid the management that was needed with CGR's route lists, and the tree is simply recomputed if needed. However, there are some trade-offs: The spanning tree computation with a node-graph doesn't find alternative routes with the same delivery rate but fewer number of hops. Even though a fix is under development, this aspect has a reduced impact in the multicast case while the multicast group size increases. Also, for a single destination, the spanning tree will be recomputed only if the residual volume for the route to this destination is exhausted. This means that SPSN would not detect if another route becomes a better match in terms of delivery delay after scheduling bundles on the first path. A recent unpublished evaluation available to the authors showed that this aspect has nearly no impact on the delivery rate for a set of satellite and public transport topologies studied. This can be due to two different aspects: Firstly, the limitation does not apply to routes having the same first hop contact, which reduces the scope where this trade-off can be encountered. Secondly, the spanning tree is also recomputed if the route volume to another destination is exhausted, updating the paths for all destinations on a regular basis in many scenarios. Even if this particularity has or is expected to have a negative impact in a given scenario, path-finding algorithms can be used interchangeably with a routing technique and conversely, and it would be also feasible to use a route management technique with route table such as CGR's one using as back-end a modified version of SPSN for the path-finding algorithm.

5.2 SPSN-Based Multicast

This section sketches a multicast forwarding approach based on SPSN and the Interplanetary MultiCast Extension Block (IMCEB) presented in [31]. The approach relies on one spanning tree computation per bundle, using SPSN's route search capabilities to find the best routes. A modification to the route search is made to leave a minimum residual volume in each route for the multicast route table management. The applied capacity-oriented search selects tree branches which are suitable for a given bundle size, thus, there is no need to select a route with a sufficient capacity from a route list like in the case of CGR. A prerequisite for SPSN-based multicast routing is that each source node knows the member list of the EID for a given multicast group before sending a bundle to this group. This set is used as the initial reception group for the multicast bundle. However, details of the used multicast group management are omitted in this section.

At each forwarding node, the list of remaining destinations should be retrieved. If the node is the source, this list is populated with the list of members of the multicast group for this multicast bundle. If the node is not the source, the list becomes the list of remaining nodes listed in the IMCEB attached to the multicast bundle. If the forwarding node is part of the list, a copy of the bundle should be delivered locally, and the node should be erased from this list for the next forwarding steps. If the list becomes empty after this step, no further forwarding is performed by this node.

After the constitution of the list of remaining destinations, SPSN comes into play: A spanning tree is computed with the local node as source and the size of the bundle as the minimal route residual volume allowed. From now on it is ensured that all reachable nodes for this size are part of the tree. The expiration time is for the moment omitted but the route search could easily be adapted to also omit routes if the projected bundle arrival time exceeds the bundle's expiration time.

The spanning tree is computed with a modified version of Dijkstra's algorithm. Therefore, only the reverse path for a destination can be extracted, requiring a post-processing step, which in the case of a spanning tree needs to construct an associative data structure, giving for each downstream branch (next hop) of a node the list of reachable downstream nodes. This post-processing has to be performed only for the nodes part of the remaining destination list, to ensure that only relevant nodes will be part of the resulting associations.

The associative data structure constructed at the source node of this tree (the local node) is particularly important, as it directly provides the information necessary for the routing decision: For each next hop (downstream branch), a new list of remaining destinations (reachable downstream nodes), which is translated by sending copies of the bundle through those branches with the remaining list of destinations being part of the IMCEBs.

5.3 Application to Example Scenario

In the scenario presented in Sect. 3.1, after taking notice of the node failure and resulting degradations of service, the operator will define a new intent for the network nodes and will select a multicast group as recipient of this intent. The multicast group can either map to the whole set of the network nodes or a subset of them. Consequently, the intent will be dispatched to the expected destinations.

In the course, SPSN will compute a spanning tree for the given bundle size, and will return the associative data structure, yielding the list of remaining destinations for each next hop. The DTN node will then be able to construct the IMCEB for each copy of the bundle, and will send those copies with the according IMCEB through the according paths.

Iteratively, the downstream nodes will receive the administrative multicast bundle with an IMCEB. The downstream nodes will first deliver the intent locally if they are members of the multicast group, and will then forward the intent using the remaining destinations listed in the IMCEB, after removing themselves from the list. If the list becomes empty, no further forwarding is performed.

It should be noted that for integrating the opportunistic sub-networks present in the example scenario, some gateways have to be expected between the scheduled network and the opportunistic sub-networks, where the applicability of SPSN with IMCEB ends and where other multicast routing schemes shall be used. Those gateways have the responsibility to be aware of the multicast members of a given group for the two regions [21].

6 Future Research Directions

As the transfer of Intent-Based Networking concepts to the domain of Delay- and Disruption-Tolerant Networking is still only a conceptual proposal, there are multiple directions of open issues that need to be addressed in future work. This section aims to provide an overview of major challenges to be tackled on the path toward productive IB-DTN deployments.

Firstly, a concept for opportunistic intent distribution is required. In an opportunistic DTN, i.e., one where transmission opportunities cannot be reliably and precisely planned ahead, a multicast technique without explicit identification of the receivers has to be used to leverage intent-based configuration. Beside the obvious option of distributing intents to all possible receivers in a broadcast manner, one can think of different techniques to reduce the distribution overhead. For example, some filter criteria may be attached to each message, specifying which types of nodes belong to the applicable receivers. Furthermore, message contents may be processed and re-forwarded at each intermediate node. Before that, a node may refine or check the intent and update it for the connected network segment.

Secondly, the network traffic as well as the computational load could also be reduced via a "divide-and-conquer" approach to intent distribution: Except of distributing an intent to all DTN routers across the network, a two-staged scheme can be imagined, which designates specific nodes to serve as *intent master nodes* for a part of the overall network. These nodes can conduct sub-network-specific intent translation and, thus, reduce the need for all nodes to perform this task individually. Such an approach may also pave the way to more flexible adaptive behavior, as the network-specific intent master nodes may leverage adaptive concepts for processing the intent themselves. By that, e.g., local (sub-)network operators may be allowed to provide part of the intent-based configuration to their respective sub-network, while keeping it integrated with the overall intent-based DTN.

Lastly, with respect to multicast forwarding leveraging SPSN with the Interplanetary MultiCast Extension Block (IMCEB, see Sect. 5.2), loop control for known bundles having an IMCEB attached is expected to differ from the unicast case. Indeed, the reception of a known bundle doesn't necessarily mean that a loop occurred, as the intersection of the already-known IMCEB (that is or was in the node's memory) with the one just received can be an empty set. With multicast forwarding, even if the replication is reduced by the use of the IMCEB, the routing algorithm is far from a single-copy approach. However, a destination listed in an IMCEB shouldn't be listed in any other IMCEB of the other copies of the bundle, in other words, it is each entry of an IMCEB that becomes the single copy amongst the existing other entries of the IMCEBs of the copies of the bundle. This should permit appropriate loop control if a node attaches the destinations listed in an IMCEB to the according entry of a known bundle list.

7 Conclusion

This chapter has provided an overview of the opportunities and challenges expected when applying Intent-Based Networking concepts to the Delay- and Disruption-Tolerant Networking domain. In that context, the strong relationship between Intent-Based Networks and the research domain of self-adaptive systems has been pointed out. On the basis of essential concepts for self-adaptivity such as the MAPE-K feedback loop, an overall approach to enable intent-based routing in a DTN deployment has been presented. The conceptual discussion shows that two central gaps need to be addressed to achieve this:

1. A self-adaptive DTN routing technique is required, addressing network- and context-specific adaptation of the routing and forwarding algorithms as well as their parametrization on each DTN node. As it was explained in Sect. 4, such a routing technique presents a feedback loop in itself and can be configured via intent.
2. For delivering intent to its appropriate receivers, delay- and disruption-tolerant multicast distribution of intent and the respective feedback on its fulfillment has to be performed throughout the network. As persistent, low-latency end-to-end connections cannot be expected in a DTN, a specific multicast approach needs to be leveraged. A possible solution for deterministic DTN was proposed in Sect. 5.

Overall, the examination of a concrete example scenario (as introduced in Sect. 3.1) with respect to the IB-DTN approach points out that intent-based management can provide major advantages in a DTN, as it can abstract from the concrete network state and local characteristics, which may change faster than new configuration could be distributed in a DTN without intent-based management capabilities. Hence, it can be concluded that Intent-based Networking is a promising option for facilitating the management and operation of large-scale DTN deployments and its application to this domain should be explored further in future research and development activities.

Acknowledgments. This work is partially funded by the German Research Foundation (DFG) withinthe Research Training Group "Role-based Software Infrastructures for continuous-context-sensitive Systems" (GRK 1907). The authors would like to sincerely thank Dr. Marius Feldmann for his contributions in preliminary discussions toward this chapter, including the proposal to distribute Intent via a DTN multicast approach, which provided the basis for the presented concepts.

References

1. Clemm, A., Ciavaglia, L., Granville, L., Tantsura, J.: Intent-Based Networking - Concepts and Definitions. draft-irtf-nmrg-ibn-concepts-definitions-03 (Internet-Draft). Network Working Group, Internet Research Task Force (2021). https://tools.ietf.org/html/draft-irtf-nmrg-ibn-concepts-definitions-03
2. Cerf, V., Burleigh, S.: Interplanetary Internet (IPN): Architectural Definition. draft-irtf-ipnrg-arch-01 (Internet Draft). IPN Research Group, Internet Research Task Force (2001). https://tools.ietf.org/html/draft-irtf-ipnrg-arch-01
3. Fall, K.: A delay-tolerant network architecture for challenged internets. In: 2003 conference on Applications, Technologies, Architectures, and Protocols for Computer Communications (SIGCOMM '03) Proceedings. ACM (2003). https://doi.org/10.1145/863955.863960
4. Cerf, V., et al.: Delay-Tolerant Networking Architecture. RFC 4838 (Informational), Network Working Group, Internet Research Task Force (2007). https://datatracker.ietf.org/doc/html/rfc4838
5. Scott, K., Burleigh, S.: Bundle Protocol Specification. RFC 5050 (Experimental). Network Working Group, Internet Research Task Force (2007). https://datatracker.ietf.org/doc/html/rfc5050
6. Birrane, E.: Asynchronous Management Protocol. draft-birrane dtn-amp-08 (Internet Draft). Delay-Tolerant Networking Working Group, Internet Engineering Task Force (2020). https://datatracker.ietf.org/doc/html/draft-birrane-dtn-amp-08
7. Cao, Y., Sun, Z.: Routing in delay/disruption tolerant networks: a taxonomy, survey and challenges. Commun. Surv. Tutor. **15**(2), 654–677 (2012). https://doi.org/10.1109/SURV.2012.042512.00053
8. Vahdat, A., Becker, D.: Epidemic routing for partially connected ad hoc networks (2000). http://issg.cs.duke.edu/epidemic/epidemic.pdf
9. Spyropoulos, T., Psounis, K., Raghavendra, C.S.: Spray and wait: an efficient routing scheme for intermittently connected mobile networks. In: 2005 ACM SIGCOMM Workshop on Delay-Tolerant Networking (WDTN '05) Proceedings. ACM (2005). https://doi.org/10.1145/1080139.1080143
10. Lindgren, A., Doria, A., Davies, E., Grasic, S.: Probabilistic routing protocol for intermittently connected networks. RFC 6693 (Experimental), Internet Research Task Force (2012). https://datatracker.ietf.org/doc/html/rfc6693
11. Consultative Committee for Space Data Systems. Schedule-Aware Bundle Routing. Recommended Standard No. CCSDS 734.3-B-1, Issue 1, CCSDS (2019). https://public.ccsds.org/Pubs/734x3b1.pdf

12. Burleigh, S., Caini, C., Messina, J. J., Rodolfi, M.: Toward a unified routing framework for delay-tolerant networking. In: International Conference on Wireless for Space and Extreme Environments (WiSEE) Proceedings. IEEE (2016). https://doi.org/10.1109/WiSEE.2016.7877309
13. Burleigh, S., Ramadas, M., Farrell, S.: Licklider Transmission Protocol - Motivation. RFC 5325 (Informational). Network Working Group, Internet Research Task Force (2008). https://datatracker.ietf.org/doc/html/rfc5325
14. Lerner, A.: Intent-based Networking. Gartner Blog Network (2017). https://blogs.gartner.com/andrew-lerner/2017/02/07/intent-based-networking/
15. Kephart, J., Chess, D.: The vision of autonomic computing. Computer **36**(1), 41–50 (2003). https://doi.org/10.1109/mc.2003.1160055
16. IBM Corporation: An architectural blueprint for autonomic computing. White Paper (2006). http://citeseerx.ist.psu.edu/viewdoc/download?doi=10.1.1.150.1011&rep=rep1&type=pdf
17. Oreizy, P., et al.: An architecture-based approach to self-adaptive software. Intell. Syst. Appl. **14**(3), 54–62 (1999). https://doi.org/10.1109/5254.769885
18. Salehie, M., Tahvildari, L.: Self-adaptive software: landscape and research challenges. Trans. Auton. Adapt. Syst. **4**(2), 1–42 (2009). https://doi.org/10.1145/1516533.1516538
19. Weyns, D., et al.: On patterns for decentralized control in self-adaptive systems. In: de Lemos, R., Giese, H., Müller, H.A., Shaw, M. (eds.) Software Engineering for Self-Adaptive Systems II. LNCS, vol. 7475, pp. 76–107. Springer, Heidelberg (2013). https://doi.org/10.1007/978-3-642-35813-5_4
20. Clare, L., Burleigh, S., Scott, K.: Endpoint naming for space delay/disruption tolerant networking. In: 2010 Aerospace Conference Proceedings. IEEE (2010). https://www.mitre.org/sites/default/files/pdf/09_5229.pdf
21. Alessi, N.: Hierarchical inter-regional routing algorithm for interplanetary networks. Master's thesis, School of Engineering and Architecture, Department of Computer Science and Engineering, Bologna, Italy (2018)
22. de Oliveira, E.C.R., et al.: Context-aware routing in delay and disruption tolerant networks. Int. J. Wirel. Inf. Netw. **23**(3), 231–245 (2016). https://doi.org/10.1007/s10776-016-0315-2
23. Dabideen, S., Ramanathan, R.: Fansy route: adaptive fan-out for variably intermittent challenged networks categories and subject descriptors. SIGMOBILE Mob. Comput. Commun. Rev. **18**(1), 37–45 (2014). https://doi.org/10.1145/2581555.2581561
24. Burleigh, S.C., Birrane, E.J.: Toward a communications satellite network for humanitarian relief. In: International Conference on Wireless Technologies for Humanitarian Relief. ACM (2011). https://doi.org/10.1145/2185216.2185280
25. Walter, F.: Prediction-enhanced routing in Disruption-tolerant Satellite Networks. Doctoral dissertation. Technische Universität Dresden, Qucosa (2020). https://nbn-resolving.org/urn:nbn:de:bsz:14-qucosa2-721622
26. Fraire, J., De Jonckère, O., Burleigh, S.: Routing in the space internet: a contact graph routing tutorial. J. Netw. Comput. Appl. **174**, 102884 (2021). https://doi.org/10.1016/j.jnca.2020.102884
27. Burleigh, S.: Interplanetary overlay network: an implementation of the DTN bundle protocol. In: 4th Consumer Communications and Networking Conference. IEEE (2007). https://doi.org/10.1109/CCNC.2007.51
28. Tripathi, A.: Space optimized multicast in delay tolerant networks. Int. J. Comput. Netw. Technol. **1**(2), 139–149 (2013)
29. Zhao, W., Ammar, M., Zegura, E.: Multicasting in delay tolerant networks: semantic models and routing algorithms. In: 2005 SIGCOMM workshop on Delay-tolerant networking Proceedings. ACM (2005). https://doi.org/10.1145/1080139.1080145

30. Boivie, R., Feldman, N., Imai, Y., Livens, W., Ooms, D., Paridaens, O.: Explicit multicast (Xcast) concepts and options. RFC 5058 (Experimental), Network Working Group, Internet Research Task Force (2007). https://datatracker.ietf.org/doc/html/rfc5058
31. De Jonckère, O.: Efficient contact graph routing algorithms for unicast and multicast bundles. In: International Conference on Space Mission Challenges for Information Technology (SMC-IT) Proceedings. IEEE (2019). https://doi.org/10.1109/smc-it.2019.00016
32. De Jonckere, O., Fraire, J.: A shortest-path tree approach for routing in space networks. China Commun. **17**(7), 52–66 (2020). https://doi.org/10.23919/J.CC.2020.07.005

QoE-Oriented Routing Model for the Future Intent-Based Networking

Andrii Pryslupskyi, Mykola Beshley(✉), Halyna Beshley, Yuliia Pyrih, and Andriy Branytskyy

Lviv Polytechnic National University, Lviv 79013, Ukraine
{mykola.i.beshlei,halyna.v.beshlei,yuliia.v.klymash}@lpnu.ua

Abstract. The Software-Defined Networking (SDN) paradigm is attracting considerable attention from industry and academia as a future Intent-Based Network (IBN) architecture. In this chapter, we developed an SDN-based IBN architecture using the northbound interface (NBI) of the SDN architecture to announce customer intentions. The proposed IBN architecture provides the ability to accept incoming data from end customers, configure networks according to customers intent, validate the correct design, implement the necessary network configurations, and then continuously monitor the execution of system intent and make changes as needed. By intentions, we propose to understand the ordering of a certain level of quality of experience (QoE). The QoE intentions in our approach are set as a score from 1 to 5, where the highest score means the best quality of service. Intents are then transmitted to the SDN controller and automatically translated by our IBN manager into pre-assembled network policies in the form of QoE-routing rules. We have proposed QoE-oriented routing for IBN. In contrast to the known ones, the proposed routing uses an adaptive QoE-oriented route metric, which is automatically calculated by the centralized network controller based on the developed mathematical model of QoS/QoE correlation, to select the optimal data path.

Keywords: Software-Defined Network (SDN) · Intent-Based Network (IBN) · Quality of experience (QoE) · Quality of service (QoS) · IBN manager · QoE-routing · QoE-monitoring

1 Introduction

Intent-based networks (IBNs) are the next generation of software-defined networks (SDNs) that will be self-aware, self-configurable, and ultimately autonomous [1–3]. Nowadays, the SDN provides the ability for programs to dynamically change and configure the network [4], but it also still needs somebody to translate the customers requirements of the enterprise into the design elements of the network [5]. The evolution of the SDN paradigm toward IBN allows decisions to be made at a higher level of the network in the form of "intent" [6]. Despite its importance for simplifying network management, the specification of intentions is not yet standardized. The intentions of future IBNs

M. Klymash et al. (Eds.): *Future Intent-Based Networking*, LNEE 831, pp. 128–144, 2022.
https://doi.org/10.1007/978-3-030-92435-5_7

should be expressed declaratively, that is, as a utility-level goal that characterizes the properties of a satisfactory result, rather than prescribing a specific solution. As a result, it allows the IBN to analyze different solutions and find the most optimal variant. This also permits the system to optimize itself by choosing its own goals that maximize utility.

Some of the advantages of expressing intentions as utility level goals is that it allows the system to deal with conflicting goals of multiple intentions. This is very important because an autonomous system must often consider multiple intentions when making a decision. For example, an IBN has one intention to provide a service with high QoE and another to minimize resource costs [7]. It can resolve such conflicts either explicitly based on weights that introduce relative importance, or implicitly based on properties of preferred outcomes defined in utility-level goals. Expectations arise from contracts or business strategies and are kept unchanged when the underlying system is replaced or changed.

In this chapter, we propose to use the northbound interface (NBI) of SDN architecture to declare user intentions. Where by intent we propose to understand the ordering of a certain level of the quality of experience (QoE). In our approach, QoE-intentions are set as a score from 1 to 5, where the highest score means the best quality of services. After that, the intentions are transmitted to SDN controller and automatically translated by the IBN manager into pre-assembled network policies in the form of routing rules. The quality of experience (QoE) in routing decisions has recently received increased attention in future networks [8–10]. When making routing decisions based on personal experience, the traditional single factor or single standard is not enough. The focus is on the method of making routing decisions based on multiple attributes, for example, QoS parameters. Furthermore, the routing decisions in an uncertain and incomplete information environment are practical but difficult to make correct routing decisions. In terms of routing decision techniques, the need to consider customers intentions and the approach of combining multiple QoE values generate many problems. It is important to make optimal decisions based on multiple experience information from multiple intentions attributes and need to be studied further.

2 QoE-Aware Intent-Based Networking Architecture

The IBN and SDN share common goals. The SDN shift the focus of the network infrastructure from hardware to software, from configurations to policies. This makes the network more programmable, improves automation, and lowers costs. Intent-based networking moves network management strategy to a higher level by combining automation with intelligence. The IBN and SDN have the potential to build on each other. The intent-based networking implementation can include an SDN controller to execute the desired policies [11–17].

Based on the above, we proposed an IBN network architecture based on SDN (Fig. 1). The aim of IBN is to create an extensible framework for determining the network requirements of customers based on natural language. Namely, in the proposed IBN, customers can make a QoE request from end to end for any service at a particular point in time. This approach is organized by establishing QoE intents by network customers. QoE intents are formed as prime numbers between 1 and 5. The higher the number of intents, the better the quality of service is guaranteed, the more expensive will be the provision of this service in the IBN network. After that the corresponding resources are allocated with the help of specified QoE-knowledge or self-learning intelligent mechanism, and then they are automatically transformed into the network equipment and interface operations.

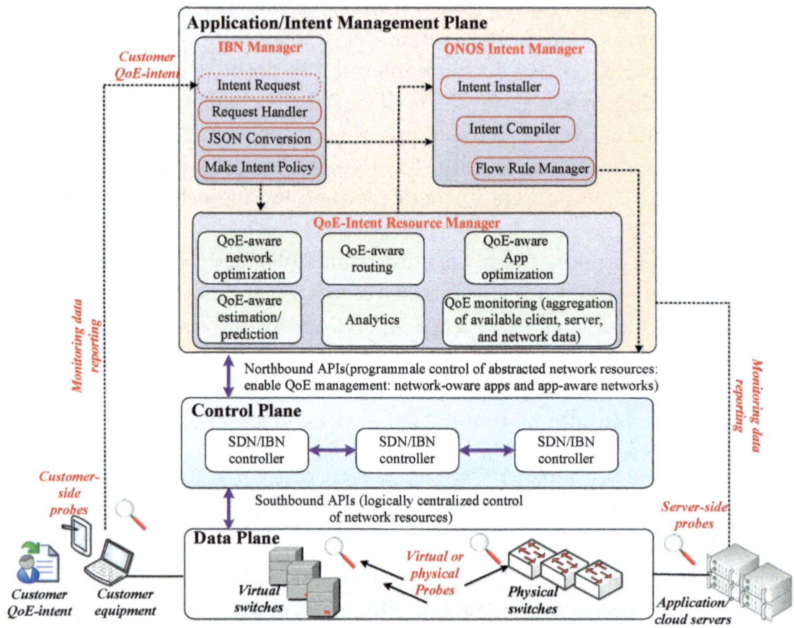

Fig. 1. QoE-aware IBN architecture based on SDN

The architecture of the proposed IBN network is somewhat different from SDN. In particular, their common feature is that the plane of control and data network is divided, which allows the flexibility to configure the network infrastructure from a central point by the software using the controller. In particular, in our work it is ONOS SDN controller. Which after our software transformation we will call SDN/IBN controller. This configuration flexibility is provided by open interfaces northbound and southbound API of the SDN architecture, which allow the exchange of information between the various functional objects of this architecture in a well-defined way. The difference between the existing SDN from the proposed IBN is that in the architecture of SDN we programmatically introduce a new plane application/intent management plane. The main elements of this plane are the well-known ONOS IBN manager, the proposed IBN manager, and the QoE-intent resource manager. The general principle of operation can be explained as follows. First, customers request QoE-intent in the IBN network. At the level of the application plane the QoE-intent is collected and analyzed by the developed IBN manager. ONOS controller interacts with IBN Manager on the management level via the Northern API. The intent manager module of the ONOS controller receives a request from IBN Manager in JSON form and, after compilation, converts it into a low-level command, which is executed on the switches. It then sends the intent (in JSON form) to the ONOS kernel installer module for installation. After successful installation, our intent installer communicates with the flow rule manager module to keep a record of the intent created for the corresponding node.

The IBN manager addresses the resource manager in order to analyze the state of the network and to reconfigure it when the QoE intention is not provided. In the reconfiguration we understand the change of routing rules. In particular, our idea is to find the optimal path through which the customer intention will be provided. In our work we will call such routing as QoE-aware routing. We will explain in detail how the proposed QoE routing works below.

The second key requirement for effective IBN is monitoring. With the shift from policies to intentions comes the need to ensure that they are effectively enforced. Policies that follow the traditional event condition model of action do not need to be monitored because no goal is specified within the policy. In the case of intent, the goal is explicit, so monitoring the network to ensure the success of that goal is critical.

For example, suppose the network defines an intent that all video conferencing streams must have 720p resolution and must not be frequently interrupted. Without monitoring the organization's video traffic, it is impossible to tell whether this intention has succeeded. In fact, this monitoring must be very specific, because the network must not only understand the quality of the video streams coming in and out of the network, but also understand their sources (e.g., Skype Business or Youtube) and possibly their purpose (a Skype call to a family member or customer). In addition, when the intent is layered (e.g., the business wants to guarantee high-speed file downloads in addition to the aforementioned video streaming intent), it becomes clear that the only way to effectively implement the intent is to have high-resolution, comprehensive monitoring of all aspects of the network. As shown in Fig. 1, without monitoring, it is impossible to report the true state of the network, making it impossible to effectively configure the network to provide the highest quality of service.

Network-level QoE monitoring by the SDN infrastructure operator is simplified because the controller autonomously generates a "global view" of the network based on its topology and performance metrics (such as throughput, delay and packet loss statistics). This allows the network operator to apply various QoE-oriented optimization strategies and make optimal decisions of routing across the entire network. Further, the SDN architecture provides open interfaces that will facilitate QoE reporting by end-customers and application servers on tracked QoE impact factors at the application/intent management level. These interfaces will provide the basis for implementing collaborative QoE management between end-customer applications and the underlying IBN. In the future, a network with high-level customers requirements will understand these QoE-intents, control itself, and be able to change the underlying infrastructure to customize the platform itself, all in real-time.

3 The QoE Video Streaming Evaluation Method Through the QoS/QoE Correlation Model

In the context of IBN ideology, the main focus is shifting from improving network performance to improving QoE perception. The level of QoE is directly proportional to the complex quality of service provisioning QoS, which is determined by the parameters of throughput, delay, packet loss and jitter. In order to implement the services with acceptable QoE it is necessary to investigate the impact of the quality characteristics of QoS provision from end to end on the service itself, allowing to describe the distribution laws of QoS characteristics and their impact on QoE parameters (connection establishment time, reaction time on command execution, image freeze, image splitting, image and voice synchronization, audio clarity and intelligibility).

In this section of the chapter, a study of the impact of technical parameters of QoS in the process of transmission of real-time video flows on the level of QoE determined by using the method of expert evaluation on a 5-point QoE scale is conducted. Various cases are presented where QoE degradation can occur during streaming of multimedia programs, conducting experiments on a dedicated SDN network topology. Quality of service degradation is observed due to constraints imposed by network conditions as well as network instability such as link failure. The experiments aim to illustrate many cases where QoE in the network suffers from degradation due to network conditions, and to indicate the need to develop a QoS parameter monitoring system to find a mathematical model of QoS/QoE correlation in order to implement QoE routing in the future software-defined intent-based networks.

An experimental scheme of the SDN topology in the Mininet environment was built to conduct research regarding the influence of QoS parameters on the quality of perception of real-time video streams (Fig. 2). Specifically, for this purpose, the real client (h1) and server (h2) run the VLC player to broadcast and view live video streams.

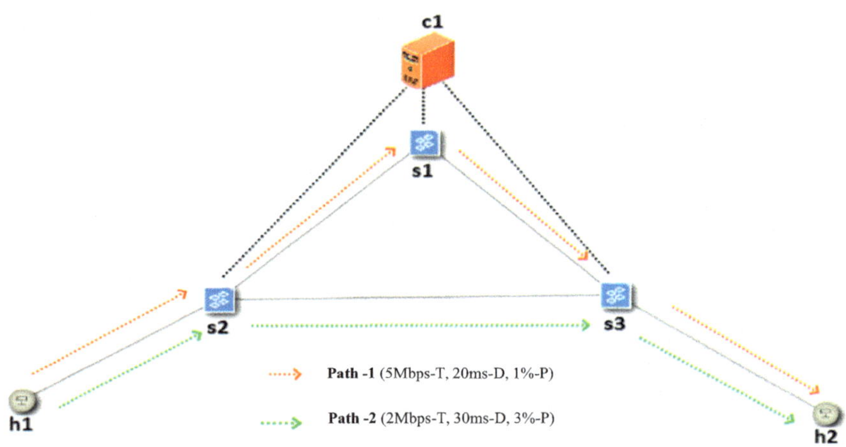

Fig. 2. SDN topology

For this study we wrote a Python script (Fig. 3), which allows us to change the parameters of the connections between the switch and the hosts. Among these QoS parameters: throughput, latency, loss, queue length. By changing these parameters it is possible to investigate their effect on the quality of service perception.

In this experimental network, there are two paths when transmitting data from host1 (h1) to host2 (h2).

```
class SingleSwitchTopo( Topo ):
    "Single switch connected to n hosts."
    def build( self, n=2 ):
        switch = self.addSwitch( 's1' )
        for h in range(n):
            # Each host gets 50%/n of system CPU
            host = self.addHost( 'h%s' % (h + 1),
                                 cpu=.5/n)

            # 100 Mbps, 1ms delay, 0% loss, 1000 packet queue
            self.addLink( host, switch, bw=100, delay='1ms', loss=0,
                          max_queue_size=1000, use_htb=True )

def myNetwork():
    topo = SingleSwitchTopo( n=1)
    net = Mininet( topo=topo,
                   host=CPULimitedHost, link=TCLink )

    print "*** Starting network"

    # Add NAT connectivity
    net.addNAT().configDefault()
    net.start()
    print "*** Hosts are running and should have internet connectivity"
    print "*** Type 'exit' or control-D to shut down network"

    CLI( net )
    # Shut down NAT
    net.stop()

if __name__ == '__main__':
    lg.setLogLevel( 'info')
    myNetwork()
```

Fig. 3. Python script code for configuring communication channels with various QoS parameters

Let us consider one example of the influence of QoS parameters on the quality of video perception. In particular, we compare the process of video streaming through two paths that provide different QoS parameters:

- Path #1 (h1-s2-s3-h2) – throughput (T) is 2 Mbps, delay (D) is 30 ms, artificial packet loss (P) is 3% and buffer size is 700 packets;
- Path #2 (h1-s2-s1-s3-h2) – throughput (T) is 5 Mbps, delay (D) is 20 ms, artificial packet loss (P) is 1% and buffer size is 850 packets.

According to the traditional routing, the shortest path 1 is chosen when transmitting a video stream. Accordingly, we obtained the following video quality, according to our own expert evaluation is described by the level of perception of the service on a scale of $QoE = 2$. Under the conditions of transmission via alternative path 2 video stream perception quality is much better and according to expert evaluation is $QoE = 5$ (Fig. 2).

Thus, by conducting a significant number of experiments it was found a certain correlation between the parameters of QoS and QoE. Based on the above we have proposed a mathematical model to determine the subjective level of user satisfaction by QoE evaluation, depending on the change of objective QoS indicators, which are provided in IBN/SDN network for real-time video. The formation of the mathematical model of QoS/QoE correlation is carried out on the basis of the obtained results of their own experimental research conducted in real SDN equipment.

In order to mathematically determine the deviation of QoS parameters in QoE estimation, it is necessary to normalize the QoS calculation procedure. For this purpose, the standard values of QoS parameters are defined in work, at which the high quality of perception of the studied video stream is provided. Also, in the process of research,

(a)

(b)

Fig. 4. Video quality obtained with different channel parameters: (a) bad quality (QoE-2) and (b) excellent quality (QoE-5)

Table 1. Estimation of QoS parameters and their influence on the QoE level when watching a real-time video stream determined by the expert evaluation method

QoS parameters	Excellent	Fair	Bad
Delay, D	<150 ms	150–200 ms	>200 ms
Packet loss, P	0–1%	1–2%	>2%
Throughput, T	>2 Mbps	1–2 Mbps	<1 Mbps
QoE level	5–4	3.5–4	<3.5

the level of importance of QoS parameters when watching video was established in the form of Table 1.

Table 2. The level of QoS parameters importance

QoS parameters	Relative importance	Weighting factor
Packet loss, P	35%	0.35
Delay, D	45%	0.45
Throughput, T	20%	0.20

The normalized value of the integral additive quality criterion is calculated by the formula (1):

$$Q = QoS(X) = 1 - (w_1(\frac{P_{\min}}{P}) + w_2(\frac{D_{\min}}{D}) + w_3(\frac{T}{T_{\max}})), \tag{1}$$

where w_1, w_2, w_3, are the weighting factors of importance of QoS parameters (X), which vary in the range from 0 to 1, and their sum must be equal to one.

Mathematical model of correlation of customer satisfaction with QoE score for video services, depending on changes in the integral criterion for QoS parameters, presented in the form of functions:

$$QoE_{video} = f_v(Q) = 5(1 - Q^2)^{15Q^5}. \tag{2}$$

Accordingly, the graph of functions (2) is shown in Fig. 5.

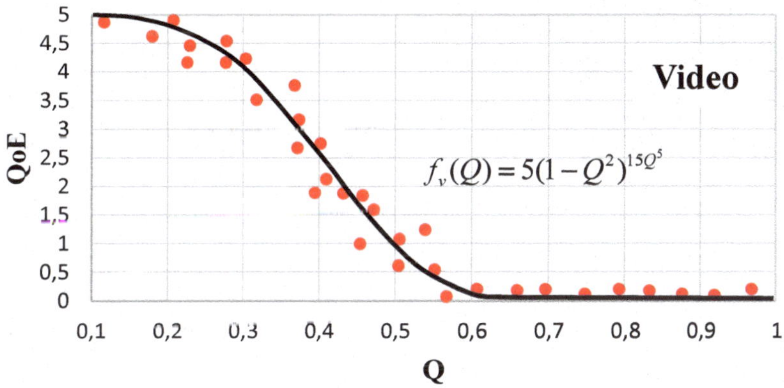

Fig. 5. Graphical QoS/QoE correlation model

Thus, the task of ensuring the ordered level of QoS perception by users on QoE evaluations, indicating the importance of the service for a particular business process, will be to find the necessary normalized value of the integral additive QoS criterion. The solution of this task can be carried out by adaptive management of network resources and their rational redistribution. In particular, one of such approaches is the development of routing data flows, the metric of which is based on the same integral additive criterion.

4 Intent-Based Software-Defined Network Testbed with QoE-Aware Routing Realization

In this part of a chapter, on the basis of hardware SDN switches ZODIAC GX [18], a prototype of IBN is developed. Within the framework of the developed prototype the QoE-aware routing model was implemented, which, in contrast to the existing ones, is based on the threshold QoE criterion metric, which characterizes the customer intention, for the choice of data transmission route. For this purpose, first of all, the system of QoE monitoring was implemented, which allows, based on QoS analysis of the network state, to identify bottlenecks in the SDN-network and automatically adjust the data transmission route to improve the QoE in conditions of its deterioration by changing the network configuration. This is achieved by using ONOS SDN controller and implementing additional functionality (QoE-intent manager) on top of the IBN controller. The IBN controller is responsible for analyzing the customers intent to provide the required level of QoE to change the path of traffic to an alternative when the QoE decreases below a specified threshold.

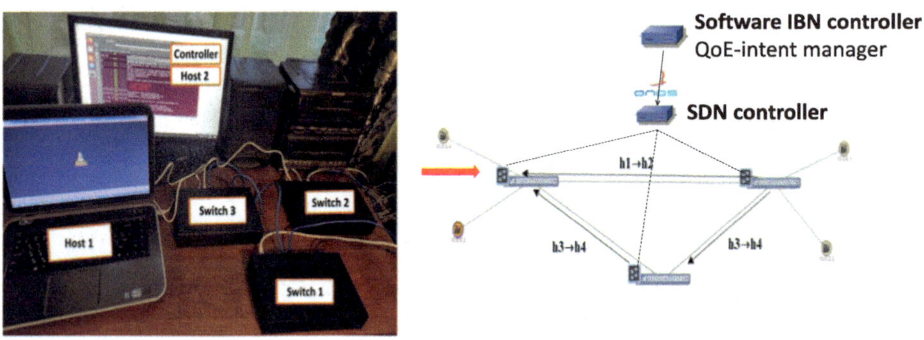

Fig. 6. Intent-based software-defined network testbed

The approach used to ensure that QoE is maintained at satisfactory levels is to monitor and assess quality based on their statistics periodically [19]. Specifically, the program calculates the shortest path between the source and destination hosts, which will be as the primary transmission path, and a second shortest path (if any), which will be the backup path in case of poor quality of service during the transmission of a real-time video stream. The quality monitoring process then begins; the IBN/SDN controller periodically collects statistics from the switches (different statistics for each type of program) and uses them to calculate the QoE level on a 5-point scale using formula 2. If the estimated value is below the specified QoE threshold, then appropriate rules are automatically set to redirect traffic to an alternate path. The operation process is shown in Fig. 7.

For each time the controller needs to measure the delay, it creates a packet with a specific source MAC address (specifically 00: 00: 00: 00: 00: 09) and forwards it to every switch in the network (except the output switch) to the source interface, so that the next switch receives it. Each switch (except the input switch, since it does not have

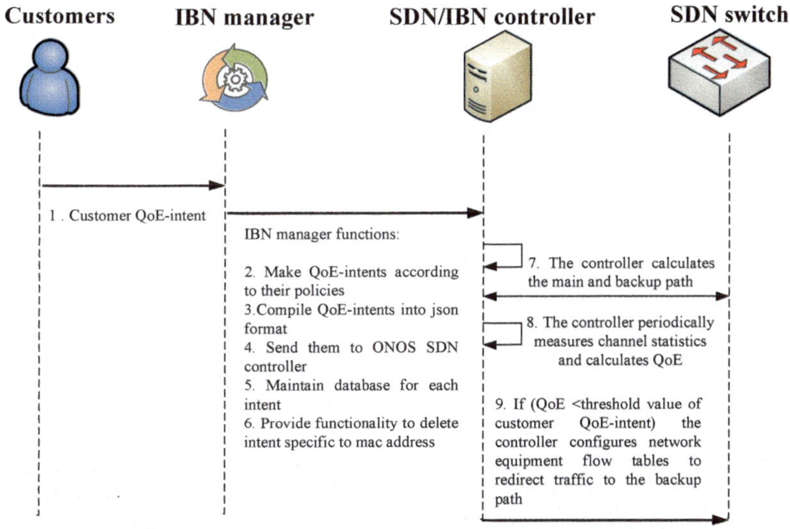

Fig. 7. The operation process of QoE-oriented routing model for IBN

an upstream switch to receive the packet) is configured with the appropriate flow rule to transmit any packet with a specific MAC address to the controller. The difference between when the switch receives the packet and when the last switch sent the packet is the delay of the specific link. Adding delays to all path links results in path delays. Each switch is configured with a rule of this format:

priority = 1000, dl_src = 00 : 00 : 00 : 00 : 00 : 09 actions = CONTROLLER : 65535.

On the basis of this research, we wrote a Python script for measuring the delay, which runs on the ONOS controller. For the delay measurement in this example, the controller creates a simple Ethernet frame. Then the controller asks the s2 switch to forward this packet through a specific port with a Packet_Out message. Switch s3 receives this packet and forwards it to the controller with a Packet_In message, since there is no responding input for this type of Ethernet. The delay is calculated using the following formula (3):

$$D = D_{total} - \frac{D_{s1}}{2} - \frac{D_{s2}}{2}, \quad (3)$$

where D_{total} is the total time of transfer,

D_{s1} is the RTT between switch1 and SDN/IBN controller,

D_{s2} is the RTT between switch2 and SDN/IBN controller.

Figure 9 shows the measured delay between s1 and s2, with a given communication channel delay of 10 ms. The process of measuring the delay on the network topology is described in Fig. 10.

To calculate packet loss, and taking into account that traffic is transmitted in UDP packets, the SDN controller periodically monitors the number of UDP packets sent by the sender (h1) and the number of UDP packets received by the receiver (h2), and

```
received_time = time.time() * 1000 - start_time

#measure T1
if event.connection.dpid == src_dpid:
  OWD1=0.5*(received_time - sent_time1)
  #print "OWD1: ", OWD1, "ms"

#measure T2
elif event.connection.dpid == dst_dpid:
  OWD2=0.5*(received_time - sent_time1)
  #print "OWD2: ", OWD2, "ms"
```

(a)

```
received_time = time.time() * 1000 - start_time

if packet.type==0x5577 and event.connection.dpid==dst_dpid:
  c=packet.find('ethernet').payload
  d,=struct.unpack('!I', c)
  print "delay:", received_time - d - OWD1-OWD2, "ms"
```

(b)

Fig. 8. Calculation of parameters Ts1 and Ts2 (a) and calculation of the delay value (b)

```
INFO:openflow.of_01:[00-00-00-00-00-01 1] connected
ConnectionUp:  00-00-00-00-00-01
INFO:openflow.of_01:[00-01-00-00-00-01 2] connected
ConnectionUp:  00-01-00-00-00-01
delay: 17.6373291016 ms
delay: 10.1243896484 ms
delay: 10.1201171875 ms
delay: 11.1755371094 ms
delay: 11.6986083984 ms
delay: 12.6079101562 ms
delay: 11.0582275391 ms
delay: 10.1243896484 ms
```

Fig. 9. The result of measuring the delay by the script

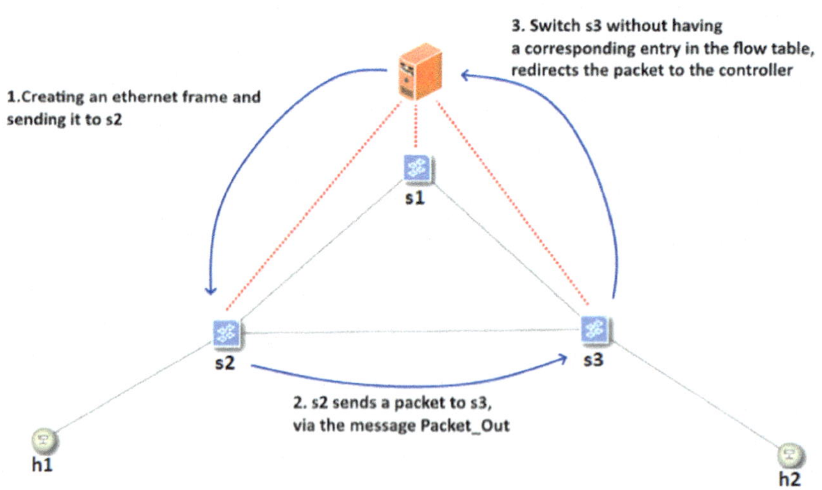

Fig. 10. The delay measurement process in IBN testbed

calculates their difference divided by the number of packets sent. To implement packet loss monitoring, the input and output switches are configured to forward to the controller other than the predefined output interface to the next node any UDP packet they receive (input receives UDP packets from the sender host and output from the previous node in the path). In turn, the controller counts the total number of incoming and outgoing UDP path packets and can determine the loss of packets.

The principle of packet loss calculation is as follows: the controller receives packet loss statistics from switches s2 and s3. Based on the received statistics of transmitted and received packets on the corresponding switches the controller calculates the total number of lost packets by the formula (4).

$$packetLoss(report) = input\_pkts - output\_pkts \quad (4)$$

The process of measuring packet loss is as follows, in the network with the above topology, artificially introduced channel losses in the path section s2-s3, packet loss level of 10%. Then h1 sends packets to h2 via ping utility. The IBN controller then sends a request for statistical data from the switches and calculates the percentage of packet loss. The result of measuring packet losses is depicted in Fig. 11.

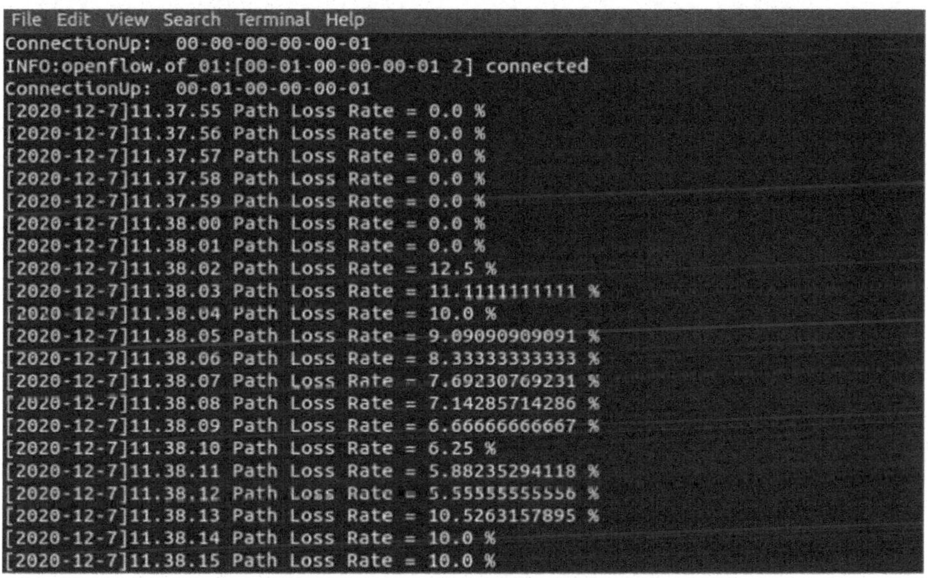

Fig. 11. Packet loss measurement result

To calculate the bitrate, the command ffmpeg -i [VIDEO_PATH] - hide_banner is executed with the Java code, and the output data is analyzed until the bitrate value is accessed. To calculate the frame rate, the command -i [VIDEO_PATH] - hide_banner is executed with Java code, and the output is parsed until the frame rate value is accessed. The results of the QoE monitoring system are shown in Figs. 13, 14 and 15.

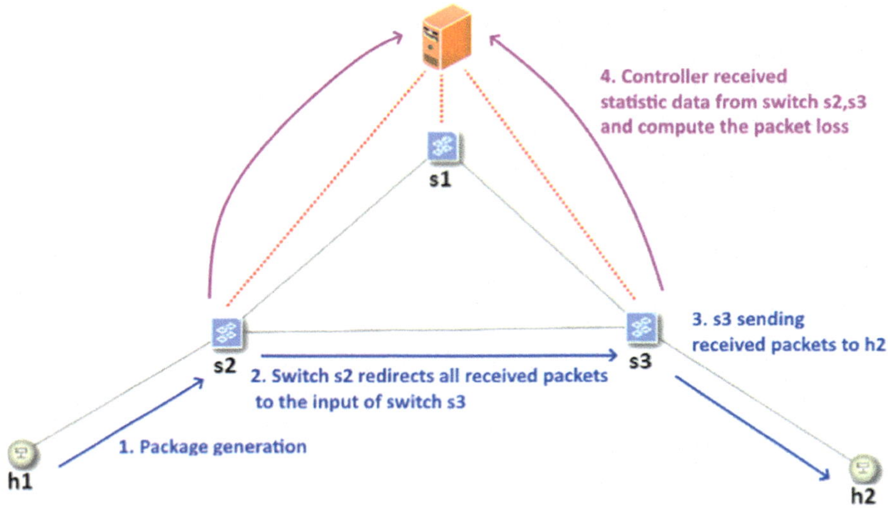

Fig. 12. Packet loss measurement process in IBN testbed

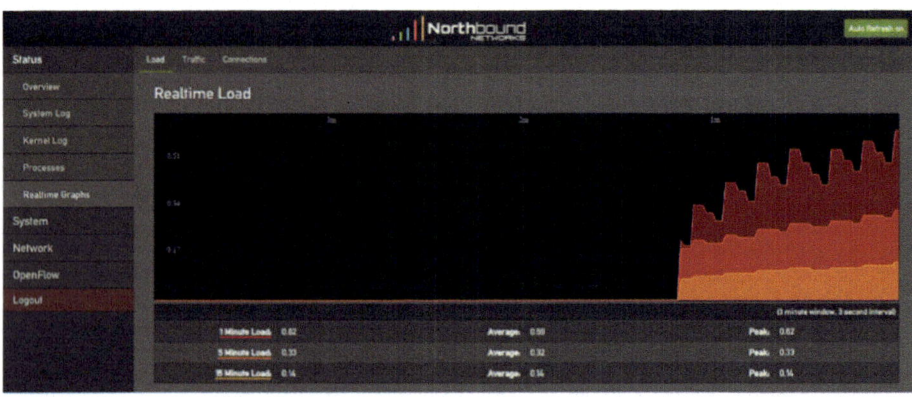

Fig. 13. ZODIAC GX switch load monitoring

We conducted a study of the developed QoE monitoring system and QoE routing. The study was done with real-time video streaming. For example, the customer QoE-intent (Threshold) is 3.92. The experiment was conducted for 12 s, the traffic itself is generated between hosts h1 and h2. Every 4 s the link parameters were degraded due to the introduction of artificial packet loss. When comparing the results, it can be seen that the proposed monitoring system can reduce the number of packet losses and generally improve the quality of service in the case of streaming video and accordingly provide a better quality of service perception for end customers based on their intentions (Fig. 16).

On the basis of the study, it was found that without the proposed monitoring system and QoE-routing, the controller does not respond to the degradation of video quality assessment, which leads to an unsatisfactory quality of service. The effectiveness of the

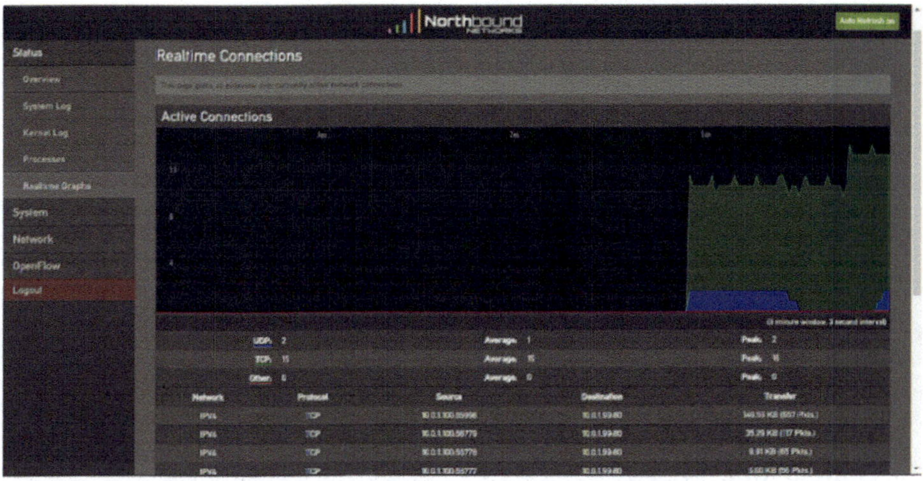

Fig. 14. Protocol-based traffic monitoring in the ZODIAC GX switch

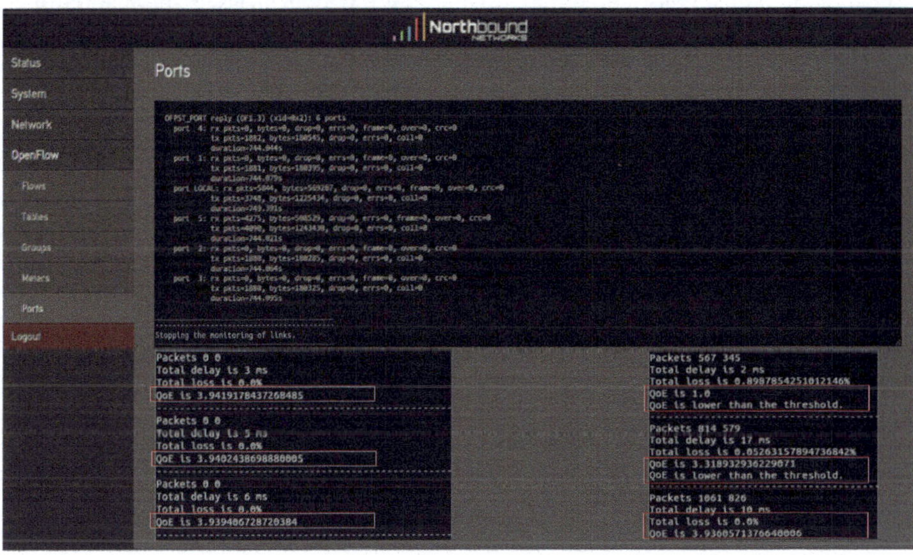

Fig. 15. QoE level monitoring and fixation of its degradation

developed QoE monitoring system and routing has been proved. According to which in the conditions of detecting degradation of video QoE level, the ordered customer intention in the communication channel is conducted by redirecting traffic to an alternative transmission path by automating the processes of the proposed QoE routing.

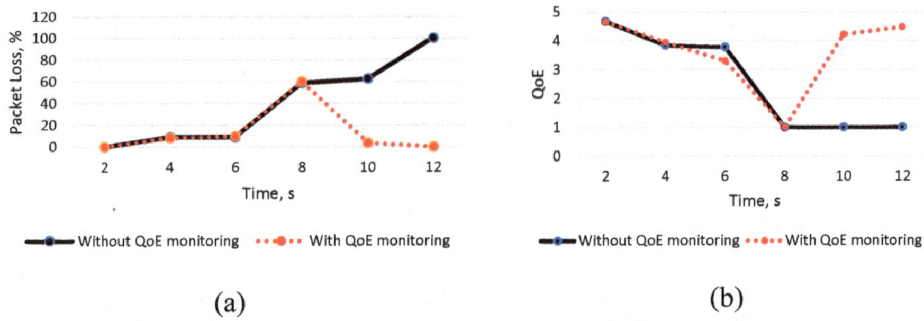

Fig. 16. Comparison of packet loss (a) and QoE level (b) in the process of video transmission without the proposed monitoring system and with the developed QoE monitoring system and routing

5 Conclusion

In our view, IBN is an addition to the already existing SDN architecture rather than a stand-alone architecture. Intent-based technology is policy-based automation of the deployment of business intentions on the network. There are three main functions in a continuous cycle: intent translation, automation of intent configuration on physical and virtual network devices, and monitoring.

In this chapter, we propose an IBN structure that provides customers with their intention on the ordered real-time quality of service in the form of 5-point QoE scores. We have implemented this by using the developed IBN manager and passing intents to the ONOS SDN controller to improve streaming routes based on users' real-time quality perception. We use a real experiment to analyze how QoE can affect different network conditions and how video stream perception degrades with dynamic parameter changes in the SDN network. A mathematical model for determining the subjective level of customer satisfaction by QoE assessment, depending on changes in objective QoS, provided in IBN/SDN network, particularly for real-time video service, has been developed. The formation of the mathematical model of QoS/QoE correlation is carried out on the basis of our own experimental research. We have programmatically implemented a QoS/QoE correlation model in the SDN/IBN controller to automatically verify the ordered intent and quality provided. This is done to identify problems and bottlenecks in the current service paths in real-time and allow network controllers to take corrective action by redirecting the streaming traffic. To do this, we have offered our own QoE-routing based on the QoE monitoring system.

The developed QoE monitoring system for the IBN corporate network prototype based on ZODIAC GX switches and the new IBN controller as a supplement to ONOS allowed identifying network bottlenecks. It also automated data routing search to ensure customer QoE-intent index by changing the network configuration.

The QoE-aware routing implemented in the IBN system was compared and evaluated with the default routing system in traditional SDN, resulting in the new system resulting in much less packet loss than the default routing and hence much higher video quality.

Acknowledgement. This research was supported by the Ukrainian government project №0120U102201 "Development of the methods and unified software-hardware means for the deployment of the energy efficient intent-based multi-purpose information and communication networks".

References

1. Rafiq, A., Mehmood, A., Song, W.-C.: Intent-based slicing between containers in SDN overlay network. J. Commun. **15**(3), 237–244 (2020). https://doi.org/10.12720/jcm.15.3.237-244
2. Singh, A., Aujla, G.S., Bali, R.S.: Intent-based network for data dissemination in software-defined vehicular edge computing. IEEE Trans. Intell. Transp. Syst. 1–9 (2020). https://doi.org/10.1109/TITS.2020.3002349
3. Rafiq, A., Afaq, M., Song, W.-C.: Intent-based networking with proactive load distribution in data center using IBN manager and Smart Path manager. J. Ambient Intell. Humanized Comput. **11**(11), 4855–4872 (2020). https://doi.org/10.1007/s12652-020-01753-1
4. Seliuchenko, M., Beshley, M., Kyryk, M., Zhovtonoh, M.: Automated recovery of server applications for SDN-based Internet of Things. In 2019 3rd International Conference on Advanced Information and Communications Technologies (AICT), pp. 149–152. IEEE (2019). https://doi.org/10.1109/AIACT.2019.8847743
5. Hyun, J., Hong, J.W.: Knowledge-defined networking using in-band network telemetry. In: 2017 19th Asia-Pacific Network Operations and Management Symposium (APNOMS), pp. 54–57 (2017). https://doi.org/10.1109/APNOMS.2017.8094178
6. Ujcich, B.E., Sanders, W.H. : Data protection intents for software-defined networking. In 2019 Conference on Network Softwarization (NetSoft), pp. 271–275. IEEE (2019). https://doi.org/10.1109/NETSOFT.2019.8806684
7. Beshley, M., Vesely, P., Prislupskiy, A., Beshley, H., Kyryk, M., Romanchuk, V., Kahalo, I.: Customer-Oriented Quality of Service Management Method for the Future Intent-Based Networking. Applied Sciences, vol. 10, no. 22, 8223–1–8223–38 (2020)
8. Rothenberg, C.E., et al.: Intent-based control loop for DASH video service assurance using ML-based edge QoE estimation. In 2020 6th Conference on Network Softwarization (NetSoft), Ghent, Belgium, pp. 353–355. IEEE (2020)
9. Wang, L., Delaney, D.T.: QoE oriented cognitive network based on machine learning and SDN. In 2019 11th International Conference on Communication Software and Networks (ICCSN), Chongqing, China, pp. 678–681. IEEE (2019)
10. Barakabitze, A.A., et al.: QoE management of multimedia streaming services in future networks: a tutorial and survey. IEEE Commun. Surv. Tutorials **22**(1), 526–565 (2019)
11. Beshley, M., Pryslupskyi, A., Panchenko O., Seliuchenko, M.: Dynamic switch migration method based on QoE- aware priority marking for intent-based networking. In 2020 15th International Conference on Advanced Trends in Radioelectronics, Telecommunications and Computer Engineering (TCSET), Ukraine, pp. 864–868. IEEE (2020)
12. Ujcich, B.E., Bates, A., Sanders, W.H.: Provenance for intent-based networking. In: 2020 6th Conference on Network Softwarization (NetSoft), Belgium, pp. 195–199. IEEE (2020)
13. Panchenko, O., et al.: Method for adaptive client-oriented management of quality of service in integrated SDN/CLOUD networks. In: 2017 4th International Scientific-Practical Conference Problems of Infocommunications. Science and Technology (PIC S&T), pp. 452–455. IEEE (2017)
14. Lewis, B., Fawcett, L., Broadbent, M., Race, N.: Using P4 to Enable Scalable Intents in Software Defined Networks. In 2018 26th International Conference on Network Protocols (ICNP), pp. 442–443. IEEE (2018)

15. Beshley, M., Pryslupskyi, A., Panchenko, O., Beshley, H.: SDN/Cloud solutions for intent-based networking. In 2019 3rd International Conference on Advanced Information and Communications Technologies (AICT), pp. 22–25. IEEE (2019)
16. Abbas, K., Khan, T.A., Afaq, M., Song, W.-C.: Network slice lifecycle management for 5G mobile networks: an intent-based networking approach. IEEE Access **9**, 80128–80146 (2021). https://doi.org/10.1109/ACCESS.2021.3084834
17. Medvetskyi, M., Beshley, M., Klymash, M.: A quality of experience management method for intent-based software-defined networks. In: 2021 16th International Conference on the Experience of Designing and Application of CAD Systems (CADSM), pp. 59–62. IEEE (2021). https://doi.org/10.1109/CADSM52681.2021.9385250
18. Northbound networks. https://northboundnetworks.com/pages/zodiac-gx-zodiac-gx-hardware-specifications
19. Romanchuk, V., Beshley, M., Polishuk, A., Seliuchenko ,M. Method for processing multi-service traffic in network node based on adaptive management of buffer resource. In 2018 14th International Conference on Advanced Trends in Radioelecrtronics, Telecommunications and Computer Engineering (TCSET), pp. 1118–1122. IEEE (2018). https://doi.org/10.1109/TCSET.2018.8336390

Complex Investigation of the Compromise Probability Behavior in Traffic Engineering Oriented Secure Routing Model in Software-Defined Networks

Oleksandr Lemeshko[1(✉)], Oleksandra Yeremenko[1], Maryna Yevdokymenko[1], Anastasiia Shapovalova[1], and Oleksii Baranovskyi[2]

[1] Kharkiv National University of Radio Electronics, Nauky Ave. 14, Kharkiv 61166, Ukraine
{oleksandr.lemeshko,oleksandra.yeremenko,marina.ievdokymenko, anastasiia.shapovalova}@nure.ua

[2] National Technical University of Ukraine "Igor Sikorsky Kyiv Polytechnic Institute", Peremohy Ave. 37, Kyiv 03056, Ukraine
o.baranovskyi@kpi.ua

Abstract. In the work, the complex investigation and analysis results of the compromise probability behavior in Traffic Engineering oriented secure routing model in software-defined networks have been presented. Within the framework of the study, the classical flow-based model based on load balancing in accordance with the principles of the Traffic Engineering concept was improved and supplemented with conditions that allow considering the network security parameters in the process of obtaining a routing solution. In this way it was obtained the secure traffic engineering routing model, the novelty of which lies in the modified conditions of load balancing considering such network characteristics as topology, features of the traffic transmitted, as well as links bandwidth and probabilities of their compromising. The use of such a model makes it possible to reduce the overload of network links with a high value of compromise probability, while more traffic will be transmitted over secure links without causing overload. Power and exponential forms of functional dependence of weighting coefficients on the link compromise probability have been used for comparison during obtaining the secure-based routing solutions. The secure traffic engineering routing flow-based model under investigation is proposed to use in a software-defined network data plane.

Keywords: Traffic engineering · Network security · Secure routing · Probability of compromise · Software-defined network

1 Introduction

In general, the main idea behind secure routing protocols is to increase the security and cyber resilience of infocommunication networks [1–6]. All secure routing protocols can be divided into proactive and reactive [4, 10–24]. Proactive secure routing protocols are

M. Klymash et al. (Eds.): *Future Intent-Based Networking*, LNEE 831, pp. 145–160, 2022.
https://doi.org/10.1007/978-3-030-92435-5_8

based on a preliminary assessment of information security risks and the use of the most secure network elements in the packet transmission process. Reactive secure routing protocols, in contrast to proactive ones, are based on the on-demand route calculation [19–21]. However, it should be noted that all secure routing protocols are an improvement on traditional routing protocols (RIP, EIGRP, OSPF) [22–24] using security metrics.

A special place is occupied by solutions for secure routing in Software-Defined Networks (SDN) [25–28]. By separating the data plane from the control plane in SDN networks, it is possible to effectively implement innovative approaches to secure routing. In turn, traditional routing protocols do not take into account the specifics of SDN networks. It should be noted that the need for security at the network level arises in different types of Software-Defined Networks: wired, wireless, embedded systems, etc.

Thus, [25] proposed a reliable RouteGuardian secure routing mechanism in the SDN network, which considers the capabilities of SDN switches in conjunction with the Network Security Virtualization framework that effectively uses distributed network protection devices to analyze abnormal traffic and isolate malicious nodes. While [26] proposed a secure routing mechanism for industrial wireless sensor SDN networks, within which internal malicious nodes were located by calculating the value of the trust indicator to the node. Whereas [27] developed a new secure SDN communication protocol based on the group key agreement approach for internal communication in the scalable architecture of the MPSoC multiprocessor system – Cloud-of-Chips (CoC).

Therefore, it is important to research and develop effective models of secure routing in SDN with load balancing [12–14] and take into account compromise probability behavior [5, 6].

2 Traffic Engineering Oriented Secure Routing Model

In this section, the notation that is used in the problem definition is explained. A network (SDN data plane) is given by a directed graph $G = (R, E)$, where $R = \{R_i;\ i = \overline{1, m}\}$ is the set nodes and $E = \{E_{i,j};\ i, j = \overline{1, m};\ i \neq j\}$ is the set of directed edges representing links between the nodes (network routers). For each edge $E_{i,j} \in E$ (network link) its bandwidth $\phi_{i,j}$ is defined. It should be noted that the number of edges (links) is determined as the set capacity $|E| = n$.

Let us suppose that in the presented basic Traffic Engineering model, each transmitted flow is unicast with a set of corresponding functional characteristics: K – set of transmitted flows of packets (k ∈ K), s_k – source router, d_k – destination router, and λ^k – k th flow average intensity (packets per second, pps).

In order to achieve the solution of the Traffic Engineering (TE) problem, the routing variables $x_{i,j}^k$ need to be calculated. Each of the variables corresponds to the portion of the k th flow intensity in the link $E_{i,j} \in E$ that is the part of the route.

The following constraints imposed on the variables when a multipath routing is used [29, 30]

$$0 \leq x_{i,j}^k \leq 1. \tag{1}$$

The flow conservation conditions are introduced in the model with the aim of ensuring the routes connectivity [8, 9]:

$$\begin{cases} \sum\limits_{j:E_{i,j}\in E} x_{i,j}^k - \sum\limits_{j:E_{j,i}\in E} x_{j,i}^k = 0;\ k \in K, \mathrm{R}_i \neq s_k, d_k; \\ \sum\limits_{j:E_{i,j}\in E} x_{i,j}^k - \sum\limits_{j:E_{j,i}\in E} x_{j,i}^k = 1;\ k \in K, \mathrm{R}_i = s_k; \\ \sum\limits_{j:E_{i,j}\in E} x_{i,j}^k - \sum\limits_{j:E_{j,i}\in E} x_{j,i}^k = -1;\ k \in K, \mathrm{R}_i = d_k. \end{cases} \quad (2)$$

Additionally, the average packet intensity of the k th flow within the link $E_{i,j} \in E$ can be calculated as follows:

$$\lambda_{i,j}^k = \lambda^k x_{i,j}^k,\ E_{i,j} \in E \quad (3)$$

To estimate the link utilization coefficient, the next expression will be used:

$$\alpha_{i,j} = \frac{\sum\limits_{k\in K} \lambda^k x_{i,j}^k}{\varphi_{i,j}}. \quad (4)$$

As shown by the analysis [15, 16], to fulfill the requirements of the Traffic Engineering concept, it is necessary to ensure balanced use of the available network link resource. This, as a rule, is implemented at the level of formulating preventing overload conditions [8, 9].

During load balancing in the network with the aim of accounting the indicators of network security, it is proposed to introduce the following conditions. Assume that each link $E_{i,j} \in E$ is associated with such an important indicator of network security as the probability of it being compromised $p_{i,j}$. The main goal of the proposed solution is to ensure maximum utilization of communication links with minimal probabilities of compromise, and vice versa – links with a high probability of compromise $p_{i,j}$ should be loaded minimally or even completely blocked.

Therefore, it is proposed to use an improved load balancing condition [5, 6]:

$$\sum_{k\in K} \lambda^k x_{i,j}^k \leq \alpha v_{i,j} \varphi_{i,j}, E_{i,j} \in E \quad (5)$$

where α is the additional control variable that numerically determines the upper bound of the network links utilization and obeys the constraints [5, 6]:

$$0 \leq \alpha \leq 1, \quad (6)$$

while $v_{i,j}$ are the weighting coefficients that in turn must comply with the following boundary conditions

$$v_{i,j} = \begin{cases} 0,\ if\ p_{i,j} = 1; \\ 1,\ if\ p_{i,j} = 0. \end{cases} \quad (7)$$

If the compromise probability $p_{i,j}$ increases from 0 to 1, the weighting coefficient $v_{i,j}$ should decrease from 1 to 0. The direct dependence of the weighting coefficient $v_{i,j}$ on $p_{i,j}$ will be given by the decreasing function on the whole interval

$$p_{i,j} \in [0; 1]. \tag{8}$$

Finally, as the optimality criterion in the problem solving of the Traffic Engineering technological task the minimum of the boundary value α is selected as it was introduced in [5, 8]:

$$\min_{x,\alpha} \alpha. \tag{9}$$

The formulation of the Traffic Engineering problem in optimization form with criterion (9) and constraints (1)–(8) is aimed at ensuring optimal load balancing with minimization of the utilization coefficient of each network link. This helps to improve network performance, time QoS indicators (average packet delay and jitter), as well as reliability indicators such as packet loss probability.

3 Link Blocking Models in Secure Based Traffic Engineering

In this work, we will consider several variants of the functional representation of the dependence $v = f(p)$ that meet the conditions (7). Within the framework of the conducted research, the set of admissible values $p_{i,j}$ will be conditionally divided into several subsets, each of which corresponds to its own scenario of network and links compromising:

- the *first scenario* covered the range of link compromise probabilities from 0 to 0.5 (the vulnerability level is "relatively low");
- the *second scenario* corresponded to the range of link compromise probabilities from 0.35 to 0.85 (the vulnerability level is "relatively average");
- the *third scenario* covered the range of link compromise probabilities from 0.5 to 1 (the level of danger is "relatively high").

Simpler blocking models have been presented in [5, 6], which may explain the operation of more complex models. For example, the link blocking model can be described as described below in comparison with functional dependence [5, 6]:

$$v_{i,j} = 1 - \frac{1}{1 + n \cdot \exp(-r \cdot p_{i,j} + b)}, \tag{10}$$

in which to fulfill conditions (7) and (8) $r = 2b$, $b \geq 7$, $n > 0$ (Fig. 1 and Fig. 2).

The use of the link blocking model (10) is explicitly focused on a certain range of values $p_{i,j}$ (compromise scenario). The average value of this range can be set using the parameter n (Fig. 2). The width of the selected range of values $p_{i,j}$ is adjusted by changing the values of parameters r and b (Fig. 1).

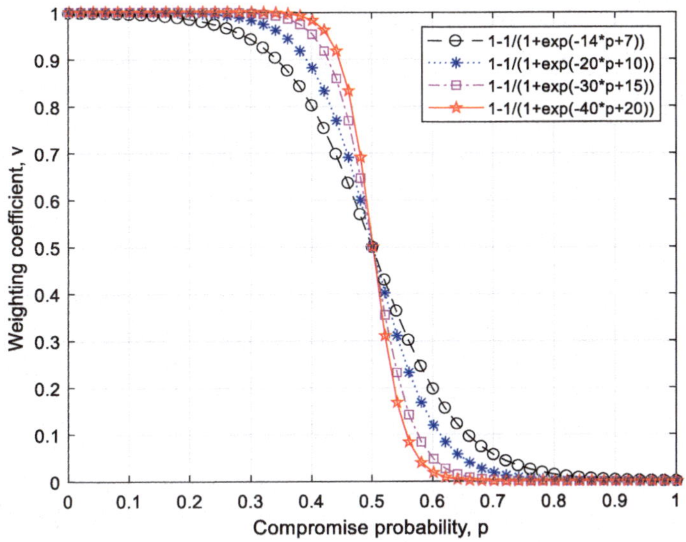

Fig. 1. Visualization of link blocking model (10) under $n = 1$.

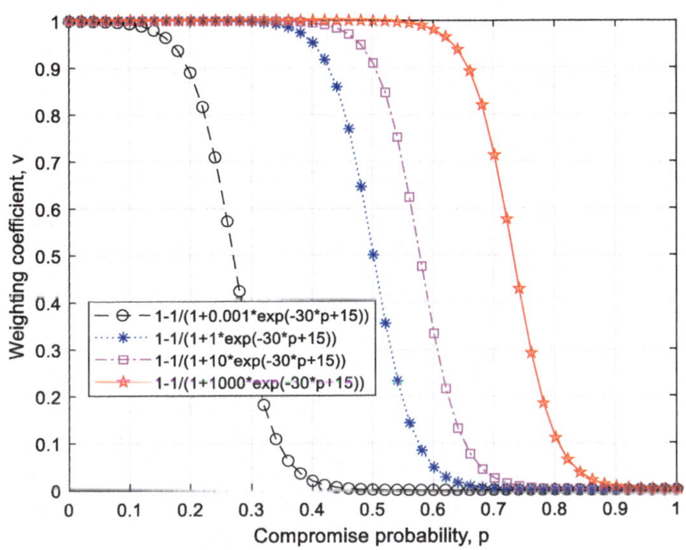

Fig. 2. Visualization of link blocking model (10) under $r = 30$, $b = 15$.

Another example of a complex blocking model is an expression

$$v_{i,j} = 1 - p_{i,j} + n\sin(2\pi p_{i,j} + \theta), \tag{11}$$

where to fulfill conditions (7) and (8) $n \leq 0.15$ and $\theta \in \{0; \pi\}$ (Fig. 3 and Fig. 4).

The blocking model (11) is generally quite versatile, as it can be used in any of the above compromise scenarios (Table 1). When $\theta = 0$ model (11) is less responsive to minimum values $p_{i,j}$ than when $\theta = \pi/2$, but more sensitive to high values $p_{i,j}$ compared to $\theta = \pi/2$. In addition, at maximum values n ($n = 0.15$) and at $\theta = \pi/2$ the blocking model (11) reacts very insignificantly to the change $p_{i,j}$ within the second compromise scenario.

Therefore, depending on the network state and the predicted scenario of compromising network elements, one or another proposed link blocking model (10), (11) can be selected for secure load balancing (5) with the fulfillment of conditions (7) and (8).

Table 1. Sensitivity of link blocking model to values of compromise probability.

Link blocking model type	First scenario $p_{i,j} \in [0; 0.5]$	Second scenario $p_{i,j} \in [0.35; 0.85]$	Third scenario $p_{i,j} \in [0.5; 1]$
(10) under $n = 1$	Very low	Average	Very high
(11) under $\theta = 0$	Low	Average	High
(11) under $\theta = \pi$	Not high	Average	Not high

4 Numerical Research of Secure Based Traffic Engineering Model on SDN Data Plane

The study analyzed the impact of network structure, compromise scenarios, and link blocking models (10) and (11) on network utilization (4) and network security level. The level of network security was evaluated by such an indicator as the probability of compromising the kth flow packets along the set of paths used:

$$p_{E2E}^{k} = \sum_{s \in S^k} \frac{\lambda_s^k}{\lambda^k} p_s, \tag{12}$$

where S^k is the set of paths (routes) used to transmit the kth packet flow between a given pair of network routers;

λ_s^k is the intensity of the kth packet flow transmitted over the sth network path;

p_s is the probability of compromising the sth path that is determined according to the formula

$$p_s = 1 - \prod_{E_{i,j} \in Path_s} (1 - p_{i,j}), \tag{13}$$

in which $Path_s = \{E_{i,j}\}$ is the set of network links that form the sth path.

Let denote by α^* the maximum value from the set of utilization coefficients (4), since when using balancing conditions (5) the value α characterizes the upper bound of network links bandwidth utilization that remained after blocking according to the values of the link compromise probability.

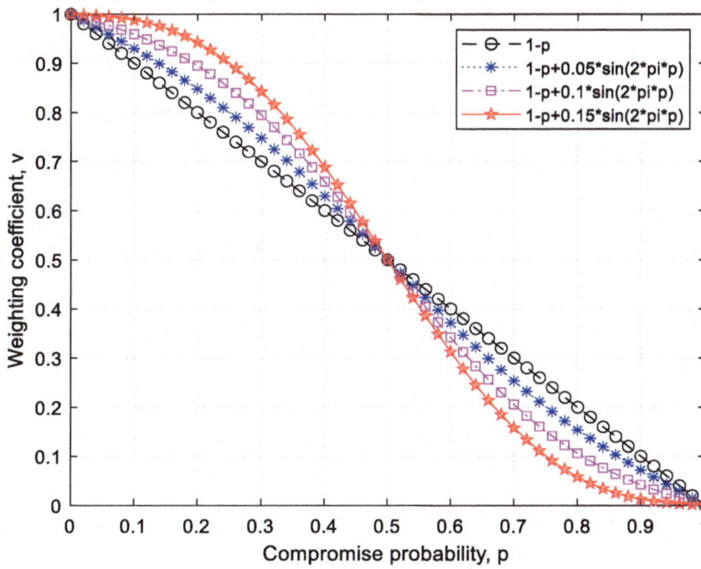

Fig. 3. Visualization of link blocking model (11) under $\theta = 0$.

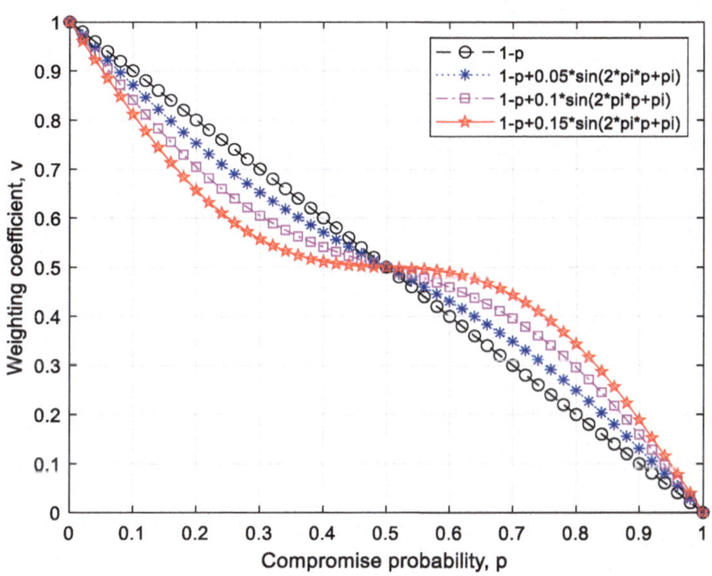

Fig. 4. Visualization of link blocking model (11) under $\theta = \pi$.

Let the structure of the studied network be presented in Fig. 5. The intensity of the input flow transmitted from the first to the fourth router was 350 pps.

Table 2 shows the bandwidth of communication links and options for the compromise probabilities in accordance with the selected strategies of network compromise.

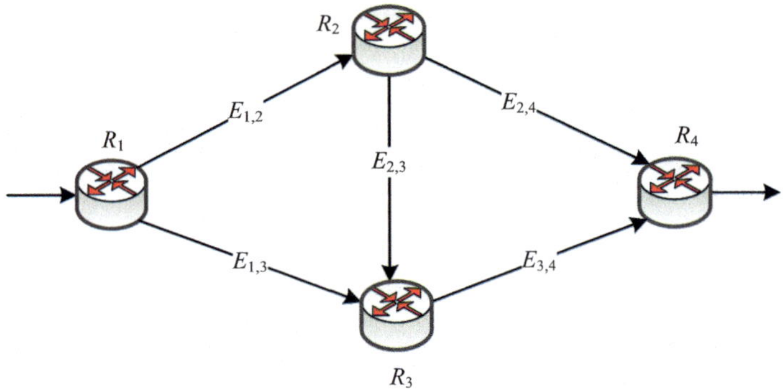

Fig. 5. Network structure under study.

Table 2. Network links characteristics.

Link	Bandwidth, pps	Link compromise probabilities		
		First scenario	Second scenario	Third scenario
$E_{1,2}$	700	0.1	0.3	0.5
$E_{2,4}$	600	0.5	0.7	0.9
$E_{1,3}$	400	0.4	0.7	0.9
$E_{3,4}$	600	0.2	0.4	0.6
$E_{2,3}$	800	0.2	0.3	0.5

Conforming to the information about the network structure (Fig. 5) and the characteristics of corresponding links, Table 3 shows data on available routes between R_1 and R_4, as well as the probability of their compromise.

Table 3. Compromise probabilities of network routes.

Route		First scenario	Second scenario	Third scenario
1	$R_1 \rightarrow R_2 \rightarrow R_4$	0.55	0.79	0.95
2	$R_1 \rightarrow R_3 \rightarrow R_4$	0.52	0.82	0.96
3	$R_1 \rightarrow R_2 \rightarrow R_3 \rightarrow R_4$	0.424	0.706	0.9

For research and comparative analysis of different link compromise scenarios (Tables 2 and 3), Table 4 shows the calculations results obtained using the classical Traffic Engineering model (1)–(6), (9) when the expression (5) $v_{i,j} = 1$.

According to the solution obtained for the TE model (Table 4), the upper bound of network link utilization was 0.3182. Using the first route, packets were transmitted with an intensity of 190.9091 pps, using the second one with an intensity of 127.2727 pps,

Table 4. Results of load balancing utilizing the Traffic Engineering model.

Link	Link compromise probabilities, $p_{i,j}$	Traffic Engineering solution	
		$\lambda_{i,j}^{1}$	$\alpha_{i,j}$
$E_{1,2}$	0.1	222.7273	0.3182
$E_{2,4}$	0.5	190.9091	0.3182
$E_{1,3}$	0.4	127.2727	0.3182
$E_{3,4}$	0.2	159.0909	0.2652
$E_{2,3}$	0.2	31.8182	0.0398

while over the third route with a rate of 31.8182 pps. Therefore, in accordance with Table 3, the probability of compromising packets in the network (12) was 0.5276.

For the same compromise scenario when using the blocking model (10) with $r = 30$ and $b = 15$, Table 5 shows the results of solving the secure routing problem with load balancing – Secure Traffic Engineering (SecTE), when the control parameter n took, for example, the values of 0.1 and 0.01.

Table 5. Comparison of obtained solutions for secure load balancing utilizing the link blocking model (10).

Link	Link compromise probabilities, $p_{i,j}$	Secure routing solution ($n = 0.1$)		Secure routing solution ($n = 0.01$)	
		$\lambda_{i,j}^{1}$	$\alpha_{i,j}$	$\lambda_{i,j}^{1}$	$\alpha_{i,j}$
$E_{1,2}$	0.1	350	0.5	350	0.5
$E_{2,4}$	0.5	29.1997	0.0487	3.4733	0.0058
$E_{1,3}$	0.4	0	0	0	0
$E_{3,4}$	0.2	320.8003	0.5347	346.5267	0.5775
$E_{2,3}$	0.2	320.8003	0.4010	346.5267	0.4332

The blocking model (10) with decreasing n provides a higher sensitivity of the link compromise probabilities. As shown in Table 5, this was reflected in the reduction of the most vulnerable links utilization, for example, link $E_{2,4}$. Table 6 shows the comparison results of the studied solutions: TE and SecTE (10) models with different values of the control parameter n, which confirmed the previous conclusions about the scope of this link blocking model and the control variables' impact on this process. In general, when the values n decreased from 0.1 to 0.01, the packet compromise probability in the network decreased in the range from 17.65% to 19.4%.

In accordance with the contents of Table 6, Fig. 6 shows the dynamics of change in the utilization and network security indicators in a network depending on the values of parameter n in the model (10). By changing the parameter n, it is possible to reduce the

Table 6. Comparison results of obtained solutions for secure load balancing utilizing the link blocking model (10).

n	α^*	α	p_{E2E}	p_{E2E} decrease
0.01	0.5775	0.5847	0.4253	19.4%
0.02	0.572	0.5756	0.4264	19.18%
0.03	0.5668	0.5691	0.4276	18.98%
0.04	0.5617	0.5634	0.4287	18.75%
0.05	0.5568	0.5581	0.4297	18.55%
0.06	0.552	0.5532	0.4308	18.36%
0.07	0.5475	0.5484	0.4317	18.17%
0.08	0.543	0.5439	0.4327	17.99%
0.09	0.5388	0.5395	0.4336	17.82%
0.1	0.5347	0.5353	0.4345	17.65%

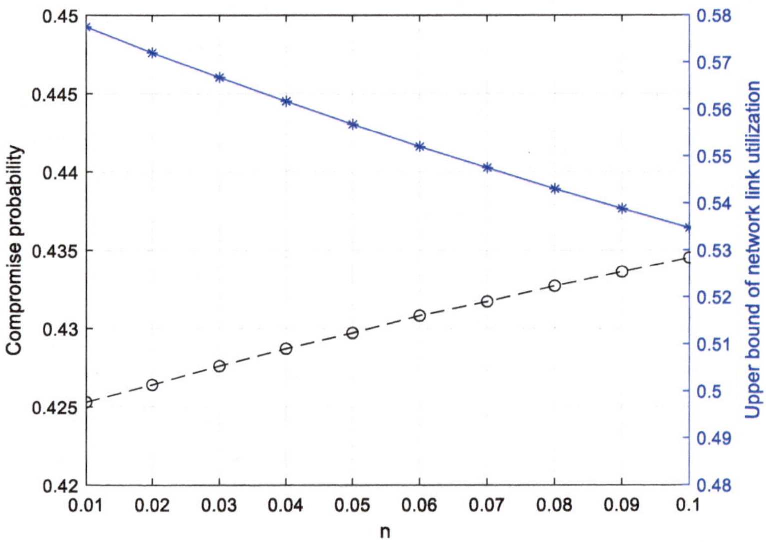

Fig. 6. Dynamics of change the utilization and network security indicators depending on parameter n values in the model (10).

packet compromise probability (from 17.65% to 19.4%) with an increase (from 68% to 81.5%) in the upper bound of network link utilization (Fig. 7).

In the study of the second scenario of communication link compromise (Table 1), the dependence (10) at $n = 1$ was chosen for the link blocking model. With the change in the compromise scenario, the use of the TE model provided a value p_{E2E} of 0.7933. Table 7 shows the comparison results of the studied solutions: TE and SecTE (10) models. The increase in the control parameters r and b values in expression (10) provided a more

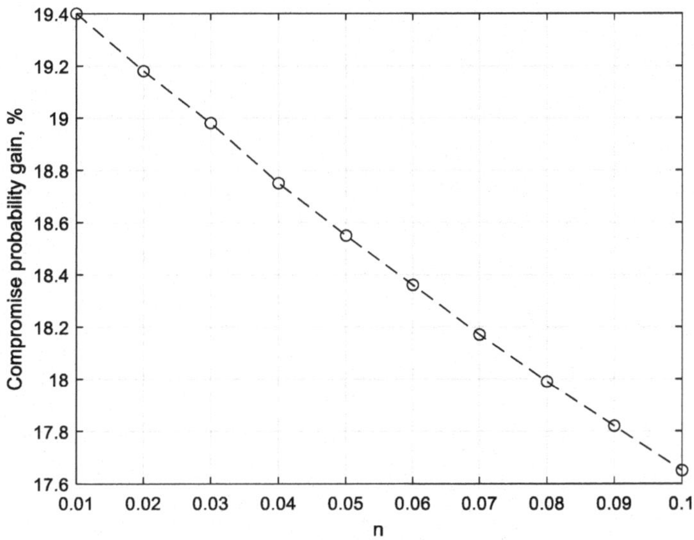

Fig. 7. Dynamics of compromise probability gain depending on parameter n values in the model (10).

intensive consideration of the link compromise probabilities under load balancing in the network.

Table 7. Comparison results of obtained solutions for secure load balancing utilizing the link blocking model (10).

r	b	α^*	α	p_{E2E}	p_{E2E} decrease
14	7	0.5444	0.6787	0.7116	10.2954%
20	10	0.5717	0.649	0.7077	10.7897%
30	15	0.5818	0.6108	0.7062	10.9742%
40	20	0.5831	0.5938	0.706	10.998%

For the same second compromise scenario, Table 8 shows the comparison results of the studied solutions: TE and SecTE (11) model at $\theta = 0$. With an increase in n from 0 to 0.15, more intensive consideration of the link compromise probabilities was provided, and the improvement (gain) of the packet compromise probability in the network ranged from 6.1413% to 9.0311% (Table 8).

In accordance with the content of Table 8, Fig. 8 and Fig. 9 show the dynamics of changes in the link utilization and network security indicators depending on the values of the parameter n in the model (11). By changing the parameter n, it is possible to reduce (from 6.1413% to 9.0311%) the packet compromise probability, but with increasing (from 42.6% to 57%) the upper bound of network link utilization (Fig. 9).

Table 8. Comparison results of obtained solutions for secure load balancing utilizing the link blocking model (11).

n	α^*	α	p_{E2E}	p_{E2E} decrease
0	0.4537	0.6481	0.7446	6.1413%
0.05	0.4945	0.6615	0.7313	7.8123%
0.1	0.5	0.6754	0.7259	8.4895%
0.15	0.5	0.6899	0.7216	9.0311%

In the study of the third scenario of communication link compromise (Table 1), the use of the TE model provided a value of p_{E2E} equal to 0.9491. The use of the link blocking model (11), when $\theta = \pi$, provided a reduction in the packet compromise probability in the network from 2.8374% to 3.7596%. The value of the control parameter n varied from 1.5 to 0.08. This small gain in packet compromise probability was due to the lack of low-compromise communication links in the third scenario (Table 1), which could be used to unload the most vulnerable network links.

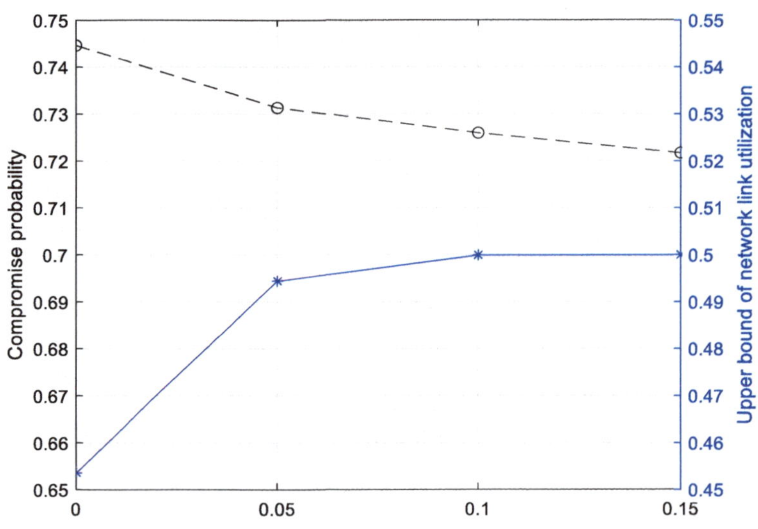

Fig. 8. Dynamics of change the utilization and network security indicators depending on parameter n values in the model (11).

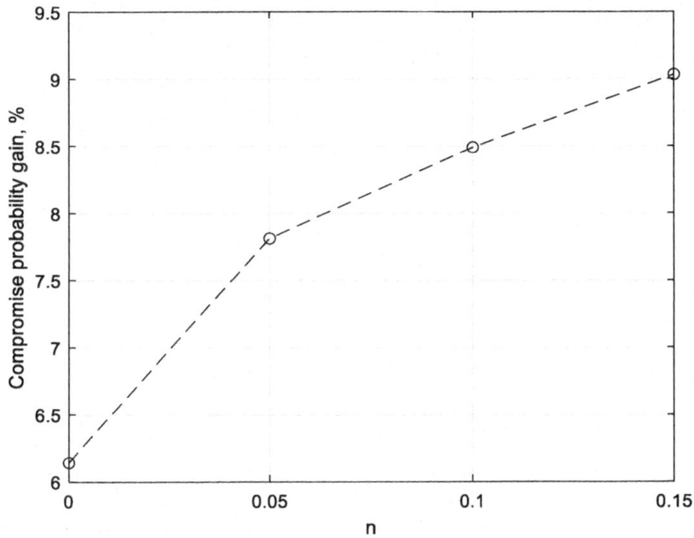

Fig. 9. Dynamics of compromise probability gain depending on parameter n values in the model (11).

5 Conclusion

Promising areas of development and enhancement of network security solutions are the improvement of traffic management and routing means. The latest solutions in the field of traffic management and routing should take into account not only parameters of network performance (bandwidth, latency, and packet loss) but also network security parameters that characterize the effectiveness of networked intrusion detection, as well as vulnerability and risk analysis.

A comprehensive study of the advanced flow-based secure routing model (1)–(8) with load balancing in accordance with the concept of Traffic Engineering based on parameters of network security in software-defined telecommunication networks has been conducted. Within this model, the solution of the technological task of secure routing with load balancing in the network was reduced to solving the optimization problem with optimality criterion (9) and constraints (1)–(8). In the implementation of multipath routing (1), the formulated optimization problem belongs to the class of Linear Programming (LP) problems.

The novelty of the studied model includes:

- modification of load balancing conditions in the network (5) that focus on minimizing the upper dynamically controlled bound of the network links utilization (9), weighted relative to the compromise probability;
- application of communication link blocking models (10) and (11) that can be used to adjust the influence of the links' compromise probabilities $p_{i,j}$ on the bound of their utilization $\alpha_{i,j}$ and the network in general.

In the process of investigating the proposed link blocking models (10) and (11), the nature of the influence of their control parameters on the sensitivity of load balancing processes and network security parameters represented as the link compromise probabilities. The reduction in the packet compromise probability transmitted by the network was provided, as a rule, by increasing the bound of the network link utilization that negatively affected the QoS level. Therefore, in each case, it is necessary to take into account the network state, predicted scenarios of compromising its elements, and packet flow requirements to the Quality of Service and network security level to choose the most appropriate link blocking model with corresponding control parameters.

References

1. Gupta, S.: Security and QoS in Wireless Sensor Networks. 1st edn. eBooks2go Inc. (2018)
2. Kiser, Q.: Computer Networking and Cybersecurity: A Guide to Understanding Communications Systems, Internet Connections, and Network Security Along with Protection from Hacking and Cyber Security Threats. Kindle Edition (2020)
3. Revathi, S., Geetha, A.: A survey of applications and security issues in software defined networking. Int. J. Comput. Network Inf. Secur. (IJCNIS) **9**(3), 21–28 (2017). https://doi.org/10.5815/ijcnis.2017.03.03
4. Yeremenko, O., Lemeshko, O., Persikov, A.: Secure routing in reliable networks: proactive and reactive approach. In: Shakhovska, N., Stepashko, V. (eds.) CSIT 2017. AISC, vol. 689, pp. 631–655. Springer, Cham (2018). https://doi.org/10.1007/978-3-319-70581-1_44
5. Lemeshko, O., Yeremenko, O., Yevdokymenko, M., Shapovalova, A., Radivilova, T., Ageyev, D.: Secure based traffic engineering model in softwarized networks. In: 2020 IEEE 2nd International Conference on Advanced Trends in Information Theory (ATIT) Proceedings, pp. 143–147. IEEE (2020). https://doi.org/10.1109/ATIT50783.2020.9349301
6. Lemeshko, O., Yeremenko, O., Shapovalova, A., Hailan, A.M., Yevdokymenko, M., Persikov, M.: Design and research of the model for secure traffic engineering fast ReRoute under traffic policing approach. In: 2021 IEEE 16th International Conference on the Experience of Designing and Application of CAD Systems (CADSM) Proceedings, pp. 23–26. IEEE (2021). https://doi.org/10.1109/CADSM52681.2021.9385253
7. Yeremenko, O.: Enhanced flow-based model of multipath routing with overlapping by nodes paths. In: 2015 Second International Scientific-Practical Conference Problems of Infocommunications Science and Technology (PIC S&T) Proceedings, pp. 42–45. IEEE (2015). https://doi.org/10.1109/INFOCOMMST.2015.7357264
8. Lemeshko, O., Yeremenko, O.: Linear optimization model of MPLS traffic engineering fast ReRoute for link, node, and bandwidth protection. In: 14th International Conference on Advanced Trends in Radioelectronics, Telecommunications and Computer Engineering (TCSET), pp. 1009–1013. IEEE (2018). https://doi.org/10.1109/TCSET.2018.8336365
9. Yeremenko, O., Tariki, N., Hailan, A.M.: Fault-tolerant IP routing flow-based model. In: 2016 13th International Conference on Modern Problems of Radio Engineering, Telecommunications and Computer Science (TCSET) Proceedings, pp. 655–657. IEEE (2016). https://doi.org/10.1109/TCSET.2016.7452143
10. Shaik, M.S., Mira, F.: A comprehensive mechanism of MANET network layer based security attack prevention. Int. J. Wireless Microwave Technol. (IJWMT) **10**(1), 38–47 (2020). https://doi.org/10.5815/ijwmt.2020.01.04
11. Shashi, R.K., Siddesh, G.K.: QoS oriented cross-synch routing protocol for event driven, mission-critical communication over MANET: Q-CSRPM. Int. J. Comput. Network Inf. Secur. (IJCNIS) **10**(11), 18–30 (2018). https://doi.org/10.5815/ijcnis.2018.11.03

12. Palani, U., Amuthavalli, G., Alamelumangai, V.: Secure and load-balanced routing protocol in wireless sensor network or disaster management. IET Inf. Secur. **14**(5), 513–520 (2020). https://doi.org/10.1049/iet-ifs.2018.5057
13. Patil, M. V., Jadhav, V.: Secure, reliable and load balanced routing protocols for multihop wireless networks. In: 2017 International Conference on Intelligent Computing and Control (I2C2) Proceedings, pp. 1–6. IEEE (2017). https://doi.org/10.1109/I2C2.2017.8321936
14. Kumar, N., Singh, Y.: Trust and packet load balancing based secure opportunistic routing protocol for WSN. In: 2017 4th International Conference on Signal Processing, Computing and Control (ISPCC) Proceedings, pp. 463–467. IEEE (2017). https://doi.org/10.1109/ISPCC.2017.8269723
15. Medhi, D., Ramasamy, K.: Network Routing: Algorithms, Protocols, and Architectures. 2nd edn. Morgan Kaufmann (2017)
16. Govindasamy, J., Punniakody, S.: A comparative study of reactive, proactive and hybrid routing protocol in wireless sensor network under wormhole attack. Electr. Syst. Inf. Technol. **5**(3), 735–744 (2018). https://doi.org/10.1016/j.jesit.2017.02.002
17. Wadhwani, G.K., Khatri, S.K., Muttoo, S.K.: Critical evaluation of secure routing protocols for MANET. In: 2018 International Conference on Advances in Computing, Communication Control and Networking (ICACCCN) Proceedings, pp. 202–206. IEEE (2018). https://doi.org/10.1109/ICACCCN.2018.8748725
18. Shashikala, R., Kavitha, C.: Secured data integrity routing for Wireless Sensor Networks. In: 2014 International Conference on Advances in Electronics Computers and Communications Proceedings, pp. 1–6. IEEE (2014). https://doi.org/10.1109/ICAECC.2014.7002419
19. Khan, S., Khan, S., Loo, J.: Cross layer secure and resource-aware on-demand routing protocol for hybrid wireless mesh networks. Wireless Pers. Commun. **62**, 201–214 (2012). https://doi.org/10.1007/s11277-010-0048-y
20. Aggarwal, A., Gandhi, S., Chaubey, N.: Trust based secure on demand routing protocol (TSDRP) for MANETs. In: 2014 Fourth International Conference on Advanced Computing & Communication Technologies Proceedings, pp. 432–438. IEEE (2014). https://doi.org/10.1109/ACCT.2014.95
21. Gu, Q., Tilborg, H.C.A., Jajodia, S.: Secure Routing Protocols, Encyclopedia of Cryptography and Security. Springer, Boston (2011). https://doi.org/10.1007/978-1-4419-5906-5_641
22. Diwan, D., Narang, V.K., Singh, A.K.: Security mechanism in RIPv2, EIGRP and OSPF for campus network. Comput. Sci. Trends Technol. **5**(2), 399–404 (2017)
23. Snihurov, A., Chakrian, V.: Improvement of EIGRP protocol routing algorithm based on information security metrics. In: 2015 Second International Scientific-Practical Conference Problems of Infocommunications Science and Technology (PIC S&T) Proceedings, pp. 263–265. IEEE (2015). https://doi.org/10.1109/INFOCOMMST.2015.7357331
24. Bhatia, M., Hartman, S., Zhang, D.: Security Extension for OSPFv2 When Using Manual Key Management, RFC 7474, 2015. https://tools.ietf.org/html/rfc7474
25. Wang, M., Liu, J., Mao, J., Cheng, H., Chen, J., Qi, C.: Route guardian: constructing secure routing paths in software-defined networking. Tsinghua Sci. Technol. **22**(4), 400–412 (2017). https://doi.org/10.23919/TST.2017.7986943
26. Li, J., Yang, Z., Yi, X., Hong, T., Wang, X.: A secure routing mechanism for industrial wireless networks based on SDN. In: 2018 14th International Conference on Mobile Ad-Hoc and Sensor Networks (MSN) Proceedings, pp. 158–164. IEEE (2018). https://doi.org/10.1109/MSN.2018.000-2
27. Ellinidou, S., Sharma, G., Rigas, T., Vanspouwen, T., Markowitch, O., Dricot, J.M.: SSPSoC: a secure SDN-based protocol over MPSoC. Secur. Commun. Networks **2019**, 1–12 (2019). https://doi.org/10.1155/2019/4869167

28. Sagare, A.A., Khondoker, R.: Security analysis of SDN routing applications. In: Khondoker, R. (ed.) SDN and NFV Security. LNNS, vol. 30, pp. 1–17. Springer, Cham (2018). https://doi.org/10.1007/978-3-319-71761-6_1
29. Lemeshko, O.V., Yevseyeva, O.Yu., Garkusha, S.V.: A tensor model of multipath routing based on multiple QoS metrics. In: 2013 International Siberian Conference on Control and Communications (SIBCON) Proceedings, pp. 1–4. IEEE (2013). https://doi.org/10.1109/SIBCON.2013.6693645
30. Lemeshko, O.V., Garkusha, S.V., Yeremenko, O.S., Hailan, A.M.: Policy-based QoS management model for multiservice networks. In: 2015 International Siberian Conference on Control and Communications (SIBCON) Proceedings, pp. 1–4. IEEE (2015). https://doi.org/10.1109/SIBCON.2015.7147124

Intelligent Traffic Engineering for Future Intent-Based Software-Defined Transport Network

Volodymyr Andrushchak[1](✉), Mykola Beshley[1], Lyubomyr Dutko[2], Taras Maksymyuk[1], and Taras Andrukhiv[1]

[1] Lviv Polytechnic National University, Lviv 79013, Ukraine
{volodymyr.s.andrushchak,mykola.i.beshlei, taras.a.maksymiuk}@lpnu.ua
[2] Lemberg Solutions LLC, Lviv 7900, Ukraine
lou@lemberg.co.uk

Abstract. This chapter addresses Traffic Engineering (TE) issues in future software-defined infrastructures using machine learning (ML) and neural networks. The Software-Defined Networks (SDN) architecture can be used to implement Intent-Based Networking (IBN) that enables the automation of network management tasks through elements of artificial intelligence (AI) and ML. The intent-based optical transport network infrastructure is proposed, adapted to the use of intelligent TE algorithms based on SDN and Optical Label Switching (OLS) technology. An algorithm for determining Intent-Based Software-Defined Transport Network (IBSDTN) states based on ML algorithms k-means and c-means is proposed. This algorithm allows the provision of an appropriate set of network parameters for training the appropriate control algorithms. A method of intelligent TE using graph neural networks to provide the necessary quality of service (QoS) parameters based on users intention during peak hours has been developed. This algorithm using the vector of network parameters, which also takes into account the parameter of energy consumption, manages network resources to provide the necessary QoS parameters.

Keywords: Software-Defined Networks · Intent-Based Networking · Intent-Based Software-Defined Transport network · Machine learning · Neural network · Traffic engineering · Graph neural network

1 Introduction

Telecommunication networks have faced the challenge of a large amount of traffic generated by various services. Universal availability and the low price of telecommunication services will only contribute to this. Thus, telecommunications operators must develop and maintain the necessary infrastructure of the optical transport network, which is the core where the bulk of traffic is transmitted [1]. Technologies and algorithms of the channel level are the primary means of efficient traffic transmission in transport networks. Any link-level technology faces typical problems of optical transport networks [2]:

M. Klymash et al. (Eds.): *Future Intent-Based Networking*, LNEE 831, pp. 161–181, 2022.
https://doi.org/10.1007/978-3-030-92435-5_9

- growth of network traffic;
- increasing the data rate in optical channels;
- evolution of WDM systems;
- network scaling;
- reduction of energy consumption of networks
- support for outdated software and hardware
- unification of services and more

Simpler blocking models have been presented in [3], which may explain the operation of more complex models. For example, the link blocking model can be described as described [4]. Accordingly, each channel layer technology uses its own or reuses existing algorithms adapted to other technologies to optimize network performance. However, each link-level technology has its characteristics. This leads to the fact that any algorithm for maximizing network resources will be limited by the configuration of the network and its technologies. Thus, it makes sense to develop more unified algorithms that would be suitable for use by different technologies of the channel layer of optical transport networks. To solve this problem, you can use algorithms based on neural networks, which will be more adaptable to changing network parameters. The domains of use of such algorithms are presented in Fig. 1.

Recently, the use of graph neural networks (GNN) is gaining popularity. A feature of these neural networks is using the graph as a contiguity matrix and data on the telecommunications network in the form of a vector of network parameters. Since a graph and many network parameters represent the telecommunication network are known, which describe the node and the communication channel, it is rational to use these neural networks to solve certain telecommunication problems.

The use of neural networks and machine learning tools in telecommunications have several disadvantages and advantages [5]. There is a high cost of error if a neural network algorithm makes the wrong decision, leading to malfunction or complete shutdown of the optical transport network [6]. However, the optical transport network transmits gigabits of information over optical channels per unit time, which provides the required number of datasets to train the appropriate models to solve this problem [7].

The use of machine learning without supervision allows you to catch atypical cases that are difficult to detect static algorithms. Such algorithms can capture many more atypical events in the network and can benefit from decisions made at the micro-level of the use of optical, time, and energy resources of the network.

The use of algorithms based on machine learning (ML) and artificial intelligence (AI) can be divided into categories of data transmission and control (see Fig. 1). For example, neural network-based algorithms at network nodes can make "recommendations" for switching and aggregating traffic. The optical transport network control plane aims to provide the necessary QoS for the relevant services. On the other hand, this plane has all the necessary information about the network, "knows" about all the system's parameters, its load at the current time, and the current configuration and architecture of the network. We can assume that the network is an autonomous system because it "owns" the necessary information and can manage itself. ITU-T refers to such networks as self-optimized networks (SON), which can respond to certain events and make the necessary decisions withot human interaction to ensure the normal operation of the network.

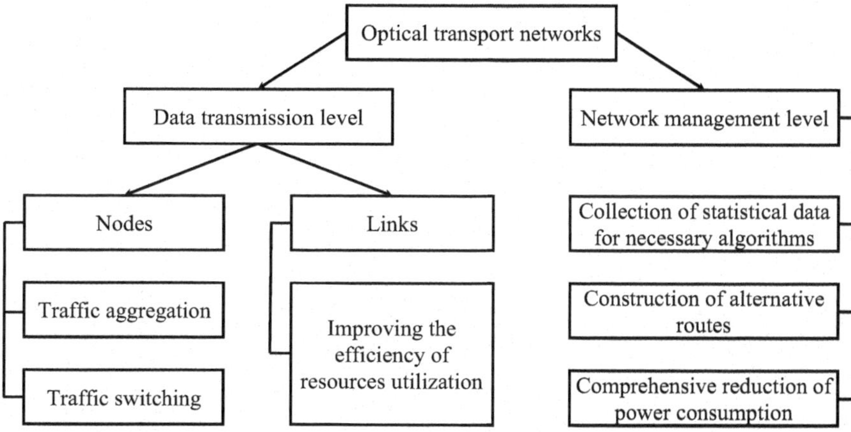

Fig. 1. Telecommunication domains for algorithms based on machine learning and artificial intelligence

In our previous works [8], we developed an algorithm for aggregating IP packets in transport modules at border nodes using neural networks, which allowed us to differ from static algorithms [9] to move away from clear aggregation conditions and more dynamically manage traffic. In addition, it was proposed to use neural networks for more intelligent switching of traffic at intermediate nodes of the optical transport network.

In general, neural network-based algorithms should not play a significant role in network management but only an ancillary one where static algorithms or a person cannot cope with sufficient speed and quality. The emergence of software-defined networks (SDN) and intent-based networks (IBN) is fundamentally changing the design of telecommunication systems [10]. The IBN is designed to solve the limitations of SDN [11]. These solutions provide connectivity to the Application Programming Interfaces (APIs) of network devices. As a result, network engineers can more simply manage, deploy and troubleshoot network equipment [12–15]. The IBN makes network programming easier by improving network automation and increasing abstraction. With this approach, companies create, implement and increase the flexibility of the network. IBN includes four elements: translation and verification; automated implementation; provisioning; awareness of network state; and dynamic self-optimization [16–18]. An IBN performs the following basic functions:

- takes input from the network engineer;
- provides network design based on enterprise intent;
- verifies that the design is correct;
- deploys network configuration;
- continuously ensures that system goal is met;
- makes changes as needed.

The IBN is still in its early stages and requires complex integration of the SDN controller, as well as some degrees of artificial intelligence, machine learning and open-source software to make it operate as required [19–21].

2 Intent-Based Software-defined Transport Network Architecture Based on Neural Networks and Machine Learning

The chapter proposes the use of intelligent algorithms based on trained models of artificial intelligence, which will be used at the channel level (transmission level) of OLS (Optical Label Switching) technology, as well as at the level of the SDN control plane (Fig. 2). The work of these algorithms focuses on the management of info-communication flows and optimizing the use of network resources. For example, neural network-based algorithms at boundary nodes can perform load aggregation to better use the spectral resources of a communication channel. Intelligent control algorithms on the SDN/IBN controller can perform network-wide optimization. However, for such algorithms that operate at the data transmission and control levels, they must have appropriate infrastructure for data collection, training, testing, and updating of relevant trained models. Many works present single algorithms based on neural networks that optimize a particular process or part of the network [3, 22, 23]. However, they do not provide data collection, neural network training, or software updates. It is not clear how a complete feedback infrastructure should work for algorithms that optimize the operation of the transport network using neural networks.

The infrastructure of such a network consists of the following elements: domain of ML algorithms on SDN/IBN controller, domain (cloud) of neural network training, the isolated domain of neural network testing, communication channels for updating neural networks and corresponding software, domain/module of algorithms using neural networks at network nodes.

The details of such an infrastructure are determined by the types of optimization algorithms used and the required accuracy of the algorithms, and the system's reliability. In the presented architecture (Fig. 2), to reduce the probability of incorrect training of models, it is proposed to use an additional cloud for testing algorithms based on neural networks - checking key metrics. This example shows how infrastructure requirements are formed using intelligent and adaptive algorithms.

First of all, we will highlight the elements of the infrastructure of data collection and model training for algorithms that are integrated into the optical transport network.

- **SRC (source)** is a node that is the source of information for machine learning algorithms and neural networks. In the proposed architecture, the source of information is the boundary and intermediate nodes and the own equipment, which operates at the data link layer;
- **C (collector)** this node is responsible for collecting data from one or more SRC nodes. In this case, this task is performed by the SDN controller;
- **PP (preprocessor)** this node is responsible for cleaning, aggregating, and performing other data processing operations. In this case, the role of this node is also performed by the corresponding software on the SDN controller;
- **M (model)** is directly a model of machine learning or neural networks;
- **P (policy)** is a node or software that describes a policy for using source models;
- **D (distributor)** is a node that determines which SRC to provide the source information from the machine learning algorithm. In this architecture, it is also an SDN controller;
- **SINK** is a node that is the target for the output of the machine learning algorithm. In the proposed architecture, it is the equipment of the channel level of OLS technology.

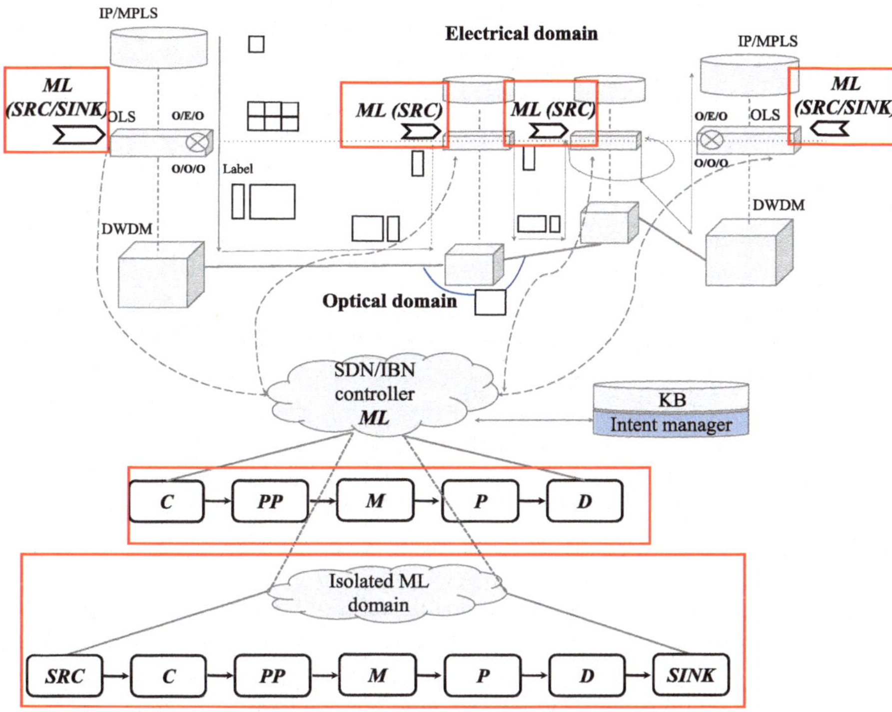

Fig. 2. Intent-Based Software-Defined Transport Network (IBSDTN) Architecture (IP/MPLS is a network layer router, OLS (equipment) is a device channel level aggregation and switching traffic, DWDM is a equipment physical layer multiplexing of optical wavelengths, O/O/O is optical domain, O/E/O is a electrical domain)

- **An intent manager** is software that parses the user's input plain text intentions into a json file using some high-level language and regular expressions. This parsing process can be improved using ML methods to permit the user to use multiple dictionary variants. The intent manager reading the intent repository from the knowledge base (KB) and trying to correlate existing vocabularies and structures with the input text. Lastly, a json file with user specifications is created and sent to the policy configurator into the SDN/IBN controller for realized intelligent TE.

In the presented architecture, it is proposed to use an isolated domain, which allows debugging intelligent control algorithms, namely the processes of preparation, testing, and evaluation before their deployment in the network. Simulated data or data from a real transport optical network can be used to teach or test such algorithms. This approach provides better preparation for the launch of such algorithms on real networks and reduces the associated risks (Fig. 3). To evaluate such algorithms, certain network metrics should be introduced, on the basis of which these algorithms will be evaluated. The isolated domain should include a simulation model of the optical transport network with appropriate components for maximum proximity to the real network. That is, such

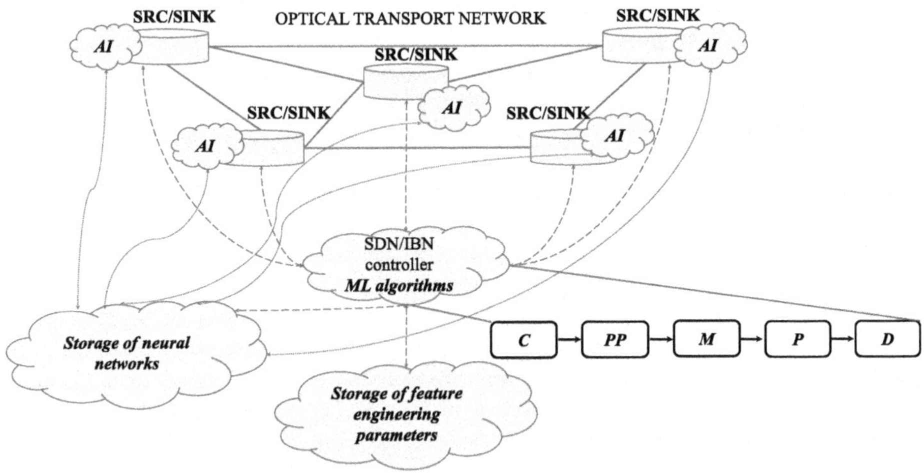

Fig. 3. Block diagram of the infrastructure of ML algorithms (NM is a neural network) for IBSDTN

things as features of data transmission protocols, rules of processing of service information, formation of data blocks, and other things should be considered. The metrics for evaluating the performance of an overtrained neural network should be the usual network parameters: end-to-end latency, energy consumption parameter, the efficiency of use of spectral resources of communication channels, CPU utilization of Layer 3 switches, etc. If the overtrained model for a particular algorithm has led to the deterioration of network parameters, this model should be returned for training or removed from the cloud storage.

Figure 3 presents a block diagram of the developed architecture of the IBSDTN with an isolated domain. For intelligent TE algorithms to give the correct result, it is necessary to collect a sufficient set of data. A sufficient data set means the optimal amount of data at which the training of models is considered complete and the process of overfitting is not observed. The optical transport network transmits huge amounts of data per unit of time, so there is more than enough data for intelligent TE algorithms.

The isolated domain performs training of the corresponding models of the related intelligent algorithms. Once the isolated domain has verified that the corresponding model ensures the operation of the corresponding algorithm, which provides the necessary parameters of quality of service, the corresponding model is loaded into the appropriate storage of models. If the model does not meet the required quality of service parameters, then the necessary network parameters are collected to overtrain the model.

3 IBSDTN State Detection Algorithm Based on ML K-Means and C-Means

To collect network data, an SDN/IBN controller is used, which according to the OpenFlow protocol, directly carries out collecting the relevant data from the nodes of the optical transport network. Each node of the network is both a source of information for

ML algorithms and can be the purpose of their application (SRC/SINK, respectively). The main data processing before training neural networks is carried out on the SDN/IBN controller. In addition, each node performs a small amount of data processing to aggregate the traffic that will be transmitted in OpenFlow packets to the SDN/IBN controller. All the collected network parameters that are needed to train neural networks are stored in the FE cloud. Only intelligent TE algorithms on the SDN/IBN controller have access to these parameters. Suppose there is a change in the state of the optical transport network, which requires retraining of the relevant models. In that case, the relevant algorithms perform the appropriate procedure of retraining them based on the new FE parameters. After that, the corresponding trained models in the cloud are replaced. When a new version of the model for the corresponding node appears, the node downloads the corresponding updated version of the model.

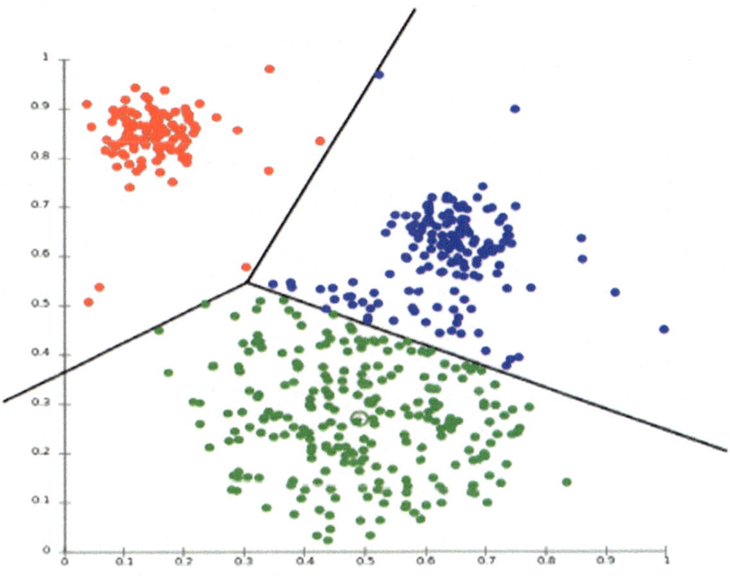

Fig. 4. Example of clustering by the k-means algorithm

The chapter uses a cluster approach to determine the states of the optical transport network with learning without supervision using ML algorithms k-means and c-means for the developed algorithm for data collection. This approach makes it possible to determine the necessary network states or the emergence of new states based on a certain period of time. In addition, this approach allows more network parameters to be taken into account with minimal changes to the software. The mathematical apparatus of the ML algorithms k-means and c-means should be briefly considered.

To implement the k-means algorithm, the data is divided into p where $(p = 1,.. k)$ is the number of clusters. Let the collected data be divided into p clusters determined by the centroid c_j where $(j = 1,.. p)$. Clustering is based on the calculation of the Euclidean distance. Given two different sets of clusters formed by two different runs of k-means, we prefer the one with the smallest error squared, because this means that the prototypes (centroids) of this clustering are the best representation of points in their cluster:

$$d = \sum_{j=1}^{k} \sum_{i=1}^{n} |x_i - c_i|^2, \tag{1}$$

$$d = \min_{j} \sum_{i=1}^{n} |x_i - c_i|^2, \tag{2}$$

where k is the number of clusters, n is the amount of data, x_i is the case i, c is the centroid for the cluster j.

Now consider the features of the c-means algorithm. This method of clustering, which involves minimizing a specific objective function [24]. When an algorithm can minimize the error function [25], it is often called c-means, which is the c - number of classes or clusters, and if the classes used to use a fuzzy technique – fuzzy c-means (FCM). The FCM approach uses fuzzy membership (probability), which determines the degree of belonging for each class. The importance of the degree of probability [26] in fuzzy clustering is similar to the pixel probability of a certain image. The advantage of FCM is the formation of new clusters from a data point that have close membership values to existing classes [27]. The FCM method has three main operators: the fuzzy membership (probability) function, the partition matrix, and the objective function.

Consider a set of n vectors for clustering into c groups:

$$(X = (x_1, x_2, \ldots, x_n) 2 \leq c \leq n). \tag{3}$$

Each vector $x_i \in R^s$ described with real dimensions that represent the features of the object x_i. A probability matrix known as a fuzzy partition matrix is used to describe a fuzzy membership matrix. Set of fuzzy matrices $(c \times n)$ is denoted by $\mathrm{M_{fc}}$ and is determined by the following relation (4):

$$M_{fc} = \{W \in R^{cn} | w_{ik} \in [0, 1], \forall i, k; \sum_{i=1}^{c} w_{ik} = 1, \forall k; 0 < \sum_{k=1}^{n} w_{ik} < n, \forall i\}. \tag{4}$$

where $1 \leq i \leq c, 1 \leq k \leq n$.

From the above definitions, it can be found that elements can fit into more than one cluster with different degrees of membership (probability). The total "membership" of an element is normalized to 1, and one cluster cannot contain all data points. The objective function of the fuzzy c-value algorithm is calculated using the probability value and the Euclidean distance (5–6).

$$J_m(W, P) = \sum_{1 \leq k \leq n} (w_{ik})^m (d_{ik})^2, \tag{5}$$

where

$$d_{ik} = |x_k - p_i|, \tag{6}$$

where $m \in (1, +\infty)$ parameter, which determines the fuzzyness of the resulting clusters, and d_{ik} is the euclidean distance from the object xk to the center of the cluster p_i.

Minimization [28] of the objective function J_m through the FCM algorithm is performed by iteratively updating the matrices w using the equations:

$$p_i = \sum_{k=1}^{n} (w_{ik})^m x_k / \sum_{k=1}^{n} (w_{ik})^m, \tag{7}$$

$$w_{ik}^{(b)} = \sum_{j=1}^{c} 1/\left[\left(d_{ik}^{(b)}/d_{jk}^{(b)}\right)^{\frac{2}{m}-1}\right]. \tag{8}$$

FCM the probability function is determined by the formula:

$$\mu_{ij} = \left[\sum_{t=1}^{c}\left(\frac{||x_j - v_i||_A}{||x_j - v_t||_A}\right)^{\frac{2}{m-1}}\right]^{-1}, \tag{9}$$

where $\mu_{i,j}$ is the probability value of the j-th value and the i-th cluster. The number of clusters is represented by c, x_j is the j-th value and vi is the centroid of the i-th cluster.

An important component of intelligent control algorithms is the collection of data for training. One of the features of the use of these algorithms in telecommunications networks is the variability of states in the network and the emergence of new and disappearance of current states, which requires additional data collection and retraining of neural networks.

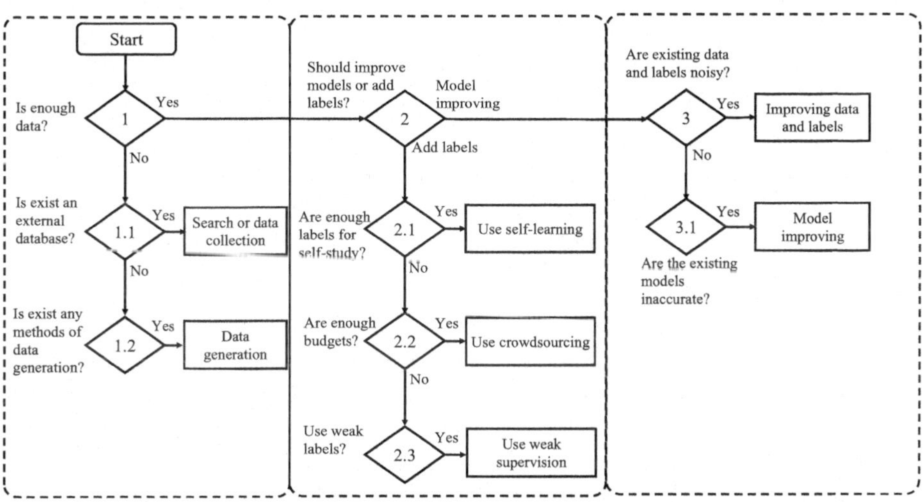

Fig. 5. Block diagram of a data collection algorithm for ML algorithms

In [29], algorithms and techniques of data collection, labeling and improvement are presented (Fig. 5). This algorithm consists of three parts:

- validity of data adequacy for training;

- improving existing data;
- improving existing models.

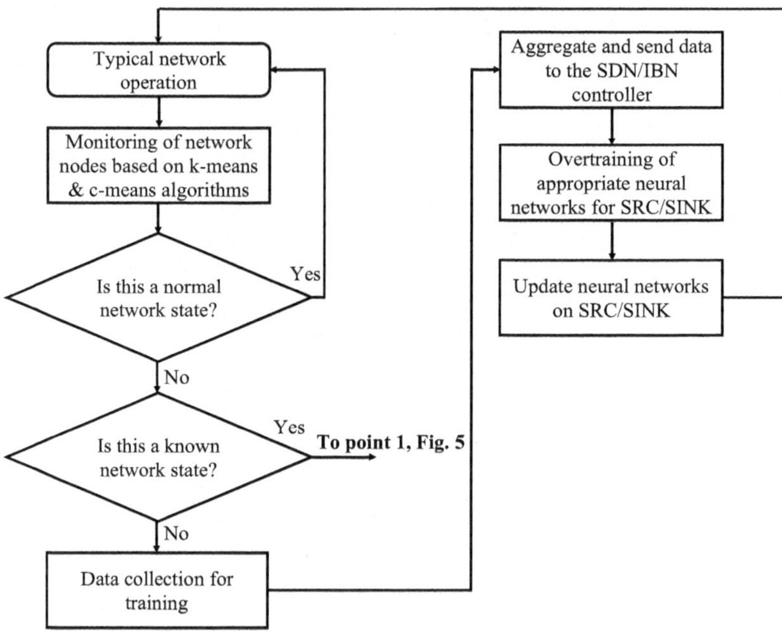

Fig. 6. Block diagram of a network state tracking algorithm for collecting and retraining the corresponding neural networks

The developed algorithm of data collection provides tracking of a condition of a network for rational data collection actually with the use of change of values of Euclidean distances in relation to a number of clusters. Thus, the algorithm monitors the state of each node. Consider the operation of the algorithm, the block diagram of which is presented in Fig. 6:

- the algorithm collects from each node of the SRC optical transport network;
- using the ML algorithm k-means, it is determined to which data cluster this state of the network belongs, and the ML algorithm c-means determines the probabilities of being in certain clusters;
- if the state of the network is known and belongs to the corresponding cluster, then accordingly the algorithm continues to work in the direction of the algorithm presented in Fig. 5;
- if the state of the network is unknown, then aggregate data is collected from the corresponding node or section of the network, which will be used for the corresponding neural networks and sent to the SDN/IBN controller;
- further, on the basis of the corresponding network data, there is a retraining of the corresponding neural networks for SINK nodes of the investigated network;

- after the overtraining of the models of the corresponding intelligent control algorithms is completed, and the test metrics in the isolated domain allow to determine that the quality of service will not deteriorate, the model is loaded into the storage cloud.

The developed algorithm for collecting and determining network states can work for both access networks and optical transport networks. This algorithm is simple in software implementation and can provide the required amount of network data for the corresponding TE algorithms.

4 An Intelligent Traffic Engineering Method Based on Graph Neural Networks for IBSDTN

The novelty of the developed intelligent TE method is that one of the FE parameters of the neural network is the energy consumption parameter, which allows to control information flows, balancing between energy consumption and latency parameter.

Graphic neural networks (GNNs) are a type of neural network that works directly with a data structure such as a graph. Each node in the column is associated with a training label, and, accordingly, the network tries to "predict" the desired node based on input data. GNN is chosen as a key control because the telecommunications network is represented as a graph to display the logical and physical connections between nodes [30]. In addition, the input parameters for such a neural network are the following elements:

- adjacency matrix, which reflects the network topology;
- network parameters of network nodes during time *dt*;
- network parameters of communication channels during time *dt*.

The purpose of GNN is to study the state of the node h_v, taking into account information about the state of neighboring nodes [30]. In the context of GNN, the state of the node v is described by a data vector of size s to generate the original data o_v. Assume that f is a parametric function that is passed through each node and updates the state of the node according to the input data. Assume that g is a local output function that describes the generation of the source data. Respectively, h_v and o_v are defined as follows:

$$h_v = f\left(x_{v}, x_{co[v]}, h_{ne[v]}, x_{ne[v]}\right) \tag{10}$$

$$o_v = g(h_v, x_v), \tag{11}$$

where x_v is the node characteristics, $x_{co[v]}$ is the link characteristics, $h_{ne[v]}, x_{ne[v]}$ are the state and characteristics of neighboring nodes, respectively.

Suppose that *H, O, X,* and X_N vectors, which are constructed by adding all the states, source data, characteristics and characteristics of all nodes, respectively, which can also be represented in the following forms:

$$H = F(H, X), \tag{12}$$

$$O = G(H, X_N), \tag{13}$$

where F is the global transition function, G is the global output function, which consists of multiply functions f and g for all nodes of the graph, respectively.

The next stage is the training of neural networks, ie learning the parameters of functions f and g. Based on target information (t_v for a specific node) for training with a supervisor, the losses l can be described as follows:

$$l = \sum_{i=1}^{p} (t_i - o_i), \tag{14}$$

where p is the number of nodes.

Algorithms using GNN can be used for the following purposes:

- optimization of global routing in the transport network;
- aggregation of traffic at border nodes;
- switching traffic at intermediate nodes;
- reduction of energy consumption of the network as a whole.

The advantage of neural networks over conventional static algorithms is that we can change the desired characteristics of the network and nodes x_v Changes are made by retraining the neural network, not by making changes to the necessary software. Accordingly, nodes or other network elements can load only an overtrained model, not the whole algorithm, which in many intelligent control algorithms is a rational approach. This approach gives more flexibility in network management, as well as the ability to quickly make changes to the required software on a real network. The developed algorithm for managing info-communication flows can work both on individual domains of the info-communication network and centrally in traffic monitoring or management systems. Actually, this algorithm will be updated with overtrained GNN, which will be pre-trained and validated in an isolated domain.

As mentioned above, one of the parameters of the input FE of the vector GNN is the energy consumption parameter. The energy consumption parameter is the amount of electricity consumption to transmit one bit of information [31]. This parameter is not a priority but may show network bottlenecks in places where there is no linear relationship between the priority network parameters. For these studies, the developed method of determining the energy consumption parameter for optical transport networks was used [32, 33].

Figure 7 presents the investigated network topology for the developed method of TE using GNN. This topology is chosen because the ring topology is a popular solution for transport optical networks.

Nodes 1–3 are located in the city's residential area, 5–6 in the business part of the city, 4 - an intermediate node between the two areas. Node color determines the load of the node - a darker color determines the load of the current node. Data transmission via rings can be performed in two directions.

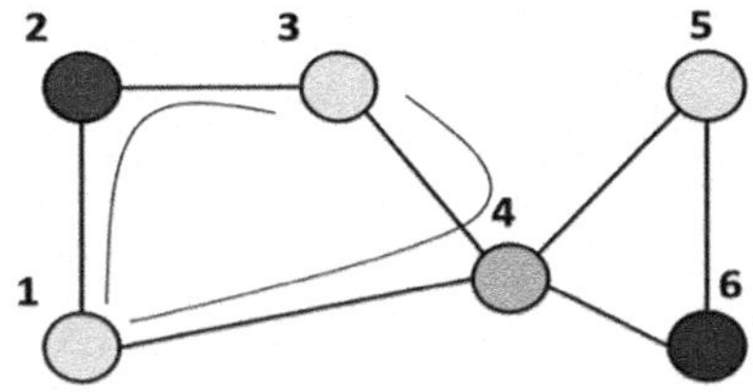

Fig. 7. Block diagram of a network state tracking method for collecting and retraining the corresponding neural networks

In Fig. 8 presents an algorithm for TE of the transport network, which performs modifications of routes on the basis of the input data of the vector FE - network parameters. Because this algorithm can run on an SDN/IBN controller, the relevant information about nodes and communication channels is transmitted to this controller for a certain time *dt*. In fact, such parameters are the dynamics of the distribution of IP packets (voice, video, M2M, etc.) for the node, CPU and RAM load, optical line path load, etc. As mentioned above, a feature of this algorithm is that as another value of the FE parameter, the energy consumption parameter is used. The energy consumption parameter is defined as the amount of energy consumed at the node or communication channel during time *dt*. This parameter is determined theoretically using the developed methodology based on the actual energy consumption of the devices. This is done in this way because the time *dt* can change. Hardware manufacturers do not provide the appropriate APIs to obtain current energy consumption.

Once the SDN/IBN controller has received the network parameters using the OpenFlow protocol, the network status data with the interval *dt*, the FE parameters are generated for GNN. There is also a stage of normalization of these FE parameters. The network adjacency matrix is known, and it is assumed that it does not change. Therefore, the input of the neural network receives three parameters:

- matrix of adjacencies of the presented network;
- FE nodes;
- FE lines.

With a sufficient set of data, GNN training takes place. If the quality of the trained GNN meets the required parameters of users intentions, then the neural network instance is deployed on the SDN/IBN controller according to the scheme of Fig. 3. The GNN labels are a specific state vector that reflects the actual state of the entire network. That is, on the one hand, the FE vector used to train the GNN reflects the state of the network at a certain point in time *dt*, and the label vector, on the other hand, reflects the effect on this state. In the developed algorithm, GNN helps to reduce the network latency parameter taking into account the energy consumption parameter.

Accordingly, the GNN returns a data vector for each node. If, for example, the original vector [0, 0, 0, 0, 0, 0] (the number of values corresponds to the number of nodes) has changed to [0, 1, 0, 0, 0, 0], it means that the state of the second node has changed and the routing should be adjusted through this node at a time when the original

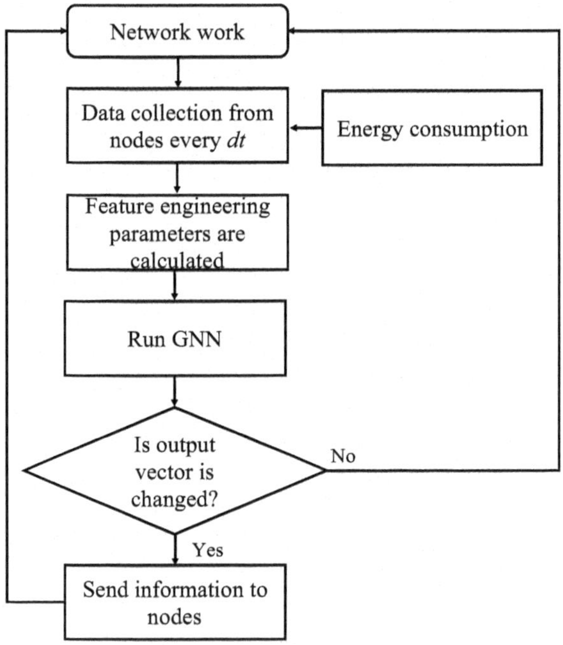

Fig. 8. Route modification algorithm in the IBSDTN

vector GNN does not change to [0, 0, 0, 0, 0, 0]. That is, the SDN/IBN controller notifies each node of the state of the network and, accordingly, the change of the output vector indicates a change of state for the recipient nodes.

5 The Development of IBSDTN Simulation Model

To study the presented methods and algorithms for TE, simulation software was developed, which is presented in Fig. 9. This model consists of the following parts:

- input simulation data;
- network modeling;
- elements of the method of TE using GNN;
- elements of the algorithm for aggregation of OLS transport blocks at the boundary node using deep neural networks.

The location of the simulation node and the time of day directly affect the dynamics of traffic distribution of the node. The network topology allows to form the routing of transmitted traffic and switching rules. In addition, the topology is important for the info-communication flow control algorithm for decision-making based on the node's location in the network.

Traffic in the network has a behavior that changes during the day, and its change depends on the location of telecommunications nodes in business or residential areas of the city. For example, in the business district, during the day, video and voice calls prevail, and there is a high activity of data transmission. At night in this area is dominated by static traffic from video cameras, security calls, etc. There are jumps in traffic during the shift, as well as during the lunch break.

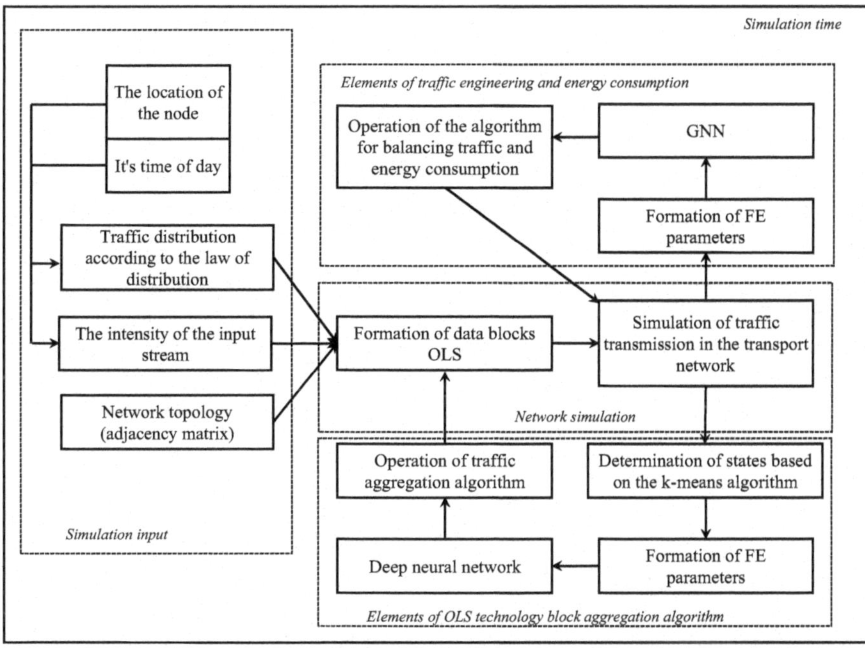

Fig. 9. The block diagram of simulation model

For a residential area, traffic changes dynamically during the day, but in the evening, there is a big jump in video content (Fig. 10). The described behavior is typical for working days for two districts of the city. These distributions were obtained from a local telecommunications operator in Lviv.

Based on the presented dynamics of traffic distribution, the studied traffic of the optical transport network for both city districts is generated. As described above, the proposed algorithm for determining network states can work with almost any set of network parameters. Thus, it allows you to control this algorithm to obtain the desired result flexibly. The following parameters are selected in work: time of day (current datestamp), the CPU load of the node, RAM load, input load of the node, the average size of the OLS block during dt, distribution of video packets, voice, data for which this node is final, energy consumption parameter and other. In Fig. 11 presents the simulated traffic of the optical transport network depending on three parameters:

- time of day;
- occupancy of the optical linear path (OLT) and the node;
- the average size of the OLS data block for time *dt*.

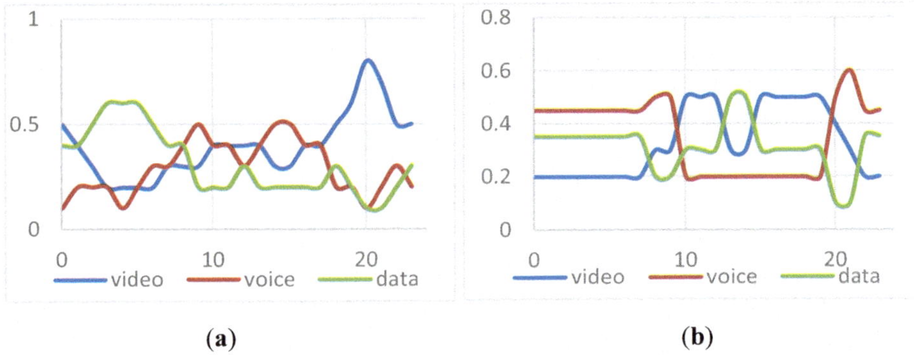

Fig. 10. Dynamics of traffic by hours for (a) residential area and (b) business area

This distribution of parameters (Fig. 11) is modeled at the macro level with an analysis time of 5 min without sharp jumps in traffic. The simulation results show how the size of the OLS block changes when the traffic dynamics change during the day. Video traffic has a significant impact - the largest size of the IP packet belongs to the actual video traffic. There is also a decline in traffic during lunch in business areas.

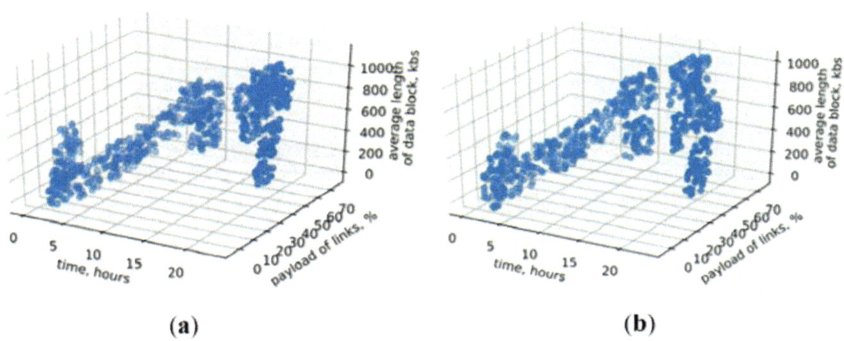

Fig. 11. Dynamics of simulated traffic (time of day, node load, average data block size) for (a) residential area (b) business area

6 Simulation Results

Based on the above data and simulated traffic, we see that the optimal number of clusters for the residential area is 4, and for the business area 2. The results of the distribution of clusters are presented in Fig. 12. The red circle indicates the centroid of the corresponding clusters.

As can be seen from the results, that for a residential area distribution is according to the hours of the day:

- late night and morning (2–8 h);

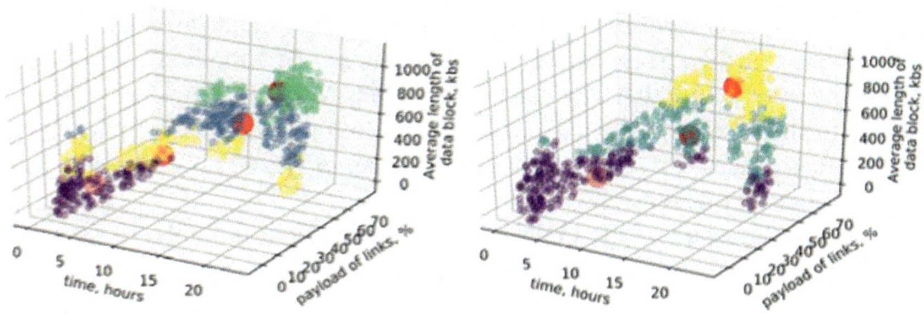

(a) (b)

Fig. 12. Distribution of clusters for (a) residential and (b) business areas

- late morning (8–10);
- lunch and late evening (10–16 and 22–2 h);
- evening (16–22).

For the business district, there is a division into two clusters, which determine the load during working and non-working hours (night and lunch). Based on the results of the ML algorithm, c-means allows for fuzzy clustering of large data sets, which allows you to more accurately identify objects at the cluster boundaries. Pre-simulated optical transport network traffic was used for more accurate validation of the results. In this study using the c-means algorithm, it is emphasized that the actual use of such a method will avoid possible errors of intelligent algorithms. That is, not only to show the affiliation of the current network state but also the probability of stay, which will allow the telecommunications operator to see the potential failures of the algorithm.

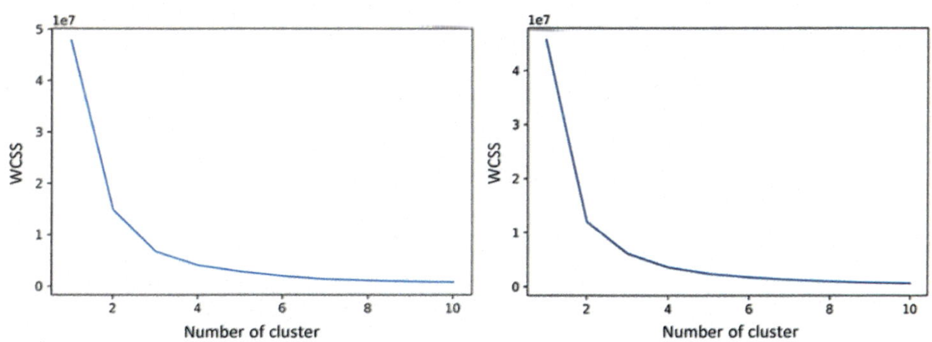

(a) (b)

Fig. 13. The result of the sum of the squares of the distances of each element of the cluster to the centroid from the number of clusters for (a) residential area (b) business area (WCSS, Within-Cluster Sum of Square)

In Fig. 13 presents the results of the sum of squares of the distances of each cluster element to the centroid from the number of clusters. This method allows to clearly determine the optimal number of clusters of k-means ML algorithm.

Based on the results presented in Fig. 13 the optimal number of clusters for the respective areas ranges from 2 to 5 units. The number of clusters is determined from the required granularity of events.

The input parameters of the GNN model are presented in Table 1. As mentioned above, the input parameters for each iteration of the workout are:

- adjacency matrix based on the studied networks;
- vector of network parameters of the node and edges;
- vector of labels for marking the trained data responsible for the corresponding state of the node.

Table 1. Input parameters for training the neural network

Parameter	Values	Parameter	Value
Learning speed	0,01	Batch size	24
Number of epochs	100	Number of simulation days	30

Based on the simulated traffic data of the optical transport network, the accuracy of the trained GNN is 0.956, which allows to correctly interpret the state of the nodes. Modeling of data transfer between nodes 1–3 is carried out (Fig. 7) (Fig. 14).

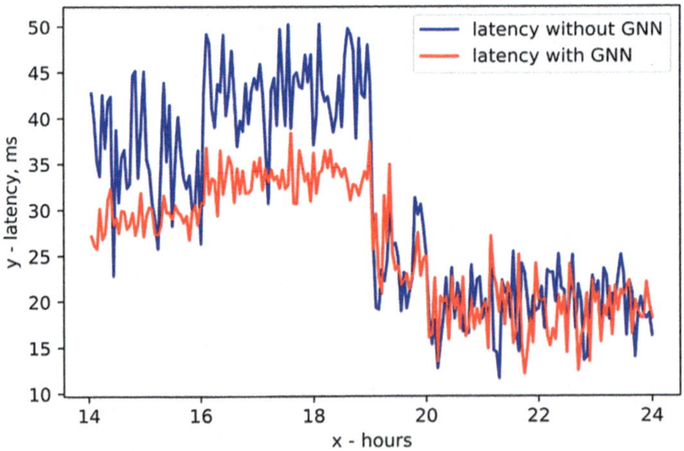

Fig. 14. Simulation of route latency between nodes 1–3

The simulation considers two data transmission routes 1-2-3 and 1-4-3. There is a higher latency between 14–19 h than between 19–24 h. In the model, this is due to the overload of node 2. The proposed intelligent TE method allows to rearrange the logical connections between nodes 1–3 and to balance traffic through node 4. As a result, there is a reduction in latency by almost 18% during 14–19 h, and the energy consumption parameter by 8.5%. In this case, it was possible to reduce both the latency parameter and the energy efficiency parameter.

7 Conclusion

The chapter proposes a conceptual model of the intent-based software-defined transport network infrastructure with elements of intelligent traffic engineering. Unlike the existing infrastructures of optical transport networks, these networks cannot provide protocol and infrastructure collection of the necessary data for intelligent algorithms. The proposed infrastructure is an isolated domain, which involves using a simulation model of the existing optical transport network to test intelligent algorithms for managing information flows.

The ML-based data collection algorithm of k-means and c-means algorithms has been developed. This algorithm allows, based on cluster approaches, to determine the states of the optical transport network and the moments of time when data should be collected from the network for the presented algorithms of TE. The method for intelligent TE using graph neural networks has been developed. The advantage of this method is the ease of implementation, since this algorithm is implemented on SDN and OpenFlow. The developed method proposes to consider the energy consumption parameter as another network parameter of the nodes and communication channels of the GNN data vector. This method allows to redistribute information flows to reduce the latency parameter during peak hours based on user intents.

References

1. Alvizu, R., Maier, G., Kukreja, N., Pattavina, A., Morro, R., Cavazzoni ,C.: Comprehensive survey on T-SDN: software-defined networking for transport networks. IEEE Commun. Surv. Tutorials **19**(4), 2232–2283 (2017). https://doi.org/10.1109/COMST.2017.2715220
2. Jia, W., Dong, X., Chen, Y., Chen, F.: A survey on All-optical IP convergence optical transport networks. In: 2019 7th International Conference on Information, Communication and Networks (ICICN), pp. 114–119 (2019). https://doi.org/10.1109/ICICN.2019.8834956
3. Musumeci, M., et al.: An overview on application of machine learning techniques in optical networks. IEEE Commun. Surv. Tut. **21**(2), 1383–1408 (2019)
4. Kaidan, M., Maksymyuk, T., Andrushchak, V., Klymash, M.: Intelligent data flow aggregation in edge nodes of optical label switching networks. In: 2019 3rd International Conference on Advanced Information and Communications Technologies (AICT), pp. 145–148 (2019)
5. Makridakis, S.: The forthcoming Artificial Intelligence (AI) revolution: Its impact on society and firms. Futures **90**, 46–60 (2017)
6. Cheatham, B., Javanmardian, K., Samandari, H.: Confronting the risks of artificial intelligence. McKinsey Quarterly (2019)
7. Yang, H., et al.: Intelligent optical network with AI and blockchain. In: 2019 18th International Conference on Optical Communications and Networks (ICOCN), pp. 1–3 (2019). https://doi.org/10.1109/ICOCN.2019.8934148

8. Maksymyuk,T., Dumych, S., Krasko, O., Kaidan, M., Strykhalyuk, B.: Study and development of next-generation optical networks. Smart Comput. Rev. **4** (2014)
9. Kaidan, M., Andrushchak, V., Maksymyuk, T., Klymash, M.: Scalability parameter in all-optical switches for optical label switching network. In: 15th International Conference on the Experience of Designing and Application of CAD Systems in Microelectronics (CADSM-2019), pp. 120–123 (2019)
10. Beshley, M., Vesely, P., Prislupskiy, A., Beshley, H., Kyryk, M., Romanchuk, V., Kahalo, I.: Customer-oriented quality of service management method for the future intent-based networking. Appl. Sci. **10**(22), 8223-1–8223-38 (2020)
11. Medvetskyi, M., Beshley, M., Klymash, M.: A quality of experience management method for intent-based software-defined networks. In: 2021 IEEE 16th International Conference on the Experience of Designing and Application of CAD Systems (CADSM), pp. 59–62 (2021). https://doi.org/10.1109/CADSM52681.2021.9385250 (2021)
12. Beshley, M., Pryslupskyi, A., Panchenko, O., Beshley, H.: SDN/Cloud solutions for intent-based networking. In: 2019 3rd International Conference on Advanced Information and Communications Technologies (AICT), pp. 22–25 (2019). https://doi.org/10.1109/AIACT.2019.8847731
13. Ujcich, B.E., Bates, A., Sanders, W.H.: Provenance for intent-based networking. In: 2020 6th IEEE Conference on Network Softwarization (NetSoft), Ghent, Belgium, pp. 195–199 (2020)
14. Faraz, M., Ismail, M.: INMTD: intent-based moving target defense framework using software defined networks. Eng. Technol. Appl. Sci. Res. **10**, 5142–5147 (2020)
15. Zeydan, E., Turk, Y.: Recent advances in intent-based networking: a survey. In: 2020 IEEE 91st Vehicular Technology Conference (VTC2020-Spring), Antwerp, Belgium, pp. 1–5 (2020)
16. Beshley, M., Kryvinska, N., Yaremko, O., Beshley, H.: A self-optimizing technique based on vertical handover for load balancing in heterogeneous wireless networks using big data analytics. Appl. Sci. **11**, 4737 (2021). https://doi.org/10.3390/app11114737(2021)
17. Beshley, M., Kryvinska, N., Beshley, H., Yaremko, O., Pyrih, J.: Virtual router design and modeling for future networks with QoS guarantees. Electronics **10**, 1139 (2021). https://doi.org/10.3390/electronics10101139(2021)
18. Beshley, M., Kryvinska, N., Seliuchenko, M., Beshley, H., Shakshuki, E., Yasar, A.: End-to-end QoS "smart queue" management algorithms and traffic prioritization mechanisms for narrow-band internet of things services in 4G/5G networks. Sensors **20**(8), 2324-1–2324-30 (2020)
19. Fadlullah, Z., et al.: State-of-the-art deep learning: evolving machine intelligence toward tomorrow's intelligent network traffic control systems. IEEE Commun. Surv. Tutorials **19**(4), 2432–2455 (2017)
20. Wu, Y.-J., Hwang, P.-C., Hwang, W.-S., Cheng, M.-H.: Artificial intelligence enabled routing in software defined networking. Appl. Sci. **10**, 656 (2020)
21. Kumar, S., Bansal, G., Shekhawat, V.S.: A Machine learning approach for traffic flow provisioning in software defined networks. In: 2020 International Conference on Information Networking (ICOIN), pp. 602–607 (2020). https://doi.org/10.1109/ICOIN48656.2020.9016529
22. Gringeri, S., Basch, B., Shukla, V., Egorov, R., Xia, J.: Flexible architectures for optical transport nodes and networks. IEEE Commun. Mag. **48**(7), 40–50 (2010)
23. Chankyun, L., June-Koo Rhee K.: Efficient Design and Scalable Control for Store-and-Forward Capable Optical Transport Networks. vol. 9, issue 8, pp. 699–710 (2017)
24. Legrand, T.: Labelled OBS test bed for contention resolution study. In: 2008 5th International Conference on Broadband Communications, Networks and Systems, pp. 82–87. IEEE (2008)

25. Lazaro, J., Arias, J., Martın, J.L., Cuadrado, C., Astarloa, A.: Implementation of a modified Fuzzy C-Means clustering algorithm for real-time applications. Microprocess, Microsyst. **29**, 375–380 (2005)
26. Icer, S.: Automatic segmentation of corpus collasum using Gaussian mixture modeling and Fuzzy C means methods. Comput. Methods Programs Biomed **112**, 38–46 (2013)
27. Asyali, M.H., Colak, D., Demirkaya, O., Inan, M.S.: Gene expression profile classification: a review. Curr. Bioinform. **1**, 55–73 (2006)
28. Runkler, T.A., Katz, C.: Fuzzy clustering by particle swarm optimization. In: 2006 IEEE International Conference on Fuzzy Systems, pp. 601–608 (2006)
29. Roh, Y., Geon H., Steven, E.: A survey on data collection for machine learning: a big data. AI integration perspective. IEEE Trans. Knowl. Data Eng. (2019)
30. Scarselli, F., Gori, M., Tsoi, A., Hagenbuchner, M., Monfardini, G.: The graph neural network model. IEEE Trans. Neural Networks **20**(1), 61–80 (2009)
31. Kaidan, M., Andrushchak, V., Pitsyk, M.: Calculation model of energy efficiency in optical transport networks. In: 2015 Second International Scientific-Practical Conference Problems of Infocommunications Science and Technology (PIC S&T), pp. 167–170 (2015)
32. Song, W., Andrushchak, V., Kaidan, M., Beshley, M., Kochan, O., Su, J.: Methodology for calculating the energy consumption of information communication systems. Techn. Electrodynamics, **4**, 80–88 (2020)
33. Przystupa, K., Beshley, M., Kaidan, M., Andrushchak, V., Demydov, I., Kochan, O., Pieniak, D.: Methodology and software tool for energy consumption evaluation and optimization in multilayer transport optical networks. Energies **13**(23), 6370-1–6370-21 (2020)

The Approach to Flow Management in Virtual Computational Environment for Up-to-Day Telecom Networks

Larysa Globa, Mariia Skulysh(✉), Dmytro Parhomenko, and Kateryna Yakubovska

National Technical University of Ukraine "Igor Sikorsky Kyiv Polytechnic Institute", Kyiv, Ukraine
lgloba@its.kpi.ua

Abstract. Data streams that are transmitted over communication networks are growing very quickly, this is facilitated by the progress of the service such as IoT, M2M, etc. The provision of high-quality communication services is increasingly influenced by the efficiency of performing calculations in the data centers of telecom operators, whose resources are limited. SDN technologies make it possible to redistribute transmitted data streams in networks to reduce their load, but they require more computing resources. Telecom operators' networks not only transmit significant amounts of information, but also process very large information flows in their computing nodes, including virtual ones. Controlling the transfer process with a large number of threads becomes a bottleneck. To service flows in SDN, an application (at the application level) call is required, which entails overloading the computational resources involved in processing incoming flows. The main approach for dealing with the computing resources overload in such systems is load balancers that should distribute the load between computing nodes efficiently. However, such approaches lead to increased latency as the flow peak rate increases and there is a computing resources limitation. In this chapter, to reduce the time for making decisions when load balancing, «the endless train» method is proposed. This method, instead of analyzing the input flow state and simultaneously the resources state, analyses the state of the computational resources only to make a decision regarding needed resources based on the current task requirements. This allows reducing the time for making a decision on the choice of a server serving the input flow. To test the efficiency of the proposed method, an implementation scheme was developed using MS Azure. The process of dynamically deploying additional virtual servers (virtual machines) to handle threads in the event of overload was tested. The testing results show the effectiveness for overload prevention but the computational recourses usage is increased.

Keywords: Software defined network · Computation in SDN · Loads balancer

1 Introduction

Today, the telecommunications telecom operator network is an organized system that includes special equipment that is serviced, monitored and managed from operational

M. Klymash et al. (Eds.): *Future Intent-Based Networking*, LNEE 831, pp. 182–196, 2022.
https://doi.org/10.1007/978-3-030-92435-5_10

data centers, where computer servers and related software are installed that serve numerous information and service flows. Rapid growth of traffic, change in its structure, the need to support a large number of mobile devices, processing IoT, M2M flows requires new solutions. Modern technologies SDN, NFV, SDR, CloudRAN and others are developing rapidly [1, 2]. Their full-scale implementation leads to the complete dependence of the telecommunications network functioning on the operation of the information and computing environment.

Service providers now virtualize parts of their network. This approach significantly influences the network performance and QoS parameters used to ensure the operation of the network. The performance of up-to-date networks significantly depends on the organization of the computing process in the systems that ensure their operation. The excessive workload in these computational systems becomes a real challenge for computer systems. Hybrid telecommunication services (TC-hybrid service) [3] are sensitive to the quality of service, especially service delays.

Planning the operation of a virtualized network, which provides maintenance of hybrid information and telecommunications services, allows a conditional infinity computing resource [4]. Its organization requires new approaches and solutions that take into account the specifics of the deployment of computing clusters and the volume of services that require maintenance [5].

Taking into account the flexibility and computational efficiency of virtualized network functions (NFV) and software-defined networking (SDN), there is considerable focus on these technologies for future network infrastructure in which operators and service providers can program network functions (e.g. routers, intrusion detection systems, load balancers, firewalls) as required by the vendor without implementing the appropriate hardware.

SDN technologies allow redistributing data streams in networks to reduce their load, but they require more computing resources, especially when the telecom operator uses its own data center. Moreover, the use of computing resources in the data center of a telecom operator is characterized by significant downtime due to uneven load during the day. The use of cloud technologies can reduce downtime and more efficient use of computing resources. So, control of the transfer process with a large number of threads becomes a bottleneck. To service flows in SDN, an application (at the application level) call is required, which entails overloading the computational resources involved in processing incoming flows. The main approach for dealing with the computing resources overload in such systems is load balancers that should distribute the load between computing nodes efficiently. However, such approaches lead to increased latency as the flow peak rate increases and there is a computing resources limitation.

The paper [4] proposes a method exploiting a Virtual Load Balancer with SDN-NFV Framework, which is based on the creation of an intelligent load management system. Recently, a lot of work has appeared in which artificial intelligence methods are used to manage QoS [3, 6, 7]. However, the work of an information and telecommunication system using templates obtained on the basis of machine learning algorithms will not give the desired results if the situation in the network changes dramatically. For such systems based on Artificial Intelligence, it will take time to retrain and create new templates. The

"Infinity train" method proposed in the article will allow you to quickly respond to any changes in the dynamics of information flows.

The chapter is structured as follows: Sect. 2 contains the processes in SDN controller description. Section 3 explains the problem to be solved by proposed approach. Section 3 describes 3 the statement of the research task and the method of organizing functions “Infinity train”. Sections 4 introduces the method for ensuring resource sufficiency at all times. Section 4 presents the mathematical model and algorithm for ensuring resource sufficiency at all times. Section 5 explains the specifics of creating virtual machines in dynamic mode and the technologies for their implementation. Section 6 includes the experiment description and Sect. 7 includes the summary and outlook on future work.

2 Description of Processes in SDN Controller

Provision of SDN controller operation associated with the implementation of a number of software modules for different purposes. The work of program modules is carried out by using a significant amount of computing resources. In Fig. 1 schematically shows the concept of SDN, according to which all control logic is carried out in the so-called controllers, which are able to monitor the operation of the entire network.

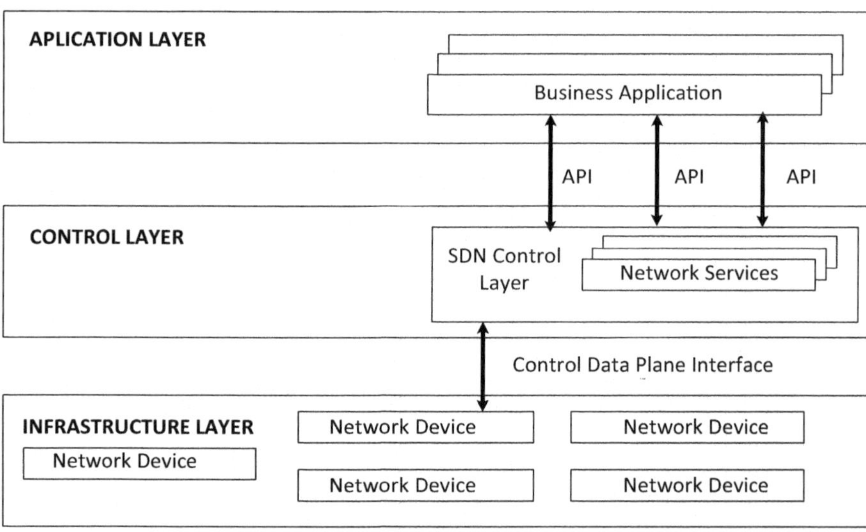

Fig. 1. The concept of software defined network

There are three types of SDN controller computational process:

- Regular computational processes associated with maintaining physical network capacity;
- Computational processes caused by a customer’s network, such as flow initialization or organization of virtual transport networks, and performing of other network functions;

- Database queries that are triggered by the subsystem controller software components.

If regular computing processes require the volume of computing resources that can be calculated, the computing processes caused by user queries are exposed only to approximate evaluation because there are many factors that affect short- and long-term trends changes of loads.

As the approach to controller service functions problem, proposed in this chapter, is designed to optimize the independent function modules computational process that can be calculated anywhere, a group of functions that works with databases was separately dedicated. This improves the process of working with databases tied to data location or large amounts of cache, which stores represent databases. Therefore, this chapter is aimed at optimizing computational processes associated with service independent functions initiated by users of the system or other applications.

The problem of load statistical analysis created by the flow of requests to date is well studied. Queueing theory provides a significant number of probabilistic models, allowing the calculation of system parameters to ensure the quality of service parameters such as delay, loss rates across the length of the queue, and so on. At the same time, today there is no need to be limited to constant service performance parameters since the use of cloud technology allows them to distance themselves from the problem of finite computing resources.

Consider in more detail the components of SDN controller architecture in order to highlight features that require a variable number of resources for implementation.

SDN controller architecture consists of three levels:

1. Southbound plugins and protocols forming the network device layer.
2. Service adaptation and network functions forming the coordination and control layer.
3. Northbound APIs and applications forming the application layer.

The controller acts like middleware in the ecosystem. It is the framework that glues together the applications requiring services of the network devices and the protocols that talk to the network devices for extracting services. The controller allows the applications to be agnostic about the network device specifications, thereby allowing the application developers to concentrate on the development of application functionality rather than writing device-specific drivers.

The operation of the controller is associated with servicing a large number of queries that trigger execution of action sequences, consisting of the launch of network applications.

Thus, the speed of processing requests in Northbound APIs and applications on the application layer directly affects the service of streams in the network as a whole. The proposed approach solves this problem.

SDN networks are a good tool for dealing with congestion, but they require improved approaches and algorithms to the problem of managing computational resources. We need improved approaches to the problem of organizing a system for managing the flow of applications and their redistribution among the available computing resources.

3 Statement of the Research Task. Method of Organizing Functions "Infinity Train"

There are different approaches to managing the volume of computing resources for each subsystem: fixed and adaptive. Today, the use of virtual servers is becoming increasingly popular, which are easily configured to change the load and allow you to implement an adaptive approach. The resource management process is a challenging task due to workload fluctuations. However, many workloads in data centers usually have periodic patterns, but there will be deviations from historical models due to unforeseen factors such as peak loads, etc. [8]. Due to the variability of the load experienced by modern systems, the placement of virtual machines must be constantly optimized in real-time [9]. Given the difficulties in predicting peak loads, the system must use a combination of dynamic resource allocation and request management to respond in a timely manner to load changes [10]. After all, the provision of resources is not free [8]; there are various associated costs and risks. Frequent implementation of the procedure of providing resources causes both losses of productivity and energy. Frequent periodic switching on and off of server power causes "wear and tear", which can lead to server failure and service outages. Therefore, providers are dealing with a compromise on energy efficiency - minimizing energy consumption while satisfying the SLA (Service Level Agreement) [9].

Dynamic resource allocation - allocating and cleaning servers for applications - has been studied in the context of single-tier applications. While multilevel Internet applications were studied, the focus was on access control issues to support the target response time, but resource allocation tasks were not considered. Extending the resources allocation mechanisms designed to service services in one step to multilevel scenarios is a non-trivial task. Classical approaches can simply shift the bottleneck to another level [10]. At the same time, systems with a single type of resources are usually considered. In this case, the tasks of resource allocation in multi-stage processing systems are considered mainly for Internet systems, while such tasks arise in various areas, including in the process of operation of the servers of the telecom operator.

In most works, the stages are considered independently or with independent resource pools [10, 11]. However, in networks, different stages can be located on the same physical machine. In such cases, it is necessary to consider a common pool of resources, and there is an additional problem - the problem of optimal distribution of total resources between different stages.

It is necessary to notice that it is more expedient to apply non-permanent decision-making about access at the receipt of the query. Such a non-permanent decision allows avoiding the losses of resources in the total the partly served queries which can be lost on later levels.

A significant volume of research has been done using queuing theory for modelling. Due to difficulties in mathematically modelling complex traffic characteristics (eg, multi-stage application processing), most research in the literature has been performed using simulation. For example, the work performed in [12] is based on a simulation model. However, careful implementation of each part of a particular system increases the complexity of the model and reduces efficiency, and such a model can usually only be used in very limited situations. Thus, the analytical model of the system will be attractive because it will be able to assess the characteristics of the system in a wide range of conditions, be calculated in a reasonable time, will allow the application of numerical optimization methods in system design.

To date, there are several approaches to the organization of the computational process in the SDN controller. Their main disadvantage is that they do not take into account the possibility of using an unlimited resource provided by cloud technology. All existing approaches are based on the fact that the number of system resources is limited and depends on the technical capabilities of the server on which the virtual machines are hosted. On the other side, the application of such approaches results in a decrease in the QoS provision in certain periods of time, which is due to the fact that the work of the monitoring system is to collect and analyse data on the quality of the network services. If the QoS values for certain services are below the threshold, but other services are provided with fairly high quality, the system may not respond to quality degradation in certain periods of time, as only the average values are taken into account.

The solution to this problem is the use of a heterogeneous cloud environment, namely certain solutions for the platform as an infrastructure (PaaS). Platform-as-Infrastructure is an isolated cluster consisting of a group of servers and services that interact as a whole system, providing the ability to easily deploy, test, maintain and scale the system. PaaS allows you to create and maintain an unlimited number of virtual machines, whose work is to maintain the computational functions of the controller. The use of an unlimited number of resources for the operation of virtual machines will avoid periods of declining quality of service.

The method proposed in this article is called the "infinity attraction effect". Its main idea is that after a certain virtual machine has received the specified number of tasks for maintenance, a new virtual machine is created, which receives the following tasks (Fig. 2).

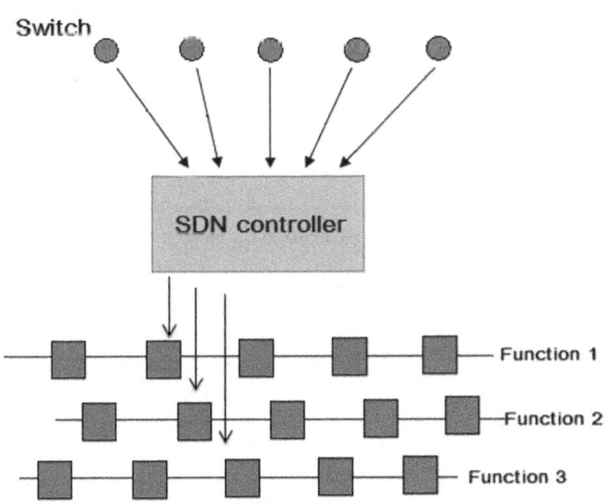

Fig. 2. "Infinity train" method

The proposed method is based on the method of dynamic migration of virtual machines, developed by Jelastic [13], which provides load balancing of cloud storage servers by creating a platform for automatic management of virtual machine containers. Also considered is the method of selecting a container for virtual machine migration, described in the paper [14], which allows you to analyse and predict the load on the network based on the assessment of resources used. Using cloud storage means that all of them are stored on a large number of servers distributed on the network. In this case, two main tasks must be performed:

- Interactive distribution of client tasks between virtual machines that are located in one or more clusters. This is, on the one hand, the task of load balancing, and, on the other - the task of ensuring the reliability of customer service;
- Control and management of the cluster of virtual machines. Cluster resources should always be sufficient for all virtual machines running on all cluster servers at the same time. The "infinity train" method assumes that all service requests are sent to the current virtual machine until it is filled. The number of requests that can be processed by a single virtual machine depends on the volume of resources allocated to it at creation. After filling this machine, a new one is created, to which all subsequent applications are redirected. The use of this method is possible if the number of system resources is relatively unlimited. The use of a heterogeneous cloud environment provides such an opportunity.

Older virtual machines continue to service their information flows until they run out. After that, the virtual machines are rolled up or waiting to be re-commissioned as "empty cars". The maximum number of applications that can be serviced in the "car" depends on the configuration of the cloud platform, or can be obtained experimentally. The number of requests is determined by the volume of resources that the controller uses to perform computational tasks. The maximum number of applications is also affected by the flexibility of migration processes of technical processes of maintenance of virtual machines. The illustration of the described method is shown in Fig. 3.

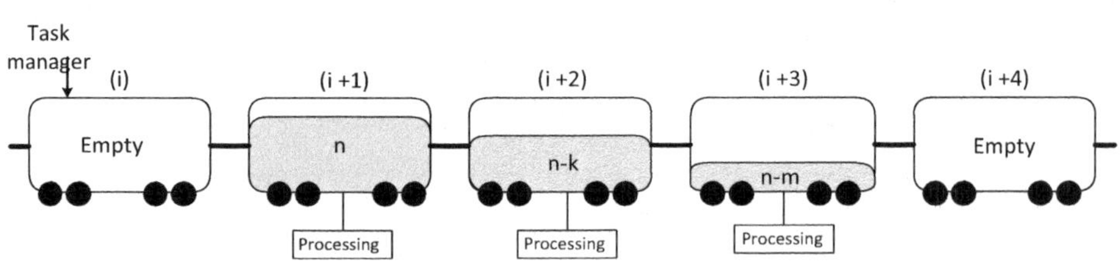

Fig. 3. The illustration of the "Infinity train" method idea

Let us introduce the following notation:

n is the maximum number of tasks for one VM;

k is the number of tasks tha t have been processed in $(i + 2)$*-th* virtual machine from the beginning of its filling to the current moment, the moment of completion of applications to the virtual machine $(i + 1)$;
m is the number of tasks that have been processed in the $(i + 3)$*-th* virtual machine from the beginning of its filling to the current moment.

According to the algorithm, incoming flows always sends to the Head VM, then when filling the Head VM, there is a generation VM by template where all new streams are sent. If after that the flow of applications decreases, the decision on dynamic removal of empty VM is made.

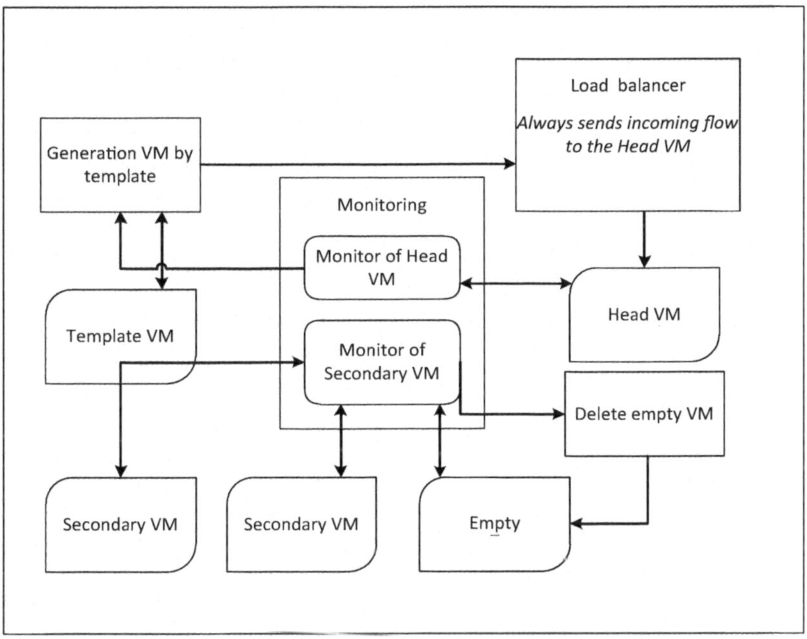

Fig. 4. The algorithm of the "Infinity train" method

Figure 4 shows the logic of the "Infinity train" method. The main structural elements, their functionality and interaction logic are shown below:

- Loadbalanser which always sends incoming flows to the Head VM. All threads of computational tasks go here. The Head VM Loadbalanser receives information about which machine is from the Generation VM by template block;
- "Generation VM by template" - a block that receives a signal from the "Monitor of Head VM" that the Head VM is full and it is necessary to create a new VM using a template. Once the virtual machine has been deployed and is ready to serve the input load, it becomes the Head VM, which is reported by the "Generation VM by template" to Loadbalanser.
- "Monitoring" block is divided into two parts:

- "Monitor of Head VM" block monitors the state of the main machine. If the volume of occupied resources of the Head VM exceeds the allowable value of R, then the "Monitor of Head VM" sends a message "Generation VM by template" about the need to initialize to create a new VM;
- "Monitor of Secondary VM" block monitors the state of the resources of all running VMs, which are called Secondary VMs. If the volume of occupied resources of the virtual machine is reduced to a level that corresponds to an empty VM, then "Monitor of Secondary VM" sends an instruction "Delete empty VM" to delete the empty VM.

"Delete empty VM" when instructed by "Monitor of Secondary VM" will delete the empty virtual machine.

4 Ensuring the Sufficiency of the Resources at All Times

To ensure sufficient resources at any time, one more function must be added to the Monitor of Head VM. The function of monitoring the sufficiency of resources. In connection with a non-uniform input flow, a problem may arise when the Monitor of Head VM block has already given a command to the Generation VM by template block, but the new Head VM has not yet been created, and the old Head VM is already full. To avoid such problems, it is necessary to determine the optimal interval for analyzing the statistics of the Head VM resource utilization, as well as monitor the dynamics of resource occupancy for a timely response to bursts. That is, it is necessary, relying on the statistics of resource occupancy, to track the moment when to initiate the creation of a new virtual machine.

The short-term planning method is an advanced method of forecasting ARIMA - autoregressive integrated moving average. However, in contrast to the known method, it is proposed to solve the problem of finding the minimum slip interval, the use of which will meet the requirements that will minimize the number of flops to perform predictions, which will ensure optimal speed of prediction.

The proposed method consists of two stages - calculation of the prediction interval based on Head VM statistics and direct periodic forecasting of Head VM resource utilization and control of their sufficiency in the next period of time T, which is determined by the time required to create a new VM by a template.

Input Data:

- T is the time interval for which the forecast is required.
- r_i is the volume of resources employed at the *i-th* moment of time, $(i \in 0,..N)$, $N = \mathrm{T_{info}} / 1_{MC}$, $r_i \in R$, where R are set of statistic values the specific amount of resources that were occupied by the Head VM during the time T_{info} (first is set, then corrected in the *2nd* stage of the method) before the forecast, $|R| = N$.
- T_{pred} is forecasting period - the time during which the statistics of the number of occupied specific resources are analyzed to build a forecast.
- M is the maximum specific volume of resources planned for one virtual machine.
- P is admissible probability of forecast error.

Short-term statistics are collected locally at the service device and stored no longer $T_{info}+ T_{pred}$, sampling time interval *1 ms*.

Output Data:

- $z \in \{0,1\}$– $z = 0$ do not create a new VM; $z = 1$ new VM have to be created.

The method algorithm:
Preparatory stage. System training based on statistics. Search for minimum T_{info} (information collection time interval):

$$T_{info} \rightarrow min,$$

for which the restriction is fulfilled:

$$r_{Tpred} + 3\sigma > M,$$

where r_{Tpred} is calculated according to the main stage, σ is the variance of employment statistics is calculated for a period of time T_{info}.

The constraint is performed for different statistical samples obtained at different time intervals.

Solution: check the values for the sequence formed on the principle $T_{info}{}^{k+1} = T_{info}{}^{k} + \Delta$; $T_{info}{}^{0} = T_{pred}$.

The main stage of dynamic control

1. Analysis of statistical data r_i, for the time interval T_{info} preceding the moment of calculation. Construction by the method of least squares regression direct dependence of occupied resources r on time i observation, calculation of coefficient estimates $\hat{a}$ and $\hat{b}$:

$$r = \hat{a}_i + \hat{b}$$

3. To calculate $r_{\bar{}Tpred} = \hat{a}\, T_{pred} + \hat{b}$.
4. If $r_{Tpred} + 3\sigma \leq M$, then $z = 0$, else $z = 1$.

Thus, the integration of the proposed functionality for monitoring the sufficiency of resources into the "Monitor of Head VM" block will provide a continuous process of servicing applications at the application level in the SDN controller with a guaranteed quality index, it is determined by the *P* parameter, admissible probability of forecast error.

5 The Specifics of Creating Virtual Machines in Dynamic Mode and the Technologies for Their Implementation

Virtualization uses software to create a level of abstraction over physical equipment. This creates a virtual computing system known as a virtual machine (VM). A virtual machine

(VM) is a virtual representation of a physical computer. This allows organizations to run multiple virtual computers, operating systems, and applications on a single physical server, essentially splitting it into multiple virtual servers. One of the main advantages of virtualization is the more efficient use of physical computer equipment, which in turn provides a greater return on investment in equipment.

Because the software is separate from the physical host computer, users can run multiple instances of the OS on a single piece of hardware, reducing management costs and physical space. Another advantage is that virtual machines can support legacy applications, reducing or eliminating the need and cost of migrating an old application to an upgraded or another operating system.

System virtual machines rely on the hypervisor as an intermediary to provide software with access to hardware resources. The most well-known hypervisors are VMware (ESX/ESXi), Intel/Linux Foundation (Xen), Oracle (MV Server for SPARC and Oracle VM Server for x86) and Microsoft (Hyper-V).

VMware vSphere - Indicates basic virtualization solutions that help you manage, track, and configure a virtual data center.

Hyper-V is a corporate data center hypervisor platform.

Hyper-V and VMware have different memory and configuration management methods available to administrators who manage both hypervisors.

The main method that Hyper-V uses to manage memory - dynamic memory management.

Dynamic memory is a method that Hyper-V uses to dynamically add additional RAM to a virtual machine running on the Hyper-V infrastructure, as well as to actually free up unused memory when memory is not in use.

Hyper-V dynamic memory component and configurations that can be changed:

- RAM at startup;
- Minimum volume of RAM;
- Maximum volume of RAM;
- Memory buffer;
- Memory weight;
- Dynamic migration of Hyper-V.

Dynamic Hyper-V migration allows you to transfer a running virtual machine, including its active memory, from one host to another. With Hyper-V Live Migration, after configuring the failures of the Windows cluster that hosts the Hyper-V role, you will be able to set up a network that will handle the transfer of virtual machines between hosts.

Scalability: VMware vs. Hyper-V.

Scalability is an important factor for any business when choosing a hypervisor that will perform its production workloads. Demand for resources may increase over time as the business grows. Additional demand may decline, and workloads too may be reduced over time. Understanding the various limitations between Hyper-V and VMware can help identify potential bottlenecks.

The proposed method of infinity train can expand the capabilities of the Hyper-V hypervisor, and provide adaptive processing of avalanche-like streams of services served by a group of virtual machines.

6 Description of the Experiment

During the experiment, an Infinity train from virtual machines was implemented. On the basis of each virtual machine the service of searching for the shortest way in a network segment by a method of Dijkstra was implemented. Virtual machines, according to the proposed method, rose under the load of the main machine and collapsed when reducing the resource used to a critical limit.

The following MSAzure services were used in the simulation process:

- Azure Virtual Machine;
- Azure Alerts from Azure Monitor;
- Azure Resource Manager Template;
- Azure Logic App.

The operation and features of using the Azure Logic App and Azure Resource Manager Template services were described above.

Azure Resource Manager - a service designed to deploy and manage Azure. It provides a level of management for creating, updating and deleting resources in an Azure account. One virtual machine is allocated to provide an account. Azure Resource Manager performs management functions such as access control, blocking, and tagging to protect and organize resources after deployment. At the same time, the infinity train method is an additional tool for managing the resources allocated to a particular account, according to the proposed method, a group of virtual machines deployed by the Hyper-V hypervisor is deployed differently for one account.

At the level of coordinated management, requests are processed from any of the Azure tools, APIs or SDKs. All requests are sent to the Resource Manager. Resource Manager performs request authentication and authorization. Resource Manager sends a request to Azure, which takes the requested action. Since all requests are processed through one API, the results and capabilities will be agreed in different means. Figure 5 shows the role of Azure Resource Manager in processing Azure requests.

Using Azure Resource Manager templates allowed you to create a sequence of virtual machines with specified characteristics of the resource group. Deployed infrastructure was managed using templates, deploy and track all resources, and manage them as a single group.

Azure Alerts from Azure Monitor has provided a system for monitoring the load of a group of resources. As part of the experiment, the CPU usage status was monitored. As described in the proposed method of infinity train, the main machine and all other secondary machines were monitored separately. The maximum load of the resource was monitored for the main machine to which the flow of requests was directed. In the event of an overload greater than the specified n%, in the experiment $n = 80\%$, the trigger to create a new virtual machine was automatically triggered. Then the new machine was marked as the main, and all new requests were sent to the new main machine. The new virtual machine has been deployed from the Azure Resource Manager Template. The old main machine was added to many secondary machines, it did not accept new requests. All secondary machines did not accept requests. The monitoring function was used for

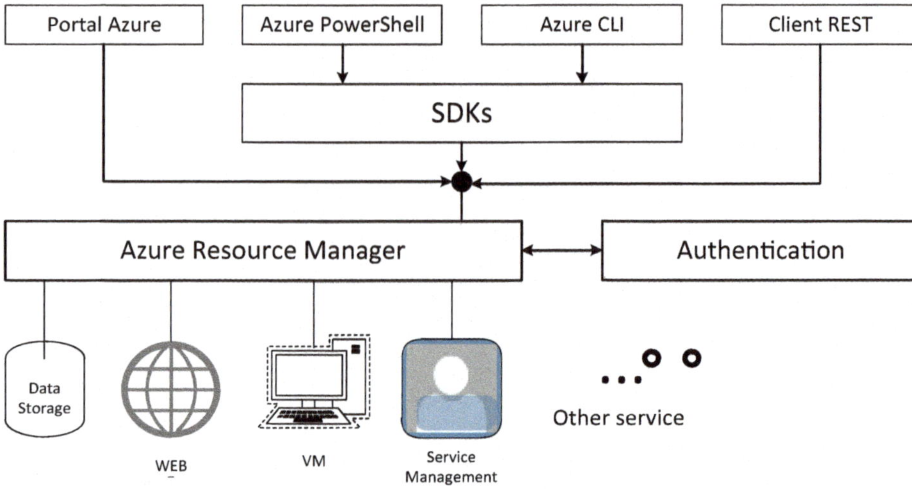

Fig. 5. Azure Resource Manager when processing Azure requests

them. In the case of reducing resource consumption to the level of an empty machine, the secondary machine was removed.

The scheme of the involved modules and the logic of their use is shown in Fig. 6.

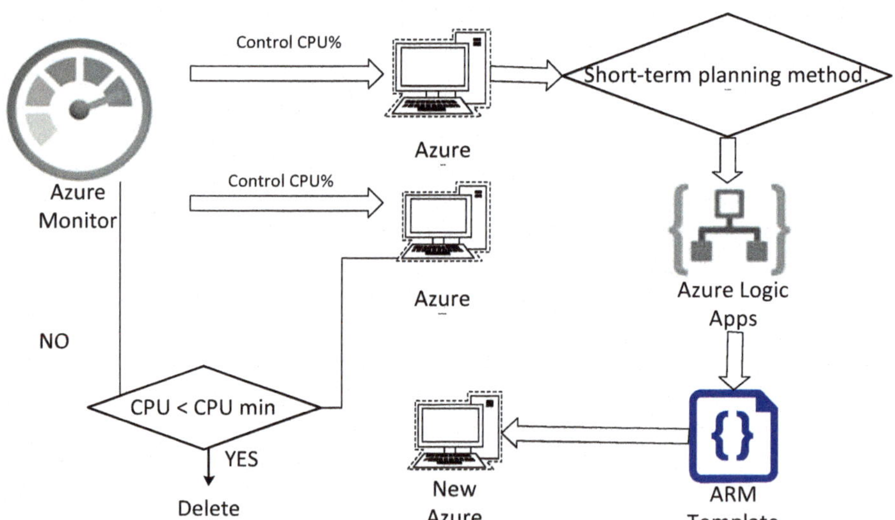

Fig. 6. Description of the experiment implemented in the environment of MS Azure

The experiment control process was carried out by the Azure Logic App, however, a similar approach to organizing a group of virtual machines can be implemented at a lower level, at the level of the Resource Hypervisor, such as Hyper-V. Such an experiment is planned in the future.

The proposed method of "infinity train" is a new approach to the use of virtual machines in systems with avalanche dynamics of flow load. This approach reduces the delay in balancing flows between computers or virtual machines. In contrast to the main technology for combating the overload of computing resources - load balancing, an approach is proposed that provides for the dynamic deployment of additional resources on demand. This reduces the time to make decisions about the choice of server for load processing. All flows are sent to one server, which is assigned to the master. In the process, the main server is replaced by a newly created according to the template virtual machine. This reduces the time to decide on the server that serves the incoming stream. The chapter explores different approaches to managing virtual machines, which are implemented in hypervisors of different vendors. The proposed solution can be an improvement for the existing virtual machine hypervisor. But to meet its needs, the company can use the tools of MS Asure, and implement for their needs the proposed algorithm of the method of infinity train. To test the effectiveness of the proposed method, an implementation scheme was developed using MS Azure. The process of dynamic deployment of additional virtual servers for processing threads in case of overloads was tested.

The testing results show the effectiveness for overload prevention but the computational recourses usage is increased.

7 Conclusions

The study showed that the existing approaches to computing resource management do not meet all the QoS requirements for TC-hybrid services, especially such as service request flows of IoT, M2M and others.

SDN networks are a good tool for dealing with congestion in the networks but require advanced approaches and algorithms to manage computing resources.

The chapter proposes a solution to the problem of organizing a flow management system of applications and their redistribution between available computing resources.

The method of "infinity train" is proposed for dealing with overloads in telecom operator networks that has the effectiveness for overload prevention but the usage of the computational resources is increased.

Future studies are expected to consider the possibility of finding a compromise between the overload prevention and the volume and time of the usage of the computational resources.

References

1. Rudyk, A.V., Semenov, A.O., Kryvinska, N., Semenova, O.O., Kvasnikov, V.P., Safonyk, A.P.: Strapdown inertial navigation systems for positioning mobile robots-mems gyroscopes random errors analysis using allan variance method. Sensors **20**(17), 4841 (2020)
2. Semenov, A.A., Semenova, O.O., Voznyak, O.M., Vasilevskyi, O.M., Maksym, Yakovlev, Yu.: Routing in telecommunication networks using fuzzy logic. In: 17th International Conference of Young Specialists on Micro/Nanotechnologies and Electron Devices EDM 2016, Erlagol, Altai - 30 June - 4 July, 2016: Conference Proceedings, 2016, pp. 173–177 (2016). https://doi.org/10.1109/EDM.2016.7538719

3. MITU-T M.3371 (10/2016). https://www.itu.int/ITU-T/recommendations/rec.aspx?rec=-13064&lang=en
4. Monir, M.F., Pan, D.: Exploiting a Virtual Load Balancer with SDN-NFV Framework
5. Globa, L., Skulysh, M., Siemens, E.: Conditionally infinite telecommunication resource for subscribers. In: Ilchenko, M., Uryvsky, L., Globa, L. (eds.) MCT 2019. LNNS, vol. 152, pp. 206–216. Springer, Cham (2021). https://doi.org/10.1007/978-3-030-58359-0_11
6. Zhang, C., Patras, P., Haddadi, H.: Deep learning in mobile and wireless networking: a survey. IEEE Commun. Surv. Tutorials **21**(3), 2224–2287 (2019)
7. Globa, L., Skulysh, M., Romanov, O., Nesterenko, M.: Quality control for mobile communication management services in hybrid environment. In: Ilchenko, M., Uryvsky, L., Globa, L. (eds.) UKRMICO 2018. LNEE, vol. 560, pp. 76–100. Springer, Cham (2019). https://doi.org/10.1007/978-3-030-16770-7_4
8. Gandhi, A., Chen, Y., Gmach, D., Arlitt, M., Marwah, M.: Minimizing data center SLA violations and power consumption via hybrid resource provisioning. In: 2011 International Green Computing Conference and Workshops (IGCC). – Orlando, USA, 2011, pp. 1–8
9. Beloglazov, A., Buyya, R.: Optimal online deterministic algorithms and adaptive heuristics for energy and performance efficient dynamic consolidation of virtual machines in Cloud data centers. Concurrency Comput. Practice Exp. **24**(13), 1397–1420 (2012)
10. Urgaonkar, B., Shenoy, P., Chandra, A., Goyal, P., Wood, T.: Agile dynamic provisioning of multi-tier Internet applications. ACM Trans. Auton. Adapt. Syst. (TAAS) **3**(1), 1–39 (2008)
11. Han, R., Ghanem, M.M., Guo, L., Guo, Y., Osmond, M.: Enabling cost-aware and adaptive elasticity of multi-tier cloud applications. Future Gener. Comput. Syst. **32**, 82–98 (2014)
12. Skulysh, M.: The method of resources involvement scheduling based on the long-term statistics ensuring quality and performance parameters. In: 2017 International Conference on Information and Telecommunication Technologies and Radio Electronics (UkrMiCo), pp. 1–4. IEEE, September 2017
13. Ye, K., Jiang, X., Huang, D.: Live migration of multiple virtual machines with resource reservation in cloud computing. In: IEEE International Symposium, pp. 267–274 (2013)
14. Pahl, C., Xiong, H.: Migration to PaaS clouds – Migra- tion process and architectural concerns. In: IEEE International Symposium, pp.86–91 (2013)
15. Luo, X., Liu, L., Shu, J., Al-Kali, M.: Link quality estimation method for wireless sensor networks based on stacked autoencoder. IEEE Access **7**, 21572–21583 (2019)

Calculation of Quality Indicators of the Future Multiservice Network

Bohdan Zhurakovskyi[1(✉)], Serhii Toliupa[2], Volodymyr Druzhynin[3], Andrii Bondarchuk[4], and Mykhailo Stepanov[2]

[1] National Technical University of Ukraine "Igor Sikorsky Kyiv Polytechnic Institute", 37 Prospect Peremogy, Kyiv 03056, Ukraine
bogdan68@ukr.net

[2] Taras Shevchenko National University of Kyiv, 60 Volodymyrska Street, Kyiv 01033, Ukraine
tolupa@i.ua

[3] National Aviation University, 1 Liubomyra Huzara Avenue, Kyiv 03058, Ukraine

[4] State University of Telecommunications, 7 Solomenska Street, Kyiv 03110, Ukraine

Abstract. This article discusses the main indicators of the quality of the multiservice network, such as: delay; delay variation (jitter); number of packets with errors; number of lost packets. Not all types of traffic are sensitive to packet transmission delays, in any case, to the values of delays that are characteristic of multiservice networks. The purpose of this study is to assess the quality of the multiservice network using simulation. The analysis of network behavior when different types of traffic enter the network is performed: real-time traffic, data traffic and mixed traffic. The obtained data showed that the most critical to changes in the parameters of the multiservice network is such a quality indicator as the delay time, i.e. a slight change in network parameters significantly changes this quality indicator.

Keywords: Time delay · Jitter · Lost packets · Queue length · Packet service intensity · Packet loss ratio

1 Introduction

1.1 Analysis of the Quality of the Multiservice Network

When analyzing the ITU Y.1540, the main indicators of the quality of the network are allocated, which are basic when providing multiservice services:

- delay;
- a delay option (jitter);
- the number of packets with errors;
- the number of packets lost.

1) Delivery delay of the IP package (IP Packet Transfer Delay, IPTD). The delay manifests itself in a number of directions, including the time required to create a specific

M. Klymash et al. (Eds.): *Future Intent-Based Networking*, LNEE 831, pp. 197–209, 2022.
https://doi.org/10.1007/978-3-030-92435-5_11

service from the initial user request and before receiving specific services (IPTD). In general, it is estimated as:

$$T_z = (t_2 - t_1), t_2 > t_1 \text{and} (t_2 - t_1) \leq T_{max}, \quad (1)$$

where t_1 is the moment of package input in the entrance point of the network; t_2 is the moment of the package's conclusion from the source network.

In general, the IPTD parameter is defined as the delivery time of the packet between the source and recipient for all packages - both successfully transmitted and affected by errors.

According to the Little's theorem, the average time of delay delivery of packages is one ratio of the middle number of packets in a queue to the intensity of the service flow of requests:

$$T_z = \frac{L_o}{\lambda}, \quad (2)$$

where L_0 is the length of the queue, λ is the intensity of packet maintenance.

Delay is a network performance parameter that is close to a network response time, but is characterized by that it always characterizes the network stages of data processing, without delay treatment with end nodes of the network.

The delay of messages is influenced by such factors as the law of distribution of intervals between messages, intensity of receipt, discipline of priority service, intensity of service. Similarly, an increase in load and a decrease in available network resources lead to an increase in queues at network nodes and, as a consequence, to an increase in packet delivery delays.

Not all types of traffic are sensitive to delays in packet transmission, in any case, to those quantities of delays that are characteristic of telecommunication networks. Packet delays generated by file service, e-mail service, or printing service, have little effect on the quality of these services in terms of network user.

On the other hand, the same delays of packages for services transmitting voice or dance can lead to a significant reduction in the quality of information - the emergence of the effect of "echo", it is impossible to disassemble some words, image vibrations, etc.

Language information and video information are examples of traffic sensitive to delays, while data annexes are mostly less sensitive to delays. Packages in which delay delivery exceeds certain values of Tmax, are discarded [1].

In real-time applications (for example, in IP-telephony) it leads to deteriorating language quality. Restrictions related to the average IP packet delay play a key role for the successful implementation of voice data for IP (VoIP), video conferences and other real-time applications. This parameter largely determines the quality of such applications.

2) Variation of the IP packet delay (IP Packet Delay Variation, IPDV) - Spread the maximum and minimum time of passage of the package from the average.

The variation of the delay (jitter) when transmitting a packet is calculated for two network nodes and is defined as the scatter of the delay of the next packet relative to the previous one. Jitter (IPDV) is characterized by the V_k parameter. For an IP packet with index K this parameter is determined between the input and output points of the network in the form of a difference between the absolute X_k delay value when delivering a packet with an index to, and a certain reference (or reference) the magnitude of the delay in delivery of the package $d_{1,2}$ for the same network points:

$$V_k = X_k - d_{1,2}. \tag{3}$$

The reference delay in the delivery of the $d_{1,2}$ packet between the source and the recipient is defined as the absolute value of delaying the delivery of the first packet between data network points.

The variation of the packet delay, or jitter, manifests itself in the fact that successive packets arrive at the recipient in irregular moments of time. In systems of IP-telephony, for example, it leads to a distortion of sound and, as a result, it becomes illegible [2].

3) The IP packet loss factor (IP Packet LOSS Ratio, IPLR) is defined as the ratio of the total number of lost packets to the total amount taken in the selected set of transmitted and received packets:

$$IPLR = \frac{\sum_t LP}{\sum_t RP}, \tag{4}$$

where LP is the number of lost packages; RP is the number of packets taken.

Losses of packets in IP networks occur in the case when the value of delays in transmission exceeds the normalized value, which is certain as Tmax. If packets are lost, then their re-transmission is possible on requesting the receiving party.

In VoIP systems, for example, packages that came to the recipient with a delay, and exceeds Tmax leads to failures in the language [3].

Among the reasons that cause packet losses, it is necessary to note the growth of queues in the nodes of the network arising from overloads.

4). The IP error packages (IP Packet Error Ratio, IPER) are defined as the total amount of errors received with errors, to the amount of successfully accepted and packets taken with errors:

$$IPER = \frac{\sum_t RPE}{\sum_t RPS + RPE}, \tag{5}$$

where RPE is the number of packets taken with errors; RPS is the number of successfully accepted packages.

Recommendation Y.1540 defines the numerical values of parameters specifically specified in it, which must be performed in networks on international tracts connecting user terminals.

2 Statement of Research Problem

2.1 Evaluation of the Quality Indicators of the Multiservice Network

For today, the main method of research of telecommunication systems as complex dynamic systems can be considered an analysis of the dynamics of the multiservice network with differential-difference equations of the state [4, 5].

The dynamics of information exchange in the multiservice network can be described by the system with $N(N-1)$ equations of the species:

$$x_{i,j} = (k+1) = x_{i,j}(k) + \sum\nolimits_{\substack{m=1\\ m\neq i}}^{N} b_{m,i}(k)u^{j}_{m,i}(k) - \sum\nolimits_{\substack{m=1\\ m\neq j}}^{N} b_{i,m}(k)u^{j}_{i,m}(k) + y_{i,j}(k), \quad (6)$$

where $k_{i,j}$ is the volume of data on the network element i and is intended for transmission of an element j at the time k ($k = 0, 1, 2, \ldots$); N is the number of network elements;

$$b_{i,m}(k) = c_{i,m}\Delta t,\ b_{m,j}(k) = c_{m,i}\Delta t,\ (i,j = 1,\ldots,N;\ \Delta t = t_{k+1} - t_k), \quad (7)$$

$c_{i,m}$ is the bandwidth of the transmission path according to the tract to the node i and from it;

$u^{j}_{i,m}$ is the part of bandwidth of the $L_{i,m}$ tract, isolated at the time of k stream with the address j;

$y_{i,j}(k) = e_{i,j}(k)\Delta t$ is the intensity of the flow of queries entering the node i for transmission to the node j for the period Δt;

$e_{i,j}(k)$ is the intensity of the flow of requestsat the time k (the total intensity of requests from users connected to the node i and the leading exchange with users connected to the node j).

Given the limitation of the queue buffers on elements of the network and bandwidth transmission paths, a number of restrictions are imposed on the change of state and control:

$$0 \leq x_{i,j}(k) = x^{max}_{i,j}, \quad (8)$$

$$0 \leq u^{j}_{i,m}(k) \leq 1;\ \sum\nolimits_{n=1}^{N} u^{n}_{i,m}(k) \leq 1, \quad (9)$$

where $x^{max}_{i,j}$ is the maximum permissible amount of data located on the network element i for traffic with the addressee j.

The system of Eqs. (6) can be recorded in a vector-matrix form:

$$X(k+1) = A(k)X(k) + B(k)U(k) + Y(k), \quad (10)$$

where $A(k)$ is a single dimension matrix $N \cdot (N-1) \times N \cdot (N-1)$;

$X(k) = \left[x_{1,2}(k), \ldots, x_{i,j}(k), \ldots, x_{N,N-1}(k)\right]^T$ is the vector of queue length on elements of the network at the time k dimension $N(N-1)$;

$U(k) = \left[u^{2}_{1,2}(k), \ldots, u^{j}_{i,j}(k), \ldots, u^{N-1}_{N,N-1}\right]^T$ is the vector partial part of the transmission tract in the moment k;

$Y(k) = \left[y_{1,2}(k), \ldots, y_{i,j}(k), \ldots, y_{N,N-1}(k)\right]^T$ is the vector intensity coming to the flow of requests at the time k dimension $N(N-1)$;

$B(k)$ is the matrix, the elements of which according to expression (6) are values $\pm b_{i,j}(k)$.

As a criteria of optimality, the maximum performance of the system achieved in the period $K\Delta t$, which in the above-mentioned model is formalized as:

$$J = \sum\nolimits_{k=0}^{K-1} \sum\nolimits_{i=0}^{N} \sum\nolimits_{\substack{j=1\\ J\neq i}}^{N} B(k)U(k) \rightarrow max, \quad (11)$$

where K is the number of intervals Δt, for which the calculation of control variables (forecast interval) is carried out.

Checking the performance of the system manageability:

$$n = rank\left[B, AB, \ldots, A^{n-1}B\right], \tag{12}$$

where n = N(N − 1).

The delay in packages and jitter of the network described by the equation system (6) are evaluated according to (2) and (3), respectively.

Based on the system of Eqs. (6) it follows that:

$$Z_{i,j}(k) = \left(x_{i,j}(k) - m\right), \tag{13}$$

where $Z_{i,j}(k)$ is the number of lost data, m is the volume of the buffer device.

Consequently, the coefficient of packet loss is calculated by the formula (4).

According to the description of the dynamics of the multiservice network, an analytical modeling was conducted, which consists of a model consisting of two network nodes associated with each other.

The bandwidth of the transmission path between the node 1 and 2 is marked as C_{12}. Y_{12} is the intensity of the flow of queries entering the node 1 for transmission to the node 2 (the index varies between 70 to 130 req/s depending on the type of traffic). The maximum volume of buffer space in each node of the considered model is 40 requests, (the indicator varies between 10 to 40 requests) (Fig. 1).

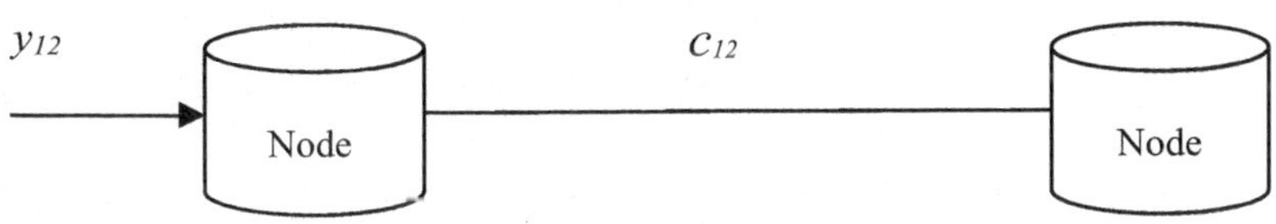

Fig. 1. The scheme of the modeled network

A fragment of the simulation model of the model that is modeled is presented in Fig. 2.

In this model, the input stream of applications plays the role of the flow of requests enters the network node in random moments of time. The random process of receiving queries is represented by the function distribution of intervals between queries. These intervals are described by Poisson distribution [6].

If at the time of receipt of the request buffer is empty and the node is free, then the request is immediately transferred to service [7].

If at the moment the buffer request is empty, but the node is busy with the service of the request received earlier, the request is expected to complete the process of processing a buffer request. The buffer is considered to be finite, that is, the requests are lost due to the exhausted buffer capacity. As soon as the node completes the service of another request, the requested request is transmitted to the exit, and the next request comes from the buffer provided that the buffer is not empty [8].

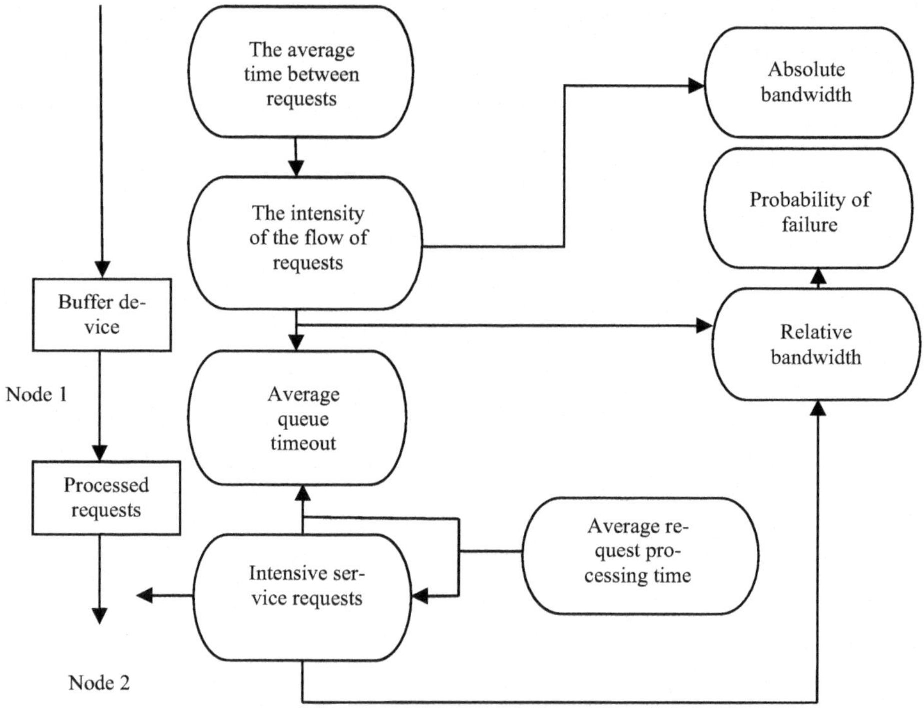

Fig. 2. Fragment of a simulation model of a multiservice network

Thus, the object that is modeled is presented in the form of an information system with feedback.

In the mathematical sense, the model represents a system of finite-difference equations solved based on a numerical integration algorithm (according to the Euler's or Runge-Kuttascheme [9]) with a constant step and given initial values.

When conducting an experiment, each network operating parameter accepts one of several values, while others are fixed. Depending on this, the value of the network quality indicators changes [10].

If the value of this indicator of the quality of the network is significantly changing when changing a certain parameter, then this quality indicator has a high sensitivity. If, with significant changes in some parameter, there are no strong changes in the values of the quality indicator, this indicates a low sensitivity to this parameter [11].

This approach provides sufficient accuracy of analysis results at a lower amount of data.

In order to determine from the number of quality indicators of those that are most sensitive to changes in the network parameters, a network behavior analysis is performed when there is a network of different traffic type: real-time traffic, data traffic and mixed traffic. 30 iterations were carried out, the averaged results of the analysis of data obtained as a result of simulation are included in Table 1.

Table 1. Changes in network quality indicators when entering the network of mixed traffic

Buffer size, requests		Bandwidth, requests/sec	Probability of losses, P_p		Latency, T_z		Jitter, V_x	
Real time traffic	Data traffic		Real time traffic	Data traffic	Real time traffic	Data traffic	Real time traffic	Data traffic
10	20	100	0,2991	0,3476	0,099	0,198	0,0083	0,0165
20	20	100	0,2972	0,3492	0,1975	0,198	0,0161	0,0165
30	20	100	0,295	0,3519	0,2955	0,198	0,0236	0,0165
40	20	100	0,293	0,3478	0,3926	0,198	0,0302	0,0165
10	10	100	0,2992	0,3696	0,099	0,099	0,0083	0,0083
10	30	100	0,299	0,3586	0,099	0,297	0,0083	0,0248
10	40	100	0,2992	0,3487	0,099	0,3959	0,0083	0,0331
10	20	70	0,5088	0,5369	0,1414	0,2829	0,0118	0,0238
10	20	80	0,4389	0,4657	0,1238	0,2475	0,0103	0,0207
10	20	90	0,3690	0,4102	0,11	0,22	0,0092	0,0184
10	20	110	0, 2293	0,2859	0,09	0,18	0,0072	0,015
10	20	120	0,1595	0,2091	0,0823	0,165	0,0068	0,0138
10	20	130	0,0897	0,1429	0,0757	0,148	0,006	0,0123

When conducting an experiment, real-time traffic has varied within 80 req/s, data traffic - from 70 to 130 req/s. The choice of partial part of the bandwidth used was based on the analysis of the provider's network.

In order to determine the quality indicators of those who are most sensitive to changes in the network parameters, the network analysis is conducted on whether the numerical values of the quality indicators are substantially changing when the assumptions about the parameters are changed in a given range. Since the parameters have the effect of interaction, that is, they act as a combined effect on quality indicators, this step is necessary.

This analysis allows you to assess the impact of the network parameters to change the value of the quality indicators selected in accordance with the point discussed above.

For the selection of critical quality indicators according to tables, a comparison of the value of the network quality indicators obtained as a result of simulation, with a limiting value set out in accordance with the selected class of service.

If the value of the quality indicator is not significantly variable when changing variable parameters, it is believed that this quality indicator has a weak sensitivity to the changing parameter, and this parameter is not critical. All uncritical parameters and corresponding quality indicators are removed from the table.

Analysis of parameters that affect, allows you to determine the indicators by different combinations of parameters and makes it possible to investigate the efficiency of each of the possible combinations.

The advantage of this method is the possibility of assessing the interaction of parameters - changes in the nature of the effect on the response function of one of the parameters depending on the value of another. This approach provides sufficient accuracy of analysis results at a lower amount of data [12].

According to the results of the analysis, a list of the most sensitive indicators of network quality [13] is made.

In the process of modeling, the main parameters of the multiservice network have changed (the intensity of the information flow, the intensity of service, etc.) and changes in the values of quality indicators were measured [14].

When transmitting real-time traffic, data must be transmitted by a uniform flow. In this case, according to the analysis of the obtained simulation results, important parameters are a delay in the package and a delay dispersion (jitter), while a partial loss of data [15] is allowed.

The data traffic, in turn, is sensitive to the integrity of the data transmitted, but it is insensitive to the time parameters. For mixed traffic, according to the results of the analysis, sensitive is as probability of losses and packet delay.

A schedule of time delay dependence, probability of losses and jitter from the basic parameters that are shown in Figs. 3, 4, 5, 6, 7 and 8.

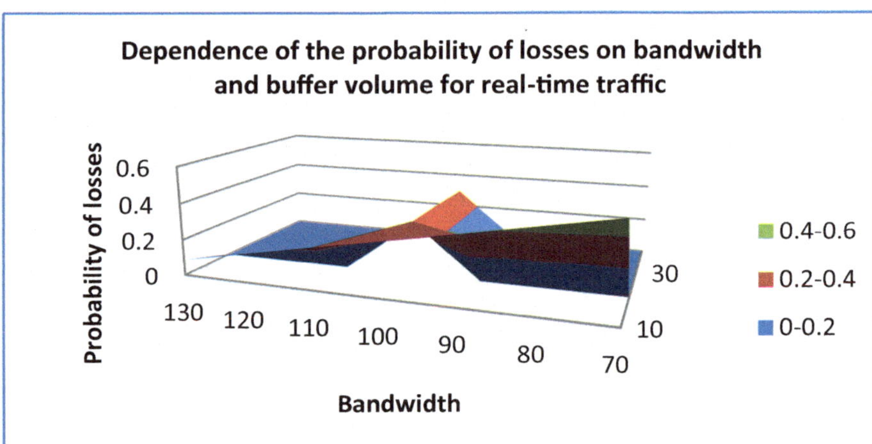

Fig. 3. Graph of the dependence of the probability of loss on the amount of buffer and bandwidth for real-time traffic

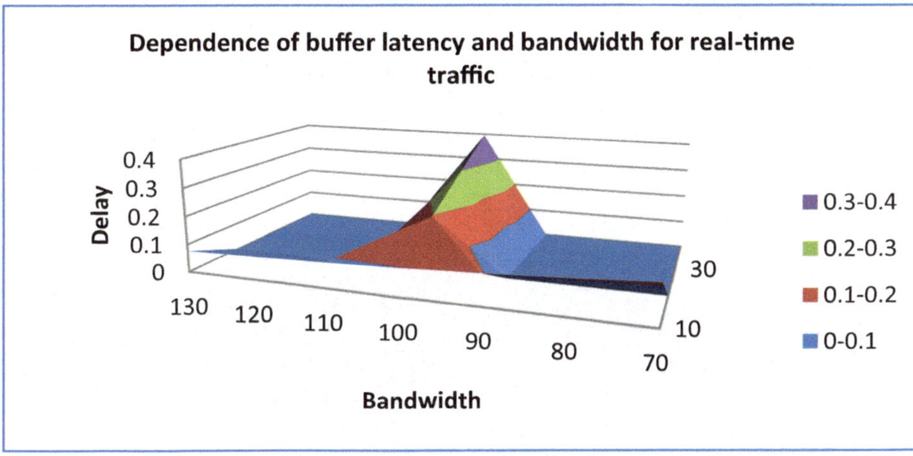

Fig. 4. Dependence of the delay on the buffer volume and bandwidth for real-time traffic

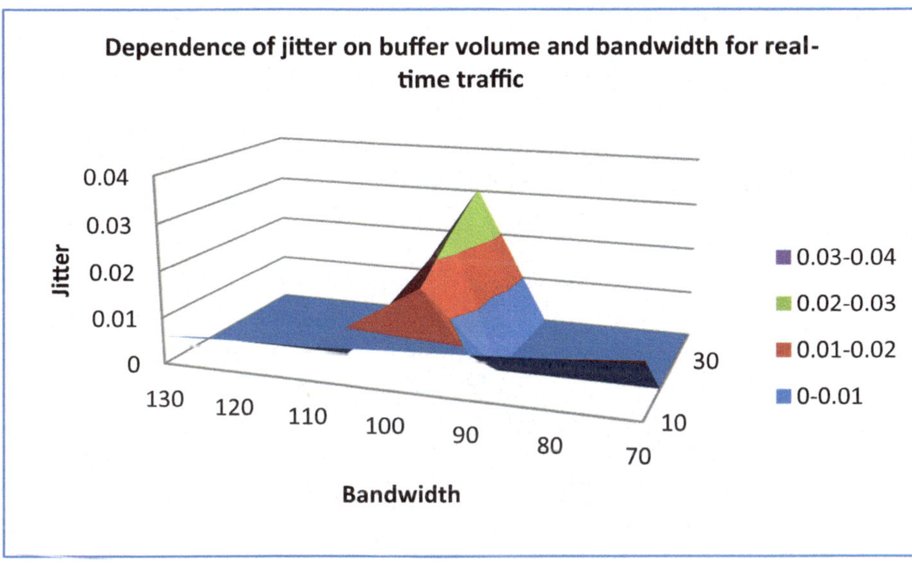

Fig. 5. Dependence of jitter on buffer volume and bandwidth for real-time traffic

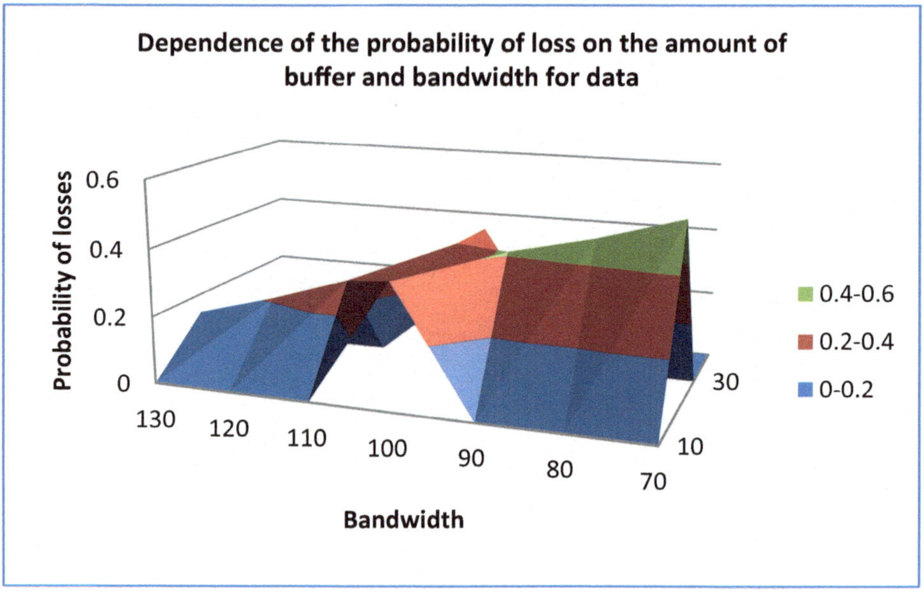

Fig. 6. Dependence of the probability of losses on the amount of buffer and bandwidth for data

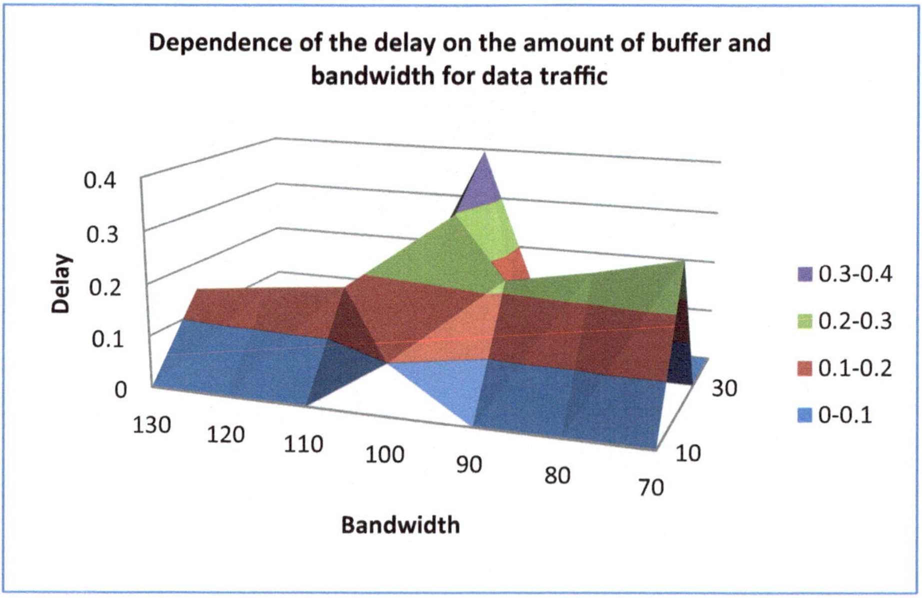

Fig. 7. Dependence of the delay on the amount of buffer and bandwidth for data traffic

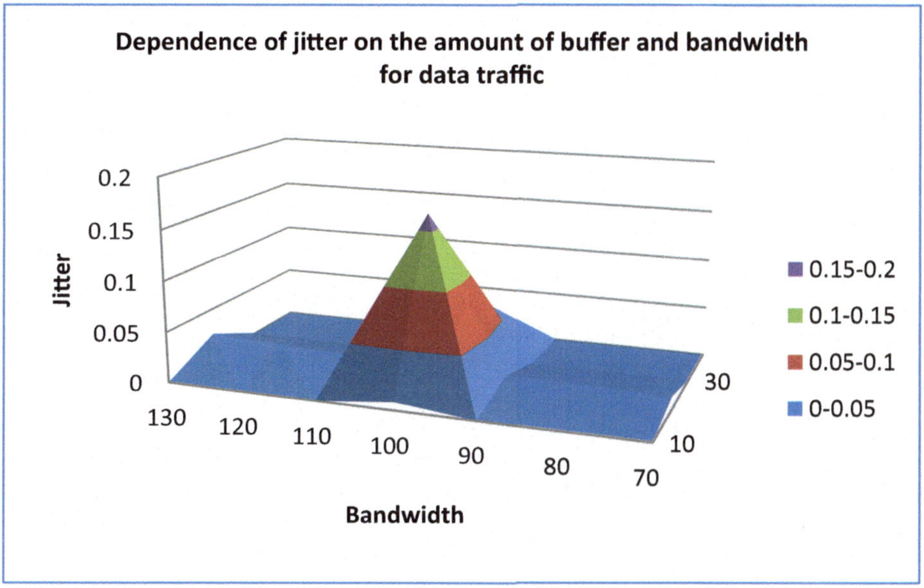

Fig. 8. Dependence of jitter on the amount of buffer and bandwidth for data traffic

3 Conclusions

The obtained data provided on the graphs (Figs. 3, 4, 5, 6, 7 and 8) showed that the most critical to changes in the parameters of the multiservice network is such a quality indicator, as a delay time, that is, a slight change in the network parameters significantly changes this quality indicator.

Depending on the sensitivity to temporary delays, the main types of services can be divided into several groups [16]:

- Asynchronous services: The delay time is practically no limited. Example: Email.
- Synchronous services: Sensitive to delays, but allow them.
- Interactive services: delays can be recorded by end users, but they do not worsen the functionality of applications. Example: a text editor that works with a remote file.
- Isochronous services: In case of delays exceeding the necessary threshold, the functionality of the program is sharply reduced. Example: The transfer of voice when it is exceeded by the threshold of delays in 100–150 ms. The quality of the reproduced voice sharply deteriorates.
- Hypersensitive to delay services. The emergence of delays in providing services reduces functionality to zero. Example: applications that control the technical object in real time. When you delay the control signal, an accident may occur.

Based on the results of the analysis to assess the dynamics of changing the state of the network, the response time, an integral part of which is the most sensitive quality indicator - delay time [17].

Network response time, according to ITU recommendations, is defined as a time interval between the user's request to any network service and receive a response to this request [18].

- the network response time consists of several components:
- time for preparing requests on user terminal;
- time transfer of requests between the user and the server through network segments and intermediate communication equipment;
- time processing requests on the server;
- time of transmission of answers from the server to the user and time of processing received from the response server to the user terminal;
- the delay time is made at each stage of the request processing.

Knowledge of constituent response time allows you to evaluate the performance of individual elements of the network, to identify bottlenecks and, if necessary, to upgrade the network to increase its overall performance.

The value of the time quality indicators of the network (in particular, the response time of the network, delay, jitter) depends on the type of service asks the user. As well as from which user and to which the instance of the service is also drawn from the current state of other elements of the network - the loading of the network elements through which the request is passed, the server loads, and so on.

Consequently, to reduce operating costs and increase the efficiency of communication network management processes and their elements require new mechanisms for providing high quality QoS service.

The solution of these problems represents an important scientific task, which determines the need for research related to the development of multiservice network management methods and an increase in the efficiency of management processes in real management systems, including the minimization of the network response time. It is also necessary to take into account that minimization of delay and response time by minimizing temporary delays in homogeneous traffic networks is provided at the design stage.

References

1. Zhurakovskyi, B., Tsopa, N., Batrakand, Y., et al.: Comparative analysis of modern formats of lossy audiocompression [electronic resource]. In: CEUR Workshop Proceedings (2020). http://ceur-ws.org/Vol-2654/paper25.pdf
2. Ibrahimov, B.G., Alieva, A.A.: Research and analysis of quality of service indicators for multimedia traffic using fuzzy logic. In: 14th International Conference on Theory and Application of Fuzzy Systems and Soft Computing, ICAFS 2020, pp. 773–780 (2020)
3. Loukodes, M.K.: Switching to VoIP, 453 p. O Relly Media Inc. (2005)
4. Druzhynin, V., Toliupa, S., Pliushch, O., Stepanov, M., Zhurakovskyi, B.: Features of processing signals from stationary radiation sources in multi-position radio monitoring systems. In: CEUR Workshop Proceedings, vol. 2746, pp. 46–65 (2020). http://ceur-ws.org/Vol-2746/
5. Zhurakovskyi, B., Boiko, J., Druzhynin, V., Zeniv, I., Eromenko, O.: Increasing the efficiency of information transmission in communication channels. Indonesian J. Electr. Eng. Comput. Sci. **19**(3), 1306 (2020)

6. Zhurakovskyi, B.Y., Moshenchenko, M.S., Zhurakovskyi, V.B.: Algorithm for detecting and troubleshooting multiservice networks. Aktualnye nauchnye issledovania v sovremennom mire, vol. 5, pp. 94–101 (2020). (in Ukrainian)
7. Rakushev, M., Kovbasiuk, S., Kravchenko, Y., Pliushch, O.: Robustness evaluation of differential spectrum of integration computational algorithms. In: Proceeding of the 2017 4th International Scientific-Practical Conference Problems of Infocommunications Science and Technology, PIC S and T 2017, January 2018, pp. 21–24 (2018)
8. Zhurakovskiy, B., Tsopa, N.: Assessment technique and selection of interconnecting line of information networks [electronic resource]. In: Proceedings of the 3rd International Conference on Advanced Information and Communications Technologies (AICT), pp. 71–75 (2019). https://doi.org/10.1109/AIACT.2019.8847726
9. Euler, L.: Integral Calculus, vol. 1, 415 p. GITTL, Moscow (1956). (in Russian)
10. Berkman, L., Tkachenko, O., Turovsky, O., Fokin, V., Strelnikov, V.: Designing a system to synchronize the input signal in a telecommunication network under the condition for reducing a transitional component of the phase error. East.-Eur. J. Enterp. Technol. **1**(9(109)), 66–76 (2021)
11. Nedashkivskiy, O., Havrylko, Y., Zhurakovskyi, B., et al.: Mathematical support for automated design systems for passive optical networks based on the β-parametric approximation formula. Int. J. Adv. Trends Comput. Sci. Eng. **9**(5), 8207–8212 (2020). https://doi.org/10.30534/ijatcse/2020/186952020
12. Berkman, L., Tkachenko, O., Kriuchkova, L., Varfolomeeva, O.: Determination of criteria for choosing the best ways and indicators of service quality in infocommunication networks. In: Proceedings of the 2019 3rd International Conference on Advanced Information and Communications Technologies, AICT 2019, pp. 223–226 (2019)
13. Kremenetskaya, Y., Makarenko, A., Markov, S., Koval, V.: Limitations of efficiency of wireless systems of telecommunications 5G and methods of their compensation. In: Proceedings of the 2019 IEEE International Scientific-Practical Conference: Problems of Infocommunications Science and Technology, PIC S and T 2019, pp. 493–496 (2019)
14. Tkachenko, O., Eremeev, Y.: Methods of measuring of loading and indexes of quality of service. In: Proceedings of the 10th International Conference on Modern Problems of Radio Engineering, Telecommunications and Computer Science, TCSET 2010, p. 254 (2010)
15. Bichkov, O.S., Nakonechnyi, V.S., Lukova-Chuiko, N.V., Zhebka, V.V., Panayotova, G.S., Dimitrov, D.A.: Measuring the effectiveness of a radio-identification system. J. Commun. **15**, 669–675 (2020). https://doi.org/10.12720/jcm.15.9.669-675
16. Olifer, N.: Service quality. A simple increase in throughput does not guarantee the performance of applications. Zhurnal setevyh reshenii/LAN (2002). (in Russian)
17. Pliushch, O., et al.: Performance study of spread spectrum systems with hard limiters. Int. J. Comput. Netw. Inf. Secur. **12**(5), 1–15 (2020)
18. Krukhmalev, V.V.: Fundamentals of building telecommunication systems and networks, 510 p. Goriachaialiniia: Telecom, Moscow (2004). (in Russian)

Intelligent Detection of DDoS Attacks in SDN Networks

Nazar Peleh, Olha Shpur(✉), and Mykhailo Klymash

Lviv Polytechnic National University, S. Bandery Street 12, Lviv, Ukraine
van_plus_k@ukr.net, {olha.m.shpur,mykhailo.m.klymash}@lpnu.ua

Abstract. In this chapter we propose intelligent detection of DDoS attacks in SDN networks based on log analysis. Due to SDN management and implementation of the self-learning element, we propose to teach the SDN controller to detect attacks using information about the state of the flow, the duration of the session and its source, using information from logs and flow tables. To do this, it is necessary to divide the total traffic flow into anomalous and normal. By identifying client requests that are the result of DDoS-attack, one can create the appropriate rules for their blocking. We propose to do this by determining the metrics of traffic behavior using the Kulbak-Leibler approach to detect flow anomalies over the session time. As a result of using machine learning, the SDN controller will block IP domains from which DDoS attacks are just starting.

Keywords: DDoS attack · SDN · Log analysis · Machine learning

1 Introduction

Increased data security from unauthorized access puts new demands on network infrastructure management. In the case of software management, this task becomes a priority. However, further increasing or improving rules management or prioritization of traffic flows, as a way of monitoring and managing network security systems do not satisfy the demand of owners of network infrastructures. This requires a monitoring system for each system that will analyze potential problems that may arise during the work, and run protection system in the service or to notify administrators about possible threats. These systems have a very short time to track changes that occur with the system – this is various injection attacks or technical failures, in which the service ceases to function normally. In addition, not only track, but also to predict in advance the possible dangers due to possible implementation of Deep Learning in the core network infrastructure.

One of the effective means of tracking and preventing this type of problem is to control the time of each session set by the user and analyze the logs as well as service information. This type of analysis should be performed at the network core level by logging all requests at the stage when the incoming packet passes the network firewall and is redirected to the Black Hole.

Many investigations have been devoted to the analysis and monitoring of IT systems and the availability of web services. Scientists Hu Q., Tang, B., and Lin, D. [1] suggested

M. Klymash et al. (Eds.): *Future Intent-Based Networking*, LNEE 831, pp. 210–222, 2022.
https://doi.org/10.1007/978-3-030-92435-5_12

that each event in the history of the log file is assigned its own weighting factor. It is determined based on the ratio between the number of unique user requests during the hour of its activity in the system, observed during n days. Thus, events were classified into normal (when the weight was weighted average) and anomalies.

Max Landauer et al. [2] suggests detecting unnormal behavior based on grouping of rows of logs by similarity to establish static cluster maps. The rows of logs are divided into existing cluster maps created in the previous and subsequent time intervals. The overlap metric is calculated, which determines the probability of transition from cluster to cluster. This is based on the relationship between two neighboring clusters of clusters (word "clusters" two times?) of cards that do not have common elements. At each subsequent stage of the system, additional clustering is performed to create new static maps.

Implementation of machine learning can solve problems monitoryng availability and protect web services, particularly in the framework of the concept of program-controlled data centers.

The authors of [3] focused their research on deep machine learning, which works on the basis of the Restricted Boltzmann Machine (RBM) method. As a result, they introduced a secure framework with SDN management and IDS structure. The studies were performed on the basis of Tensorflow with KDD99 input set. The proposed algorithm showed 94% accuracy. Another variant of IDS is presented in [4]. Researchers simulate secured network by introducing identifiers of attacks types and parallel neural cross-training context. Implementation of IDS on the basis of machine learning and software control was carried out in [5, 6], in particular, the authors [6] proposed an alternative scenario for the operation of intelligent transport SDN networks.

Lin and Wang researchers proposed a mechanism for detecting and protecting DDoS attacks [7]. Their research is based on a method that separates Openflow and sFlow controls for anomaly detection. As a result the work and deployment of such protection is a rather difficult task. A more accurate method of detecting attacks is proposed in [8]. Yang et al. proposed a method which is based on the value of the entropy between the flow information and the average value of the entropy of the flow. Even though the information entropy is more accurate for detecting anomalies, it still needs to be combined with other technologies in determining the threshold and multi-element weight distribution. Said et al. [9] investigated that based on the analysis of the characteristics of each TCP/UDP/ICMP protocols, the method of detecting abnormal traffic should distinguish between the packet protocol, and using the ANN training algorithm to detect DDoS attacks. However, this is quite difficult, as it is necessary to constantly analyze all packet headers.

On the other hand the use of statistical approaches to anomaly detection, are not out. The authors of paper [10] for detect DDoS-attacks suggest using the SOM algorithm, which works by extracting flow statistics in a certain time interval. But, the disadvantage of this method is that the behavior of the attack is not timely and inaccurate. In [12] the authors proposed the mechanism for detecting DDoS attacks based on the analysis of anomalous characteristics of the source IP address and destination IP address. However, the method does not take into account and does not adjust a certain threshold of IP address anomaly values.

Underlying reviewed scientific papers are mainly statistical learning methods of protection. In addition, some of them require additional analysis of network traffic. In this chapter, we would like to present the system concept that allows to detect and predict DDoS attacks through the introduction of Deep learning, SDN management and log analysis.

2 Concept of Intelligent Detection of DDoS Attacks in SDN Networks Using Machine Learning

2.1 General Concept

Today, it is difficult to imagine a system that works with web applications without software control. Transmission, management and collection of statistical information is carried out by the SDN controller. It searches for records in group flow tables from one or/more interfaces or applications. When it comes to applications, the controller redirects request flows to the required application. In this case, the controller will be responsible for the security of these applications.

Web applications are most vulnerable to DDoS attacks. Failure of certain parts of a web application can lead to loss of system performance and loss of confidential user information. Today there is no universal tool to counteract DDOS-attacks. To counter distributed denial-of-service attacks, there are two main tasks to be followed:

1. DDOS attack should be detected as soon as possible.
2. The overall traffic flow should be identified as normal and abnormal. By knowing which of the client requests are the result of an attack, we can create appropriate rules to block them.

To detect DDoS attacks as soon as possible, we suggest using machine learning: to teach the SDN controller to detect attacks using information about the state of the flow, the duration of the session and its origin. We propose to obtain this information on the basis of log analysis and data from the controller's flow table.

To solve the problem of identifying normal and anomalous traffic, we propose to determine the metrics of traffic behavior using the Kulbak - Leibler approach to detect flow anomalies over the duration of the session. The basis of the proposed concept is shown in Fig. 1.

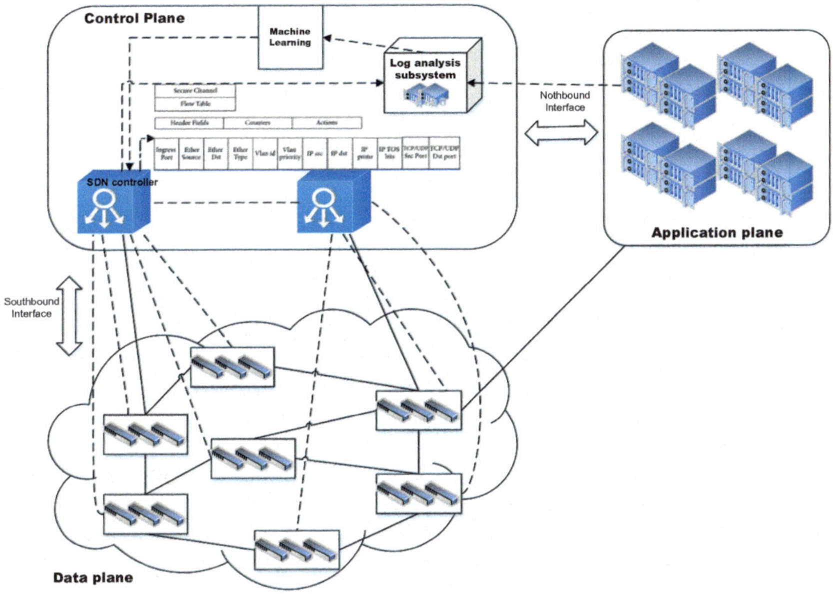

Fig. 1. The proposed concept of SDN architecture with DDoS attacks detection

2.2 The Principle of Log Analysis

Most security systems operate at the network core. Some incoming requests are redirected to the Black Hole. The only possible way to increase the effectiveness of system protection is to log all requests in the firewall. By splitting the log file, you can find out all the necessary information. Therefore, the developed system analyzer will operate after the firewall.

Continuous reading from the file will be quite costly in terms of resources. That is why we create a database and transfer to it all log files. This will increase the efficiency of search and filtering logs. The most ideal option for this is to use a so-called in-memory database, as it holds the data being analyzed in RAM. This allows to quickly perform recording and search operations.

Figure 2 shows the algorithm for writing/reading the log to the database. For some simplification and study sample was taken on the log form, that:

$$[{}''yyyy'/'MM'/'dd' : 'HH' : 'mm' : 'ss']\_ipaffress\_ID\ port. \qquad (1)$$

This log records the main parameters: date and time of connection, IP address of the client and port number (socket), the amount of time the client spent connected to the server (in seconds). To recognize DoS-attacks we offer an algorithm that can identify the attacker among others clients (Fig. 3).

Log analysis subsystem analyzes all requests to web services during the last 24 h from the moment when suspicious request was identified. We separate the IP address from which the largest number of requests have been sent in the last 24 h. To check whether the sequence ofrequests is typical and does not differ from the number of requests of

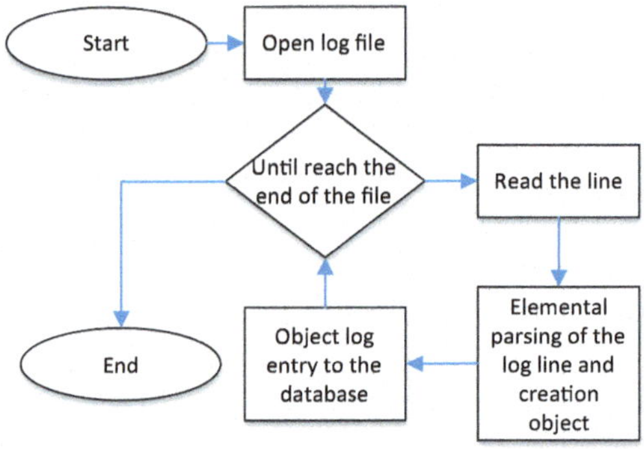

Fig. 2. Simplified steps for transferring logs to the database

other users, we determine the average value of the number of requests from all other users, except the maximum number of requests. Let M_{23} denote an average number of requests from all other users. After that, we will compare whether the maximum number of requests will be greater than the average. To do this, we introduce a correction factor that will be equal to 10. This will cut off bursts of traffic during peak hours from real DoS attacks. Another important parameter for us is the time of this session.

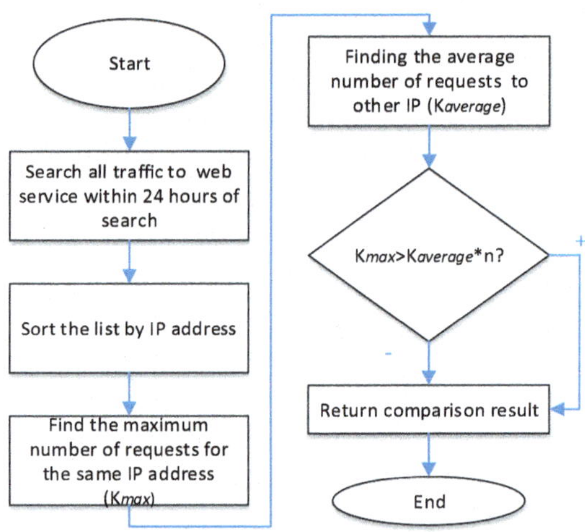

Fig. 3. Traffic analysis algorithm of the DDoS attack [13]

Let's define $\{T_{access}\}$ as the time to access the web server from multiple IP addresses $\{P\}$. Based on this information, we will train the system. To do this, we determine the relative entropy of Kulbak-Labler. We will compare the time of the session with the time to access the web server from specific IP addresses, which were sorted as a result of the algorithm presented in Fig. 3. To determine the Kullback-Leibler (KL) divergence, which is a measure of how much one probability distribution differs from another reference probability distribution, we calculate the information entropy of both distributions and find their difference. In our case it is determined based on the expressions 2 and 3.

$$KL(T_{aver}||T_{access\,last\,hour}) = \sum T_{aver}(P)\log\frac{T_{aver}(P)}{T_{access\,last\,hour}(P)}, \tag{2}$$

$$H(T_{aver}) = \log T_{aver}(P) - KL. \tag{3}$$

The calculation results are recorded in a database.

The machine learning algorithm "Decision Tree" was chosen for the analysis of logs because of its simplicity, reliability, and relevance. In this case, clients should be classified into categories as follows: if the value of the information entropy of both distributions and their difference is quite large then the attacker is detected and the controller creates a rule that blocks the corresponding IP address and port. If it is not possible to determine whether there is an attack or not, there is a comparison of time to access the service for the last seven days. The accumulation of KL values in the ML database will allow to detect anomalies in the receipt of request flows, based on the analysis of time to access the service and prescribe the rules of the controller. As a result of using such training and constant monitoring of sessions, the SDN controller will block IP domains from which DDoS attacks are just beginning.

The general scheme of this algorithm is shown in Fig. 4.

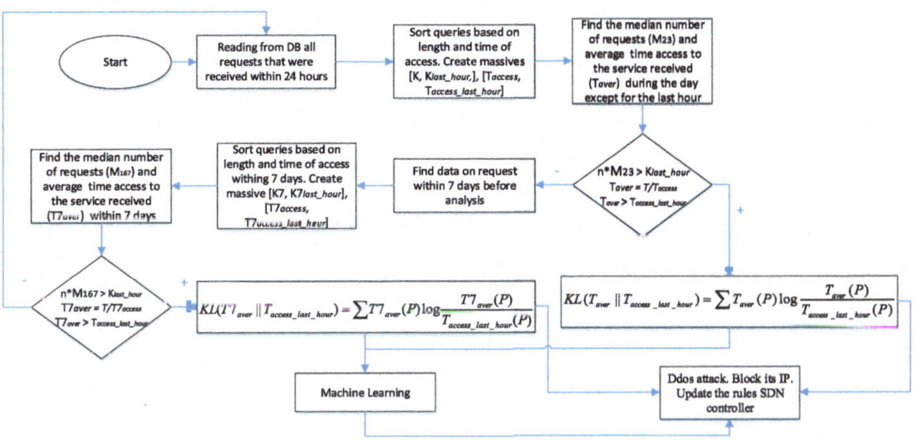

Fig. 4. The algorithm that recognizes the DDoS attack

3 Implementation of Protection from DDoS Attacks Based on Analysis of the Service Information

3.1 Web Applications Security Monitoring Using Log Analysis Subsystem

For investigation of efficiency of the developed system for detecting DDoS attacks we performed simulation using Java and Python programming languages. The socket module of the Python programming language is used to establish a connection between clients and the server. This module provides a user-friendly and consistent interface for API Berkeley sockets. Log data from the file generates data according to the normal distribution law. Before starting the web server, random data is generated and written to a log file. Data in file is not consistent over time, so they are further sorted before recording to the database.

For correct work 150 000 entries are written into the event Logs.txt file. For them, the IP address and time are generated according to the Gaussian distribution law and in case of error it is recorded. The contents of the log file will appear as shown in Fig. 5.

```
[2021-05-24 22:22:20] 192.168.56.1 62758 n3      12
[2021-05-24 22:48:04] 192.168.56.1 63074 n1 af   52
[2021-05-24 22:49:54] 192.168.56.1 63108 n4 af   22
[2021-05-24 23:16:57] 192.168.56.1 63478 n1 af   10
[2021-05-24 23:18:29] 192.168.56.1 63489 n7 af   12
[2021-05-24 23:21:46] 192.168.56.1 63674 n1 af   12
[2021-05-24 23:23:29] 192.168.56.1 63705 n9 af   41
[2021-05-24 23:24:00] 192.168.56.1 63718 n2      11
[2021-05-25 00:27:31] 192.168.56.1 64388 n4      12
[2021-05-25 00:29:12] 192.168.56.1 64415 n7      42
[2021-05-25 00:40:25] 192.168.56.1 64556 n2      12
[2021-05-25 00:40:42] 192.168.56.1 64562 n3      11
[2021-05-25 00:40:56] 192.168.56.1 64563 n1      11
[2021-05-25 14:41:38] 192.168.56.1 49975 n1      55
```

Fig. 5. An example of the records of logs generated in a file

After starting the server, a non-blocking "incoming server" socket is created on a predefined port. The data of this socket is written to the INPUTS and OUTPUTS lists. From there, they are written to the ReadList and WriteList lists using the select module of the socket library. In the infinite loop the polling of events and processing of the results of events is happening. In this step, information about the new connection from the client is written to the lists. A special function parses the data coming to the main descriptor lists INPUTS and OUTPUTS, and writes the ports to special lists for ports, the InputPorts and OutputPorts. After that, the program begins to process the request from the client, identifieswhich client has connected and which file the client needs, adds client data to the OUTPUTS and OutputPorts list, generates a message for the client, records the time when it was generated and sends it to the client. This time will be considered as the connection time and logged along with the IP, port number and customer ID. After these procedures, the client can initiate repeated requests to retrieve other files, or disconnect. If the client re-initiates the request, the steps are repeated. If the client is disconnected, the server records the disconnection time and calculates the total amount of time of this connection, and writes it to the log. As a next step all client data is deleted from all lists (Fig. 6).

```
_____________________________ after handling_OutputEvents ### 2
  client
<socket.socket [closed] fd=-1, family=AddressFamily.AF_INET, type=SocketKind.SOCK_STREAM, proto=0>
INPUTS:
[<socket.socket fd=600, family=AddressFamily.AF_INET, type=SocketKind.SOCK_STREAM, proto=0, laddr=('192.168.56.1', 9753)>]
List of InputPorts: ['9753']
readList:
[]
OUTPUTS:
[]
List of OutputPorts: []
writeList:
[<socket.socket [closed] fd=-1, family=AddressFamily.AF_INET, type=SocketKind.SOCK_STREAM, proto=0>]
messageForSocket:
{}

Time of Connect:  2021-05-29 18:24:06.888116
Time of Disconnect:   18:24:12.959
Time on server: 6
Connection closed, data removed.
```

Fig. 6. An example of the clear lists and record time

To simplify the task, the MAC address was not taken into account. In order to distinguish a regular client from an attacker, the number of files that the client received was entered as an additional feature.

Logback was selected as the analyzer for logging messages. This wrapper allows to configure different options for recording logs in different directions. In this case, the file entry can be configured to split logs into different files. If the log is out of date, it will be automatically overwritten. Logging can be divided into 2 stages: load test day compared to the average level of the test and the number of requests in the current hour and an average of 7 days. It will trace the attack, which lasted more than an hour to keep under attack as a web server for a long time - costly process. Figure 7 shows the result of how the system is tested for attack using the proposed method. As shown in the figure at the moment there is no attack.

```
: if number of visits for current hour more than average number of visits for 7 day  * 10: false
: number of visits for current number 0
: average number of visits for 7 day 222
: Ddos false
```

Fig. 7. Logging for DDoS attack

To verify the work, we will change the data in the database that will allow to understand if analyzer operates correctly. We will conduct 2 simulations. The first simulation will be responsible for checking the presence of the attack in the current hour, when checking with requests for one day. The second test will correspond to the situation when the first test gave a negative result. After that, we check whether the attack lasts a long time.

The result of the first check is presented in Fig. 8. With 260 requests for the current hour and an average of 25 requests, our web server is attacked. The logic of the analyzer took requests from the database and counting them gave the answer that the attack is happening.

```
number of visits for current hour 260
average number of visits for day 25
Number of visits for last hour more than average number of visits for a day
Ddos true
```

Fig. 8. The first step of the attack test

The result of the second test is shown in Fig. 9. In this case, the data for the current hour and the average value of requests for the week were taken into account. The results show that the average number of requests for the current day is not very different from the maximum, which allows you to pass the test with a negative result for the current day and go to the test for the week.

```
number of visits for current hour 3000
average number of visits for day 2500
if number of visits for current hour more than average number of visits for 7 day  * 10: true
number of visits for current number 3000
average number of visits for 7 day 221
Ddos true
```

Fig. 9. The second step of the attack test

3.2 Attack Detection Using Kulbak-Leibler Approach

We simulated the determination of the divergence between different session time using the Kulbak-Leibler approach. For this purpose we used the Python programming language and math library. Figure 10 shows the probability of events (note as P and Q) depending on session duration.

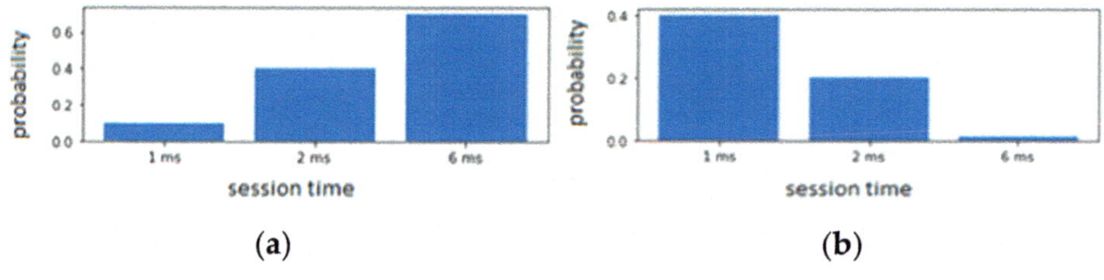

Fig. 10. Probability of events P (a) and Q (b) depending on session duration

We calculated the average value of the probability of each event and the divergence between them using the Kulbak-Leibler algorithm (Table 1).

In another experiment the probability of two events is almost the same for each session time (Fig. 10).

When we calculated the average value and the divergence between P and Q, we can see that the divergence has decreased (Table 2) (Fig. 11).

The obtained results confirm the effectiveness of using the proposed approach to differentiate between normal client requests and DDoS-attacks.

Table 1. Value and divergence between P and Q using the Kulbak-Leibler algorithm

P = 1.200 Q = 0.610	
KL(P ǁ Q):	4.490 bits
KL(Q ǁ P):	0.539 bits

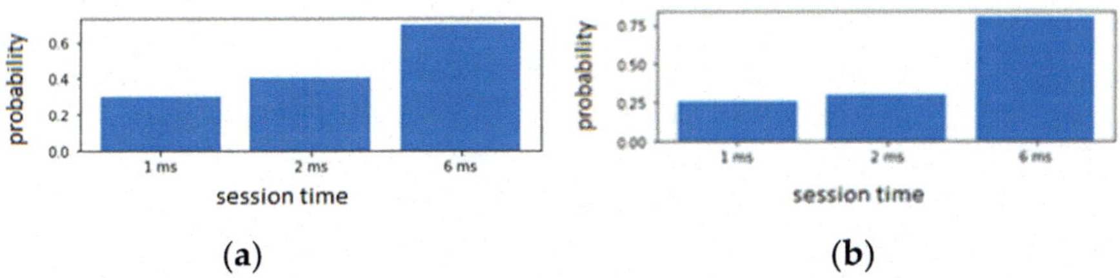

Fig. 11. Probability of events P (a) and Q (b) depending on time duration provided a similar probability

Table 2. Value and divergence between P and Q provided a similar probability

P = 1.400 Q = 1.350	
KL(P ǁ Q):	0.110 bits
KL(Q ǁ P):	−0.036 bits

3.3 Machine Learning and Attack Detection

As noted above, all information received is sent to the database for machine learning where algorithm is taught based on "Decision tree". A separate edited dataset was created for the training sample, which is located in the Log-learn.txt file. The model is constructed using DecisionTreeClassifier function of the sklearn library (Fig. 12).

In this dataset, the key column, with yes and no, indicates whether the client is attacker or not.

After building the model, test predictions were made to assess its quality. The first two predictions from the dataset (4 files and 13 s of time and 1 file and 57 s of time) are correct. The following tested data that are not in the dataset (1.60; 1.65; 1.70). All of them are characteristics of malicious clients and all of them were correctly identified by the program.

Because the log analysis has been successfully tested and proven to be effective, it can be used to detect malicious customers. The program parses the log and substitutes the data obtained from each connection into the model, which checks whether the client is malicious or not. If the client is a snapshot, its ID is stored in the Check.txt file (Fig. 13).

To test the effectiveness of the proposed solutions, we will attack the appropriate web service. The results obtained are presented in Fig. 14 which shows that the traffic

4	13	no
4	12	no
4	60	no
4	16	no
1	57	yes
1	70	yes
1	12	no
1	22	no
1	61	yes
1	73	yes
4	12	no
1	60	yes
4	16	no
1	47	no
1	70	yes
1	12	no
1	82	yes
1	61	yes

	Files	Time
0	4	13
1	4	12
2	4	60
3	4	16
4	1	57
5	1	70
6	1	12
7	1	22
8	1	61
9	1	73
10	4	12
11	1	60
12	4	16
13	1	47
14	1	70
15	1	12
16	1	82
17	1	61

	Key
0	no
1	no
2	no
3	no
4	yes
5	yes
6	no
7	no
8	yes
9	yes
10	no
11	yes
12	no
13	no
14	yes
15	no
16	yes
17	yes

Fig. 12. Training dataset and result of attack detection

```
['[2021-05-25', '14:41:38]', '192.168.56.1', '49975', 'n1', 'of', '', '55\n']
Answer is:  yes
['[2021-05-25', '14:41:59]', '192.168.56.1', '49983', 'n3', 'of', '', '12\n']
Answer is:  no
['[2021-05-25', '14:45:41]', '192.168.56.1', '50068', 'n3', 'of', '', '11\n']
Answer is:  no
['[2021-05-25', '14:46:27]', '192.168.56.1', '50093', 'n1', 'of', '', '12\n']
Answer is:  no
['[2021-05-25', '14:46:48]', '192.168.56.1', '50099', 'n5', 'of', '', '11\n']
Answer is:  no
['[2021-05-25', '14:47:07]', '192.168.56.1', '50105', 'n2', 'of', '', '12\n']
Answer is:  no
['[2021-05-25', '14:55:31]', '192.168.56.1', '50217', 'n4', 'of', '', '70\n']
Answer is:  yes
```

Fig. 13. The analysis of the log and identify malicious clients.

analyzer, which works on the basis of our proposed algorithm, tracked the moment when the DDoS attack occurred.

Each time a client connects to the server, the server checks the client ID with the malicious client IDs stored in the Check.txt file. If they match, the controller blocks the client connection (Fig. 15).

Fig. 14. Dependences of service traffic (a) and the hourly load on web service (b) relative to the number of hours

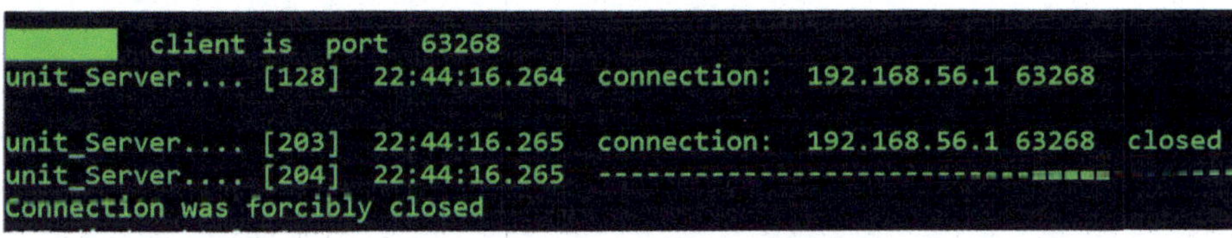

Fig. 15. Blocking by the controller of the malicious client

4 Conclusion

In this chapter we continue to explore the availability of web services in software-defined networking and detecting/predicting DDoS attacks based on log analysis. Due to SDN management and implementation of the self-learning element, we propose to teach the SDN controller to detect attacks using information about the state of the flow, the duration of the session and its source, using information from logs and flow tables. To do this, it is necessary to divide the total traffic flow into anomalous and normal. Understanding which client requests are the result of DDOS attack, we can create the appropriate rules for their blocking. We propose to do this by determining the metrics of traffic behavior using the Kulbak-Leibler approach to detect flow anomalies over the session time. In our case, we will compare the average session time with time to access the server from specific IP addresses. The obtained values will be recorded in the Machine Learning database. If the result of the comparison did not bring results, the duration of access to the service during the last seven days is compared. Similarly, the value of KL is determined and written to the ML database. KL accumulation values in a ML will identify anomalies in the flow admission requests by analyzing the length of service and access to prescribed rules of controller.

References

1. Hu, Q., Tang, B., Lin, D.: Anomalous user activity detection in enterprise multi-source logs. In: Proceeding of the IEEE International Conference on Data Mining Workshops (ICDMW), New Orleans, pp. 797–804 (2017)

2. Landauer, M., Wurzenberger, M., Skopik, F., Settanni, G., Filzmoser, P.: Dynamic log file analysis: an unsupervised cluster evolution approach for anomaly detection. Comput. Secur. **79**, 94–116 (2018). https://doi.org/10.1016/j.cose.2018.08.009
3. Dawoud, A., Shahristani, S., Raun, C.: Deep learning and software-defined networks: towards secure IoT architecture. Internet Things **3–4**, 82–89 (2018). https://doi.org/10.1016/j.iot.2018.09.003
4. Smith, R., Zincir-Heywood, A., Heywood, M., Jacobs, J.: Initiating a moving target network defense with a real-time neuro-evolutionary detector. In: Proceedings of the 2016 on Genetic and Evolutionary Computation Conference Companion, New York, pp. 1095–1102 (2016)
5. Zhang, H., Wang, Y., Chen, H., Zhao, Y., Zhang, J.: Exploring machine-learning-based control plane intrusion detection techniques in software defined optical networks. Opt. Fiber Technol. **39**, 37–42 (2017). https://doi.org/10.1016/j.yofte.2017.09.023
6. Raj, A., Truong-Huu, T., Mohan, P., Gurusamy, M.: Crossfire attack detection using deep learning in software defined ITS networks. In: Proceedings of 89th Vehicular Technology Conference (VTC2019-Spring), Kuala Lumpur, Malaysia (2019). https://doi.org/10.1109/VTCSpring.2019.8746594
7. Lin, H., Wang, P.: Implementation of an SDN-based security defense mechanism against DDoS attacks. In: Proceedings of the 2016 Joint International Conference on Economics and Management Engineering (ICEME 2016) and International Conference on Economics and Business Management (EBM 2016), Pennsylvania (2016). https://doi.org/10.12783/dtem/iceme-ebm2016/4183
8. Yang, J.G., Wang, X.T., Liu, L.Q.: Based on traffic and IP entropy characteristics of DDoS attack detection method. Appl. Res. Comput. **33**(4), 1145–1149 (2016)
9. Saied, A., Overill, R., Radzik, T.: Detection of known and unknown DDoS attacks using artificial neural networks. Neurocomputing **172**, 385–393 (2016). https://doi.org/10.1016/j.neucom.2015.04.101
10. Braga, R., Mota, E., Passito, A.: Lightweight DDoS flooding attack detection using NOX/OpenFlow. In: Proceedings of the 35th Annual IEEE Conference on Local Computer Networks, LCN 2010, Denver, pp. 408–415 (2010)
11. Bawany, N., Shamsi, J., Salah, K.: DDoS attack detection and mitigation using sdn: methods, practices, and solutions. Arab. J. Sci. Eng. **42**(2), 425–441 (2017). https://doi.org/10.1007/s13369-017-2414-5
12. Wang, X., Chen, M., Xing, C., Zhang, T.: Defending DDoS attacks in software-defined networking based on legitimate source and destination IP address database. IEICE Trans. Inf. Syst. **E99D**(4), 850–859 (2016)
13. Klymash, M., Peleh, N., Shpur, O., Hladun, S.: Monitoring of web service availability in distributed infocommunication systems. In: Proceedings of the 15th International Conference on Advanced Trends in Radioelectronics, Telecommunications and Computer Engineering, TCSET-2020, Lviv-Slavske, Ukraine, pp. 723–728 (2020)

Mathematical Methods of Reliability Analysis of the Network Structures: Securing QoS on Hyperconverged Networks for Traffic Anomalies

Nina Kuchuk[1](✉), Andriy Kovalenko[2], Heorhii Kuchuk[1], Vitaly Levashenko[3], and Elena Zaitseva[3]

[1] National Technical University «KhPI», Kyrpychova str., 2, Kharkiv, Ukraine
{nina_kuchuk,kuchuk56}@ukr.net

[2] Kharkiv National University of Radio Electronics, Nauky Ave., 14, Kharkiv, Ukraine
andriy.kovalenko@nure.ua

[3] University of Zilina, Univerzitna 8215/1, Zilina, Slovakia
{vitaly.levashenko,elena.zaitseva}@fri.uniza.sk

Abstract. The chapter considers the approach to providing QoS in hyperconverged networks with traffic anomalies. Analytical dependences for calculation of statistical characteristics of traffic on its samples are offered. It is proved that all considered statistical characteristics are unambiguously determined by means of only three parameters: fractal exponent; the intensity of the traffic flow process; fractal setup. A mathematical model of anomalous traffic is proposed. The model is adequate to real traffic and takes into account the fractal nature of the anomaly. The model uses the properties of scale invariance. Packet losses will be compensated by an increase in message transmission time, which leads to the formation of long statistically temporary dependencies. In the obtained model, the influence of losses and the cause of extended dependences are formally taken into account by introducing the fractional integration operation. The model of anomalous traffic of a hyperconvergent system was used to construct a short-term forecast. The forecast is received by the system hypervisor and used to quickly reallocate resources. As a result, QoS performance improves. The provision of QoS in the hyperconvergent network of the Cisco HyperFlex HX220c M4 Node system in the event of traffic anomalies using the proposed approach has been shown experimentally.

Keywords: Hyperconverged stratification · Anomaly · Fractal traffic · Hurst parameter

1 Introduction

1.1 Motivation

Currently, the popularity of hyperconvergent platforms is growing rapidly [1, 2]. Such a platform involves combining memory, computing, software and network resources into a single cluster. [1]. The cluster is managed by a hypervisor [2]. The advantages of a hyperconvergent infrastructure are:

M. Klymash et al. (Eds.): *Future Intent-Based Networking*, LNEE 831, pp. 223–241, 2022.
https://doi.org/10.1007/978-3-030-92435-5_13

1) simplification of the management infrastructure, it allows to optimize day-to-day maintenance;
2) large-scale storage capacity, this makes it easier and faster to add additional gigabytes;
3) fast preparation and resource allocation;
4) faster response of information technology;
5) simplified cloud migration makes it easier to implement private or hybrid clouds;
6) enhanced control provides simultaneous control of multiple functions and devices.

However, with all the advantages, hyperconverged platforms have a number of common disadvantages. They make them much more difficult to use in practice. The main disadvantage is the decrease in QoS values due to centralized management. Therefore, in case of traffic anomalies, the probability of QoS provision decreases, sometimes significantly.

One of the approaches to solving this problem is the operational redistribution of system resources. The basis for redistribution can be a forecast of traffic behavior based on its reports, which are carried out in real time.

1.2 Analysis of Related Works

Many scientific works were devoted to the study of the issues of fulfilling the QoS requirements [3–17]. They considered different approaches. So, works [3–5] suggest optimal resource allocation. In works [6, 7] information flows are redistributed. In work [8] autonomous clusters are considered. In work [9] mobile components of computer systems are considered. A number of works are focused on using the results of short-term traffic forecasting [10–17]. But in all the studies reviewed, the features of abnormal traffic in the hyperconverged network are not taken into account.

1.3 Goals and Structure

The goal of this chapter is to develop an approach to ensuring QoS in hyperconverged networks in case of traffic anomalies. The method should be based on a model that describes traffic anomalies in the hyperconverged network. Based on the simulation results, the network hypervisor must perform an operational reallocation of resources.

In the second section, analytical dependencies are proposed for calculating the statistical characteristics of traffic based on its counts. Section 3 defines the boundaries of changes in the main parameters in case of traffic anomalies of the hyperconverged network. In the 4th section, a mathematical model of abnormal traffic is proposed. Section 5 analyzes the results of the experiment on the HyperFlex HX220c M4 Node network.

2 Statistical Characteristics of Hyperconverged Network Traffic

Investigation of the statistical characteristics of traffic in a hyperconverged network should be carried out using a point model processes [2]. When describing traffic as a point process, such statistical characteristics can be used as the intensity of the point process of packet transmission, moment function of the second order; spectral density; correlation function of the number of samples; normalized variance of the number of samples; variance of the number of samples; correlation coefficient and others.

Traffic $f(x)$ consider as a sampling function x. If the traffic is fractal in nature, then it has the property of scale invariance:

$$f(ax) = g(a)f(x), \tag{1}$$

where a is the scale parameter; $g(a)$ is the scaling function.

To identify the model of such traffic, consider a stationary random point process $\{\xi_k\}$, in which the intervals between events are independent random variables. To describe them, an indicator variable is introduced R. The distribution function of such a process is discrete. The moments of its change are random values:

$$N_\tau = \sum_{j \in P_\tau} e(t - \tau_j), \tag{2}$$

where τ_j is the moment of arrival j-th package; P_τ are sets of packages that have arrived over time τ; $e(t - \tau_j)$ is the boolean function, which is equal to one only for $t \geq \tau_j$.

The information flows of a hyperconverged system are controlled by its hypervisor. Let us describe these flows using the characteristic $\theta(V, T)$ and correcting $L(u, T)$ functionals [15]. The characteristic functional is a generalization of the Fourier transform of the probability densities $\{\xi(t_i),\ i = 1, 2, ...n\}$ with an increase in the number of its counts, when $n \to \infty$ [15]:

$$\theta(V, T) = M\left[\exp\left(j\int_0^T V(t)\xi(t)dt\right)\right], \tag{3}$$

where $M\{\cdot\}$ is the mathematical expectation; $V(t)$ is the helper function.

Moment $m_n(\cdot)$ and correlation $k_n(\cdot)$ functions or density functions and order density correlations n are used to describe the local characteristics of point processes. There are such algebraic dependencies between them [15]:

$$\begin{aligned} m_1(t) &= k_1(t); \\ m_2(t_1, t_2) &= k_2(t_1, t_2) + k_1(t_1) \cdot k_1(t_2); \quad ... \ ; \end{aligned} \tag{4}$$

$$\begin{aligned} k_1(t) &= m_1(t); \\ k_2(t_1, t_2) &= m_2(t_1, t_2) - k_1(t_1) \cdot k_1(t_2); \quad ... \ . \end{aligned} \tag{5}$$

Function systems $m_n(\cdot)$ and $k_n(\cdot)$ is statistically orthogonal. With their help, you can easily obtain the numerical characteristics of random processes. So the function $k_2(t_1, t_1) = D$ is the variance and $k_1(t)$ – the intensity of the renewal flow.

The correcting functionality can be calculated as [12]

$$L(u, T) = M\left[\prod_{i=1}^{n}(1 + u(t_i))\right], \tag{6}$$

where $u(t_i)$ is the auxiliary function calculated at the points of occurrence of events t_i. This allows to get the decomposition of traffic according to the system of basic functions $f_n(\cdot)$ and $g_n(\cdot)$ for which

$$\begin{gathered} f_1(t) = g_1(t); \\ f_2(t_1, t_2) = g_2(t_1, t_2) + g_1(t_1) \cdot g_1(t_2); \quad \ldots \; ; \end{gathered} \tag{7}$$

$$\begin{gathered} g_1(t) = f_1(t); \\ g_2(t_1, t_2) = f_2(t_1, t_2) - f_1(t_1) \cdot f_1(t_2); \quad \ldots \; , \end{gathered} \tag{8}$$

These functions are density functions and order density correlations n respectively. So, the function $f_1(t)$ is the average rate of a point process. If there are statistical relationships between the moments of occurrence of events, then the second-order density correlation function is used to describe them $f_2(t_1, t_2)$. These functions characterize the joint probability of the appearance of points near the moments t_1 and t_2. Functions $f_1(t_1)$ and $f_2(t_2)$ characterize the probabilities of independent events. This helps to study the properties of random traffic rates. It should also be borne in mind that the implementation of a random intensity is a stream of delta-pulses as a result of differentiation of a random point process N_τ, i.e.

$$\xi(t) = \frac{dN_\tau}{dt} = \sum_i \delta(t - t_i), \tag{9}$$

where t_i is the coordinate of the appearance of the point (packet and) on the time axis; $\delta(\cdot)$ is Dirac delta-function.

Using the filtering properties of the delta function, one can obtain a relation between the characteristic $\theta(V, T)$ and corrective $L(u, T)$ functionals:

$$\theta(V, T) = L(u, T) \cdot e^{jV(t) - 1}. \tag{10}$$

Let us write down the function series for these functionals:

$$\theta(V, T) = \exp\left(\sum_{n=1}^{\infty} \frac{j^n}{n!} \int_0^T \ldots \int_0^T k_n(t_1, \ldots t_n) \prod_{r=1}^{n} V(t_r) dt_1 \ldots dt_n \right); \tag{11}$$

$$L(u, T) = \exp\left(\sum_{n=1}^{\infty} \frac{1}{n!} \int_0^T \ldots \int_0^T g_n(t_1, \ldots t_n) \prod_{r=1}^{n} u(t_r) dt_1 \ldots dt_n \right). \tag{12}$$

We obtain the relations connecting the characteristics of the flows:

$$\begin{gathered} k_1(t) = g_1(t); \\ k_2(t_1, t_2) = g_1(t_1)\delta(t_1 - t_2) + g_2(t_1, t_2); \\ k_3(t_1, t_2, t_3) = g_1(t_1)\delta(t_1 - t_2)\delta(t_1 - t_3) + g_2(t_1, t_3)\delta(t_1 - t_2) + \\ + g_2(t_2, t_3)\delta(t_2 - t_1) + g_2(t_1, t_2)\delta(t_1 - t_3) + g_3(t_1, t_2, t_3); \quad \ldots \; . \end{gathered} \tag{13}$$

The identification of such a traffic model will be carried out taking into account (9). Consider only second-order statistics, despite the fact that they do not depend on the current time. Their values are determined by the variable $\tau = t_2 - t_1$, therefore with (13) the following ratios occur:

$$k_1 = g_1 = f_1 = const;\ k_2(\tau) = \lambda\delta(\tau) + g_2(\tau), \tag{14}$$

where λ is the intensity of point process.

It is known that

$$f(t_2|\,t_1) = f(t_2 - t_1) = f(\tau)f(t_2|\,t_1) = f(t_2 - t_1) = f(\tau), \tag{15}$$

$$\text{then } g_2(\tau) = f_2(t_1, t_2) - f_1^2 = \lambda(f(t_2|t_1) - \lambda) = \lambda(f(\tau) - \lambda). \tag{16}$$

This function $f(\tau)$ can be determined from the integral recovery equation:

$$f(\tau) = \psi(\tau) + \int_0^t \psi(\tau - t)f(t)dt, \tag{17}$$

where $\psi(\tau)$ is the probability density of time intervals between points.

Applying to the Eq. (14) Fourier transform, we obtain the expression for the spectral density of the centered component of the random intensity:

$$S(\omega) = \int_{-\infty}^{\infty} k_2(\tau)\exp(-j\omega\tau)\,d\tau = \lambda + \int_{-\infty}^{\infty} g_2(\tau)\exp(-j\omega\tau)\,d\tau = \lambda + S_1(\omega). \tag{18}$$

Let's write an expression for a correlation function $G_N(\tau)$:

$$G_N(\tau) = m_2(\tau) = k_2(t_1, t_2) + k_1(t_1) + k_2(t_2).$$

Taking into account (14), we obtain

$$G_N(\tau) = k_2(\tau) + k_1^2 = k_2(\tau) + \lambda^2 - \lambda\delta(\tau) + g_2(\tau) + \lambda^2 = \lambda\delta(\tau) + R_1(\tau), \tag{19}$$

where $R_1(\tau) = g_2(\tau) + \lambda^2$ is the modulating component of the moment function; $\tau = t_2 - t_1$ is the correlation interval of events.

For fractal processes, the second-order correlation function has the form:

$$g_2(\tau) = \lambda^2(\tau/\tau_0)^{\alpha-1}, \tag{20}$$

where $\tau_0 = k\frac{\lambda\cdot\Gamma(1-\alpha/2)}{\Gamma(\alpha/2)\Gamma(1-\alpha)}$; $\Gamma(\cdot)$ is a gamma function; k is the normalizing factor, fractional exponent $\alpha < 1$ is the fractal exponent linearly related to Hurst exponent H correlation $H = \frac{\alpha+1}{2}$. Other sampling statistics are used to determine the many characteristics of the scale invariance properties. For example, the Fano factor:

$$F(T) = \frac{D(T)}{\lambda T}.$$

From expressions (17) and (19) [12]:

$$F(T) = D(T)(\lambda T)^{-1} = 2(\lambda T)^{-1}\int_0^T (T-\tau)k_2(\tau)d\tau =$$
$$= 2(\lambda T)^{-1}\left(\int_0^T (T-\tau)\lambda\delta(\tau)d\tau + \frac{\lambda^2}{\tau_0^{\alpha-1}}\int_0^T (T-\tau)\tau^{\alpha-1}d\tau\right). \quad (21)$$

The first integral in the right part of the expression (21) based on the filtering properties of the delta function is equal to $\lambda T/2$. After calculating the second integral, we obtain the value $\lambda^2 T^{\alpha-1}\Big/\left(\alpha(1+\alpha)\tau_0^{\alpha-1}\right)$. Then the expression for the Fano factor:

$$F(T) = 1 + (T/T_0)^{\alpha};\ T_0^{\alpha} = \tfrac{1}{2}\alpha(1+\alpha)\Big/\left(\lambda\tau_0^{1-\alpha}\right), \quad (22)$$

where T_0 is the fractal time.

Similarly, we can obtain an expression for the correlation function of the number of samples $C(k;T)$ when $k \geq 1$ $(\tau_1 = -\tau)$:

$$C(k;T) = \int_{-T}^{T} (t-|\tau|)\,k_2(kT-\tau)d\tau = C(k;T) =$$
$$= \lambda\int_0^T (T-\tau_1)\delta(kT-\tau_1)d\tau_1 + \lambda\int_0^T (T-\tau_1)\delta(kT+\tau_1)d\tau_1 + \frac{\lambda^2}{\tau_0^{\alpha-1}}\int_0^T (T-\tau)(kT-\tau)^{\alpha-1}d\tau + \quad (23)$$
$$+\frac{\lambda^2}{\tau_0^{\alpha-1}}\int_0^T (T-\tau_1)(kT+\tau_1)^{\alpha-1}d\tau_1 = J_1 + J_2 + J_3 + J_4,$$

where $\tau_1 = -\tau$; $J_1 + J_2 = 0$ (considering the filtering properties of the delta-function), and integrals J_3 and J_4 are equal:

$$J_3 = \frac{\lambda^2}{\tau_0^{\alpha-1}}\int_0^T (T-\tau)(kT-\tau)^{\alpha-1}d\tau =$$
$$= \frac{\lambda^2 T^{\alpha+1}}{\tau_0^{\alpha-1}}\left(\left(\frac{1}{\alpha}k^{\alpha} - (k-1)^{\alpha}\right) + \frac{1}{\alpha}(k-1)^{\alpha}\right) + \frac{1}{\alpha(1+\alpha)}\left((k-1)^{\alpha} - k^{\alpha+1}\right),$$

$$J_4 = \frac{\lambda^2}{\tau_0^{\alpha-1}}\int_0^T (T-\tau_1)(kT+\tau_1)^{\alpha-1}d\tau_1 =$$
$$= \frac{\lambda^2 T^{\alpha+1}}{\tau_0^{\alpha-1}}\left(\left(\frac{1}{\alpha}(k+1)^{\alpha} - k^{\alpha}\right) - \frac{1}{\alpha}(k+1)^{\alpha} + \frac{1}{\alpha(1+\alpha)}\left((k+1)^{\alpha+1} - k^{\alpha+1}\right)\right),$$

$$i.e\ C(k;T) = J_3 + J_4 = \tfrac{1}{2}\lambda T\left(\tfrac{T}{T_0}\right)^{\alpha}\left((k+1)^{\alpha+1} - 2k^{\alpha+1} + (k-1)^{\alpha+1}\right). \quad (24)$$

In accordance with (19) the spectral density of the studied traffic is equal to

$$S_N(\omega) = \int\limits_{-\infty}^{\infty} G_N(\tau)\exp\{-j\omega\tau\}d\tau = 2\pi\lambda^2\delta(\omega) + \lambda\left(\omega/\omega_0\right)^{-\alpha} + \lambda, \tag{25}$$

where $\omega_0^\alpha = 2\lambda\cos\left(\frac{\pi\alpha}{2}\right)\Gamma(\alpha)\cdot\tau_0^{1-\alpha}$; $\Gamma(\cdot)$ is a gamma function, or

$$S_N(\omega) = S_1(\omega) + \lambda, \tag{26}$$

where $S_1(\omega) = \int\limits_{-\infty}^{\infty} R_1(\tau)\exp\{-j\omega\tau\}d\tau = 2\pi\lambda^2\delta(\omega) + \left(\omega/\omega_0\right)^{-\alpha}$ is spectral density of the modulating signal. We'll pretend (25) how

$$S_N(\omega) = 2\pi\lambda^2\delta(\omega) + S(\omega),$$

where $S(\omega)$ is spectral density of the centered component of the random intensity of the point process. Can be calculated:

$$T_0^\alpha = \frac{1}{2}\frac{\alpha(1+\alpha)}{\lambda\tau_0^{1-\alpha}};\ \omega_0^\alpha = 2\lambda\cos\left(\frac{\pi\lambda}{2}\right)\Gamma(\alpha)\cdot\tau_0^{1-\alpha},$$

and also their connecting ratio

$$\omega_0^\alpha T_0^\alpha = \cos\left(\pi\lambda/2\right)\Gamma(\alpha+2).$$

Calculate the traffic correlation function estimate:

$$r(k;T) = \frac{C(k;T)}{D(T)} = T^\alpha\left((k+1)^{\alpha+1} - 2k^{\alpha+1} + (k-1)^{\alpha+1}\right)\Big/\left(2\left(T^\alpha + T_0^\alpha\right)\right). \tag{27}$$

When $k >> 1$ we can write the following approximate equalities:

$$(k+1)^{\alpha+1} \approx k^{\alpha+1} + (\alpha+1)k^\alpha + \frac{1}{2}\alpha(\alpha+1)k^{\alpha-1}; \tag{28}$$

$$(k-1)^{\alpha+1} \approx k^{\alpha+1} - (\alpha+1)k^\alpha + \frac{1}{2}\alpha(\alpha+1)k^{\alpha-1}, \tag{29}$$

Therefore, we can write the following asymptotic equality:

$$r(k;T) \sim \left(\alpha(\alpha+1)/\left(2\left(1+(T_0/T)^\alpha\right)\right)\right)k^{\alpha-1}. \tag{30}$$

Stepwise character $r(k;T)$ indicates the fractal nature of the correlation dependence. Weighted average traffic counts have a similar property:

$$x^{(m)} = \{x_k^{(m)} : k = 0,\ 1,\ \ldots n,\ \ldots\} - x^{(m)} = \{x_k^{(m)} : k = 0,\ 1,\ \ldots n,\ \ldots\} =$$
$$= \left\{\frac{x_1 + \ldots + x_m}{m}, \ldots, \frac{x_{km+1} + \ldots + x_{(k+1)m}}{m}\right\} = \frac{1}{m}\sum_{i=km+1}^{(k+1)m} x_i,$$

where m and k are the aggregation and offset parameters respectively. For such an aggregated process, second-order statistics are:

$$G^{(m)}(k;T) = m^{-2} \int_{-mT}^{mT} (mT - |\tau|)\left(G(kTm - \tau) - \lambda^2\right) d\tau =$$
$$= m^{-2} c(k, mT); \quad D^m(T) = m^{-2} C(0, mT);$$
$$r^{(m)}(k;T) = \left(1 \Big/ \left(2\left(1 + \left(\frac{T_0}{mT}\right)^{\alpha}\right)\right)\right) \times \left((k+1)^{\alpha+1} - 2k^{\alpha+1} + (k-1)^{\alpha+1}\right). \tag{31}$$

When $m \to \infty$ correlation coefficient $r^{(m)}(k;T)$ no longer depends on the aggregation method and therefore retains its structure.

The correlation coefficient is independent of the scaled parameter m and has the form of a power dependence

$$r^{(m)}(k;T) \sim \tfrac{1}{2}\left((k+1)^{\alpha+1} - 2k^{\alpha+1} + (k-1)^{\alpha+1}\right). \tag{32}$$

At big m the following asymptotic expression for the variance is valid:

$$\begin{gathered} D^{(m)}(T) = \tfrac{\lambda mT}{m^2}\left(+1(mT/T_0)^{\alpha}\right) = \\ \left(m^{-1} + (T/T_0)^{\alpha} m^{\alpha-1}\right) \sim \lambda T (T/T_0)^{\alpha} m^{\alpha-1} \end{gathered} \tag{33}$$

All the statistical characteristics considered above are uniquely determined using only three parameters:

- α is the fractal exponent;
- λ is the intensity of the traffic process;
- T_0 is the fractal setting time.

Therefore, the identification of these parameters is sufficient for building a traffic model of a hyperconverged system. To prepare for traffic modeling, it is necessary to determine the boundaries of changes in its frequency and spatial properties when anomalies occur.

3 Defining the Limits of cChanges in Traffic Properties

Consider space $L^2(R)$. This is a function space $z(t)$, which are defined on the entire valid axis $R(-\infty, \infty)$ and have a finite square norm.

$$\|z(t)\| = \int_{-\infty}^{\infty} |z(t)|^2 dt < \infty. \tag{34}$$

Let us construct an orthogonal wavelet basis in it

$$L^2(R) = V_0 \oplus \left\{ \bigoplus_{j=-0}^{\infty} W_j \right\}, \tag{35}$$

where W_j is the orthogonal spaces; V_0 is the subspace, which is a normal circuit $L^2(R)$ shifts of the scaling function $\psi(x)$:

$$V_0 = [\psi_{0n}(t) = \psi(t-n)]_{n\in Z} = \left\{ \sum_{n\in Z} C_{0n}\psi_{0n} \middle| \sum_{n\in Z} |C_{0n}|^2 < \infty \right\}; \tag{36}$$

This wavelet basis consists of integer shifts $\{\psi_{0n}\}_{n\in Z}$ and bursts $\{\phi_{jn}\}_{j\in Z, n\in Z, j\geq 0}$. Hyperconverged network traffic is a function of $L^2(R)$. Any function with $L^2(R)$ can be expanded in a series in this wavelet basis:

$$z(t) = \sum_{j\in Z}\sum_{n\in Z} W_{jn}\phi_{jn}(t) = U_0 + \sum_{j=0}^{\infty}\sum_{n\in Z} W_{jn}\phi_{jn}(t), \tag{37}$$

where $U_0 = \sum_{n\in Z} U_{0n}\psi_{0n}(t)$ is the function from the subspace of functions of unit scale; coefficients U_{0n} is a decomposition of traffic with a resolution of the rule "one point on 2^n points of the analyzed traffic». After renormalization of the time argument ($t \in [0,\ 1]$) we have

$$z(t) = U_0 + \sum_{j=0}^{\infty}\sum_{n\in Z} W_{jn}\varphi_{jn}(t). \tag{38}$$

After a large-scale transformation of the basis, the norm $\varphi(2^j t)$ will be equal to

$$\left\|\varphi(2^j t)\right\|_2 = \sqrt{\int_{-\infty}^{\infty} \varphi(2^j t)\cdot\varphi(2^j t)dt} = \sqrt{\int_{-\infty}^{\infty} 2^{-j}\cdot\varphi^2(2^j t)d(2^j t)} = 2^{-j/2}\|\varphi(t)\|_2. \tag{39}$$

Taking into account that the displacement does not change the value of the function norm, we obtain:

$$\left\|\varphi(2^j t - k)\right\|_2 = 2^{-j/2}\|\varphi(t)\|_2. \tag{40}$$

So if $\varphi(t) \in L^2(R)$ has a unit norm, then all functions $\{\varphi_{jk}\}$ of the form:

$$\varphi_{jk}(t) = 2^{j/2}\varphi\left(2^j t - k\right),\ j, k \in Z \tag{41}$$

characterized by the fact that $\|\varphi_{jk}\|_2 = \|\varphi\|_2 = 1$.

Moreover, if the functions $\{\varphi_{jk}\}$ form an orthogonal basis of the functional space $L^2(R)$, then each function $f(t) \in L^2(R)$ can be represented as a series

$$f(t) = \sum_{j,k=-\infty}^{+\infty} W_{jk}\varphi_{jk}(t). \tag{42}$$

Haar function $\varphi^{H}(t)$ meets all of the above requirements,

$$\varphi_{jk}^{H}(t) = \varphi^{H}\left(2^{j}t - k\right), j, k \in Z. \tag{43}$$

Then any two functions φ_{jk}^{H} and $\varphi_{\ell m}^{H}$ form a basis in $L^2(R)$.

Note that the series (37) for any values t is positive if the condition $\left|W_{jk}\right| \leq U_{jk}$ [12]. To simplify the calculation of values W_{jk} it is possible to determine the adjacent expansion coefficients by the formulas:

$$U_{(j+1)(2k)} = \left(1 + a_{jk}\right) \cdot 2^{-1/2} U_{jk};\ U_{(j+1)(2k+1)} = \left(1 - a_{jk}\right) \cdot 2^{-1/2} U_{jk}.$$

In this case, the simulated traffic counts are calculated as

$$z^{(n)}(k) = 2^{-n/2} \cdot U_{nk},\ k = \overline{0,\ 2^{n} - 1},$$

and the value n defines the highest accuracy or the smallest scale of representation of the simulated traffic.

Let's start the calculation from the coefficient U_{00}. Then, proceeding from the binary structure of the coefficient tree, we determine

$$U_{j,k_j} = 2^{-j/2} U_{00} \prod_{i=0}^{j-1} \left(1 + (-1)^{k_j'} a_{ik_j}\right), \tag{44}$$

$$\text{where } k_j = \sum_{i=0}^{j-1} k_i' 2^{j-1-i}. \tag{45}$$

Setting the value k_i', where $i = \overline{0,\ j}$, can be determine the value k_i and all coefficients of the corresponding binary tree from U_{00} to U_{j,k_j}. We get such equality:

$$x^{(n)}(k) = 2^{-n} U_{00} \prod_{i=0}^{n-1} \left(1 + (-1)^{k_i'} a_{ik_i}\right) = 2^{-n} U_{00} \prod_{j=0}^{n-1} \left(1 + (-1)^{k_i'} a_{(j)}\right), \tag{46}$$

where $a_{(j)}$ is the random variable whose distribution corresponds to the generated model.

Note that the moments of order q for adjacent zoom levels $(j - 1)$ and j algebraically related

$$M\left[U_{j-1,k}^{q}\right] = M\left[U_{j,k}^{q}\right] \cdot 2^{q/2} M\left[\left(1 - a_{(j-1)}\right)^{q}\right]^{-1}. \tag{47}$$

This makes it possible to use the properties of statistical moments at various aggregation intervals to form a traffic model.

An approach to the study of traffic is considered that makes it possible to quickly determine the boundaries of changes in the scale and frequency properties of the process under consideration.

4 Mathematical Model of Abnormal Traffic in a Hyperconverged Network

To build a model, we introduce the distribution density function $f(t)$, It is time of the packet transition from the node with the coordinate η to the node $\eta + 1$ for the time t. Abnormal traffic is fractal in nature, therefore

$$f(t) = \theta \cdot (1+\theta)^{-(\theta+1)}, \quad 0 < \theta < 1, f(t) > 0; \quad \int_0^\infty f(t)\,dt = 1. \tag{48}$$

Introduce the function $F(\tau)$, characterizing the probability that the packet will not make the transition to the next node before the moment of time τ, i. e

$$F(\tau) = 1 - \int_0^\infty f(t)\,dt = (1+\tau)^{-\theta}, \tag{49}$$

where τ is the time of packet stay in the node of the virtual connection with coordinate η. Let's calculate the most probable number of packets in a node η in time t:

$$\ell(\eta, t) = \int_0^t \ell(\eta - 1; t-\tau) \cdot f(\tau)\,d\tau + \ell_0(\eta)\,F(t), \tag{50}$$

where $\ell_0(\eta)$ is the number of packets in the node η before packets arrive from the host $(\eta - 1)$, and the difference $\Delta L = \ell(\eta, t) - \int_0^t \ell(\eta; t-\tau) \cdot f(\tau)\,d\tau$.

Then, taking into account (50)

$$\Delta L = \int_0^t (\ell(\eta - 1; t-\tau) - \ell(\eta; t-\tau))\,f(\tau)\,d\tau + \ell_0(\eta) \cdot F(t). \tag{51}$$

Decompose $\ell(\eta - 1\,;\ t - \tau)$ in Taylor's row:

$$\ell(\eta - 1; t-\tau) = \ell(\eta; t-\tau) + \frac{\partial \ell(\eta; t-\tau)}{\partial \eta} \cdot ((\eta - 1) - \eta) +$$
$$+ R_2(\eta) = \ell(\eta; t-\tau) - \frac{\partial \ell(\eta; t-\tau)}{\partial \eta} + O\left(\eta^2\right).$$

Neglect of order terms $O(\eta^2)$ we get

$$\delta L = \ell(\eta, t) - \int_0^t \ell(\eta; t-\tau) \cdot f(\tau)\,d\tau = -\int_0^t \frac{\partial \ell(\eta; t-\tau)}{\partial \eta} \cdot f(\tau)\,d\tau + \ell_0(\eta) \cdot F(t)\,. \tag{52}$$

We apply to the obtained equation of the Laplace transform in time [15]. With considering (48) we get:

$$L\left(\int_0^t \frac{\partial \ell(\eta; t-\tau)}{\partial \eta}\cdot f(\tau)\,d\tau + \ell_0(\eta)\cdot F(t)\right) =$$

$$= \frac{\partial \ell_p(\eta; p)}{\partial \eta}\int_0^\infty \exp(-p\tau)\cdot f(\tau)\,d\tau = \frac{\partial \ell_p(\eta; p)}{\partial \eta}\int_0^\infty \exp(-p\tau)\cdot \theta\cdot(1+\tau)^{-(\theta+1)}d\tau, \tag{53}$$

where p Ta $\ell_p(\eta; p)$ are the parameter and Laplace transform of a function $\ell(\eta; t)$.

To calculate the integral on the right side (53) let's make the change of variables $y = \tau + 1$. After replacing obtain:

$$\theta\int_0^\infty \exp(-p(y-1))y^{-(\theta+1)}dy = \theta\cdot\exp(p)\int_0^\infty \exp(-py)\cdot y^{-(\theta+1)}d\tau\,. \tag{54}$$

If $py = t$, that is $dy = dt/p$, then from (54) after transformations we obtain

$$\theta\cdot\exp(p)\int_0^\infty \exp(-py)\theta\cdot y^{-(\theta+1)}d\tau = 1 - e^p\cdot p^\theta\int_p^\infty \exp(-t)\cdot t^{-\theta}dt.$$

Using the expression for the incomplete gamma function [15]

$$\Gamma(a;\eta) = \Gamma(a) - g(a;\eta) = \int_\eta^\infty \exp(-t)\cdot t^{a-1}dt,$$

where $\Gamma(a)$ is the Euler's gamma function, we obtain

$$p(a;\eta) = \Gamma^{-1}(a)\int_0^\eta \exp(-t)\cdot t^{a-1}dt$$

Next, we write down the right side (53)

$$\frac{\partial \ell_p(\eta; p)^\infty}{\partial \eta}\int_0^\infty \exp(-p\tau)\cdot\theta\cdot(1+\tau)^{-(\theta+1)}d\tau = \frac{\partial \ell_p(\eta; p)}{\partial \eta}\big(1 - p\cdot F_p(p)\big),$$

where $\Gamma(1-\theta; p) = \int_\eta^\infty \exp(-t)\cdot t^\theta dt\Gamma(1-\theta; p) = \int_\eta^\infty \exp(-t)\cdot t^\theta dt$,

and consider the Laplace transform the left-hand side of (52), integral term of which

$$L\left(\int_0^t \ell(\eta; t-\tau)\cdot f(\tau)d\tau\right) = \ell_p(\eta; p)\cdot f_p(p),$$

where $f_p(p)$ is the Laplace transform of the distribution density function $f(t)$.

For Laplacetransform $F(t)$ $F_p(p) = (1 - f_p(p))/p$. For the time component of the transformation:$F(t) = 1 - \int_0^t \theta \cdot (1-\tau)^{\theta+1} d\tau = (1+t)^{-\theta}$ compute explicitly the Laplace image of the function $F(t)$:

$$F_p(p) = \int_0^\infty \exp(-p\tau) \cdot (1+\tau)^{-\theta} d\tau = F_p(p) = \int_0^\infty \exp(-p\tau) \cdot (1+\tau)^{-\theta} d\tau =$$

$$= \int_1^\infty \exp(-p(y-1)) \cdot y^{-\theta} dy = \exp(p) \int_1^\infty \exp(-py) \cdot y^{-\theta} dy.$$

Considering that $py = t$, calculate the resulting integral expression

$$e^p \int_1^\infty \frac{e^{-py}}{y^\theta} dy = e^p \int_p^\infty \frac{e^{-1}}{t^\theta p^{-\theta}} \frac{dt}{p} = e^p p^{\theta-1} \int_p^\infty t^{-\theta} e^{-t} dt = e^p p^{\theta-1} \Gamma(1-\theta, p) .$$

Thus, the Laplacetransform of the function $F(t)$ has the form $F_p(p) = e^p p^{\theta-1}\Gamma(1-\theta, p)$. Therefore, the left side of the Eq. (53) is such:

$$\ell_p(\eta; p) - \ell_p(\eta; p) \cdot (1 - pF_p(p)) = \ell_p(\eta; p) \cdot pF_p(p),$$

namely, the equation in the image domain of the Laplace transform takes the form

$$pF_p(p) \cdot \ell_p(\eta; p) = -\frac{\partial \ell_p(\eta; p)}{\partial \ell} \cdot (1 - pF_p(p)) + \ell_0(\eta) \cdot F_p(p) . \quad (55)$$

Define the solution (55) for the case when $p << 1, \theta < 1$:

$$\Gamma(1-\theta) \cdot p^\theta \ell_p(\eta; p) = -\frac{\partial \ell_p(\eta; p)}{\partial \eta} + \Gamma(1-\theta) \cdot p^{\theta-1} \ell_0(\eta) . \quad (56)$$

Moving on to the originals, we get

$$\frac{\partial}{\partial t} \int_0^t \frac{\ell(\eta, \tau)}{(t-\tau)^\theta} d\tau = -\frac{\partial \ell(\eta, t)}{\partial \eta} + \frac{\ell_0(\eta)}{t^\theta}. \quad (57)$$

In (57) the left side is the fractional derivative of the function $\ell(\eta; t)$:

$$D_t^\theta[\ell(\eta; t)] = \frac{1}{\Gamma(1-\theta)} \cdot \int_0^t \frac{\ell(\eta; \tau)}{(t-\tau)^\theta} d\tau.$$

Therefore, the packet propagation equation is:

$$\Gamma(1-\theta)\, p^\theta \ell(k; p) = -\ell(k+1; p) + \ell(k; p) + \Gamma(1-\theta)\, p^{\theta-1} \ell_0(k).$$

From the Eq. (56), taking into account the discrete nature of the coordinate change η at $\eta = k$ and the finite difference approximating the partial derivative, we obtain

$$\Gamma(1-\theta)\cdot D_t^{\theta}[\ell(\eta;t)] = -\frac{\partial \ell(\eta;t)}{\partial \eta} + \frac{\ell_0(\eta)}{t^{\theta}}.$$

After arithmetic conversions, we write

$$\ell(k+1;p) = \left[1-\Gamma(1-\theta)\cdot p^{\theta}\right]\cdot \ell(k;p) + \Gamma(1-\theta)\cdot p^{\theta-1}\ell_0(k)\,. \qquad (58)$$

To obtain a solution to this equation, consider the case when the second term is absent on the right-hand side, that is

$$\ell(k+1;p) = \left[1-\Gamma(1-\theta)\cdot p^{\theta}\right]\cdot \ell(k;p).$$

The solution to this equation has the form

$$\ell(k;p) = \left\{\prod_{m=0}^{k-1}\left[1-\Gamma(1-\theta)\cdot p^{\theta}\right]\right\}\cdot \ell(0;p),$$

$$or\ \ell(k;p) = \left[1-\Gamma(1-\theta)\cdot p^{\theta}\right]\cdot \ell(0;p).$$

The inhomogeneous equation can be written as

$$\ell(k+1;p) - \ell(k;p) + \Gamma(1-\theta)\cdot p^{\theta}\ell(k;p) = \Gamma(1-\theta)\cdot p^{\theta-1}\cdot \ell_0(k)\,.$$

Presenting expression $\ell\,(k\,;\,p)$ at the form $\ell(k;p) = u(k)\cdot v(k)$ and designating $\Delta\,\ell\,(k\,;\,p) = \ell\,(k+1\,;\,p) - \ell\,(k\,;\,p)$, we get

$$\Delta\ell(k;p) + \Gamma(1-\theta)\cdot p^{\theta}\ell(k;p) = \Gamma(1-\theta)\cdot p^{\theta-1}\cdot \ell_0(k)$$

or $u(k+1;p)\cdot \Delta v(k) + v(k)\left[\Delta u(k) + \Gamma(1-\theta)\cdot p^{\theta}u(k)\right] = \Gamma(1-\theta)\cdot p^{\theta-1}\cdot \ell_0(k)\,.$

From the last relation, applying Bernoulli's method [12], we get:

$$v(k) = \frac{\Gamma(1-\theta)\cdot p^{\theta-1}}{u(0)}\sum_{m=0}^{k-1}\frac{\ell_0(m)}{\left[1-\Gamma(1-\theta)\cdot p^{\theta}\right]^m} + v(0)v(k) =$$

$$= \frac{\Gamma(1-\theta)\cdot p^{\theta-1}}{u(0)}\sum_{m=0}^{k-1}\frac{\ell_0(m)}{\left[1-\Gamma(1-\theta)\cdot p^{\theta}\right]^m} + v(0).$$

As a result of solving the Eq. (58)

$$\ell(k;p) = \left[1-\Gamma(1-\theta)\cdot p^{\theta}\right]^k\left[\Gamma(1-\theta)\cdot p^{\theta-1}\sum_{m=0}^{k-1}\frac{\ell_0(m)}{\left[1-\Gamma(1-\theta)\cdot p^{\theta}\right]^m} + u(0)\cdot v(0)\right].$$

This solution can be written in the following form:

$$\ell(k;p)=\left[1-\Gamma(1-\theta)\cdot p^{\theta}\right]^{k}\left\{\Gamma(1-\theta)\cdot p^{\theta-1}\sum_{m=0}^{k-1}\frac{\ell_0(m)}{\left[1-\Gamma(1-\theta)\cdot p^{\theta}\right]^{m}}+\ell(0;p)\right\}. \tag{59}$$

When $\ell_0(0)=\ell_0$ Ta $\ell_0(k)=0\,,\quad k=1,2,\ldots,$

$$\ell(k;p)=\left[1-\Gamma(1-\theta)\cdot p^{\theta}\right]^{k}\left\{\Gamma(1-\theta)\cdot p^{\theta-1}\ell_0+\ell(0;p)\right\}.$$

Using solution (59), we write down the obtained solution under the conditions $\ell\,(0\,;\,t)\;=\;\ell_0\cdot\delta(t)$, simulating the generation of a series of packets of volume ℓ_0 packets at time t, has the form.

$$\ell(k;p)=\left[1-\Gamma(1-\theta)\cdot p^{\theta}\right]^{k}\left\{\Gamma(1-\theta)\cdot p^{\theta-1}+1\right\}\cdot\ell_0$$

Considering the right-hand side of the obtained expression for adjacent nodes, we can write down the following approximate equality

$$\ell(k;p)=\ell_0\left[1-\Gamma(1-\theta)\cdot p^{\theta}\right]\cdot\left\{\Gamma(1-\theta)\cdot p^{\theta-1}+1\right\}=$$
$$=\ell_0\left\{\Gamma(1-\theta)\cdot p^{\theta-1}+1-k\Gamma^2(1-\theta)\cdot p^{2\theta-1}-k\Gamma(1-\theta)\cdot p^{\theta}\right\}.$$

After switching from images to originals

$$\ell(k;t)=\ell_0\left\{\frac{1}{t^{\theta}}-k\frac{\Gamma^2(1-\theta)}{\Gamma(1-2\theta)\cdot t^{2\theta}}-k\Gamma(1-\theta)\cdot\frac{1}{\Gamma(-\theta)\cdot t^{\theta+1}}\right\}$$

or $\ell(k;t)=\ell_0\left\{\frac{1}{t^{\theta}}-k\left[\frac{\Gamma^2(1-\theta)}{\Gamma(1-2\theta)}\cdot\frac{1}{t^{2\theta}}+\frac{\Gamma(1-\theta)}{\Gamma(-\theta)}\cdot\frac{1}{t^{\theta+1}}\right]\right\}.$

Taking into account that the solution obtained above is asymptotic, for $k=0$ the solution for the initial nodes of the virtual connection can be written in the form $\ell(0;t)=\ell_0\{1/t^{\theta}\}$, and for:

$$\ell(1;t)=\ell_0\left\{\frac{1}{t^{\theta}}-\left[\frac{\Gamma^2(1-\theta)}{\Gamma(1-2\theta)}\cdot\frac{1}{t^{2\theta}}+\frac{\Gamma(1-\theta)}{\Gamma(-\theta)}\cdot\frac{1}{t^{\theta+1}}\right]\right\}\ell(1;t)=$$
$$=\ell_0\left\{\frac{1}{t^{\theta}}-\left[\frac{\Gamma^2(1-\theta)}{\Gamma(1-2\theta)}\cdot\frac{1}{t^{2\theta}}+\frac{\Gamma(1-\theta)}{\Gamma(-\theta)}\cdot\frac{1}{t^{\theta+1}}\right]\right\}.$$

Thus, calculations can be continued for an arbitrary k. This allows us to calculate the correlation function of the obtained solution.

For the case when the initial conditions are of the form $\ell\,(0\,;\,t)=\ell_0\cdot\delta(t)$, you can write that

$$\ell(k;p)=\left[1-\Gamma(1-\theta)\cdot p^{\theta}\right]\left\{\Gamma(1-\theta)\cdot p^{\theta-1}+1\right\}\ell_0. \tag{60}$$

According to the definition, the expression for the correlation function is calculated in relation to the process $\ell\,(k\,;\,p)$, has the form

$$c(m;p) = \lim_{N\to\infty}\frac{1}{N}\sum_{i=1}^{N}\ell(i+m;p)\cdot\ell(i;p).$$

Then, taking into account (40) you can write such an expression

$$c(m;p) = \ell_0^2\Big\{\Gamma(1-\theta)\cdot p^{\theta-1}+1\Big\}^2\times$$
$$\times\Big[1-\Gamma(1-\theta)\cdot p^{\theta}\Big]^m\cdot\lim_{N\to\infty}\frac{1}{N}\sum_{i=1}^{N}\Big[1-\Gamma(1-\theta)\cdot p^{\theta}\Big]^{2i}.$$

On condition $p << 1$, for additional nodes, the asymptotic part of the correlation function has the form

$$c(m;p)\approx\ell_0^2\cdot\Gamma(1-\theta)\cdot p^{2(\theta-1)}\cdot\Big[1-m\Gamma(1-\theta)\cdot p^{\theta}\Big]=$$
$$=\ell_0^2\cdot\Gamma(1-\theta)\cdot\Big[p^{2\theta-2}-m\Gamma(1-\theta)\cdot p^{3\theta-2}\Big].$$

Moving on to the originals, we get

$$c(m;p)\approx\ell_0^2\cdot\Gamma(1-\theta)\times\left[\frac{1}{\Gamma(1-2\theta+1)}\cdot\frac{1}{t^{2\theta-1}}-m\frac{\Gamma(1-\theta)}{\Gamma(1-3\theta+1)}\cdot\frac{1}{t^{3\theta-1}}\right]=$$
$$=\ell_0^2\cdot\Gamma(1-\theta)\cdot t^{1-2\theta}\cdot\left[\frac{1}{\Gamma(2-2\theta)}-m\frac{\Gamma(1-\theta)}{\Gamma(2-3\theta)}\cdot\frac{1}{t^{\theta}}\right].$$

The expression for the variance will be obtained for $m = 0$ and has the form

$$D(t)=c(0;t)=\frac{\ell_0\cdot\Gamma(1-\theta)}{\Gamma(2-2\theta)}\cdot t^{1-2\theta}.$$

This expression is characteristic of processes with long-term statistical dependences and the property of asymptotic self-similarity.

5 Discussion

The developed mathematical model of anomalous traffic of the hyperconverged system is used to construct a short-term forecast. The system hypervisor receives the forecast and uses it to reallocate resources quickly. As a result, QoS performance is improved.

The proposed approach was used when operating the hyperconverged network of the Cisco HyperFlex HX220c M4 Node. The system is powered by Intel Xeon processors. The deployment provides pre-integrated clusters that include networking fabric, data optimization, unified servers, and VMware ESXi/vSphere. This solution delivers consistently high performance in today's most popular hyper-converged VMware environments.

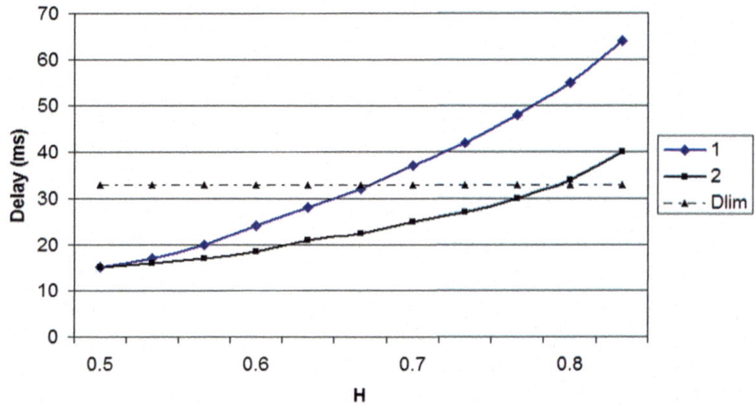

Fig. 1. Analyzing average packet latency: 1 – standard approach; 2 – proposed approach

Cisco sees this configuration option as a simple system for small projects. Under heavy load, traffic anomalies often occur.

The experimental results are shown in Fig. 1.

As seen from Fig. 1, the use of the proposed approach gives a greater advantage in case of greater traffic anomalies. The degree of abnormality is determined by the Hurstparameter H. One of the main components of QoS is the average packet delay time. When $H = 0.8$ the average packet delay time can be reduced to 40%.

Usually, the quality of service is assessed by an objective indicator. This is the probability of meeting all QoS requirements. This indicator significantly decreases with traffic anomalies. However, the proposed approach made it possible to raise this indicator to the required level (Fig. 2).

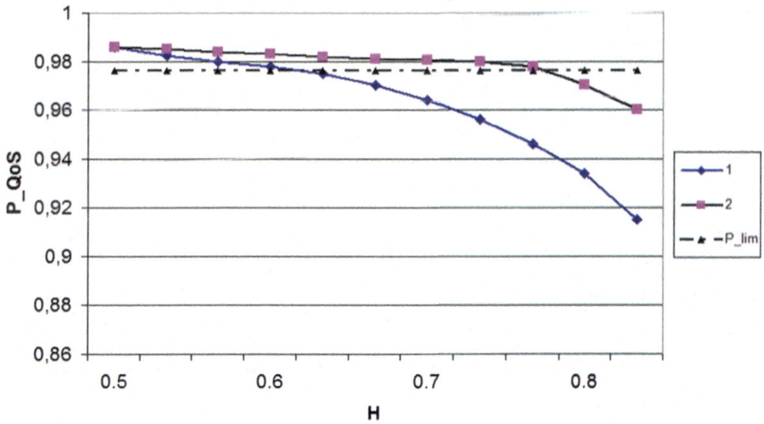

Fig. 2. Probability of meeting requirements QoS: 1 – standard approach; 2 – proposed approach

6 Conclusion

The chapter considers the approach to providing QoS in hyperconverged networks with traffic anomalies. Analytical dependences for calculation of statistical characteristics of traffic on its samples are offered. It is proved that all considered statistical characteristics are unambiguously determined by means of only three parameters: fractal exponent; the intensity of the traffic flow process; fractal setup. A mathematical model of anomalous traffic is proposed. The model is adequate to real traffic and takes into account the fractal nature of the anomaly. The model uses the properties of scale invariance. Packet losses will be compensated by an increase in message transmission time, which leads to the formation of long statistically temporary dependencies. In the obtained model, the influence of losses and the cause of extended dependences are formally taken into account by introducing the fractional integration operation. The model of anomalous traffic of a hyperconvergent system was used to construct a short-term forecast. The forecast is received by the system hypervisor and used to quickly reallocate resources. As a result, QoS performance improves. The provision of QoS in the hyperconvergent network of the Cisco HyperFlex HX220c M4 Node system in the event of traffic anomalies using the proposed approach has been shown experimentally.

References

1. Semenov, S., Kuchuk, N., Lukova-Chuiko, N.: Method of determining optimal batch capacities of hyperconverged network. Adv. Inf. Syst. **3**(4), 28–32 (2019). https://doi.org/10.20998/2522-9052.2019.4.03
2. Kuchuk, N.: The method of calculating the maximum intensities of information flows in hyperconvergent system. Control Navig. Commun. Syst. **4**(56), 53–56 (2019)
3. Merlac, V., Smatkov, S., Kuchuk, N., Nechausov, A.: Resourses distribution method of university e-learning on the hypercovergent platform. In: Conference Proceedings of 2018 9th Int. Conference on Dependable Systems, Service and Technologies, DESSERT'2018, pp. 136–140. IEEE (2018). https://doi.org/10.1109/DESSERT.2018.8409114
4. Svyrydov, A., Kovalenko, A., Kuchuk, H.: The pass-through capacity redevelopment method of net critical section based on improvement ON/OFF models of traffic. Adv. Inf. Syst. **2**(2), 139–144 (2018). https://doi.org/10.20998/2522-9052.2018.2.24
5. Davydov, V., Hrebeniuk, D.: Development the resources load variation forecasting method within cloud computing systems. Adv. Inf. Syst. **4**(4), 128–135 (2020). https://doi.org/10.20998/2522-9052.2020.4.18
6. Semenov, S., Sira, O., Gavrylenko, S., Kuchuk, N.: Identification of the state of an object under conditions of fuzzy input data. East.-Eur. J. Enterp. Technol. **1**(4), 22–30 (2019). https://doi.org/10.15587/1729-4061.2019.157085
7. Mukhin, V., et al.: Decomposition method for synthesizing the computer system architecture. In: Hu, Z., Petoukhov, S., Dychka, I., He, M. (eds.) ICCSEEA 2019. AISC, vol. 938, pp. 289–300. Springer, Cham (2020). https://doi.org/10.1007/978-3-030-16621-2_27
8. Fränti, P.: Efficiency of random swap clustering. J. Big Data **5**(1), 1–29 (2018). https://doi.org/10.1186/s40537-018-0122-y
9. Ye, Q., Zhuang, W.: Distributed and adaptive medium access control for internet-of-things-enabled mobile networks. IEEE Internet Things J. **4**(2), 446–460 (2017). https://doi.org/10.1109/JIOT.2016.2566659

10. Tkachov, V., Hunko, M., Volotka, V.: Scenarios for implementation of nested virtualization technology in task of improving cloud firewall fault tolerance. In: 2019 International Scientific-Practical Conference Problems of Infocommunications, Science and Technology (PIC S&T), Kyiv, Ukraine, pp. 759–763, IEEE (2019). https://doi.org/10.1109/PICST47496.2019.9061473
11. Pliushch, O., Vyshnivskyi, V., Berezovska, Y.: Robust telecommunication channel with parameters changing on a frame-by-frame basis. Adv. Inf. Syst. **4**(3), 62–69 (2020). https://doi.org/10.20998/2522-9052.2020.3.07
12. Kuchuk, G., Kovalenko, A., Komari, I.E., Svyrydov, A., Kharchenko, V.: Improving big data centers energy efficiency: traffic based model and method. In: Kharchenko, V., Kondratenko, Y., Kacprzyk, J. (eds.) Green IT Engineering: Social, Business and Industrial Applications. SSDC, vol. 171, pp. 161–183. Springer, Cham (2019). https://doi.org/10.1007/978-3-030-00253-4_8
13. Svyrydov, A., Kuchuk, H., Tsiapa, O.: Improving efficienty of image recognition process: approach and case study. In: 2018 IEEE 9th International Conference on Dependable Systems, Services and Technologies, DESSERT 2018 Proceedings, pp. 593–597. IEEE (2018). https://doi.org/10.1109/DESSERT.2018.8409201
14. Ruban, I., Martovytskyi, V., Lukova-Chuiko, N.: Approach to classifying the state of a network based on statistical parameters for detecting anomalies in the information structure of a computing system. Cybern. Syst. Anal. **54**(2), 302–309 (2018). https://doi.org/10.1007/s10559-018-0032-1
15. Raskin, L., Sira, O., Parfenyuk, Y.: Selection of the optimum route in an extended transportation network under uncertainty. Adv. Inf. Syst. **5**(1), 62–68 (2020). https://doi.org/10.20998/2522-9052.2021.4.08
16. Mozhaiev, M.: Modeling nonlinear elements of critical computer network. Adv. Inf. Syst. **4**(4), 27–32 (2020). https://doi.org/10.20998/2522-9052.2020.4.04
17. Kovalenko, A., Kuchuk, H.: Methods for synthesis of informational and technical structures of critical application object's control system. Adv. Inf. Syst. **2**(1), 22–27 (2018). https://doi.org/10.20998/2522-9052.2018.1.04

Parametric Analysis of Statistical and Correlation Characteristics of Discrete Processes in Dynamic Systems with Non-stationary Nonlinearities in Time for the Secure Intent-Based Networks

Vasil Dubrouski[1(✉)], Anatolii Semenko[2], Mykola Kushnir[3], and Mohammed M. Steita[4]

[1] Belarusian State Academy of Communications, F. Skaryny Street 8/2, 220076 Minsk, Republic of Belarus
[2] Open International University of Human Development "Ukraine", 23 Lvivs'ka Street, Kyiv 04071, Ukraine
setel@ukr.net
[3] Yuriy Fedkovych Chernivtsi National University, Kotsyubynsky 2, Chernivtsi 58012, Ukraine
[4] Belaruskaya Street 4, 220030 Minsk, Republic of Belarus

Abstract. The chapter presents the results of numerical simulation of dynamic systems with nonlinear feedbacks based on first-degree polynomials with restrictions on the dynamic range of possible values and non-stationary in time nonlinearities with two and three degrees of freedom. The conditions for generating sequences with acceptable auto- and inter-correlation functions for use in information transmission and security systems are determined. Practically realizable structural and functional schemes of signal generation devices are proposed. The regions of parameters of a nonlinear system at which discrete processes with the given spectral-time and statistical characteristics are formed are determined.

Keywords: Signal generation · Nonlinear dynamics · Information securing · Probability distribution density · Autocorrelation function · Statistical dependence · Nonlinear mixing · Dynamic range · Digital processing

1 Introduction

One of the directions of the evolutionary development of network technologies in the last decade is the phased implementation of Intent-based Networking (IBN). IBN networks are the technological basis for the digital transformation of businesses and enterprises. The relatively complex infrastructure of switches and wireless access points required to create self-organizing IBN networks adapting to external conditions and tasks requires, among other things, ensuring the following quality characteristics:

M. Klymash et al. (Eds.): *Future Intent-Based Networking*, LNEE 831, pp. 242–256, 2022.
https://doi.org/10.1007/978-3-030-92435-5_14

1) reliability (fidelity) of information flow transmission;
2) strong information security when transmitted through open communication channels.

The fulfillment of these conditions will allow IBN networks to function according to the built-in scenario of deployment and intelligent adaptation to multi-factor intentions. One of the possible ways to achieve a given degree of reliability of information transmission and its confidentiality is the methods of nonlinear dynamics implemented in software-defined networks.

[1] presents an extensive class of random-like sequence generators based on algorithms with nonlinear dynamics that provide a significant degree of stochastization of oscillations, a relatively low resource consumption and high noise immunity of information transmission systems based on them. In the context of the discussion of this class of generators, the question concerning the estimation of the degree of unpredictability of the generated sequences, their auto- and inter-correlation properties within a wide range, remains open to the present time changes in the system parameters, its initial and boundary conditions.

Some results of research in this area with the adoption of a number of restrictions that simplify the calculations are presented in [2]. Similar studies conducted for other types of generators are given in [3–5]. Since the class of nonlinear dynamic systems with time-distributed, and more broadly – non-stationary nonlinearities, are convenient for implementation by the simplest signal processors and microcontrollers, it is important to comprehensively investigate the behavior of the above-mentioned systems for subsequent practical application by communication engineers in real systems for transmitting confidential information over open channels. The complexity of the class of systems under study requires the determination of a number of basic characteristics. In this chapter, we will limit ourselves to the following:

1) for the correlation characteristics, we will investigate single-ring dynamical systems with nonlinearities described by first-degree polynomials with two and three degrees of freedom;
2) as in [2], the main condition for determining the nonlinear forming function (NFF) will be its limiting tangent of the slope angle-no more than 0.18, which approximately corresponds to 10°;
3) the quality of the probability distribution density of the generated sequences will be evaluated by its proximity to the uniform law;
4) the signals at the output of the main ring are subject to mandatory normalization generator, and auxiliary generators that generate reports of NFF parameters the main generator.

The second condition is indicated in order to preserve the noise immunity [1] of the information transmission systems in which the generated signals will be used.

An increase in this parameter has a fruitful effect on the correlation properties of the generated sequences, but it significantly reduces the noise immunity of the system, which limits their applicability in practice.

The fourth condition is determined by the need to perform calculations in devices with a specified bit depth and exceptions to register overflow situations.

The most important impact of this research for so-called IBN consists of the deployment of the below described methods within up-to-date mobile and wireless networks in two possible ways:

1) on OSI layer 2 for efficient coding based on spread-spectrum techniques (like FHSS, DSSS, Chirp) and aimed to fault tolerance optimization,
2) on the OSI layers 5–7 for securing of password generation routines aimed to authenticated access to multiple servers, desktop applications and mobile apps. (extended by Editors).

2 Algorithms for Generating Random-Like Processes

Based on condition 1, we define two types of generators with non-stationary time-forming functions of the type:

$$f(x,\ y) = p_0 + p_1 x + p_2 y.\ f(x,\ y,\ z) = p_0 + p_1 x + p_2 y + p_3 z, \tag{1}$$

where p_i are arbitrary parameters; $x,\ y,\ z$ are independent arguments.

The concept of «non-stationarity in time» in this case means changing the values of the p_i parameters at each cycle of the generator. The simplest case of the generating function $f(x) = p_0 + p_1 x$ is not considered in this chapter, since the basic characteristics of systems based on it are considered in [1], and from the point of view of ensuring the confidentiality of information transmission, this is not the best choice. Long-term observation of the implementation of an encrypted signal at high signal-to-noise ratios (S/W), formed on the basis of a function of a single argument, under certain conditions allows us to identify the structure and logic of changing the parameters of the information stream encoder. Taking into account (1), the algorithm for generating processes for a system with two degrees of freedom will be described by the expression:

The ribbon consists of buttons for the available style elements: title, authors, affiliations, headings, normal text, etc. To use one of these styles, first enter text and then click the button. The style will then be assigned to the paragraph that currently has the cursor in it. The headings H1 and H2 will be created with automatic numbering. For regular text please use "Normal text" button which will format the paragraph without indentation of the first line if it immediately follows a heading. As a general rule, required amounts of space before and after various elements (headings, equations, figures, etc.) are included in respective styles so do not use empty lines for this purpose. Also, please do not insert Word's index, table of contents or extra page numbers.

$$\begin{cases} h_k = F\{p_{0,k} + p_{\max} p_{1,k} h_{k-1} + p_{\max} p_{2,k} h_{k-2}\} \\ p_{0,k} = f_0(p_{0,k-1}) \\ p_{1,k} = f_1(p_{1,k-1}) \\ p_{2,k} = f_2(p_{2,k-1}) \end{cases}, \tag{2}$$

where k is the number of the signal sample, $k \in \mathbb{N}$; h_{k-1}, h_{k-2} are states at the output of the sequence generation system at time points separated by 1 and 2 working cycles, respectively $p_{0,k}, p_{1,k}, p_{2,k}$ are parameters that change at each clock cycle during the entire observation session.

The nonlinear functions $f_i(\cdot)$, $i = 0, 1, 2$ determine the operation of auxiliary generators that form the values of the parameters of the main ring of the system. No special conditions are imposed on the functions $f_i(\cdot)$. From the point of view of the convenience of practical calculations, it is desirable that they do not contain the problematic operations of division, square root extraction, logarithm, etc.

In our study, to determine $f_i(\cdot)$, we will use polynomials of degree $q = 2$ and higher, defined by the general expression:

$$f_i(p) = F\{\mu(p - \delta)^q + \varepsilon\}. \tag{3}$$

Here μ, δ, ε are arbitrary constants that determine the characteristics of the system.

The $p_{\max}$ is the parameter is a constant that determines the maximum possible tangent of the angle of inclination of the plane in the three-dimensional state space. As mentioned above, the value of the $p_{\max}$ parameter should not be set higher than 0.18.

We pay special attention to the fulfillment of condition 4, mentioned in the «Introduction» section. The value $p_{0,k} + p_{\max}p_{1,k}h_{k-1} + p_{\max}p_{2,k}h_{k-2}$ is the argument of the normalizing function $F(\cdot)$, defined as follows (for more information, see [2]):

$$F(h) = \begin{cases} 2 - h, & \text{if } h > 1 \\ -2 - h, & \text{if } h < -1 \end{cases}. \tag{4}$$

The functional scheme implementing algorithm (2) with consideration for (4) is shown in Fig. 1.

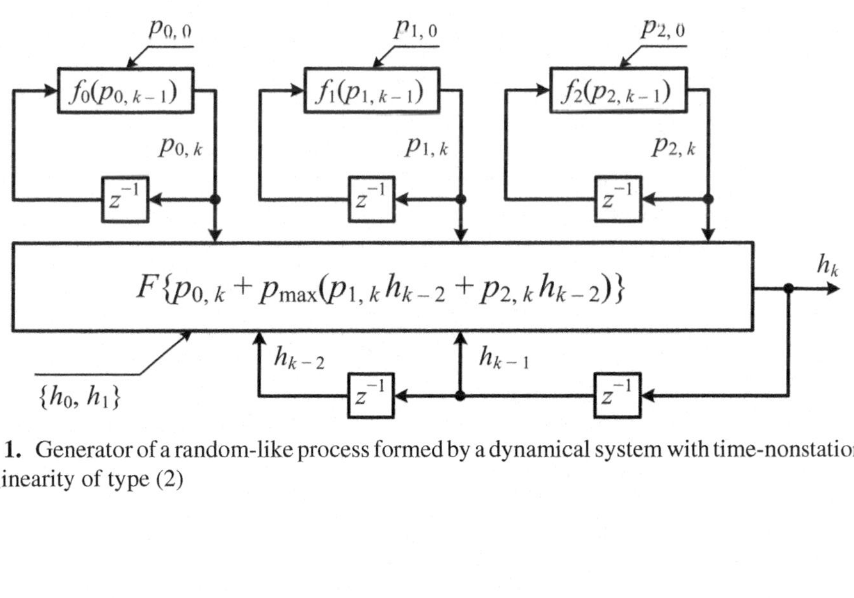

Fig. 1. Generator of a random-like process formed by a dynamical system with time-nonstationary nonlinearity of type (2)

Similarly, we define an algorithm for generating a random-like process for a nonlinear system with three degrees of freedom:

$$\begin{cases} h_k = F\{p_{0,k} + p_{\max}p_{1,k}h_{k-1} + p_{\max}p_{2,k}h_{k-2} + p_{\max}p_{3,k}h_{k-3}\} \\ p_{0,k} = f_0(p_{0,k-1}) \\ p_{1,k} = f_1(p_{1,k-1}) \\ p_{2,k} = f_2(p_{2,k-1}) \\ p_{3,\,k} = f_3(p_{3,\,k-1}) \end{cases} . \tag{5}$$

The functional scheme implementing algorithm (5) with consideration for (4) is shown in Fig. 2.

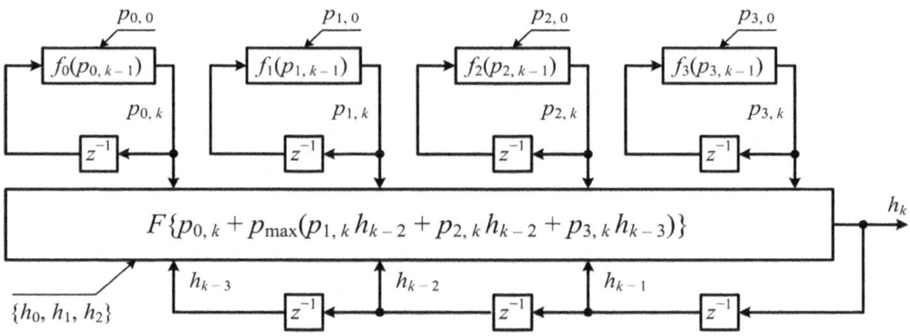

Fig. 2. Generator of a random-like process formed by a dynamical system with time-nonstationary nonlinearity of the type (5)

3 Estimation of Statistical and Spectral-Temporal Characteristics of Processes Formed by a System with Two Degrees of Freedom

The quality of the probability distribution function (PDF) of the generated random-like processes, taking into account remark 3, indicated in the introduction, will be evaluated by the mean square deviation σ_h of the states of the histogram of the instantaneous values of the process h from the average:

$$\sigma_h(X) = \sqrt{\frac{1}{N-1}\sum_{n=1}^{N}(X_i - \mu_X)^2}, \tag{6}$$

where N is the number of intervals for constructing the histogram; X_i is the number of hits of the process h on the i-th interval of values; μ_X is the mathematical expectation of the vector X, defined by the expression:

$$\mu_X = \frac{1}{N}\sum_{n=1}^{N}X_n. \tag{7}$$

Thus, the closer to zero the value $\sigma_h(X)$ is for a given parameter vector $[p_{\max},\ q,\ \mu,\ \delta,\ \varepsilon]$, the better the generated process is in terms of the uncertainty of the generated values. The quantitative indicator Ξ_h for this quality criterion is defined by the following expression:

$$\Xi_h(p_{\max},\ q,\ \mu,\ \delta,\ \varepsilon) = \frac{\sigma_h(p_{\max},\ q,\ \mu,\ \delta,\ \varepsilon)}{\mu_X(p_{\max},\ q,\ \mu,\ \delta,\ \varepsilon)} \tag{8}$$

The quality of the correlation (spectral-time) characteristics of the generated processes will be determined by the value of the maximum values of the lateral outliers $R_h(\tau)$ of the autocorrelation function (ACF). The inter-correlation functions (VCF) of the processes should be subjected to additional verification.

Since the functional Ξ_h has five independent parameters as arguments, we define their physically reasonable boundary conditions for algorithm (2), taking into account (3) and (4):

1) $p_{\max} \in (0;\ 0{,}18]$;
2) $q = 2$;
3) $\mu \in (0;\ 100]$, $\delta \in [-10;\ 10]$, $\varepsilon \in [-20;\ 20]$.

The boundary conditions in clause 3 are determined based on the behavior of a polynomial of the 2nd degree on the domain of its definition, in the class of problems under consideration $[-1;\ 1]$. We will fix the $p_{\max}$ parameter at the maximum recommended level of 0,18 and will not vary it. Taking into account (3), the task of this study lies in a very wide area, therefore, fixing the parameters of two of the three functions $f_i(p;\ \mu,\ \delta,\ \varepsilon)$, we evaluate the quality indicators of the generated sequences by variations of the third function Ξ_h and $R_h(\tau)$.

3.1 Statistical Indicators of the Generated Process Quality

For values

$$\begin{cases} \mu_0 = \text{var} \\ \delta_0 = 10 \\ \varepsilon_0 = 20 \end{cases},\quad \begin{cases} \mu_1 = 50 \\ \delta_1 = -10 \\ \varepsilon_1 = 20 \end{cases},\quad \begin{cases} \mu_2 = 25 \\ \delta_2 = 10 \\ \varepsilon_2 = -20 \end{cases}, \tag{9}$$

we define the quality parameter Ξ_h of the PDF of the process formed at the output of the system with two degrees of freedom.

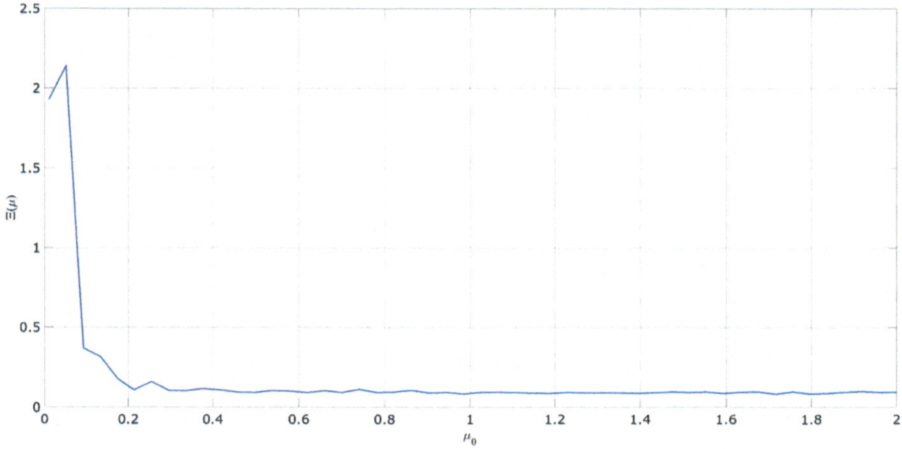

Fig. 3. Estimation of the uniformity of the PDD process at the output of the generator (2) when the parameter μ_0 is changed

The graph shows that at the parameters (9), already starting from $\mu_0 = 0{,}3$ the PDF of a random−like process at the output of the generator (2) becomes almost uniform with a quality coefficient $\Xi_h = 0{,}08...0{,}09$. It follows that unacceptable distributions can be called distributions at $\Xi_h > 0{,}15$.

The nature of the PDF changes from localized at $\mu_0 = 0{,}01$ (Fig. 4) to displaced at $\mu_0 = 0{,}17$ (Fig. 5).

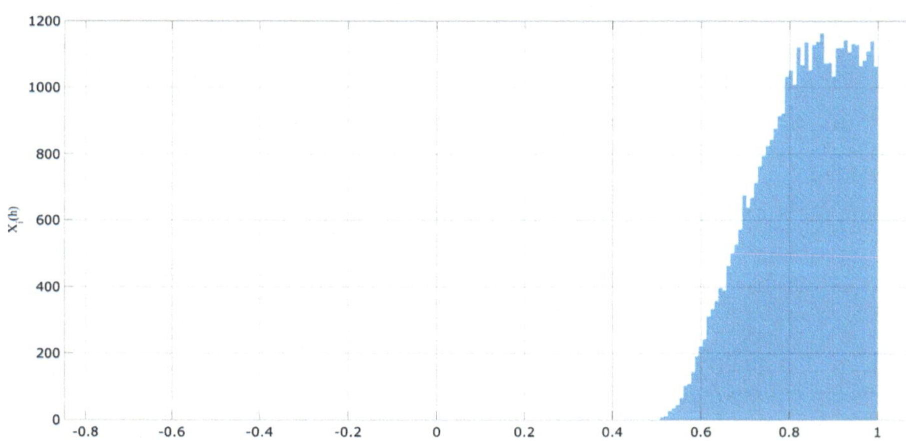

Fig. 4. PDF of the process at the output of the generator (2) with the parameter $\mu_0 = 0.01$

The general dependence of the quality coefficient of PDF Ξ_h on the interval of change of the parameter $\mu_0 = (2;\quad 100)$ is shown in Fig. 6.

To understand the nature of the change in the PDF of a random-like process at the output of the generator (2), we estimate the quality coefficient Ξ_h at lower values of

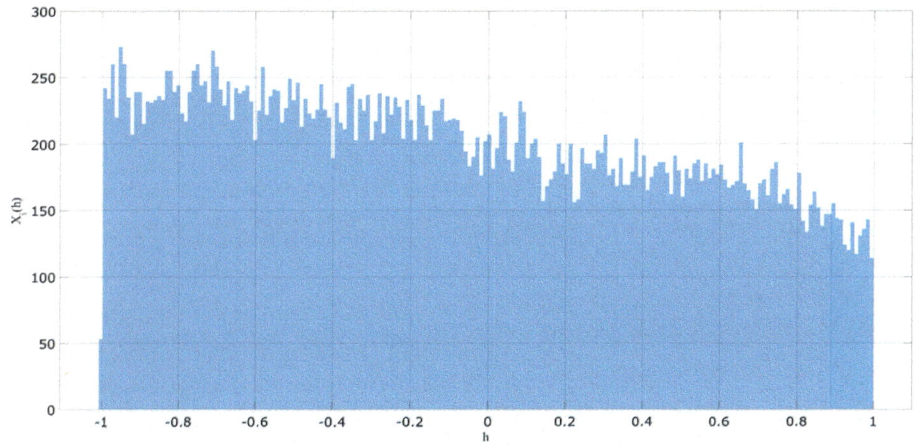

Fig. 5. PDF of the process at the output of the generator (2) with the parameter $\mu_0 = 0.17$

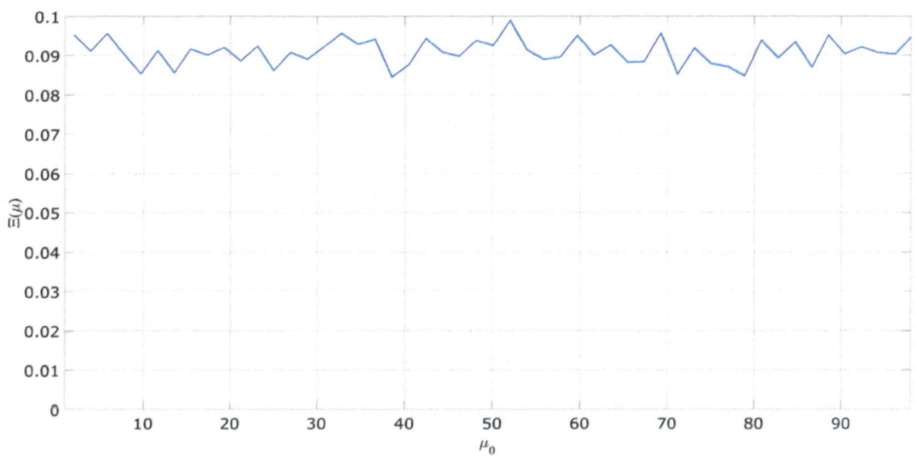

Fig. 6. Estimation of the uniformity of the PDF process at the output of the generator (2) when the parameter μ_0 to 100 is changed

other related parameters:

$$\begin{cases} \mu_0 = \text{var} \\ \delta_0 = 1 \\ \varepsilon_0 = 2 \end{cases}, \begin{cases} \mu_1 = 5 \\ \delta_1 = -1, \\ \varepsilon_1 = 2 \end{cases} \begin{cases} \mu_2 = 2,5 \\ \delta_2 = 1 \\ \varepsilon_2 = -2 \end{cases}. \tag{10}$$

The calculation results are shown in the figure below.

The quality factor of the PDF generated by a random-like process significantly worse, it is approximately 0,12–0,20 and stabilizes at the level of 0,16, starting from $\mu_0 = 6$. For $\mu_0 > 18$, the coefficient $\Xi_h \approx 0{,}12$. Studies on the influence of variations of the parameters $\delta_1,\ \delta_2,\ \varepsilon_1,\ \varepsilon_2$ at fixed values of $\mu_0,\ \mu_1,\ \mu_2$ exceeding 10 have shown good and very good values of the quality coefficient PDF $\Xi_h \simeq 0{,}08...0{,}11$.

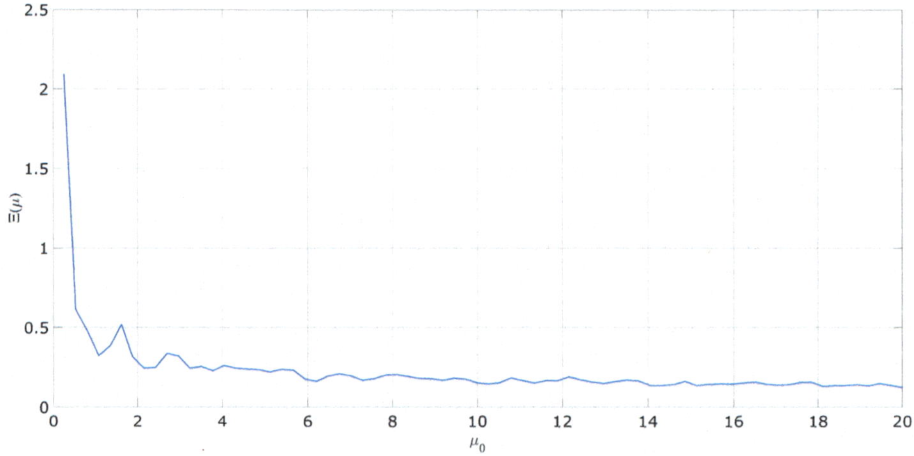

Fig. 7. Estimation of the uniformity of the PDF process at the output of the generator (2) at different μ_0 under conditions (10)

3.2 Correlation Properties of the Generated Process

The maximum value of the ejection of the side lobe of the ACF of the process is evaluated, followed by normalization by the value $\sqrt{N}$, where N – is the number of samples of the generated process. The emission values for the level $(3\ldots5)/\sqrt{N}$ can be considered acceptable for use in secure information transmission systems. If the denominator exceeds the value of 5, this indicates a significant statistical relationship between some states of a random-like process. The results of the study for conditions (9) are presented below.

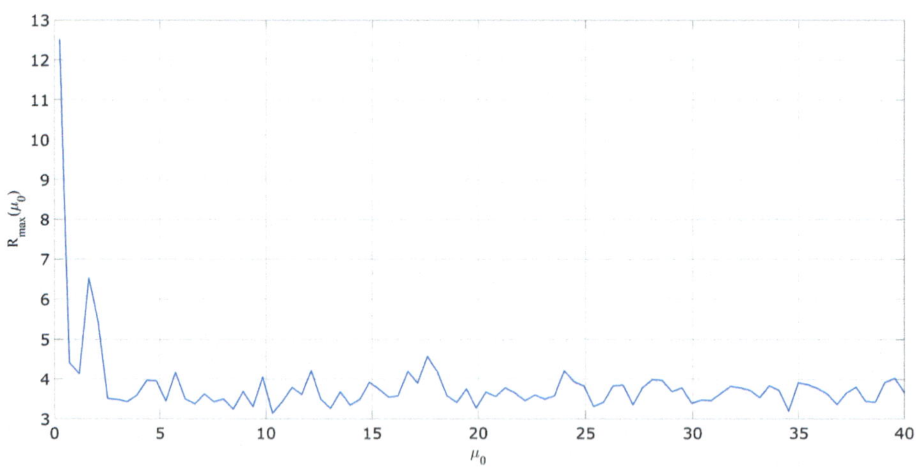

Fig. 8. Estimation of the maximum value of the side lobes of the ACF of the process at the output of the generator (2) for the conditions (9)

Note in particular: the maximum value of the ejection of the side lobes of the ACF of the generated sequences is postponed along the axes of the ordinates of the graph: $R_{\max}(\tau) \Big/ \sqrt{N} \Big|_{\tau>0}$. Analysis of the graph shows that for all values of $\mu_0 > 3$, the system (2) forms sequences with very good ACF. Less expected results are observed in the case of a 10 fold reduction in the parameters of the auxiliary ring functions forming the parameters p_1 and p_2. The figure shows the wave-like nature of the deterioration of the ACF emissions of processes at some values of the variable parameter.

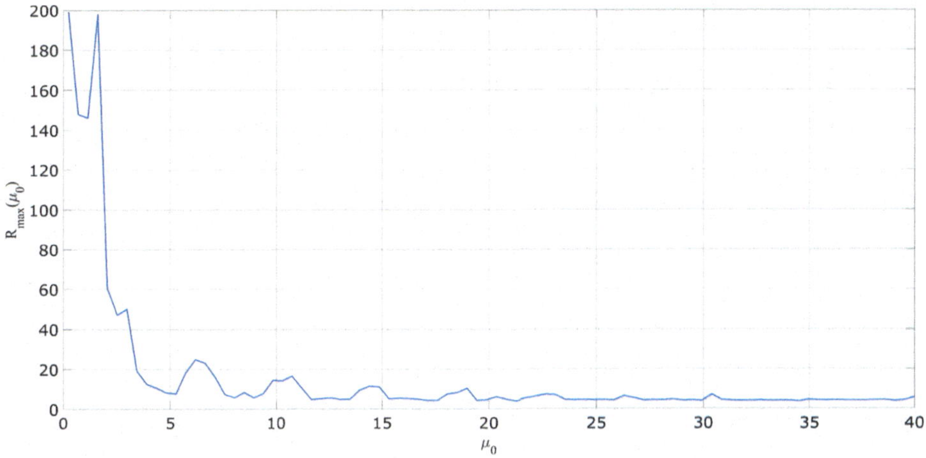

Fig. 9. Estimation of the maximum value of the side lobes of the ACF of the process at the output of the generator (2) for the conditions (10)

As in clause 3.1, we estimate the value of the maximum ejection of the side lobes of the ACF of the generated process near the lower limit of the parameter μ_0.

Thus, in the region of small values of μ_0, the outliers of the side lobes under conditions (10) show very poor results.

4 Estimation of Statistical and Temporal Characteristics of Processes Formed by a System with Three Degrees of Freedom

4.1 Statistical Indicators of the Quality of the Generated Process

According to (5) taking into account (3) for the values

$$\begin{cases} \mu_0 = \text{var} \\ \delta_0 = 10 \\ \varepsilon_0 = 20 \end{cases}, \begin{cases} \mu_1 = 50 \\ \delta_1 = -10 \\ \varepsilon_1 = 20 \end{cases}, \begin{cases} \mu_2 = 25 \\ \delta_2 = 10 \\ \varepsilon_2 = -20 \end{cases}, \begin{cases} \mu_3 = -25 \\ \delta_3 = 10 \\ \varepsilon_3 = -20 \end{cases}, \tag{11}$$

we determine the quality parameter Ξ_h of the PDF of the process formed at the output of the system with two degrees of freedom.

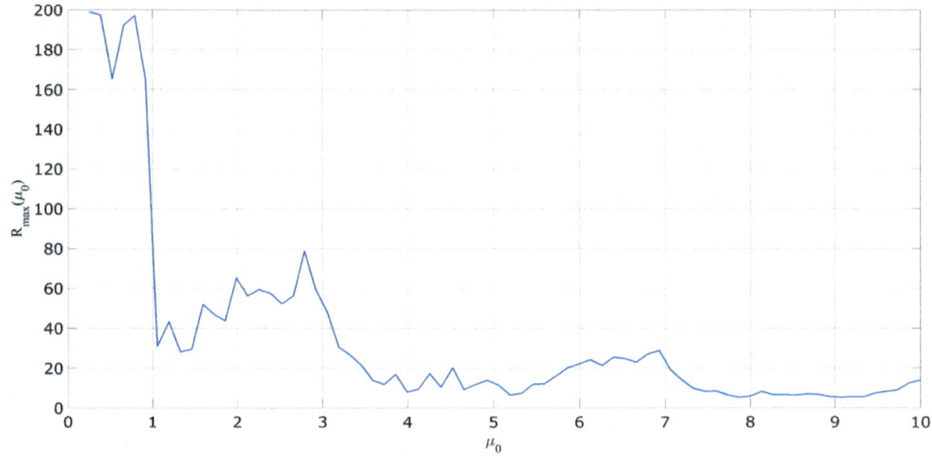

Fig. 10. Estimation of the maximum value of the side lobes of the ACF of the process at the output of the generator (2) under conditions (10)

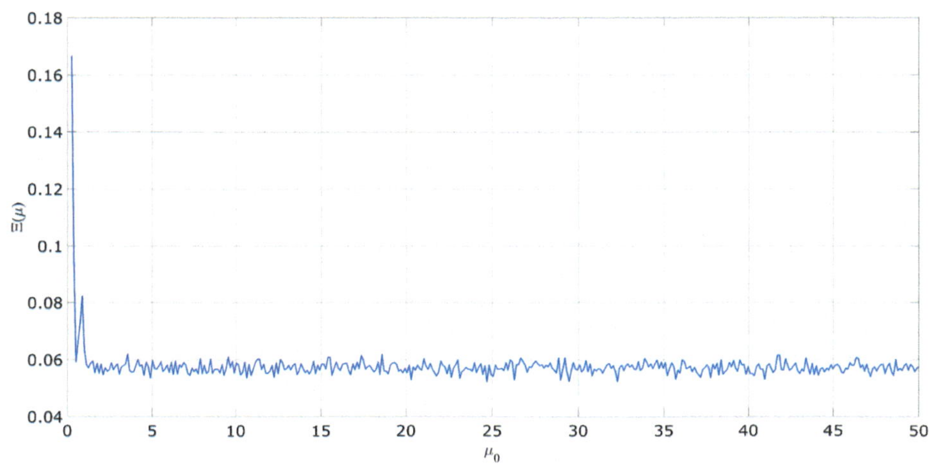

Fig. 11. Estimation of the uniformity of the PDF process at the output of the generator (5) when the parameter μ_0 is changed

In this case, the quality of the PDF of the generated process at $\mu_0 > 2$ is estimated by the value $\Xi_h \in (0{,}05; \quad 0{,}06)$, which is noticeably better than the results given in Sect. 2 (see Fig. 6).

Reducing the values of the parameters (11) of the auxiliary functions (3) for the algorithm (5) 10 times gives the following result.

The quality index Ξ_h deteriorated to values of 0,09 0,13, and at values $\mu_0 < 5$, areas with an abnormal deterioration in the statistical properties of the generated sequence were identified.

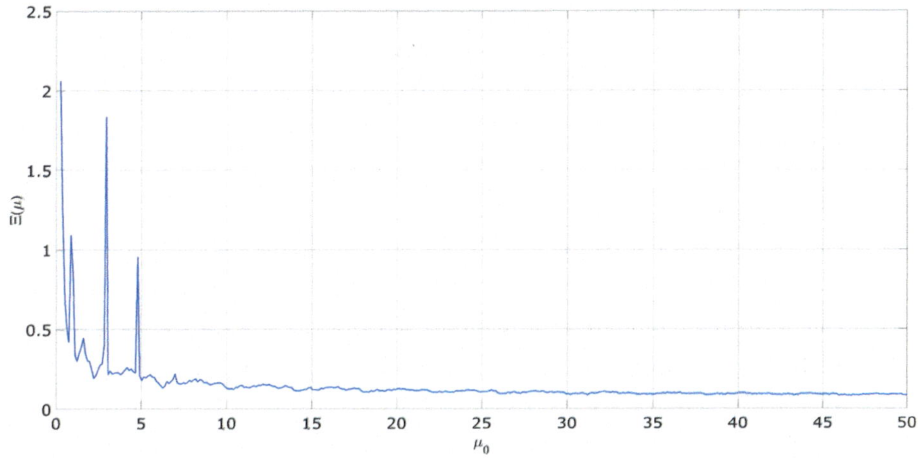

Fig. 12. Estimation of the uniformity of the PDF process at the output of the generator (5) after reducing the parameters (11) by 10 times

4.2 Correlation Properties of the Generated Process

An estimate of the $\sqrt{N}$ normalized value of the lateral ACF emissions of a randomly generated process for conditions (11) is shown in the graph below.

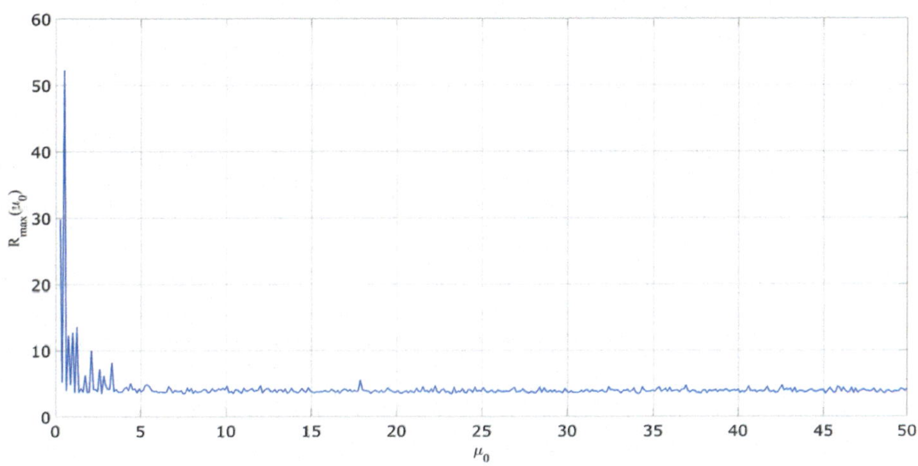

Fig. 13. Estimation of the maximum value of the side lobes of the ACF of the process at the output of the generator (5) for the conditions (11)

The value of the side lobe emissions in this case is 3.4–3.9. In the case of a 10 fold decrease in the parameters (11), a predictable deterioration in the correlation properties is observed.

The value of the outliers of the side lobes is in this case the value of 9.5–13.0, which, as in the results shown in Figs. 9 and 10, is a low indicator and is unacceptable for use in

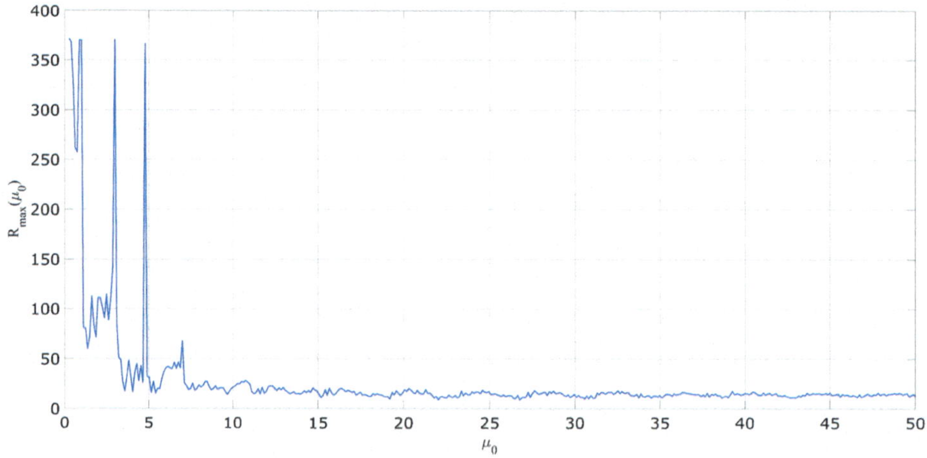

Fig. 14. Estimation of the maximum value of the side lobes of the ACF of the process at the generator output (5) for the conditions (11)

information transmission systems. Thus, reducing the parameters of auxiliary functions preserves the quality of the law of distribution of instantaneous values, but leads to low quality indicators of the ACF. The intercorrelation properties of the generated sequences with the variation of the parameters of systems (2) and (5), as well as the initial generation conditions, in all considered cases show good results at the level of $(2,5\ldots3,5)/\sqrt{N}$.

5 Conclusions and Practical Recommendations

1) The basic element of a nonlinear dynamical system described by expression (2) is a «piecewise linear plane» in a three-dimensional state space: two variables h_{k-1} and h_{k-2} are independent arguments, on the basis of which a polynomial of the first degree is first calculated, and then this value is normalized by the function $F(\cdot)$. Thus, the system has two degrees of freedom, but the phase space of states is characterized by a certain figure in three-dimensional space, and this figure is modified at each clock cycle of the circuit, which leads to a significant entanglement of the phase trajectories of the generated process.

   Similarly, the basic element of the system described by expression (5) is a «piecewise linear figure» in a four-dimensional state space. In this case, the three variables h_{k-1}, h_{k-2} and h_{k-3} are independent arguments defining a system with three degrees of freedom. The appearance of the four-dimensional figure also changes at each clock cycle of the circuit, which further complicates the assessment of the characteristics of the systems by an outside observer – an unauthorized user.
2) Throughout the text of this chapter, the term «random-like process» is used to determine the nature of the generated sequences. From the point of view of the countability and finiteness of the states of a digital automaton implementing algorithms (2) and (5), such a definition is valid. But as soon as we introduce an information stream into the generators, which is random in nature, the phase trajectories of the process

at the output acquire an unpredictable structure due to the nonlinear nature of the dynamic system. Thus, in real-world applications, the process output of such at he «generator-modulator» will have a random nature, which leads us to a completely new entity: a completely deterministic system that generates a random signal. In this regard, two important properties of such an entity should be particularly noted:

- the random signal generated by systems (2) and (5) by nonlinear mixing of the information stream is not correlated with the latter, in contrast to classical modulation systems;
- if you know the initial conditions and the structure of the generators, it is always possible to extract the information process from a random process.

3) The function $p_{0,k} = f_0(p_{0,k-1})$ in expressions (2) and (5) largely determines the statistical characteristics of the generated processes, since it makes the greatest contribution to the generated sample. However, the spectral-temporal properties are largely defined auxiliary functions $p_{1,k} = f_1(p_{1,k-1})$, $p_{2,k} = f_2(p_{2,k-1})$, $p_{3,k} = f_3(p_{3,k-1})$ because they determine the «fine structure» of phase transitions of the system from state to state. The correlation properties are also significantly affected by the $p_{\max}$ parameter.
4) The vector of parameters $p_{\max}, \mu, \delta, \varepsilon$ in the formation algorithms (2) and (5), as well as the initial conditions $[p_0(k),\ \ p_1(k),\ \ p_2(k),\ \ p_3(k)]$ for $k = 0$ and $[h(0),\ \ h(1)]$ для (2); $[h(0),\ \ h(1),\ \ h(3)]$ for (5), are encryption keys for information transmission systems. The structure of the generator and the nature of the feedbacks in it may be known to third parties.
5) A quality factor has been introduced that allows an objective assessment of uniformity PDF within a fixed dynamic range of process values at the output of the generator (8). It is shown that distributions that are unsuitable for use in information transmission systems, we can call distributions having a quality coefficient $\Xi_h > 0{,}15$. Exceeding this value indicates that the generated process can be delayed within a certain range of instantaneous values, or more often be pulled together there, thereby forming a certain quasi-regular trajectory or a predictable set of states.
6) The correlation properties of the generated sequences under conditions (9) and (11) show good results at the level $(3\ldots4)/\sqrt{N}$, indicating a small statistical relationship between the samples of the sequences.
7) The peculiarity of the methods and algorithms presented in the chapter is the relatively simple possibility of software implementation in the form of additional modules used in wireless communication lines of IBN networks. In contrast to the known means of ensuring information security, systems based on nonlinear dynamics have the property of self-synchronization and recovery of operability in the event of an intentional or unintentional failure in the IBN network.

References

1. Polovenia, S., Dubrouski, V.: Ensuring the secrecy of the information chaotic signals based on the mappings distributed in time. Vestnik of the BSU **3**, 51–56 (2012)

2. Dubrouski, V., Lavshuk, O.: Temporal and statistical properties of sequences generated by nonlinear parametric systems. Probl. Commun. Minsk **2**(142), 50–55 (2017)
3. Anishchenko, V., Vadivasova, T., Okrokvertshov, G., Strelkova, G.: Correlation analysis of dynamic chaos. Phys. A Stat. Mech. Appl. **325**, 199–212 (2003)
4. Paulson, J.A., Bühler, E.A., Mesbah, A.: Arbitrary polynomial chaos for uncertainty distribution of correlated random variables in dynamic systems. IFAC-Papers On Line **50**(1), 3548–3553 (2017)
5. Queens, V., Dogra, S., Lakshmi Narayan, A.: Quantum correlations as a probe of chaos and the ergodicity of the. Opt. Commun. **420**, 189–193 (2018)

Methodology of ISMS Establishment Against Modern Cybersecurity Threats

Vitalii Susukailo, Ivan Opirsky(✉), and Oleh Yaremko

Lviv Polytechnic National University, Stepan Bandera Street, 12, Lviv 79013, Ukraine
{vitalii.a.susukailo,ivan.r.opirskyi,oleh.m.yaremko}@lpnu.ua

Abstract. The chapter addresses the Information Security Management System (ISMS) establishment approach, ensuring necessary controls to avoid widespread cybersecurity threats nowadays. The most common attack vectors and techniques of the last three years were analyzed in this chapter to define a set of information security practices, which can minimize risks related to modern cybersecurity threats. Conducted analysis of cybersecurity frameworks such as ISO 27001/2, CIS Top 18, NIST 800-53, and their differentiators. An ISMS establishment algorithm is proposed in this chapter with a detailed explanation of each phase and controls required for system implementation. The document defines cybersecurity technologies for management systems are determined based on the infrastructure type. The documentation management framework and risk management methodology are proposed and analyzed modern awareness strategies and defined education roadmap for ISMS roles.

Keywords: Information Security Management System (ISMS) · Security controls · Cybersecurity threats · Education roadmap · ISO 27001/2 · CIS Top 18 · NIST 800-53

1 Introduction

Information security is a set of technical and organizational measures and developed documents in the broadest sense of the word. The primary purpose is to protect and preserve the information owned by the organization. However, information security remains an integral part of cybersecurity, a much broader category, and includes protecting information and data and protecting systems, networks, and more [1–3]. The main goals of information security also involve creating a set of business processes that will protect information assets regardless of how information is formatted, whether it is in transit, processed, or at rest, i.e., stored in appropriate databases. According to experts, the value of an is determined primarily by what information is owned by the company and how this information is stored. Information security is a crucial factor in ensuring the effective conduct of business operations and maintaining and gaining the trust of customers, both future and existing [4].

To ensure the most effective information security within the company, it is first necessary to determine the basic strategy that the company must follow. Such a strategy

M. Klymash et al. (Eds.): *Future Intent-Based Networking*, LNEE 831, pp. 257–271, 2022.
https://doi.org/10.1007/978-3-030-92435-5_15

should be determined by the company's management, with the simultaneous involvement of information security specialists, who can be both employees of the company and the external contractors involved. The result of the developed strategy is usually an approved project on information security and ways and methods of its implementation. After defining the global strategy system, many usually creates a specialized team of information security specialists. Typically, this group is led by the Chief Information Security Officer (CISO). Other team members should be selected based on their level of competence and information security skills [5].

The information security team is responsible for risk management, the implementation of processes that constantly assess vulnerabilities and threats to information held by the company, as well as for the adoption and application of appropriate safeguards. However, the information security team must respond to all violations of information security and promptly decide to eliminate such violations, as well as to minimize risks to the interests of the company or fundamental rights, freedoms and interests of individuals, if the violation in any way concerns personal data of such persons [6].

The information security project should build around the three main aspects of security, which are to maintain the confidentiality, integrity and accessibility of information, including within the company's IT systems and databases. The implementation of such principles will help ensure the disclosure of confidential information only to authorized parties (confidentiality), prevent unauthorized alteration of data (integrity) and ensure that information can only be accessed by authorized parties (if any) [7].

Nowadays, the quantity of cybercriminals is rapidly growing every day. As more specialists join their ranks, more malware is being launched daily, with approximately 230,000 new malware samples per day according to the information from PandaLabs security researchers statistics. During this time growing quantity of cybersecurity threats should be analyzed, and applicable security measures should be defined. New techniques and technologies supports threat actors with exploits development. Vulnerabilities and flaws in OS, services and applications occurs on a daily basis. That is why, it became hard for cybersecurity experts to detect threats in timely manner [8].

2 Analysis of Attack Vectors

The most important part of attack vectors analysis is to understand the motivation of threat actors. The motivation of modern attacks is the financial gain which caused by the financial crisis in the world. Hackers can compromise organizational infrastructure or get unauthorized access to data or information, which can be used for financial gain. Also, hackers disrupt or sabotage manufacturing, electric power generation etc. to create chaos and anarchy [14–16]. When motivation is defined, it is necessary to understand how the attack is performed. The review of latest attacks is provided in this chapter. The popularization of cloud computing without proper hardening measures would almost always lead to organizational compromise. Data breaches as a threat retains its number one for cloud environments. Breaches can cause great reputational and financial damage. They could potentially result in loss of intellectual property (IP) and significant legal liabilities. Inadequate access management, as a cloud environment, not threat can lead to cloud system compromise. To avoid this threat, cloud customers should protect credentials, ensure automated rotation of cryptographic keys, passwords and certificates, ensure scalability, require cloud service administrators to use multi-factor authentication, define password policy for management plane and each service deployed in the cloud.

In 2019 more than 540 million records of Facebook users were compromised and were published on Amazon's cloud computing service. The root cause of this incident were insecure backups that were publicly available on AWS without any access control mechanism.

There were two separate instances. First involved Mexico City-based digital platform - named Cultura Colectiva, which openly stored 540 million (approx. 146 GB) records of Facebook users, including identification numbers, account names, comments, and reactions. The records were accessible and can be downloaded by anyone who could find them online.

The second instance contained a backup from a Facebook-integrated application – namely, "At the Pool," which was exposed to the public internet via an Amazon S3 Bucket. The backup contained data of 22,000 users that further consists of user I.D.s, frienI.D.slist, likes, music, movies, books, groups, check-ins, passwords (in plaintext) and more. These passwords were likely for the "At the Pool" app rather than the Facebook account of the user [8].

December 8 2020 - the leading cybersecurity company FireEye announced that it had been hacked by a group of government hackers. As part of this attack, the attackers even stole the tools of the so-called red team - a group of FireEye experts who conduct as close as possible to real cyberattacks to check the security systems of their customers. It was not known how the hackers gained access to the FireEye network until December 13, when Microsoft, FireEye, SolarWinds and the U.S. governmU.S.t released a coordinated report that SolarWinds had been hacked by a group of government hackers - FireEye was just one of SolarWinds' customers affected. In January 2021, representatives of the US DepartmU.S.t of Justice confirmed that the Ministry of Justice also suffered from the hacking of SolarWinds. Worse, the agency was one of the few victims in which hackers continued to attack and eventually gained access to internal mailboxes.

Based on the fact that the staff of the Ministry of Justice is estimated at about 100,000–115,000 people, the number of victims ranges from 3,000 to 3,450 people. The attackers gained access to the SolarWinds Orion build system and added a backdoor to the SolarWinds.Orion.Core.BusinessLayer.dll file. This DLL was then distributed to SolarWinds customers through an automatic update platform. After downloading, the backdoor connects to a remote management and control server in the avsvmcloud [9] Com subdomain to receive "jobs" to run on the infected computer (Fig. 1).

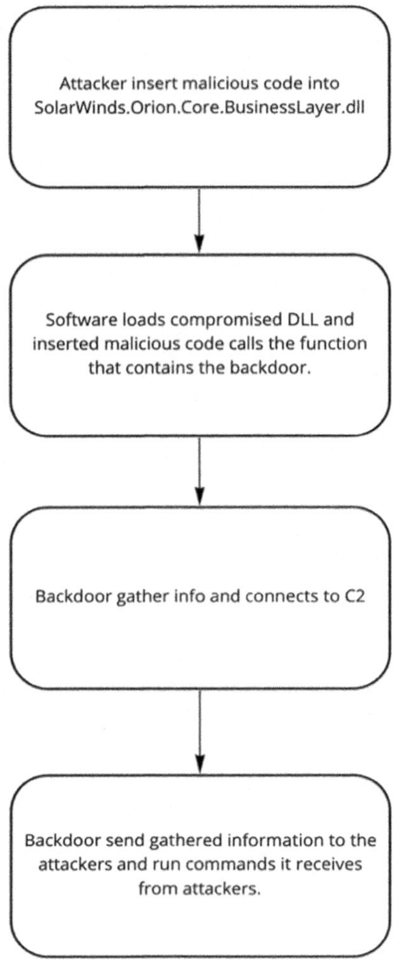

Fig. 1. SolarWinds attack flow

May 2021 - Colonial Pipeline was shut down after the ransomware attacks. The company delivers 100 million gallons of fuel a day.

Colonial closed its operations, due to the ransomware infection found on its computer systems. The shutdown affected the supply of gas in parts of the East Coast, with some people waiting an hour or more at filling stations or not finding gas at all [7]. State and federal officials had warned against hoarding and panic buying that could exacerbate the problem. The ransomware infection at Colonial highlighted the vulnerability of the country's critical infrastructure, which has been the target of an increasing number of cyberattacks [10].

Based on the mentioned above attacks it's highly important to highlight the following cybersecurity challenges:

Insufficient protection of critical infrastructure. The following challenge can lead to an increased quantity of private and government Security Operations Centres. Also, it can affect the quality of services which provides Security Operations Centres. Cybersecurity experts need to provide high-level security monitoring, vulnerability management and incident response to ensure that there will not be interruptions in critical infrastructure and personal patient data will be protected.

An insufficient number of Cyber Security specialists is an actual problem nowadays, and the pandemic situation shows how important it is to have qualified experts, which ensures the protection of information in government and private organizations. Following challenge can lead to improving the educational system in Cyber Security and increasing quantity institutions, which can prepare qualified specialists.

Incorrect determination of cybersecurity frameworks that are relevant to ISMS. Information security experts within organization must determine appropriate cybersecurity framework, which should be followed by organization.

Supply chain attacks as a result of poor implementation of information security controls due to lack of information security requirements, which suppliers are required to followed. This challenge can lead to more comprehensive and detailed controls definment within supplier management process.

3 Analysis of Cybersecurity Frameworks

While Information Security Management System establishment organization must determine appropriate Cybersecurity Framework. Cybersecurity frameworks are comprehensive, and are designed to achieve a specific objective. The most commonly used are ISO 27001/2, NIST Cybersecurity Framework and CIS top 18.

NIST Cybersecurity Framework is the cybersecurity framework established by the National Institute of Standards and Technology (NIST). The following framework offers detailed guidance on everything from risk assessment and continuous monitoring to incidence response and awareness training. NIST offers not only a comprehensive plan for data protection and risk mitigation but also a methodology for limiting the impact of adverse events.

Like the NIST, the ISO is designed to provide an approach for achieving a certified level of data security compliance that meets external assessment standards. Instead of NIST, which is designed by the U.S. federal government, the ISO 27001 is developed by the International Organization for Standardization (ISO) and the International Electrotechnical Commission (IEC). ISO 27001 can be implemented in any organization: commercial or non-commercial, private or public, small or large. It was written by the world's leading experts in the field of information security and offers a methodology for implementing information security management in the enterprise. It also allows companies to obtain certification, which means that an independent certification body will confirm that the organization has implemented information security in accordance with ISO 27001.

The Center for Internet Security is an American public nonprofit organization best known as the creator and main contributor to CIS Controls and CIS Benchmark, the world's recognized best practices for securing I.T. systems I.T.nd data. CIS TOP 18

Controls is a basic set of actions that collectively aim to protect against the most common attacks on businesses and their security systems. The framework helps to ensure security at a minimum acceptable level, taking into account the likelihood and realism of existing threats. All measures are divided into basic, fundamental and organizational. The description of each of the 20 measures (controls) begins with an overview of its importance and why it is critical. Examples of common attacks and an explanation of the success of attackers in the absence of such control are given. Also, it defines list of specific actions procedures and tools that organizations should take to build protection, as well as examples of diagrams and charts to illustrate approaches to implementing measures.

To help organizations define appropriate Cybersecurity framework for Information Security Management System implementation it was created Cybersecurity Frameworks comparation matrix (Table 1).

Table 1. Cybersecurity frameworks comparation matrix

Objective	ISO 27001/2	CIS Top 18	NIST
Does the cybersecurity framework has implementation guidance for its controls?	Yes	Yes	Yes
Is there available mapping of its controls to other frameworks?	Yes	No	Yes
Is it required to have specific knowledge to implement the following framework?	Yes	No	Yes
Is it publicly available?	No	Yes	Yes
Does the cybersecurity framework provide technical security recommendations, which the organization can implement?	Yes	Yes	Yes
Can the organization be certified after the implementation of the cybersecurity framework?	Yes	No	No
Can a person be a certified implementor of the following framework?	Yes	No	No

4 Educational Roadmap for Cyber Security Specialists

The most important thing during ISMS implementation is to have the necessary knowledge. There are many courses, which can be taken by cybersecurity specialists, such as ISO 27001:2013 Foundation, Lead Implementor or Lead auditor. Those would allow can professionals develop skills in ISMS establishment. But to implement cybersecurity controls, which can ensure protection against modern security threats specialists must develop necessary professional skills. This scientific work propose possible educational roadmaps for different roles within organization and qualification levels. Proposed below educational roadmap can help develop cybersecurity skills for different specialists within organization, which can ensure security controls establishment as well as risk oriented mindset development at different organizational levels (Table 2).

Table 2. Educational roadmap for cyber security specialists

Role	Junior	Middle	Senior
Information Security Engineer	Comptia Security+, CSX Cybersecurity Nexus	CCSK, CCSP, AWS	CASP+. CISSP
Information Security Analyst	Comptia Security+, CSX Cybersecurity Nexus	CCSK, Comptia CySA+, CASM, CSM	CASP+. CISSP
Information Security Administrator	Comptia Security+, CSX Cybersecurity Nexus	ITIL MP, CRISC	CISM, CISA
Penetration Tester	CEH	ComptiaPentester+ OSCP	OSCE, OSWE, OSEE
Quality Assurance Engineer	CEH	ComptiaPentester+	CSSLP
DevOps Engineer	Comptia Security+, CSX Cybersecurity Nexus	CCSK	CSSLP
Software Engineer	GIAC Secure Software Programmer-Java, GIAC Secure Software Programmer-.Net	CASE	CSSLP

5 ISMS Implementation Model

ISMS implementation model is defined and explicitly described in this chapter, which could be adopted per any organization and IT infrastructure type (Fig. 2).

The schematic ISMS implementation model, which can ensure protection against modern cybersecurity threats provided above. At the first stage of ISMS establishment it is necessary to conduct gap assessment. To conduct gap assessment effectively organization should divide

1. Determine audit scope.
2. Define is it necessary to include subject matter experts or consulting company.
3. Select framework, which will be used as an audit criteria.
4. Determine auditees.

All this information should be specified in gap assessment plan. Also, before gap assessment it is necessary to know the organization mission, vision, organizational structure, business models, and infrastructure to apply organization specific security controls.

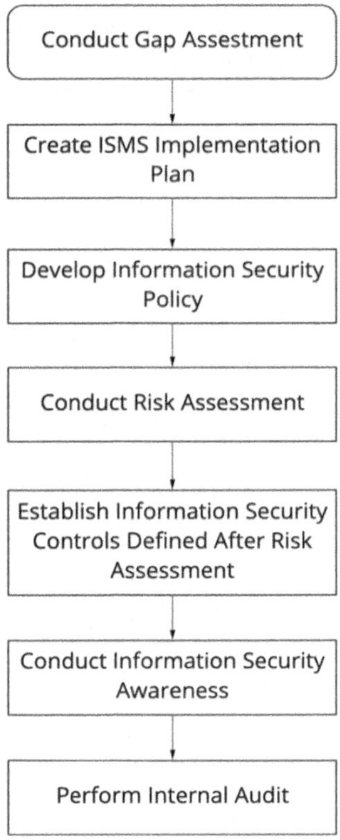

Fig. 2. ISMS Implementation Model

To conduct gap assessment effectively it is necessary to prepare Checklist of security controls, which would be used to consolidate gap assessment results. As a checklist can be used controls from proposed in the previous chapter frameworks, such as NIST, ISO 27001/2 or CIS top 18.

Based on gap assessment results, it is necessary to define corrective actions for detected gaps. If for gap assessment, was used ISO 27001/2, the best approach would be to use task management solution to track progress on corrective actions implementation.

After the gap assessment stage the first and the most important item that is essential to implement is an Information Security Policy that meets business objectives. Information Security Policy is ISMS's main document that highlights the scope, objectives, responsibilities, and information security improvement framework. The following objects may be always included into the policy:

1 Ensure compliance with relevant information security laws, regulations, and agreements to satisfy applicable requirements related to information security.
2 Prevent information security incidents from occurring.
3 Educate personnel on the confidentiality, availability, and integrity of information and information assets.

4 Ensure appropriate handling of information.

In this research we also propose KPI for the following objectives provided in Table 3. Objective-KPI Mapping.

Table 3. Objective-KPI mapping

Objective	KPI
Ensure compliance with relevant information security laws, regulations, and agreements to satisfy applicable requirements related to information security	Quantity of lawsuits, caused by information security incidents is 0
Prevent information security incidents from occurring	Percentage of information Security incidents resolved within required time period is X%
Educate personnel on the confidentiality, availability, and integrity of information and information assets	Percentage of specialists passed yearly information security awareness is more that X%
Ensure appropriate handling of information	Percentage of documentation shared with specialists labeled based on Information Classification requirements is 100%

The most critical task during ISMS implementation is Risk Assessment. The Management framework for modern ISMS proposed on Fig. 3.

The first stage of Risk Assessment is Asset identification. During this stage it is necessary to define all valuable assets, within organization [8]. Once assets are identified it is necessary to define risks for these assets. The approach that we propose in this research rely on the commonly used practice: define vulnerability and threat that are relevant to the asset. For example, execution of unexpected commands to malicious software by remote hackers due to lack of antimalware application modules updates on end-user PC. The threat in this risk is malicious software and the vulnerability - lack of antimalware application modules updates. Multiple vulnerabilities can be applicable to this threat as well as multiple treats can be analyzed for this vulnerability [12]. Those should be evaluated for each asset type and specified within Risk Management phase.

For each risk and gap defined during assessment stage it is necessary to define appropriate control. The administrative and organizational controls are provided in cybersecurity frameworks and must be adopted by each organization individually. For this research we defined set of cybersecurity tools that can be used to mitigate risks defined during risk assessment and gap assessment phases [26] [28,29] (Table 4).

The effectiveness and performance of each management system should be evaluated from time to time. The information security management system is not an exclusion. It is necessary perform monitoring and measurement of the ISMS regularly. It's an essential part of the ISMS implementation and improvement, which needs to be controlled [19] (Fig. 4).

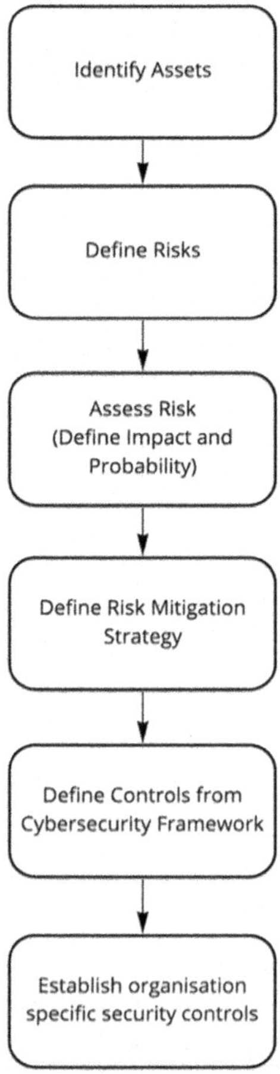

Fig. 3. Risk management framework

Evaluation if ISMS is achieving KPIs and objectives. If ISMS is not achieving some KPIs or objectives, it's time to review the weak process and improve it. Security specialists also regularly check your information security goals to ensure that they are still actual for organization. Otherwise, ISMS team need define other objectives to improve an ISMS continuously [18–22].

Similar to gap assessment, it is necessary to conduct an internal audit. But ISMS team need to select auditors and conduct audits that ensure objectivity and the impartiality of the audit process. This process would help Security Team evaluate ISMS progress periodically and prepare organization for external audits.

The modern ISMS's can be also evaluated gathering anonymous feedback from your colleagues to know what they think about the Information Security processes within your organization.

Table 4. Educational roadmap for cyber security specialists

Security Control	Solution	Purpose	ISO 27001/2 control	NIST CSF Control
VPN	OpenVPN	Ensures secure remote connection to on-remise infrastructure	A6.2.2	PR.AC-3
HIDS	OSSEC, Wazuh	Host-based intrusion detection system should be used to detect threats on endpoints	A16.1.2	DE.DP-4
Backup and recovery	Bacula, Urbackup	Backup software is necessary to restore critical business infrastructure during the pandemic	A12.3.1	PR.DS-4
Infrastructure monitoring	Zabbix, Nagios	Infrastructure monitoring software should be used to monitor the on-premise infrastructure state	A16.1.2	DE.DP-4
Endpoint protection	Armadito, Clam AV	Endpoint protection software should be used to avoid malware infection	A12.2.1	PR.DS-2
Vulnerability management	Wazuh, OpenVas	Wazuh can be used to detect vulnerabilities on endpoints	A12.6.1	PR.IP-12
Patch Management	OPSI	Patch management tools must be used to update endpoints and infrastructure assets	A12.6.1	PR.IP-12
Security orchestration	Patrowl Demisto	Security Orchestrations platforms need to be used to automate security operations activities during a pandemic	A16.1.2	RS.CO-2

(*continued*)

Table 4. (*continued*)

Security Control	Solution	Purpose	ISO 27001/2 control	NIST CSF Control
SAST	SonarQube, DerScanner	Source code analysis tools, also referred to as Static Application Security Testing (SAST) Tools, are designed to analyze source code or compiled versions of code to help find security flaws	A14.2.1	PR.IP-2
DAST	OWASP ZAP, Arachni NIkto	Dynamic Application Security Testing (DAST) examines applications for vulnerabilities like these in deployed environments	A14.2.1	PR.IP-2
SCA	Npm-audit, OWASP Dependency check	Software composition analysis (SCA) software enables users to analyze and manage the open-source elements of their applications	A14.2.1	PR.IP-2
Cloud configuration audit	Scout-suite	Security-auditing tool, which enables security posture assessment of cloud environments	A18.2.3	ID.RA-1

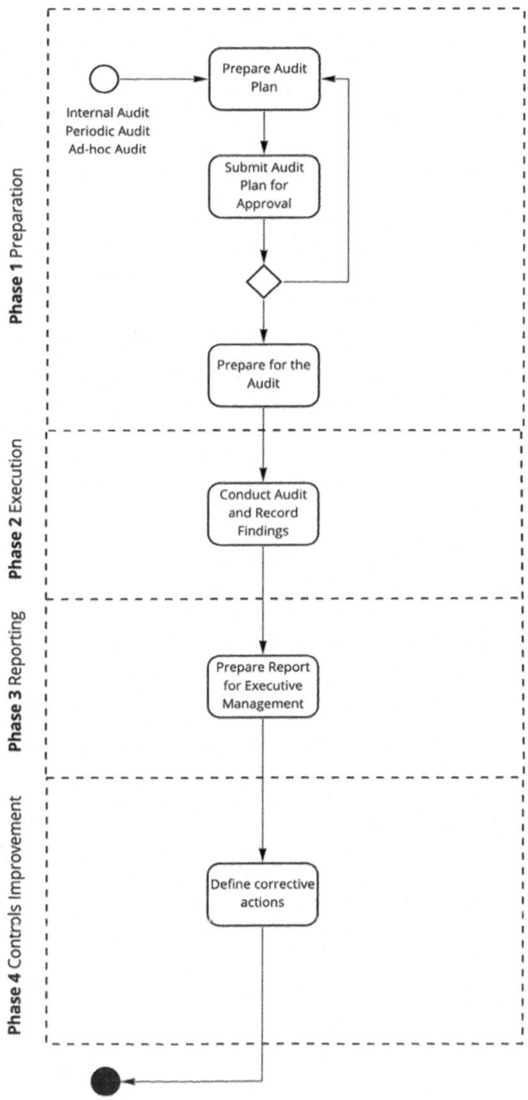

Fig. 4. Internal audit business process diagram

6 Conclusion

Modern threats forced organizations and individuals to embrace new practices such as remote working. It provides new challenges for cybersecurity area. New approaches, techniques and solutions, should be developed during the next years to ensure secure remote working conditions and more advance controls for critical infrastructure. The defined ISMS establishment framework in the following research work can be applied to ensure appropriate security controls for modern threats. As it is described in this research the organization can use template solutions that are popular in the market. For example, the ISO 27001/2 standard or NIST is an extremely popular and well-known basis for the ISMS developed by the company. However, the organization must take into account all aspects of its activities, as well as other features of the information held by the organization.

Information security guidelines, processes, and policies typically involve the development and implementation of physical and technical security measures to protect information from unauthorized access, use, distribution, or destruction. Also, an important measure that should be included in the ISMS is the audit of existing documents and systems of the organization for compliance with the requirements that ensure maximum protection of information. A security audit should be performed by the company to assess the company's ability to maintain the security of the system based on and within established criteria. Those controls as well as qualified specialists can minimize risks related to current cybersecurity threats.

References

1. Lakhno, V., Kozlovskii, V., Boiko, Y., Mishchenko, A., Opirskyy, I.: Management of information protection based on the integrated implementation of decision support systems. East. Eur. J. Enterp. Technol. **5**(9(89)), 36–42 (2017). https://doi.org/10.15587/1729-4061.2017.111081
2. Dudykevych, V., et al.: A multicriterial analysis of the efficiency of conservative information security systems. East. Eur. J. Enterp. Technol. **3**(9 (99)), 6–13 (2019). https://doi.org/10.15587/1729-4061.2019.166349
3. Susukailo, V., Opirskyy, I., Vasylyshyn, S.: Analysis of the attack vectors used by threat actors during the pandemic. In: Proceedings of the 2020 IEEE 15th International Scientific and Technical Conference on Computer Sciences and Information Technologies, CSIT 2020, vol. 2, pp. 261–264 (2020)
4. Ahmed, Z., Mahmood, S., Shah, H., Ahmed, J.: Information security management needs more holistic approach: a literature review. Int. J. Inf. Manage. **36**(2), 215–225 (2016). https://doi.org/10.1016/j.ijinfomgt.2015.11.009
5. McLaughlin, M.-D., Gogan, J.: Challenges and best practices in information security management. MIS Q. Exec. **17**(3), 237–262 (2018)
6. Bongiovanni, I.: The least secure places in the universe? A systematic literature review on information security management in higher education. Comput. Secur. **86**, 350–357 (2019)
7. Tu, C.Z., et al.: Strategic value alignment for information security management: a critical success factor analysis. Inf. Comput. Secur. **26**, 150–170 (2018)
8. Topa, I., Karyda, M.: From theory to practice: guidelines for enhancing information security management. Inf. Comput. Secur. **27**, 326–342 (2019)
9. Bahuguna, A., Bisht, R.K., Pande, J.: Roadmap amid chaos: cyber security management for organisations. In: 2018 9th International Conference on Computing, Communication and Networking Technologies (ICCCNT). IEEE (2018)
10. Stanton, J., et al.: Behavioral information security: defining the criterion space (2021)
11. Shamala, P., Ahmad, R., Zolait, A., Sedek, M.: Integrating information quality dimensions into information security risk management (ISRM). J. Inf. Secur. Appl. **36**, 1–10 (2017)
12. Bergström, E., Lundgren, M., Ericson, Å.: Revisiting information security risk management challenges: a practice perspective. Inf. Comput. Secur. **27**, 358–372 (2019)

13. Lundgren, M., Bergström, E.: Dynamic interplay in the information security risk management process. Int. J. Risk Assess. Manage. **22**(2), 212–230 (2019)
14. Morris, D., Madzudzo, G., Garcia-Perez, A.: Cybersecurity threats in the auto industry: tensions in the knowledge environment. Technol. Forecast. Soc. Chang. **157**, 120102 (2020)
15. Narayanan, S.N., et al.: Early detection of cybersecurity threats using collaborative cognition. In: 2018 IEEE 4th International Conference on Collaboration and Internet Computing (CIC). IEEE (2018)
16. Zong, S., et al.: Analyzing the perceived severity of cybersecurity threats reported on social media. arXiv preprint arXiv:1902.10680 (2019)
17. Dashti, S., Giorgini, P., Paja, E.: Information security risk management. In: Poels, G., Gailly, F., Asensio, E.S., Snoeck, M. (eds.) The Practice of Enterprise Modeling: 10th IFIP WG 8.1. Working Conference, PoEM 2017, Leuven, Belgium, 22–24 November 2017, Proceedings, pp. 18–33. Springer, Cham (2017). https://doi.org/10.1007/978-3-319-70241-4_2
18. Al-Darwish, A.I., Choe, P.: A framework of information security integrated with human factors. In: Moallem, A. (ed.) HCII 2019. LNCS, vol. 11594, pp. 217–229. Springer, Cham (2019). https://doi.org/10.1007/978-3-030-22351-9_15
19. Steinbart, P.J., Raschke, R.L., Gal, G., Dilla, W.N.: The influence of a good relationship between the internal audit and information security functions on information security outcomes. Acc. Organ. Soc. **71**, 15–29 (2018). https://doi.org/10.1016/j.aos.2018.04.005
20. Stafford, T., Deitz, G., Li, Y.: The role of internal audit and user training in information security policy compliance. Manag. Auditing J. **33**(4), 410–424 (2018). https://doi.org/10.1108/MAJ-07-2017-1596
21. Hina, S., Dominic, P.D.D.: Information security policies' compliance: a perspective for higher education institutions. J. Comput. Inf. Syst. **60**(3), 201–211 (2018)
22. Asghar, M.R., Qinwen, H., Zeadally, S.: Cybersecurity in industrial control systems: issues, technologies, and challenges. Comput. Netw. **165**, 106946 (2019). https://doi.org/10.1016/j.comnet.2019.106946
23. Sanguino, T.D.J.M., Domínguez, J.M.L., Baptista, P.D.C.: Cybersecurity certification and auditing of automotive industry. In: Policy Implications of Autonomous Vehicles, vol. 5, p. 98 (2020)
24. Baskerville, R., Rowe, F., Wolff, F.-C.: Integration of information systems and cybersecurity countermeasures: an exposure to risk perspective. ACM SIGMIS Database DATABASE Adv. Inf. Syst. **49**(1), 33–52 (2018)
25. Souppaya, M., et al.: Critical cybersecurity hygiene: patching the enterprise. National Institute of Standards and Technology (2018)
26. Kelo, T., Eronen, J.: Experiences from development of security audit criteria. In: Proceedings of the 16th European Conference on Cyber Warfare and Security, ECCWS 2017 (2017)

QoE Estimation Methodology for 5G Use Cases

Roman Odarchenko(✉) and Tetiana Dyka

National Aviation University, Liubomyra Guzara 1, Kyiv 03058, Ukraine
{odarchenko.r.s,tanya_dyka}@ukr.net

Abstract. Promising development of information technology in the modern world completely affects all areas of human activity. From now on, the concept of "Quality of Experience" (QoE) is becoming increasingly popular, and great efforts have been made to improve and provide reliable and additional services with a high level of experience for users. This chapter considers the problems of functioning of quality assurance systems in the fifth generation networks. As well as the importance of service quality in wireless and mobile networks and analyzed the main features of the use of 5G technologies. Providing standard definitions and the most important developed measurement methods. Demonstrates significant improvements and approaches for service quality control to meet user expectations.

Keywords: 5G · Quality of Experience (QoE) · Quality of Service (QoS) · IoT · Network · Technology · URLLC · eMBB · V2X

1 Introduction

Despite the advances in the development of fourth-generation cellular networks, the new demands arising from new communication needs require the creation of a fifth-generation mobile network. The 5G standard promises to be a breakthrough. Unlike previous generations, it has much higher data rate and power, as well as much better reliability. It is claimed that thanks to 5G Internet of Things (IoT), unmanned vehicles and virtual reality will move rapidly from the pages of technological media to our reality.

The cellular network will operate from low to high frequencies, which can transmit large amounts of data. The developers of this standard are also trying to reduce delays and reduce electric power consumption. This is a very important solution for mobile devices and IoT devices compared to 4G. The fifth generation network will be able to transmit data on the unlicensed frequencies currently used for Wi-Fi, without creating a conflict with existing Wi-Fi networks. This is very similar to T-Mobile's LTE-U technology [1].

5G technology provides for the presence of an extensive telecommunications infrastructure (a system of powerful terrestrial data transmission channels - fiber-optic communication channels). In this regard, 5G requires a much larger number of base stations (gNb) than required by 4G or any other mobile standard. For the full functioning of the network, stations ("towers") must be located every few hundred meters.

M. Klymash et al. (Eds.): *Future Intent-Based Networking*, LNEE 831, pp. 272–287, 2022.
https://doi.org/10.1007/978-3-030-92435-5_16

New applications, such as high-quality video streaming, touch Internet, traffic safety, remote real-time control and management, include new requirements for bandwidth, end-user latency (E2E) and network reliability. In addition, services include the provision of periodic or ongoing communication for machine-type communications, covering a variety of services, such as connected cars, homes, mobile works and sensors, which must be supported in efficient and scalable ways. Moreover, some new trends, such as "smart" devices, 3D immersion technology and virtual reality, affect the behavior of end users and directly affect the requirements for the network.

Also, the 5G network is expected to significantly improve Quality of Service (QoS) communication. Despite the variety of Quality of Experience (QoE) requirements, providing low latency and high bandwidth generally improves QoE. Thus, most of the previously mentioned tools can improve QoE. Traffic optimization methods can be used to meet QoE expectations. Also, installing cache and computing resources at the edge of the network allows the operator to place content and services close to the end user. This can provide very low latency and high QoE for critical interactive services such as video editing and augmented reality.

Big data, including information from sensors (such as on a device) and user statistics, can be used wisely using such models to more accurately estimate the user's expected QoE and determine the optimal resources to use to meet the expected QoE.

2 Previous Research Analysis

The benefits of 5G technology are needed to maintain quality of service. As in the previous generation, such as 3G and 4G, a quality known as QoE was identified. The 5G generation has additional qualities that the satisfaction service material focuses on QoE.

The notion, practicalities, and applications of QoE have considerably evolved since its inception in the telecommunication context [2]. Defined at the time as the "the overall acceptability of an application or service, as perceived subjectively by the end-user", QoE was understood to include "end-to-end system effects" and that it "may be influenced by user expectations and context". The evolution of QoE inference and use have included improvements on the manually aggregated scores, e.g., the mean opinion score (MOS); standardizing QoE mappings to network measurements of service delivery; association/correlation studies with user-end observations and responses; and more [3].

QoE can be defined as a process of measuring or evaluating quality for a set of program or service users with a special procedure and taking into account impact factors (possibly controlled, measured, or simply collected and reported). For example, quality assessment based on SDN/NFV [4–6] methods is an important step towards quality-based monitoring and management. Depending on the purpose and direction of the study in Rec. ITU-T describes various methods and guidelines for subjective assessment [7–9].

Regarding the measurement of QoE in 5G technology, a number of studies have conducted different approaches to measurement. The lack of measurement standards and QoS and QoE values for 5G provides opportunities and variations in measurement objects, methods and data acquisition. For example, in studies [10] and [11, 12] they used mathematical approaches, while in [13] they used a virtual approach. An approach that uses statistical analysis is also performed, as in studies [14]. The location and type of device also affect QoS and QoE measurements, as in studies [15] and [16].

It is expected that 5G technology with all the advantages meets the value of QoE, reliability and high security. Simulation to obtain QoS values from previous technologies may not be suitable for 5G technology. This is due to the value of quality present in this 5G era. Parameters such as packet loss, loss rate, network delay, PSNR and travel time are considered less effective in 5G, mainly for video media, because in assessing the quality of video media is the value of satisfaction presented in QoE. Therefore, despite the fact that the QoE parameter is still considered vital, it is not enough for the value of user satisfaction [10].

Performing a measurement process to obtain QoE values using 5G technology, a number of researchers have performed it. Various approaches, methods and objects are used to obtain results, which can illustrate the advantage of 5G technology in ensuring the quality of service to its users. Transport communication is the most widely used object, as in studies [12, 15, 17–23]. Parameters in traffic, such as data rate and reliability, are measured to obtain values from QoS and QoE. In a study [24], the value of traffic success is a measure to assess the QoE in obtaining the value of quality in terms of security.

Objects involving users are conducted to assess the QoE, such as research [25], which suggests which 5G scenario is used and where the location is located.

The article [26] proposes a dynamic approach to resource allocation by VBS and RRH, aimed at improving resource efficiency and energy, while ensuring a high level of QoE.

3 Problem Statement

5G networks has brought a lot of new possibilities to the vertical industries. And now it is necessary to provided needed level of user's satisfaction. There are a lot of new use cases appeared. And now it is necessary to measure QoE for each of them. Existing approaches are not able to provide correct measurements of this very important parameter. That's why the aim of the research is to develop the novel unified methodology for assessing the QoE in 5G networks for different use cases (UCs). To achieve the goal of the research the next tasks have to be solved:

- 5G use cases classification development;
- QoE estimation methodology development;
- Development of the approach of QoE estimation methodology implementation;
- Experimental studies of the developed methodology.

4 QoE-QoS Estimation Methodology for the 5G Use Cases

4.1 5G Use Cases Classification

Wireless networks have improved their capabilities, trying to keep up with the development of technology. Different generations of wireless cellular networks were developed before the advent of 5G. The development of the new network promises to provide extremely high data rates, much less latency and high integrity.

The fifth generation is the basic technology that is necessary for society, the state as a whole and the digital transformation of business. Not taking into account the fact that the specifications for fifth-generation broadband access are still under development. But we can say that today it is obvious that the effect of the application of this technology will go far beyond the telecommunications business.

Consider the possibility of using 5G technology in various spheres of life. For this purpose the analysis of characteristics is executed, features of a structure and characteristics of 5G technologies are considered. The main problems that can be solved using 5G technologies are highlighted below (see Fig. 1) [27].

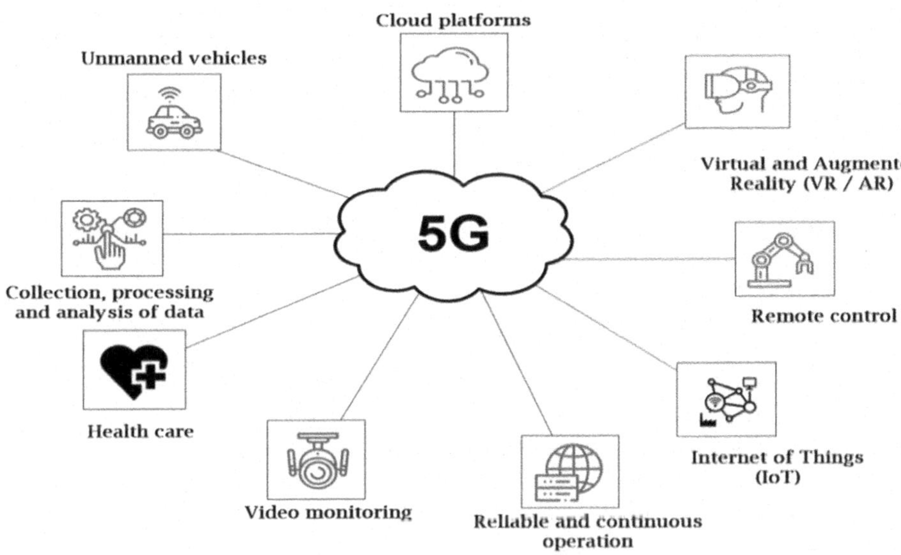

Fig. 1. Use of 5G technologies

Thanks to the basic ability of 5G to connect a large number of digital devices, such networks can implement a wide range of services called Internet of things (IoT) - finally open the door to a world where gastronomic preferences will be clear to home appliances – ordering, delivery and payment will be automated. Such 5G-based systems open up incredible opportunities for any industry and industry – from automatic humidity control and field irrigation (non-critical systems) to automatic design and production of new products in unmanned production (critical systems).

In smart city projects, 5G will allow real-time transmission of information from a much larger number of sensors at various sites. It will be possible to deploy a thousand sensors instead of hundreds, for the maintenance of which there will be enough fewer base stations than with existing networks. These can be, for example, sensors for monitoring the condition of housing and communal services, sensors of “smart lighting” or sound sensors installed for security and order in the city. In the latter case, the sensors can detect suspicious or too loud sounds, and this information will be automatically transmitted to law enforcement.

Self-governing vehicles are another area that requires a new generation of communication networks. Cars can be equipped with sensors that read all sorts of information about the road situation: the nearest vehicles, weather conditions, the condition of the asphalt, road signs. Based on this data, the trip can be managed automatically. In addition to the convenience for the driver of the car, equipping cars with sensors opens up new opportunities to improve road safety. On 5G networks, cars will be able to communicate with each other and make instant decisions about what to do in a given situation based on information received from other vehicles on the road. For example, a car could send a signal about its sudden braking, so that the car, which is in danger of a collision, could also brake sharply in automatic mode. It is impossible to implement such a service in 4G networks [27].

Augmented and virtual reality (AR/VR) services have been evolving for several years, but have not yet become widespread. The growing use of technology is constrained by the lack of a collaboration platform in an AR/VR environment. The proliferation of 5G will make AR/VR massive, allowing you to deliver 3D content, 3D video at high speed, providing low latency and interactivity.

High bandwidth of 5G channels and cloud computing power will minimize wearable AR/VR devices, eliminating the need for local video processing. Due to the high data transfer speeds and 5G synchronization, they will provide easy joint AR/VR immersion of several remote users. Third-party developers of "heavy" AR-/VR-content will have access to all the technical capabilities and services of 5G.

New services using 5G can be implemented in medicine. One of the areas of application is remote monitoring of patients. The doctor will be able to quickly receive information from special sensors and monitor the condition of patients around the clock. Due to the very low data transfer delays, 5G will also open up more opportunities for remote operations using robots. This service is especially relevant for small settlements where there are no local surgeons. Due to 5G, such a service can be deployed in wireless networks [28].

The 5G network will become a universal platform for remote control of equipment in remote and closed areas, transmission of process information to other production units, partners and regulatory authorities.

Remote control uses platform services of video streams, AR/VR, but requires cooperation with manufacturers of machinery and equipment, development of special applications. Therefore, it is logical to allocate remote manipulations in a separate platform service for scaling and distribution in different industries for a variety of types of controlled equipment.

Clinical and outpatient care is becoming radically more accessible and effective with the use of special devices for collecting and transmitting biological and medical indicators from sensors on patients, as well as with intelligent machine analysis of these data in the diagnosis and evaluation of treatment.

New technologies allow to transfer a huge amount of data without delay and are in demand in pediatrics, psychotherapy, dermatology, neurology and even in resuscitation: if the patient can not be transported to another clinic, an urgent video call from a more competent specialist can save lives. A highly qualified surgeon can remotely monitor what happens during the operation through a 5G video session and correct the actions of colleagues, or control auxiliary devices. Secure sharing ensures that data is kept confidential. Because smart devices are deployed and run in a peripheral infrastructure rather than through a centralized node, the data collected elicits immediate feedback, improving the response of the entire system.

Real-time machine analysis of data and UHD images from a remote patient builds a model from which the doctor works over the network through a connected AR-/VR-headset. Comparing the digital model of a particular patient with numerous digitized images and information from a powerful cloud MIS simplifies diagnosis, identifies future health disorders, studies their causes, suggests ways of prevention and treatment, helps to implement them remotely. Even more valuable is the use of AR tools in surgery. Often the maximum speed of completion of intervention is important, manipulations are complicated by difficulties of distinction of fabrics and bodies. Augmented reality images provide visual accentuation of diseased tissues, operational recommendations and online assessment of actions, significantly reduce the risk of medical errors and speed up the work of the surgeon. Real and VR images, 3D models are broadcast to other doctors for advice.

Thus, we can conclude that 5G will allow you to transfer practical skills over your networks, developing a new direction, not just information. Mobile networks will become an important part of the infrastructure for the development of key industries. There is a statement, according to Ericsson, that by 2026 the launch of the fifth generation mobile network standard will lead to a completely new market, reaching about $582 billion globally. Remote control of heavy industry will become a reality due to almost zero delay. This will reduce production costs and increase safety for employees.

Absolutely every user of technology will witness the development of intelligent transport systems. Unmanned automotive future awaits users in the near future. In total, 10 million smart vehicles will travel on most of the world's roads.

Ericsson Mobility Report predicts that by 2022 there will be 29 billion connected gadgets worldwide. The Internet of Things will account for 18 billion of the total. These results mean that each active consumer of technology will have several smart things at once [29].

4.2 QoE Estimation Methodology

In recent years, the technical community has shifted some attention from one related gauge, quality of service (QoS), to a more consumer-centric metric, quality of experience (QoE). Network operators and service providers from the very advent of telecommunications wanted to know, what is the level of service quality which is provided to the end users. This is because that knowledge can be extremely useful when trying to manage network topology, optimize its capacity and operating costs, introduce new services or plan investments and expansion of a network.

International Telecommunication Union (ITU) defines QoE as the overall acceptability of an application or service, as perceived subjectively by the end-user QoE can be considered as an extension of the traditional QoS in the sense that QoE provides information about the delivered service from an end-user point of view.

Whereas QoS stands between the network and an application, QoE is centred on the subscriber. In particular, QoE focuses on person-as-user who interacts with an application and person-as-customer who deals with a service provider, see Fig. 2 [29, 30].

Fig. 2. QoE model

Thus, the service quality has some objective and subjective properties. Obviously that the user will hardly be satisfied if a network performance (QoS) is poor. For instance, if the re-buffering of a video is frequent during the streaming session a user will most certainly be annoyed and unsatisfied. But it was also showed that achieving the QoS targets does not necessarily ensure satisfied users. Something was still missing.

The difference between QoE and QoS is underlined below [31]:

QoS – Quality of Service:

- network characteristics/behavior;
- performance guarantees given by network provider based on measurements;

QoE – Quality of Experience:

- impact of network behavior on end user;
- some imperfections may go unnoticed;
- some imperfections may render application useless;
- not captured by network measurements.

QoE is not directly depending of radio channel conditions, but the expectation will increase with higher performance. Increasing expectation changes QoE but it happens for all technologies then. QoE considers a user's expectation, QoS is more rational based on technical measurements (Fig. 3).

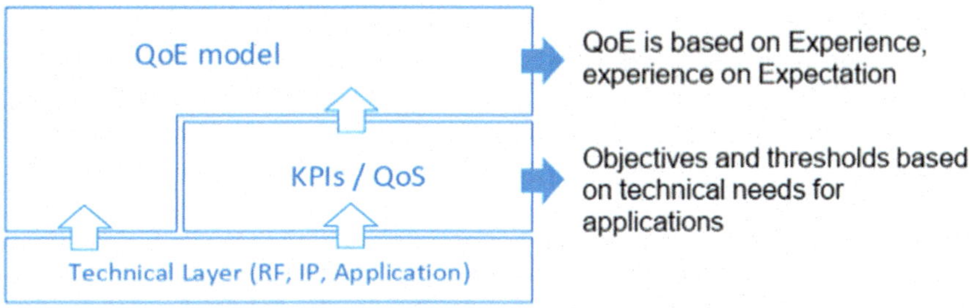

Fig. 3. Relationships between QoE, QoS and KPIs

Based on the above, the following approach is proposed for evaluating the overall QoE using QoS metrics, that can be estimated in more objective way.

To implement this approach, a set of services (UCs) was introduced that should be analyzed:

$$\left\{\bigcup_{i=1}^{n} \mathbf{S_i}\right\} = \{\mathbf{S_1}, \mathbf{S_2}, \ldots, \mathbf{S_n}\}, \tag{1}$$

where $\mathbf{S_i} \subseteq \mathbf{S}$, $(\mathrm{i} = \overline{1, \mathrm{n}})$, n is a number of services, and

$$\mathbf{S_i} = \left\{\bigcup_{\mathrm{j}=1}^{\mathrm{m_i}} \mathbf{S_{ij}}\right\} = \{\mathbf{S_{i1}}, \mathbf{S_{i2}}, \ldots, \mathbf{S_{im_i}}\}, \tag{2}$$

with S_{ij} $(\mathrm{j} = \overline{1, \mathrm{m_i}})$ is a subset of the elements of the quality assurance system.

The Subsets of QoE metrics $\mathbf{S_{ij}} \subseteq \mathbf{S_i}$ can be represented as:

$$\mathbf{S_{IJ}} = \left\{\bigcup_{\mathbf{p}=1}^{\mathbf{r_{ij}}} \mathbf{S_{IJP}}\right\} - \left\{\mathbf{S_{ij1}}, \mathbf{S_{ij2}}, \ldots, \mathbf{S_{ijr_{\mathit{j}}}}\right\}, \tag{3}$$

where $\mathbf{S_{jp}}$ $(\mathrm{p} = \overline{1, \mathrm{r_{ij}}})$ are QoE indicators that characterize the QoE for $\mathbf{S_{ij}}$; $\mathrm{r_{ij}}$ is the number of such indicators.

At the second stage, QoS and QoE indicators are selected $\mathbf{S_{ijp}}$, using multi-factor correlation-regression analysis. To construct a multi-factor regression model, the following steps have to be completed:

Step 1. Select all possible QoS factors that affect the QoE indicator (or process) that is being investigated. For each factor it is necessary to determine its numerical characteristics. If some factors can not be quantitatively or qualitatively determined or statistics are not available to them, then they are removed from further consideration.

Step 2. Choose the form of a regression or multivariate model, that is, finding an analytic expression that best reflects the relationship of factor characteristics with the resultant, that is, the choice of function:

$$\hat{\mathrm{Y}} = \mathrm{f}(\mathrm{x_1}, \mathrm{x_2}, \mathrm{x_3}, \ldots, \mathrm{x_n}), \tag{4}$$

where $\hat{Y}$ is the effective sign-function; $x_1, x_2, x_3, \ldots, x_n$ are factor signs.

On the next stage subsets of QoS indicators have to be calculated using corresponding algorithms and formulas for their calculations [32]. QoE has to be calculated using i.e. MOS, DSCQR, ACR [33] or other appropriate methods/techniques. For example, in 5G-TOURS project [34], the special questionnaires were developed to estimate the QoE for each use case.

On the last stage, the obtained values are compared with the maximum permissible, possible to ensure the normal functioning of the network and achieved KPIs.

To compare the values obtained as a result of calculations with the maximum allowable was introduced the logical function of equivalence:

$$E(x, y) = \begin{cases} 1, & \text{if } x > y, \\ 0, & \text{if } x \leq y. \end{cases} \tag{5}$$

QoE is almost the most important parameter, estimating which it is possible to determine user experience and compare it with users' expectations. Respective approach based on the principles of machine learning has been developed to assess and optimize the state of the network in order to improve the QoS provision to users.

That's why was developed the QoE evaluation methodology in order to evaluate the level of satisfaction of end-users and verticals' players with the deployed use cases. This includes users' QoE as well as the feedback from the vertical players on how the technology provided can improve their business operations [31].

In addition to the validation of the QoS results which illustrates mainly the performance of the network KPIs and can be compared against the 5G PPP targets, it is of paramount importance to validate the actual satisfaction of i) end-users and ii) the

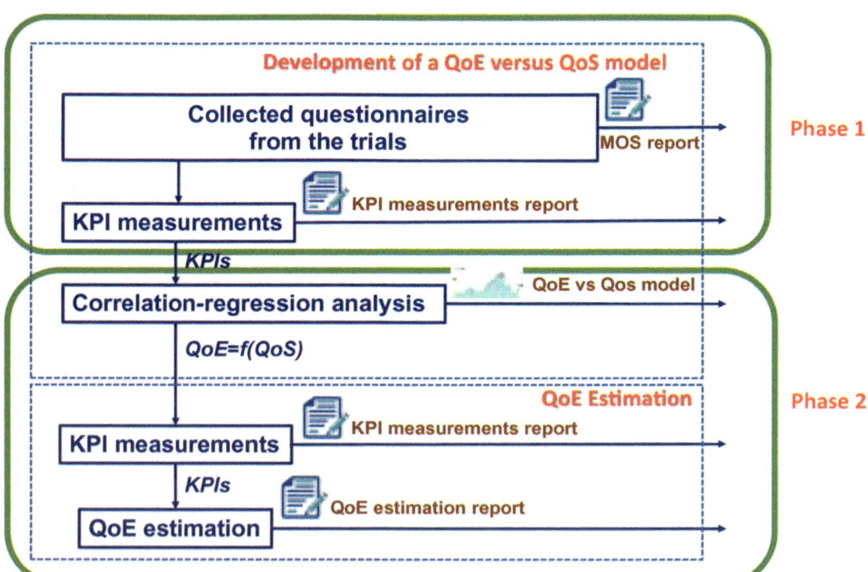

Fig. 4. . General approach for evaluation methodology

vertical players (either as service providers or users of secondary service flows). In this direction, was developed the QoE evaluation methodology, the high-level architecture of which is presented in Fig. 4.

It was discriminated between two phases. The first phase is realized during the trials execution and collects both the QoS metrics, automatically collected from the infrastructure, and the QoE metrics (and vertical satisfaction) collected using appropriate questionnaires. The second phase is realized after the trials executions and by using correlation-regression analysis which tries to create a model for QoS-QoE correlation.

Every approach taken in previous research to measure QoS and QoE has its respective potential. This potential is expected to meet the balance between QoS and QoE, an easy, fast, and accurate process, and flexibility between ideal conditions and reality. It should also remind that in 5G technology, there are three major scenarios in which there are multi-technologies in implementation so that the measurement of QoS and QoE has the potential to be very diverse.

The measurement of traffic and user response in the object approaches group has potential from the balance of QoS and QoE, ease, and flexibility as in the sub-material that becomes the measurement target. This is because the traffic has various sub-material as well as sub-material in the user response that is very flexible. User traffic and response also meet the value requirements of highly technical and systematic QoS and QoE that is very attached to users.

Quality of experience is essentially a human related experience that is difficult to measure using quantitative techniques, being the ones related to videos the most famous ones. QoE is traditionally measured through Mean Opinion Scores and questionnaires. These questionnaires include a set of multiple questions with a specific weight usually defined with a specific scale, as originally proposed by, extended with some open questions.

The rationale behind this decision is to exploit the questionnaire filling procedure to also achieve some insights behind the ones already obtained by the questions. Clearly those open questions cannot be mathematically evaluated, as discussed next, but they can provide further useful feedback.

The final version of the mentioned questionnaires tried to address four critical requirements:

a) Validate both the user satisfaction as well as the vertical satisfaction (in each UC questionnaires are generated for both users and main shareholders).
b) Cover aspects that will become useful during the QoE-QoS correlation process.
c) Share some commonalities between UCs.
d) Deal with cost and pricing aspects.

For the estimation of the overall QoE level we rely on the hypothesis function:

$$h(\theta) = \theta_0 \cdot x_0 + \theta_1 \cdot x_1 + \ldots + \theta_n \cdot x_n, \tag{6}$$

where n is the number of features in the data set; x is the QoS parameters, θ are the weight coefficients.

For estimation of the θ was proposed to use the next Normal Equation:

$$\theta = \left(X^T X\right)^{-1} \cdot \left(X^T y\right). \tag{7}$$

In the above equation,

- θ (Estimated QoE): hypothesis parameters that define it the best.
- X (Input QoS parameters): Input feature value of each instance.
- Y (Output QoE): Output value of each instance.

4.3 QoE Estimation Methodology Implementation

In this chapter will be described the process of developed methodology implementation applying to the 5G-TOURS UC5 "Remote and distributed video production". The main objective of the use case is to exploit the 5G-TOURS network features for remote television production and analyze how 5G networks could support various scenarios in which high-quality video (e.g., in 4K, HD/HDR or Video 360°) is generated and transmitted. In a typical TV production environment, video contents is delivered from cameras located in places where an event is taking place to a TV studio in the broadcasting center or to a remote studio facility on the event location itself. Such video contents could be used both for immediate live broadcasting of the event or recorded to be further edited and used in TV programs later on [35].

There was proposed to use the next implementation algorithm on practice:

Step 1 – Definition of the most relevant for the UC KPIs. For the mentioned above UC these KPIs are (Table 1):

Table 1. Most relevant for the UC5 KPIs

KPI_1	KPI_2	KPI_3
Latency	Throughput DL	Throughput UL

Step 2– Definition of the most relevant for the UC QoE parameters, which are the next for the UC (Table 2):

Table 2. The most relevant for the UC QoE parameters

	QoE_1	QoE_2	QoE_3	QoE_4	QoE_5
	Video quality	Audio quality	Stability of 5G signal	Breaks in the video/audio	Impact of the artifacts
Weight coefficient (*K*)	0,2	0,2	0,2	0,2	0,2

Step 3– Definition of the weight coefficients for each QoE parameter (see table above).
Step 4– KPI measurements according to pre-defined methodology [32].
Step 5– Collection of the QoE questionnaires. The example of QoE related part of the developed questionnaire is represented on the Fig. 5.
Step 6– Processing of the obtained experimental data (QoE and QoS). This data has to be represented in the table form (Table 3).

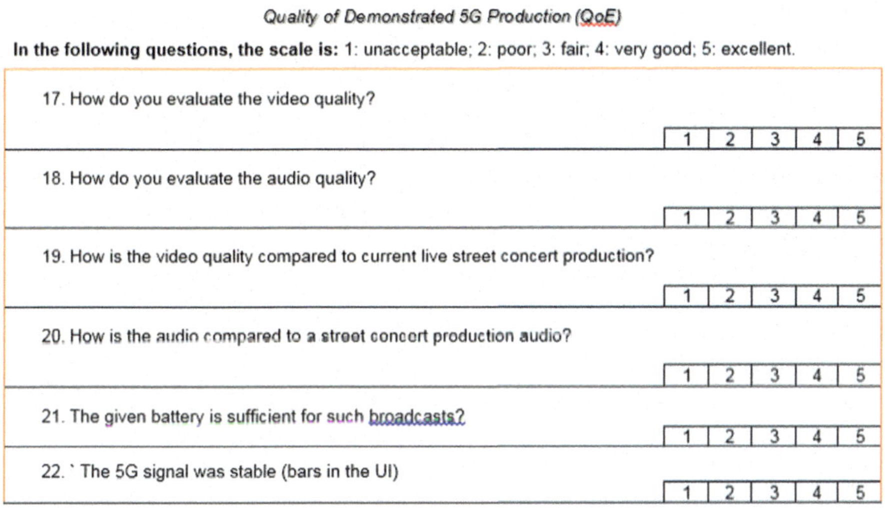
Quality of Demonstrated 5G Production (QoE)

In the following questions, the scale is: 1: unacceptable; 2: poor; 3: fair; 4: very good; 5: excellent.

17. How do you evaluate the video quality?	1	2	3	4	5
18. How do you evaluate the audio quality?	1	2	3	4	5
19. How is the video quality compared to current live street concert production?	1	2	3	4	5
20. How is the audio compared to a street concert production audio?	1	2	3	4	5
21. The given battery is sufficient for such broadcasts?	1	2	3	4	5
22. ` The 5G signal was stable (bars in the UI)	1	2	3	4	5

Fig. 5. QoE questionnaires

For this described above UC was proposed the next QoE-QoS mapping function:

$$QoE = A0 + A(Thr) \cdot Throughput + A(Delay) \cdot Delay, \tag{8}$$

where weight coefficients can be calculated using the normal equation (Table 4):
To estimate the MOS was proposed the next formula:

$$MOS = K1 \cdot Parameter\_1 + \ldots Kn \cdot Parameter\_n, \tag{9}$$

where $K1 + \ldots + Kn = 1$ and $K1, \ldots Kn > 0$.
During the data obtaining should be estimated the next parameters (Table 5).

Table 3. Collected during trials QoE and QoS values

QoE	QoS parameters	
MOS	Troughput, Mbps	Delay, ms
80	10	1
90	12	1
88	11	2
…	…	…

Table 4. Calculated values of weight coefficients

Coefficients		
A(Delay)	A(Thr)	A0
−4,55314	3,107015	57,3893

Table 5. Calculated parameters related to the correlation analysis

Correlation coefficient_Throughput	−0,961191417
Correlation coefficient_Throughput	0,961047169
Estimated t-criterion t	4,442136041
The table value of the t-criterion trh	2,776445105
Tabular value standard. normal Distr. zy	1,959963985
Fisher transform value z’	−1,538344072
Left interval estimate for z	−2,669929806
Right interval estimate for z	−0,406758338
Left interval estimate for rxy	−0,990452706
Right interval estimate for rxy	−0,38571676
Standard deviation for rxy	0,205270997

5 Conclusions

5G network is the first and so far the only technology that allows to flexibly combine platform services on a single technological basis, eliminates the need for the corporate consumer to build their own network infrastructure. These qualities make 5G the basis of scalable services, which significantly reduces the time of their development and implementation in various sectors of the economy. 5G technology provides more advantages than its predecessors, especially in terms of speed and power. The possibility of 5G technology in each scenario opens up different possibilities and approaches for calculating the value of quality that this technology will give.

Compared to previous generations of mobile networks, the architecture of the 5G system has been significantly improved, thanks to the introduction of network analytics functions and enhanced capabilities for interaction with third-party application functions. Combining these capabilities, new experience quality assessment (QoE) features can be developed and implemented in the next generation network. However, it is unclear how 5G networks can collect monitoring data and application metrics, how they relate to each other, and what methods can be used in 5G systems to assess quality. That is why the problems of functioning of quality assurance systems in the fifth generation networks were considered in the chapter. As well as the importance of service quality in wireless and mobile networks and analyzed the main features of the use of 5G technologies. Were provided standard definitions and the most important developed measurement methods. Were demonstrated significant improvements and approaches for service quality control to meet user expectations.

Acknowledgements. This work was supported in part by the European Commission under the 5G-TOURS: SmarT mObility, media and e-health for toURists and citizenS (H2020-ICT-2018-2020 call, grant number 856950). The views expressed in this contribution are those of the author and do not necessarily represent the project.

References

1. ITU–T Recommendation G.1080: Quality of Experience Requirements for IPTV Services; International Telecommunication Union Telecommunication Standardization Sector, Geneva, Switzerland (2008)
2. Porcu, S., Floris, A., Voigt-Antons, J.-N., Atzori, L., Moller, S.: Estimation of the quality of experience during video streaming from facial expression and gaze direction. IEEE Trans. Netw. Serv. Manag. **17**, 2702–2716 (2020)
3. Kourtis, M.-A., Liberal, F., Koumaras, H., Xilouris, G., Trouva, E.: Exploiting NFV techniques towards future VQA methods. In: IEEE 22nd International Workshop on Computer Aided Modeling and Design of Communication Links and Networks (CAMAD), Lund, Sweden, June 2017 (2017)
4. Kourtis, M.-A., Koumaras, H., Xilouris, G., Liberal, F.: An NFV based video quality assessment method over 5G small cell networks. IEEE Multimedia **24**, 68–78 (2017)
5. Koumaras, H., Kourtis, M., Sakkas, C., Xilouris, G., Kolometsos, S.: In-service video quality assessment based on SDN/NFV techniques. In: 2016 23rd International Conference on Telecommunications (ICT), Thessaloniki, Greece, pp. 1–5, May 2016 (2016)
6. ITU-R, BT.500.11: Methodology for the subjective assessment of the quality of television pictures, January 2012
7. ITU-T Rec. P.910: Subjective video quality assessment methods for multimedia applications, April 2008
8. ITU-T Rec. P.913: Methods for the subjective assessment of video quality, audio quality and audiovisual quality of Internet video and distribution quality television in any environment, March 2016
9. Ouali, K., Kassar, M., Nguyen, T.M.T., Sethom, K., Kervella, B.: Modeling D2D handover management in 5G cellular networks. In: 2017 13th International Wireless Communications and Mobile Computing Conference (IWCMC), June 2017 (2017)

10. Pedreno-Manresa, J.-J., Khodashenas, P. S., Siddiqui, M.S., Pavon-Marino, P.: Dynamic QoS/QoE assurance in realistic NFV-enabled 5G access networks. In: 2017 19th International Conference on Transparent Optical Networks (ICTON), July 2017 (2017)
11. Pierucci, L.: The quality of experience perspective toward 5G technology. IEEE Wirel. Commun. **22**(4), 10–16 (2015)
12. Tikhvinskiy, V., Bochechka, G., Gryazev, A., Aitmagambetov, A.: Comparative analysis of QoS management and technical requirements in 3GPP standards for cellular IoT technologies. J. Telecommun. Inf. Technol. **2**, 41–47 (2018)
13. Tao, X., Liu, Y., Jiang, C., Wang, Z., Qin, X.: QoE-oriented multimedia assessment: a facial expression recognition approach. IEEE Multimedia **26**(1), 41–50 (2019)
14. Andriyanto, F., Suryanegara, M.: The QoE assessment model for 5G mobile technology. In: 2017 International Conference on Broadband Communication, Wireless Sensors and Powering (BCWSP), November 2017 (2017)
15. Koumaras, V., Foteas, A., Foteas, A., Kapari, M., Sakkas, C., Koumaras, H.: 5G performance testing of mobile chatbot applications. In: 2018 IEEE 23rd International Workshop on Computer Aided Modeling and Design of Communication Links and Networks (CAMAD), September 2018 (2018)
16. Agiwal, M., Roy, A., Saxena, N.: Next generation 5G wireless networks: a comprehensive survey. IEEE Commun. Surv. Tutor. **18**(3), 1617–1655 (2016)
17. Dighriri, M., Alfoudi, A.S.D., Lee, G.M., Baker, T., Pereira, R.: Comparison data traffic scheduling techniques for classifying QoS over 5G mobile networks. In: 2017 31st International Conference on Advanced Information Networking and Applications Workshops (WAINA), March 2017 (2017)
18. Mebarkia, K., Zsoka, Z.: QoS modeling and analysis in 5G backhaul networks. In: IEEE 29th Annual International Symposium on Personal, Indoor and Mobile Radio Communications (PIMRC), September 2018 (2018)
19. Cominardi, L., Contreras, L.M., Bcrnardos, C.J., Berberana, I.: Understanding QoS applicability in 5G transport networks. In: IEEE International Symposium on Broadband Multimedia Systems and Broadcasting (BMSB), June 2018 (2018)
20. Ali, M.A., Esmailpour, A., Nasser, N.: Traffic density based adaptive QoS classes mapping for integrated LTE-WiMAX 5G networks. IEEE International Conference on Communications (ICC), May 2017 (2017)
21. Ye, Q., Li, J., Qu, K., Zhuang, W., Shen, X.S., Li, X.: End-to-end quality of service in 5G networks: examining the effectiveness of a network slicing framework. IEEE Veh. Technol. Mag. **13**(2), 65–74 (2018)
22. Lee, S.-H., Yoon, M.-S., Cho, S.-H., Cho, H.-K.: Study on simplified test bench for QoS analysis using traffic models of Pre5G service. In: 2018 10th International Conference on Ubiquitous and Future Networks (ICUFN), July 2018 (2018)
23. Shuminoski, T., Janevski, T., Risteski, A., Bogdanoski, M.: Security and QoS framework for 5G and next-generation mobile broadband networks. In: 17th International Conference on Smart Technologies, IEEE EUROCON 2017, July 2017 (2017)
24. Andriyanto, F., Suryanegara, M.: The QoE assessment model for 5G mobile technology. In: International Conference on Broadband Communication, Wireless Sensors and Powering (BCWSP), November 2017 (2017)
25. Pompili, D., Hajisami, A., Tran, T.X.: Elastic resource utilization framework for high capacity and energy efficiency in cloud RAN. IEEE Commun. Mag. **54**(1), 26–32 (2016)
26. Makarenko, Y.V.: The analysis of the possibilities of using 5G technology in internet systems of things. Igor Sikorsky Kyiv Polytechnic Institute, National Technical University of Ukraine (2019)
27. Top 5 applications 5G - [Electron resource] - Mode of access: https://kursiv.kz/news/hi-tech/2016-09/top-5-sfer-primeneniya-5g

28. Marativna, H.V.: Rational choice of the size of the MIMO antenna array to increase the efficiency of the 5G system. Igor Sikorsky Kyiv Polytechnic Institute, National Technical University of Ukraine, p. 108 (2018)
29. QoE and QoS: Definitions and implications - [Electron resource] - Mode of access: https://www.v-net.tv/2012/08/27/qoe-and-qos-definitions-and-implications/
30. BRAS solutions market for ISPs and telecom operators - [Electron resource] - Mode of access: https://vasexperts.ru/blog/bras/rynok-reshenij-bras-dlya-internet-provajderov-i-operatorov-svyazi/
31. Hassan Osman, Roman Odarchenko et all Deliverable D7.2 First Integrated 5G-TOURS Ecosystem. 5G-TOURS - ICT-19-2019 – G.A:856950
32. 5G-TOURS, D7.1 (2020)
33. ITU, Subjective video quality assessment methods for multimedia applications
34. 5G-TOURS project official website. http://5gtours.eu/
35. 5G-TOURS D4.1 Robotic, Media and Smart Cities solutions for Touristic Cities

Software Implementation Research of Self-similar Traffic Characteristics of Mobile Communication Networks

I. Strelkovskaya[2], I. Solovskaya[2](✉), J. Strelkovska[2], and A. Makoganiuk[1]

[1] Odesa, Ukraine
a.makoganyuk@onat.edu.ua
[2] National University «Odesa Law Academy», Odesa, Ukraine
{strelkovskaya,i.solovskaya}@onat.edu.ua

Abstract. A study of the characteristics of self-similar traffic for a queuing system (QS) of the form $W_B/M/1/\infty$, which simulates the service of self-similar traffic using a two-parameter Weibull distribution. Using the Laplace-Stieltjes transform, an analytical expression is obtained to find the quality characteristics of self-similar traffic, for which software solutions based on PYTHON are proposed. The obtained results will allow to choose the configuration of connections between base stations according to the criterion of the average waiting time at the stage of planning, design and further operation of mobile networks, while in real processes of network operation to take into account its construction.

Keywords: Self-similar traffic · Queuing system · QoS characteristics · Weibull distribution · The Laplace-Stieltjes transformation · The average of time delay · The average queue length

1 Introduction

Modern mobile networks, such as 4G/LTE (Long Term Evolution), LTE-Advanced (Rel. 10–12), LTE-Advanced Pro (Rel. 13–14) are based on packet traffic technologies. By its nature, the traffic served in mobile networks is heterogeneous, as it is formed by many different sources of services, services and applications, providing a range of broadcasting, data and video services (YouTube, Video Surveillance, streaming video, OTT-services, M2M (Machine to machine), IoT (Internet of Things)) [1].

The rapid growth of subscriber traffic in the 4G/LTE mobile network, the change in its nature and structure, and the significant increase in bandwidth may contribute to possible congestion of network facilities, their buffers and, consequently, lead to delays and packet losses. Therefore, when serving self-similar traffic, special attention is paid to supporting the QoS (Quality of Service) characteristics of the service.

In the design and further optimisation phase of a 4G/LTE mobile network, when selecting the network structure and network node performance, it is important to use calculation methods that will take into account the self-similarity characteristics of the traffic. This raises a number of challenges, among them the choice of distribution to

M. Klymash et al. (Eds.): *Future Intent-Based Networking*, LNEE 831, pp. 288–304, 2022.
https://doi.org/10.1007/978-3-030-92435-5_17

describe a particular type of traffic and the method of calculating the QoS characteristics that will match the distribution. The use of "classical methods" of teletraffic theory based on Poisson distribution for which calculation methods are already known does not allow today to obtain reliable results, and generally known and recognised calculation methods for quality of service characteristics of self-similar traffic, practically do not exist.

The QoS performance of self-similar traffic has been the subject of a considerable amount of work by various authors [2–6]. Most work is based on experimental data or derived simulation results, but reliable and recognised analytical solutions for estimating the QoS characteristics of self-similar traffic for different distributions have not yet been obtained. And only for some cases of traffic description by means of different distributions, for example, gamma-distribution, approximate solutions for some kinds of queuing systems (QS) have been obtained [4]. In [6–10], to estimate QoS characteristics of self-similar traffic, the Pollachek-Hinchin formulas for M/G/1 QS, the Norros formulas for fBM/D/1/∞ were used.

This chapter examines the characteristics of self-similar traffic, the need to assess which is due to the impossibility of existing solutions to adequately assess the required amount of traffic in the mobile network and, accordingly, to obtain decisions on its allocation and efficient use of available network resources of QoS (Quality of Service). The difference between the proposed solutions from those already known is that other approaches are used to study the characteristics of traffic, such as Laplace-Stieltjes transformation. In general, each of the considered methods allows to increase the accuracy of estimates of traffic characteristics and service quality indicators, as well as to provide a number of practical recommendations for their application [6–16].

This chapter considers self-similar traffic, which is described by the Weibull distribution [2] and has the following properties:

1) the Weibull distribution, when the distribution curve parameter H = 0.5, becomes exponential, which corresponds exactly to the value of the Hurst parameter, which determines the degree of self-similarity, at H = 0.5 there is no sign of self-similarity for the traffic;
2) the waveform parameter of the Weibull distribution curve is able to represent the intensities of traffic changes on the timescale and in doing so take into account the peaks of traffic intensity increases, which are most characteristic of self-similar traffic;
3) the asymptote of the Weibull distribution, which has a "tail" implies a significant variance.

A study of the characteristics of self-similar traffic for a queuing system (QS) of the form W_B/M/1/N, which simulates the service of self-similar traffic using a two-parameter Weibull distribution. For this study, traffic modelling was carried out using the PYTHON software [17].

Using the Laplace-Stieltjes transformation, an analytical expression was obtained to find the quality characteristics of self-similar traffic, for which software solutions based on MATLAB were proposed [18]. The obtained results allow at the stage of planning, design and further operation of mobile networks to optimally choose the configuration of connections between base stations according to the criterion of the average waiting time, while in real processes of network operation take into account the peculiarities of its construction.

The purpose of this chapter is to find the QoS characteristics of self-similar traffic described by a Weibull distribution in a W_B/M/1/N QS.

2 The Self-similar Traffic in Mobile Networks

It is known [2, 3] that packet traffic served by 4G/LTE network has a self-similar (fractal) nature, the main reason for which is the long-term relationship between the arrival of packets, which is determined by the correlation function at different times, the presence of after effects and high pulsation.

Definition of self-similar process, according to [3]. The real process *X(t)*, $t \in R$ is self-similar to the exponent $H > 0$ if for all $a > 0$, the finite-dimensional distributions for $\{X(at), t \in R\}$ are identical to the finite-dimensional distributions $\{a^H X(t), t \in R\}$, i.e. if for any $k \geq 1, t_1, t_2, ..., t_k \in R$ and any $a > 0$

$$(X(at_1), X(at_2), ..., X(at_k)) \equiv \left(a^H X(t_1), a^H X(t_2), ..., a^H X(t_k)\right), \tag{2.1}$$

or

$$\{X(at), t \in R\} \equiv \left(a^H X(t), t \in R\right).$$

Based on formula (2.1), we can conclude that the process is automatically repeated with the preservation of statistical properties, due to the fact that the statistical characteristics do not change during scaling. As a quantitative assessment of the degree of self-similarity used Hurst parameter $0.5 \leq H < 1$.

In addition to statistical similarity in scaling, these processes have certain quantitative properties. Self-similar processes can be evaluated by several equivalent features [3].

1) Hyperbolically decaying correlation function of the form

$$R(k) = k^{(2H-2)} L(t), \text{ if } k \to \infty,$$

where $L(t)$ is the slowly variable infinity function,

$$\lim_{t \to \infty} \frac{L(tx)}{L(t)} = 1, \text{ for all } x > 0.$$

That is, the correlation function is un summed, and the series formed by the research values of the correlation function diverges $\sum_k R(k) = \infty$. This infinite sum is another definition of long-term dependence.

2) The dispersion of the sample mean decays slower than the reciprocal of the sample size. If you enter a new time sequence $\left\{X_i^{(m)}; i = 1, 2, \ldots\right\}$, obtained by averaging the initial sequence $\{X_i; i = 1, 2, \ldots\}$ on non-intersecting successive blocks of size *m*, then for self-similar processes will be characterized by a slower decrease in variance according to the law

$$\sigma^2(X^m) m^{(2H-2)}, \text{ if } m \to \infty..$$

3) If we consider the processes in the frequency domain, the phenomenon of long-term dependence leads to the power nature of the spectral density near zero. X is long term dependent if

$$S(\omega) - \omega^{-\gamma} L_2(\omega), \text{ if } w \to 0.$$

where $0 < \gamma < 1$, L_2 is slow variable in 0 and $S(\omega) = \sum_k R(k)e^{ik\omega}$ is the spectral density.

The characteristics of the packet traffic that is serviced are affected by a significant number of high-speed services, services and applications used by the mobile subscriber. Therefore, to describe self-similar traffic, given that the moments of arrival of packets have a distribution with "heavy tails", most often use the Pareto, Weibull distribution or lognormal [3].

In frequency-spatial planning at the stage of design and further optimization, when the choice of network structure and production of network nodes (e-NodeB, S-GW, P-GW, etc.), it is necessary to use calculation methods that take into account the similarity of packet traffic. It should be noted that to describe self-similar traffic, the use of classical methods of teletraffic theory based on the description of traffic using the Poisson distribution is not appropriate.

Self-similar traffic has a more complex nature, and the distribution that describes this traffic is significantly different from Poisson. Therefore, the use of classical methods to calculate the main characteristics of self-similar traffic gives incorrect, unreasonably optimistic results, and traffic processing algorithms based on the use of the simplest flow are inefficient for flows with self-similarity. Statistical characteristics (average value, spectral density, autocorrelation function, etc.) of self-similar traffic have a character that is very different from exponential. Despite the long period of studying the self-similarity of teletraffic, a significant class of problems remains unsolved:

1) in fact, there is no strict theoretical basis that would replace the classical theory of teletraffic in the design of modern mobile networks that use self-similar traffic;
2) there is no single generally accepted model of self-similar traffic;
3) there is no reliable and recognized method of calculating the quality characteristics of QoS for systems and networks serving self-similar traffic;
4) there are no algorithms and mechanisms that ensure the quality of service in terms of self-similar traffic.

It is known that the solution of these problems has not only theoretical but also practical significance. For example, in TCP/IP-based packet switching networks, when using the protocol without guaranteed UDP delivery, a significant reduction in packet latency is achieved. However, it is difficult to provide increased connection quality requirements only with the help of the transport protocol (UDP or TCP), because the reasons that lead to long delays are more at the network level.

The current situation in modern mobile networks, the presence of a significant number of routes of self-similar traffic, which periodically occur sharp fluctuations in traffic intensity or delays in packet data transmission, accompanied by packet loss, manifestations of self-similar traffic, require the required level of quality QoS service, taking into account the requirements for various applications, make relevant the task of researching self-similar traffic. All the above allows us to conclude that the interest in the study of the characteristics of the quality of QoS self-similar traffic for mobile networks.

2.1 Research of Quality Characteristics of Self-similar Traffic

2.1.1 Modeling of Self-similar Traffic for QS $W_B/M/1/N$

To model self-similar traffic for QS $W_B/M/1/N$ will makes the program in PYTHON [17].

Solve implementation code, a block diagram is shown in Fig. 1 [17]:

- math – represents a three-dimensional functionality for working with numbers. Provides access to some popular mathematical functions and constants that can be used in code for more complex mathematical calculations. The library is a built-in Python module, so no additional installation is required.
- scipy.special – the SciPy library depends on NumPy, which provides easy and fast manipulation of the N-dimensional array. The SciPy library is designed to work with NumPy arrays and provides many convenient and efficient numerical methods, such as numerical integration and optimization procedures. SciPy consists of subpacks covering various areas of scientific computing. In this case/.special - connects any special mathematical functions.
- random – random value generator. With their help, you can quickly create sequences of different numbers or symbols that are impossible to predict. To this end, Python uses a built-in library with many methods for managed generation.
- matplotlib.pyplot – one of the most popular Python packages used for data visualization. This is a cross-platform library for creating various graphs from data in arrays. Matplotlib is written in Python and uses NumPy, a numerical mathematical extension of Python.

The implementation of modeling using the PYTHON software environment is presented: In Fig. 1 the results of the simulation, namely simulation traffic models for QS of the form $W_B/M/1/N$ with different Hurst parameters.

According to Fig. 2, we can see that for the obtained self-similar traffic on the interval [0;3000] ms there is a scale invariance, a significant number of "bursts" of traffic intensity and a long-term dependence between the moments of requests for service.

2.1.2 Research of Quality Service Characteristics of Self-similar Traffic for QS $W_B/M/1/\infty$

Consider a queuing system, type $W_B/M/1/N$, which simulates traffic service by the base station NodeB (e-NodeB) and serves the flow of applications, the intervals between which are described by the Weibull distribution, service time has an exponential distribution M, QS is single-line [10–14].

Consider the Weibull distribution, which is most characteristic of data, services and applications traffic served by the e-NodeB base station in the 4G/LTE mobile networks.

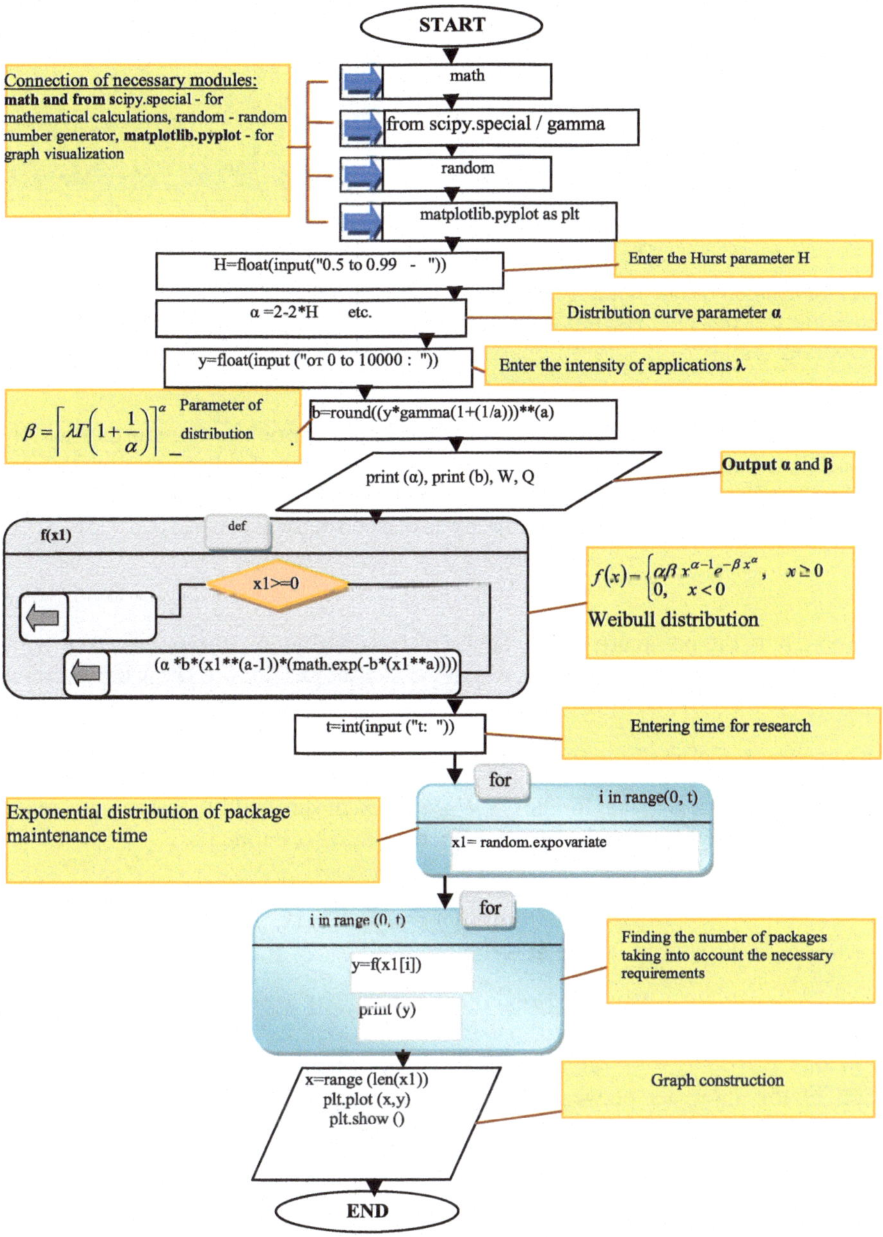

Fig. 1. Block diagram of the algorithm for a simulation model of traffic

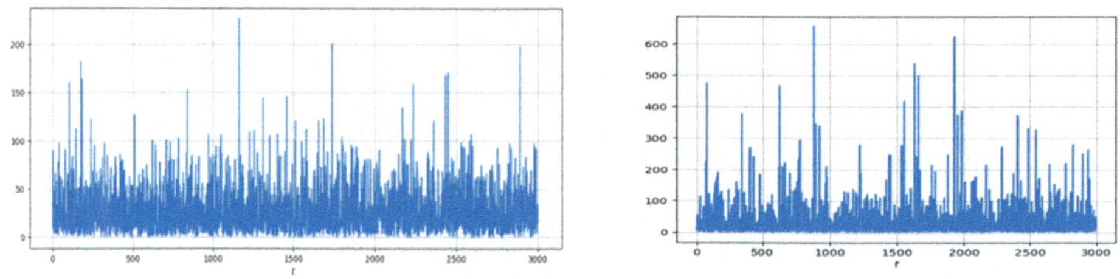

a) *b)*

Fig. 2. Simulation of the traffic model for QS of the form WB/M/1/N with parameters: a) H = 0,6; λ = 50, α = 0,8; β = 25.267, t ∈ [0; 3000] ms, b) H = 0,7; λ = 50; α = 0,6; β = 13,361; t ∈ [0; 3000] ms

We use another approach, namely: we find the Laplace-Stieltjes transformation for the Weibull distribution in order to obtain QoS characteristics of self-similar traffic [7, 8].

Consider the Weibull distribution given by the differential distribution function [2]:

$$f(x) = \begin{cases} \alpha\beta\, x^{\alpha-1} e^{-\beta x^{\alpha}}, & x \geq 0 \\ 0, \quad x \leq 0 \end{cases} \tag{2.2}$$

where α is the parameter of the shape of the distribution curve, $0 < \alpha < 1$; $\alpha = 2 - 2H$, H is the Hurst parameter $0,5 \leq H < 1$, $\beta = \left[\lambda\Gamma\left(1 + \frac{1}{\alpha}\right)\right]^{\alpha}$ is the distribution parameter, $\beta > 0$, λ is the intensity receipt of requests for service in the QS, $\Gamma(k)$ is the Euler gamma function of the form $\Gamma(k) = \int_0^{+\infty} t^{k-1} e^{-1} dt$.

It is known from [19] that for the QS $W_B/M/1/\infty$ the probability that a newly received application will be found in the QS, n service requests is defined as:

$$r_n = (1-\sigma)\sigma^n, 0 \leq \sigma < 1, \tag{2.3}$$

where σ is the root of the equation

$$\sigma = F(\mu - \mu\sigma), 0 \leq \sigma < 1. \tag{2.4}$$

In this case, the F is the Laplace-Stieltjes transformation of the density distribution of the intervals between the applications $f(t)$ in the QS and has the form $F(s) = \int_0^{+\infty} e^{-st} f(t) dt$, s is the complex variable, μ is the intensity of service of applications in QS.

Finding the roots of σ Eq. (2.4), we can determine the following quality characteristics for QS $W_B/M/1/N$ [19]:

- the average waiting time W of the application in the system, defined by the formula

$$W = \frac{\sigma}{\mu(1-\sigma)}, \tag{2.5}$$

where μ is the intensity of service of applications in QS,

– the average number of applications Q in the queue, determined by the formula

$$Q = \frac{\rho\sigma}{\mu(1-\sigma)}, \tag{2.6}$$

where ρ is the load factor of QS.

In the QS $W_B/M/1/\infty$ the probability density of the distribution of the duration τ intervals of applications receipt in the QS has the form [10–14]:

$$f(t) = \alpha\beta t^{\alpha-1}e^{-\beta t^{\alpha}}, t \geq 0, 0 < \alpha < 1. \tag{2.7}$$

Write the formula (2.7) in the form [7, 8]:

$$\sigma = \alpha\beta \int_0^{+\infty} e^{-(\mu-\mu\sigma)t}t^{\alpha-1}e^{-\beta t^{\alpha}}dt \tag{2.8}$$

Using the schedule function e^x in Maclaurin series [15] $e^x = \sum_{n=0}^{\infty} \frac{x^n}{n!}$, where $x \in (-\infty, +\infty)$, we will receive [10–14]:

$$f(t) = \alpha\beta\, t^{\alpha-1} \sum_{n=0}^{\infty} \frac{(-\beta t^{\alpha})^n}{n!},$$

or

$$f(t) = \sum_{n=0}^{\infty} \frac{\alpha(-1)^n \beta^{n+1} t^{(n+1)\alpha-1}}{n!}.$$

Thus,

$$f(t) = \sum_{n=0}^{\infty} \alpha \frac{(-1)^{n-1}}{(n-1)\,!} \beta^n t^{n\alpha-1}. \tag{2.9}$$

For the original $f(t)$, which is defined by formula (2.9), it is necessary to find the image F. It is known that the Laplace transform has the property of linearity for a finite number of terms [20, 21], i.e. if $f_1(t), f_2(t), \ldots, f_N(t)$ – originals, then for any constants $c_i, i = \overline{1, N}$, function $f(t) = \sum_{k=1}^{N} c_k f_k(t)$ is also the original and equality is valid [10–14]:

$$L\left[\sum_{k=1}^{N} c_k f_k(t)\right] = \sum_{k=1}^{N} c_k L[f_k(t)], \tag{2.10}$$

where L is the Laplace operator.

The Laplace transform for an infinite number of terms of the series is found in [10–14].

Then, using formula (2.10) for the original $f(t)$, defined by formula (2.9), and the uniform convergence of the series (2.9) with the above restrictions on the region of convergence of this series, we obtain an image $F(p)$, that takes the form [10–14].

$$F(p) = \sum_{n=1}^{\infty} \alpha \frac{(-1)^{n-1}}{(n-1)\,!} \beta^n \frac{\Gamma(n\alpha)}{\mathrm{p}^{n\alpha}},\ n\alpha > 0, \tag{2.11}$$

where Γ is the gamma function.

Then formula (2.11), given the fact that $\mathrm{p} = \mu - \mu\sigma$, has the form [10–14]:

$$F(\mu - \mu\sigma) = \sum_{n=0}^{\infty} \alpha \frac{(-1)^{n-1}}{(n-1)\,!} \beta^n \frac{\Gamma(n\alpha)}{(\mu - \mu\sigma)^{n\alpha}},\ n\alpha > 0 \tag{2.12}$$

Equation (2.4) will take the form [10–14]:

$$\sigma = \sum_{n=1}^{\infty} \alpha \frac{(-1)^{n-1}}{(n-1)\,!} \beta^n \frac{\Gamma(n\alpha)}{(\mu - \mu\sigma)^{n\alpha}},\ n\alpha > 0 \tag{2.13}$$

Solving Eq. (2.13), we find the root of equation σ. This will define the required quality characteristics for QS form $W_B/M/1/N$ [10–14]: the value of the average waiting time W of the application in the QS (2.5), the average number of applications Q in the QS queue (2.6).

2.1.3 Analysis of Program Results Determine the Characteristics of Service Quality Self-similar Traffic QS $W_B/M/1/\infty$

To obtain solutions of the transcendental equation and search for the root, we use two software environments, the MATLAB application package [18] and the object-oriented Python programming language [17]. To solve the problem using the software environment MATLAB we will solve Eq. (2.8), which is equivalent to Eq. (2.9) [10–14].

We solve Eq. (2.8) by the graphical method, ie by plotting the functionsy $= \sigma$, $y = \alpha\beta \int_0^{+\infty} e^{-(\mu-\mu\sigma)t} t^{\alpha-1} e^{-\beta t^{\alpha}} dt$. The point of intersection of both graphs of the above functions and will be the root σ of the Eq. (2.8). To find σ, make a program MATLAB [18].

For this purpose we find the solution of the transcendental Eq. (2.8), namely the roots for which condition $0 \leq \sigma < 1$, is fulfilled, by means of the given values:

- Hurst parameter H ∈ [0,6; 0,95] with a 0.05 step,
- values of the intensity λ of requests to the $W_B/M/1/N$ QS [0.1; 0.8] with a step of 0.1;
- value of the intensity of service of requests $\mu = \text{const} = 1{,}5$.

The results of the solution of Eq. (2.8) are listed in Table 1 and are shown in the Fig. 3.

According to the obtained in Fig. 2 graph of the dependence of the obtained root of the equation σ = σ(λ, H) on the intensity λ of service requests in the QS WB/M/1/N QS and values of the Hurst parameter H, the following conclusions can be made:

Table 1. The results of solving Eq. (2.8) for different Hurst coefficients based on the MATLAB software environment, (μ = const = 1,5)

Indicator	Value							
λ	0.1	0.2	0.3	0.4	0.5	0.6	0.7	0.8
$H = 0.6, \alpha = 0.8$								
Value of the root σ, σ (H = 0.6)	0.118	0.204	0.282	0.354	0.423	0.486	0.531	0.537
$H = 0.65, \alpha = 0.7$								
Value of the root σ, σ (H = 0.65)	0.160	0.258	0.341	0.415	0.483	0.545	0.593	0.611
$H = 0.7, \alpha = 0.6$								
Value of the root σ, σ (H = 0.7)	0.220	0.330	0.417	0.491	0.556	0.615	0.661	0.682
$H = 0.75, \alpha = 0.5$								
Value of the root σ, σ (H = 0.75)	0.309	0.428	0.515	0.586	0.646	0.697	0.735	0.752
$H = 0.8, \alpha = 0.4$								
Value of the root σ, σ (H = 0.8)	0.439	0.560	0.641	0.703	0.752	0.789	0.811	0.819
$H = 0.85, \alpha = 0.3$								
Value of the root σ, σ (H = 0.85)	0.626	0.729	0,791	0.833	0.860	0.872	0.874	0.876
$H = 0.9, \alpha = 0.2$								
Value of the root σ, σ (H = 0.9)	0.855	0.903	0.916	0.926	0.941	0.956	0.965	0.977
$H = 0.95, \alpha = 0.1$								
Value of the root σ, σ (H = 0.95)	0.850	0.892	0.946	0.977	0.981	0.986	0.995	0.997

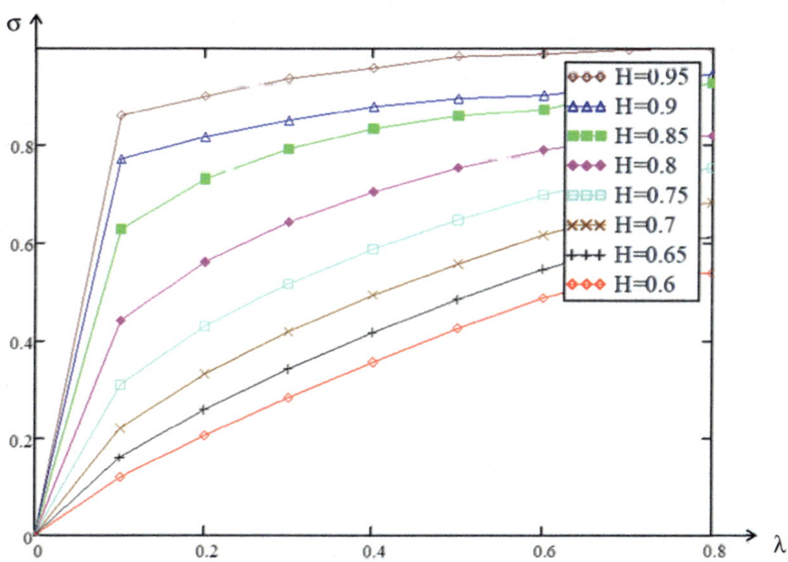

Fig. 3. Dependence graph $\sigma = \sigma(\lambda, H)$ for WB/M/1/N QS

- the value of the root increases with increasing Hurst coefficient in a given region of values H ∈ [0.6;0.95] and affects the degree of self-similarity of the traffic, it should be noted that the root of the equation a maximum value σ = 0.997 at H = 0.95;
- with the increase in the intensity of applications to the WB/M/1/N QS in the given range of values λ ∈ [0; 0.8] with step 0.2 at a fixed value of the intensity of applications service μ in the QS μ = 1.5 the value of the root of the equation almost doubles.

The obtained values of the root of the Eq. (2.8) allow us to obtain values of the traffic service quality characteristic, such as the average delay time W = W (H,μ) in the WB/M/1/N QS, which are shown in Table 2.

Table 2 The results of the obtained characteristics of the quality of self-similar traffic for the value of the Hurst parameter H = 0.9, μ = const = 1.5

Indicator	Value							
Intensity of the receipt of applications to QS, λ	0.1	0.2	0.3	0.4	0.5	0.6	0.7	0.8
H = 0.6, α = 0.8								
Value of the root σ, σ (*H* = 0.6)	0.118	0.204	0.282	0.354	0.423	0.486	0.531	0.537
Average delay time, W	0.089	0.171	0.262	0.366	0.488	0.630	0.754	0.773
H = 0.65, α = 0.7								
Value of the root σ, σ (*H* = 0.65)	0.160	0.258	0.341	0.415	0.483	0.545	0.593	0.611
Average delay time, W	0.127	0.232	0.345	0.473	0.622	0.798	0.972	1.047
H = 0.7, α = 0.6								
Value of the root σ, σ (*H* = 0.7)	0.220	0.330	0.417	0.491	0.556	0.615	0.661	0.682
Average delay time, W	0.188	0.328	0.476	0.643	0.836	1.063	1.298	1.432
H = 0.75, α = 0.5								
Value of the root σ, σ (*H* = 0.75)	0.309	0.428	0.515	0.586	0.646	0.697	0.735	0.752
Average delay time, W	0.298	0.498	0.708	0.944	1.218	1.534	1.848	2.027
H = 0.8, α = 0.4								
Value of the root σ, σ (*H* = 0.8)	0.439	0.560	0.641	0.703	0.752	0.789	0.811	0.819
Average delay time, W	0.523	0.848	1.190	1.576	2.020	2.496	2.869	3.019

(*continued*)

Table 2 (*continued*)

Indicator	Value							
$H = 0{,}85$, $\alpha = 0{,}3$								
Value of the root σ, σ ($H = 0.85$)	0.626	0.729	0.791	0.833	0.860	0.872	0.874	0.876
Average delay time, W	1.118	1.798	2.527	3.329	4.079	4.529	4.644	4.568
$H = 0.9$, $\alpha = 0.2$								
Value of the root σ, σ ($H = 0.9$)	0.855	0.803	0.816	0.826	0.841	0.856	0.865	0.877
Average delay time, W	3.933	6.213	7.210	7.212	7.245	8.113	8.253	8.355
$H = 0.95$, $\alpha = 0.1$								
Value of the root σ, σ ($H = 0.95$)	0.890	0.899	0.936	0.957	0.981	0.986	0.995	0.997
Average delay time, W	4.933	6.513	7.920	8.630	9.450	10.113	12.253	13.355

The results of calculations of the average waiting time $W = W(H,\mu)$ for requests in the W_B/M/1/N QS are shown in Fig. 4.

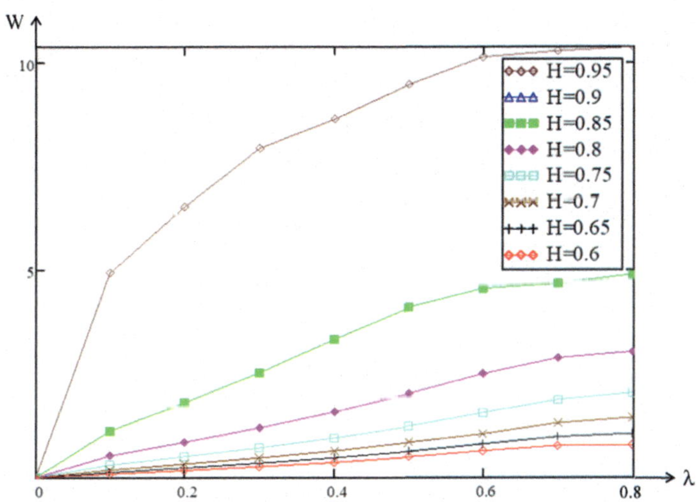

Fig. 4. The average time of the $W = W$ (H, μ) in the WB/M/1/N QS

The obtained values of the root of the Eq. (2.8) and formula (2.6) allow us to obtain values of the characteristics of traffic service quality, such as the average length of the service order line Q for W_B/M/1/N QS, which are shown in Table 3; the graph of dependence $Q = Q$ (H, ρ) is shown in Fig. 5.

Table 3. The results of the obtained characteristics of the average number of applications Q in the queue for the value of the Hurst parameter H = 0,9, μ = const = 1,5

Indicator	Value							
Load factor of QS, ρ	0,1	0,2	0,3	0,4	0,5	0,6	0,7	0,8
H = 0,6, α = 0,8								
Value of the root σ, σ (H = 0,6)	0,118	0,204	0,282	0,354	0,423	0,486	0,531	0,537
Average number of applications in the queue, Q	0.009	0.034	0.079	0.146	0.258	0.378	0.528	0.619
H = 0,65, α = 0,7								
Value of the root σ, σ (H = 0,65)	0,160	0,258	0,341	0,415	0,483	0,545	0,593	0,611
Average number of applications in the queue, Q	0.013	0.046	0.103	0.189	0.311	0.479	0.680	0.838
H = 0,7, α = 0,6								
Value of the root σ, σ (H = 0,7)	0,220	0,330	0,417	0,491	0,556	0,615	0,661	0,682
Average number of applications in the queue, Q	0.019	0.066	0.143	0.257	0.417	0.639	0.910	1,144
H = 0,75, α = 0,5								
Value of the root σ, σ (H = 0,75)	0,309	0,428	0,515	0,586	0,646	0,697	0,735	0,752
Average number of applications in the queue, Q	0.030	0.083	0.212	0.377	0.608	0.920	1.294	1.617
H = 0,8, α = 0,4								
Value of the root σ, σ (H = 0,8)	0,439	0,560	0,641	0,703	0,752	0,789	0,811	0,819
Average number of applications in the queue, Q	0.052	0.170	0.357	0.631	1.011	1.428	2.002	2.413
H = 0,85, α = 0,3								
Value of the root σ, σ (H = 0,85)	0,626	0,729	0,791	0,833	0,860	0,872	0,874	0,876
Average number of applications in the queue, Q	0.112	0.359	0.757	1.330	2.048	2.725	3.237	3.838
H = 0,9, α = 0,2								
Value of the root σ, σ (H = 0,9)	0,855	0,863	0,876	0,878	0,880	0,886	0,890	0,892
Average number of applications in the queue, Q	0.393	0.840	1.413	1.919	2.444	3.109	3.776	4.405
H = 0,95, α = 0,1								
Value of the root σ, σ (H = 0,95)	0,895	0,899	0,936	0,947	0,956	0,966	0,975	0,981
Average number of applications in the queue, Q	0.568	1.187	2.925	4.765	7.242	11.365	18.200	27.537

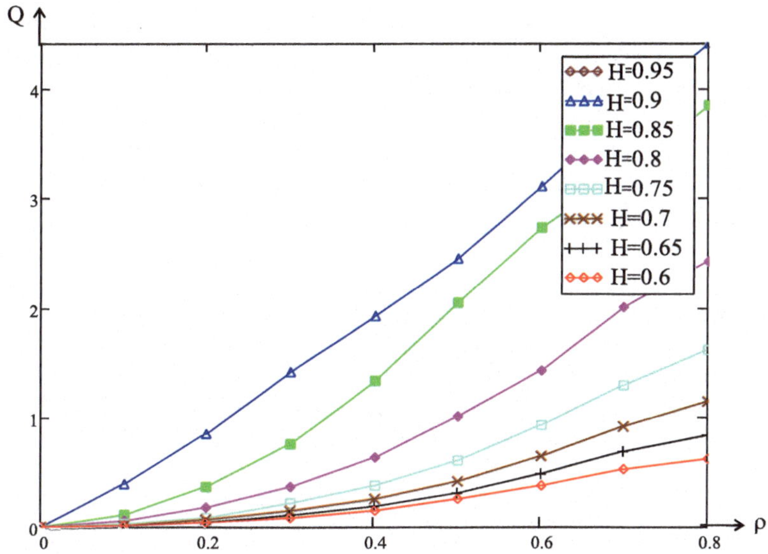

Fig. 5. Average length of the requests Q = Q (H,ρ) to the WB/M/1/N QS

3 Practical Recommendations for Using the Results of Research Quality Characteristics in Mobile Networks

The study was devoted to solving the complex problem of assessing the characteristics of the quality of service of self-similar traffic, modeled using the Weibull distribution for QS $W_B/M/1/N$ [10–14]. The presented solution allows to determine the main characteristics of QoS, such as:

- the average waiting time for the application in the QS $W_B/M/1/N$,
- the average number of applications in the QS $W_B/M/1/N$,
- the average length of the queue of applications in the QS $W_B/M/1/\infty$.

According to the obtained values of these characteristics, it is possible to obtain a solution of a number of practical problems:

1) at the stage of planning, design and further optimization of mobile networks, when the choice of network structure and performance of network nodes (e-NodeB, S-GW, P-GW, etc.), the use of calculation methods that take into account the similarity of packet traffic, allows get more accurate estimates of QoS characteristics and based on them select the necessary equipment;
2) when choosing the characteristics of hardware and software of the mobile network for $W_B/M/1/N$ traffic serviced in QS, using the proposed method of determining the quality characteristics, it is possible to predict the required size of buffer equipment of network objects, thereby providing high-speed services, services and applications in the network of the mobile operator;

3) an important step in designing a mobile network of any technology or standard is to solve the problem of choosing the optimal configuration of connections between e-NodeB base stations in the service area, according to the obtained values of the average time of application, taking into account the queue length. Taking into account the peculiarities of construction and structure of e-NodeB base stations in E-UTRAN radio neteork, as well as the obtained values of the average number of applications, predict the configuration of connections in 4G/LTE network in real processes of their operation.

Given the fact that the coefficient σ dependent parameter α, which determines the shape of the curve Weibull distribution and Hurst coefficient H is obvious that the value of the average waiting time W depends on the above values. Thus, confirming that the value $\sigma \in [0;1)$ and, accordingly, form Weibull distribution curve α with increasing Hurst coefficient increases the average latency W of packets in QS $W_B/M/1/N$.

4 Conclusions

1. The simulation of self-similar traffic in PYTHON for the queuing system of the form $W_B/M/1/N$, which simulates the service of self-similar traffic using the two-parameter Weibull distribution, is carried out.
2. Applying the Laplace-Stieltjes transform, an analytical formula is obtained to find the characteristics of the quality of self-similar traffic, such as the average waiting time of the application in the QS, the average number of applications in the QS. The obtained values allow at the design stage to predict the required size of the buffer equipment of network objects, thereby providing the ability to provide high-speed services, services and applications in the network of the mobile operator.
3. The software solutions of the problems of determining the quality characteristics for QS of the form $W_B/M/1/N$ on the basis of PYTHON are given. The results allow the planning phase and design of mobile network optimally choose the configuration of connections between base stations in real conditions of the functioning of the network.

References

1. 3GPP TS 23.002 3rd Generation Partnership Project; Technical Specification Group Services and Systems Aspects; Network architecture (Release 8-12) (2018). 94 pages.
2. Krylov, V.V., Samokhvalov, S.S.: Teletraffic theory and its applications. SPb. BHV-Petersburg (2005). 288 p.
3. Shelukhin, O.I, Osin, A.V., Smolsky, S.M.: Self-similarity and fractals. Telecommunication applications. Fizmalit, Moscow (2008). 368 p.
4. Ponomarev, D.Yu.: About servicing in the system with the input gamma flow. The materials of V Russian conference of young scientists on mathematical modeling and information technologies. [Electronic resource]. http://www.sbras.ru/ws/YM2004/8510/

5. Chaurasiya, P.K., Siraj Ahmed, V.: Comparative analysis of Weibull parameters for wind data measured from met-mast and remote sensing techniques. Renewable Energy **115**, 1153–1165 (2017). https://doi.org/10.1016/j.renene.2017.08.014
6. Millán, G.A., Lefranc, G.: Fast multifractal model for self-similar traffic flows in high-speed computer networks. Procedia Comput. Sci. **17**, 420–425 (2018). https://doi.org/10.1016/j.procs.2013.05.054
7. Lemeshko, O., Yeremenko, O.: Linear optimization model of MPLS traffic engineering fast ReRoute for link, node, and bandwidth protection. In: 2018 14th International Conference on Advanced Trends in Radioelecrtronics, Telecommunications and Computer Engineering (TCSET), pp. 1009–1013 (2018). https://doi.org/10.1109/TCSET.2018.8336365
8. Yeremenko, O.: Enhanced flow-based model of multipath routing with overlapping by nodes paths. In: 2015 Second International Scientific-Practical Conference Problems of Infocommunications Science and Technology (PIC S&T), pp. 42–45 (2015). https://doi.org/10.1109/INFOCOMMST.2015.7357264
9. Yeremenko, O., Tariki, N., Hailan, A.M.: Fault-tolerant IP routing flow-based model. In: 2016 13th International Conference on Modern Problems of Radio Engineering, Telecommunications and Computer Science (TCSET), pp. 655–657 (2016). https://doi.org/10.1109/TCSET.2016.7452143
10. Strelkovskaya, I.V., Grygoryeva, T.I., Solovskaya, I.N.: Self-similar traffic in G/M/1 queue defined by the Weibull distribution. Radioelectron. Commun. Syst. **61**(3), 173–180 (2018). https://doi.org/10.20535/S0021347018030056
11. Strelkovskaya, I., Solovskaya, I., Grygoryeva, T., Paskalenko, S.: The solution to the problem of the QoS characteristics definition for self-similar traffic serviced by the W/M/1 QS. In: Problems of Infocommunications Science and Technology: Conference Proceedings of the 2016 Third International Scientific-Practical Conference Proceedings (PICS&T 2016), Kharkiv, Ukraine, 4–6 October 2016, pp. 40–42. Kharkiv National University of Radio Electronics, Kharkiv (2016). https://doi.org/10.1109/infocommst.2016.7905330
12. Strelkovskaya, I., Siemens, E., Solovskaya, I., Fedotova, I.: Estimation of QoS characteristics of self-similar traffic for the W/M/1 queuing system. Збірник наукових праць ОНАЗ ім. О.С. Попова. Вип. 1. С. 27–33 (2018)
13. Strelkovskaya, I., Solovskaya, I., Makoganiuk, A., Balyk, A.: Research of the quality characteristics of self-similar traffic of a mobile communication network on the basis of software release. Int. Res. J. Inf. Telecommun. Sci. **11**, **2**(21), 51–57 (2020)
14. Strelkovskaya, I., Solovskaya, I., Makoganiuk, A.: Finding some QoS characteristics of self-similar traffic serviced by a mobile network. In: Proceedings 2th IEEE International Conference Advanced Information and Communication Technologies 2017 (AICT-2017), Lviv, Ukraine, 4–7 July 2017, pp. 146–149 (2017). https://doi.org/10.1109/AIACT.2017.8020086
15. Uryvsky, L., Maptynova, R.: Complex analytical model of priority requires service on cloud server. In: 2019 International Conference Radio Electronics & Info Communications (UkrMiCo) (2019). https://doi.org/10.1109/UkrMiCo47782.2019.9165323
16. Uryvsky, L., Pilipenko, A., Trach, B.: Asymptotic properties of self-similar traffic models based on discrete-time and continuous-time martingales. J. Telecommun. Sci. **4**(2), 19–21 (2013)
17. https://www.python.org
18. https://www.mathworks.com/products/matlab.html
19. Kleinrock, L., Grushko, I.I.: Theory of queuing. Mashinostroenie, Moscow (1979). 432 p. (1976)

20. Strelkovskaya, I.V., Paskalenko, V.M.: The higher mathematics for the specialists in the sphere of communications: P.IV.: Integral on oriented area. Vector analysis. Series. Differential equations. Textbook [for the students of higher education institutions] edited by of P.P. Vorobienko. BMB, Odessa (2015). 668 p.
21. Beytman, G., Erdeyi, A.: Tables of integral transformations. P. 1. Fourier, Mellin Laplace, transformations. Transl. from Engl. Science (1969). 343 p.

Universal Method of Multidimensional Signal Formation for Any Multiplicity of Modulation in 5G Mobile Network

Lyubov Berkman, Larysa Kriuchkova, Viktoriia Zhebka(✉), and Svitlana Strelnikova

State University of Telecommunications, Solomyanska Street, 7, Kyiv 03680, Ukraine
{090289,scianalyst,viktoria_zhebka}@ukr.net,
s.strelnikova@dut.edu.ua

Abstract. A new method of forming a multidimensional signal with amplitude-phase difference modulation (MAPDM signal) of OFDM technology for 5G mobile networks, which makes it possible to increase the noise immunity of reception by 2 times in comparison with two-dimensional OFDM signals has been proposed in the chapter. The informational parameters of the MAPDM signal are the amplitude, phase and time distance between the boundaries of the parcels and the signal integration interval. An increase in noise immunity is provided by increasing the equivalent signal energy, determined by the distance between the two nearest signal points, and thus increasing the resolution of the receiver. The use of the MAPDM signal allows the information transfer rate to be brought closer to the channel capacity, which is necessary for the implementation of 5G mobile networks.

Keywords: MAPDM signal · OFDM · 5G mobile networks · Noise immunity

1 Introduction

The rapid growth in the number of devices connected to the Internet and the ever-increasing requirements of subscribers to the speed of mobile Internet access make it necessary to increase the speed of information transfer and reduce network delays at a given reliability.

Many real-time applications will require reliable communication links with extremely low latency and extremely high bandwidth to avoid signal distortion when watching videos, controlling drones or industrial robots.

It is well-known from the theory of potential noise immunity; noise immunity is primarily determined by the equivalent energy of the signals. The greater the signal energy, defined as the distance between adjacent signal points, the higher the noise immunity of the system, all other things being equal.

With the same reception method, different constellations provide different noise immunity. This is due to the peculiarities of the placement of the boundaries of the signal regions. In the geometric representation of the signal, the placement of signal points at equal distances ensures a minimum probability of error. This arrangement

M. Klymash et al. (Eds.): *Future Intent-Based Networking*, LNEE 831, pp. 305–321, 2022.
https://doi.org/10.1007/978-3-030-92435-5_18

provides the same probability of error in receiving any signal (the signal areas are the same) and the minimum average signal power (areas of the densest packing).

The known signals of the densest packing are realized, as a rule, by placing points at the nodes of spatial networks that have a regular structure. In one-dimensional space, the densest packing is the placement of signal points on a straight line. In two-dimensional space, variants of close packing on a plane are considered.

For multi-position signals, the information transfer rate is determined by the modulation rate and elementary signaling period.

$$V = \frac{\kappa}{\tau} \text{ Kbit/s}, \quad (1)$$

where κ– modulation rate, τ – elementary signaling period.

To increase the information transmission rate, it is necessary to reduce the duration of the signaling period and increase the modulation rate. However, a decrease in the duration of a signaling period by more than a certain value τ causes linear distortion of the signal, and with an increase in the modulation rate, the distance between two adjacent signal points decreases and, accordingly, the equivalent signal energy decreases, and also noise immunity decreases.

An effective means of increasing the noise immunity is the formation of the OFDM signal in such a way that for a given ratio of the signal energy to the spectral power density of the interference, the distance between adjacent signal points is maximal. This is provided by shaping the signal in three-dimensional space.

2 Analysis of the Literature Data and the Problem Statement

In the studies of the world and domestic authors, it is shown that the main task in the implementation of 5G is to ensure the speed of information transfer, close to the throughput of the communication channel [1–7].

The maximum allowable speed is provided with an increase in the modulation rate and a decrease in the duration of the signaling period. However, a decrease in the duration of a signaling period leads to intersymbol distortions, and an increase in the modulation multiplicity leads to a decrease in the equivalent signal energy, and, consequently, to a decrease in noise immunity [8–10].

When introducing 5G technology, ultra-dense networks should be created based on new types of signal-code structures [11–13]. When synthesizing a signal-code structure, it is necessary to ensure the maximum possible modulation rate, and, consequently, the number of signal positions, while maximizing the equivalent signal energy at a given ratio of signal energy to interference power spectral density.

In modern studies [14–16] it is shown that an increase in spectral efficiency in 5G networks can be achieved through the use of non-orthogonal signals (for example, FTN, F-OFDM, etc.) [17]. However, the literature does not provide calculations of noise immunity, namely, the magnitude of the inter-channel interference power.

The wireless industry has identified three main directions for 5G development:

- improved mobile broadband network;

- IoT (Carther predicts more than 20 billion Internet objects by 2020);
- highly reliable networks with low latency.

An important criterion of quality is the indicator of the probability of errors, in fact, this indicator determines the possibility of providing high quality services. In [18], signal-code structures are proposed for use in modern highly efficient telecommunication systems. However, all the methods proposed in the literature do not allow providing the required information transfer rate in the radio access section, since interference and distortions decrease by an order of magnitude the noise immunity, and, accordingly, the information transfer rate.

3 The Purpose and Objectives of the Study

The aim of this work is to synthesize a multidimensional signal with amplitude-phase-difference modulation (MAPDM), in which an additional information parameter is the time distance between the boundaries of the parcels and the boundaries of the integration interval, which makes it possible to ensure, under given conditions and constraints, the maximum reliability and information transfer rate.

To achieve this goal, it is necessary to solve the following tasks of:

- a synthesized multidimensional sixteen-position signal, where the signal points are located in three-dimensional space;
- the development of an algorithm for optimal coherent reception with synchronization of a multidimensional signal according to the working signal;
- the development of a method for calculating the noise immunity of the information transmission system on the basis of simulation modeling.

4 Special Elements

OFDM signals, which are used in 4G mobile networks, are formed in two-dimensional space, that is, the information parameters are the amplitude and phase of the transmitted signal. An example of such a 16-position signal is shown in Fig. 1.

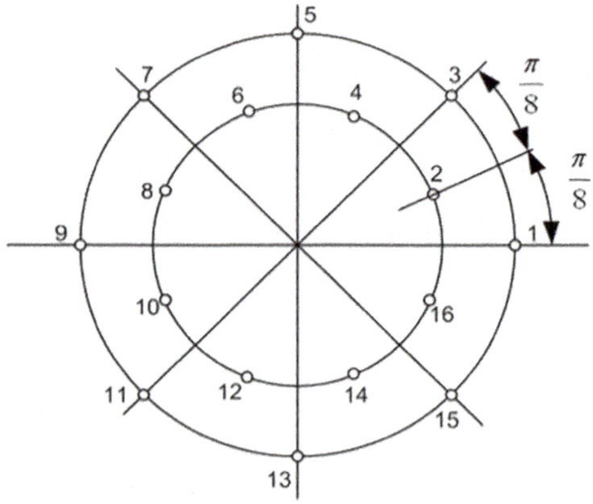

Fig. 1. Multi-position signal with amplitude-phase modulation

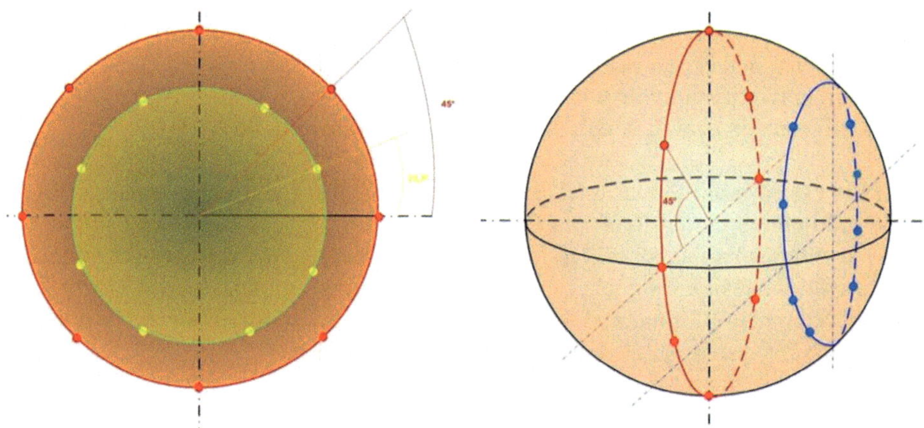

Fig. 2. Sixteen-position multidimensional signal

The signal points are located in two circles, and the ratio of the radii of these circles is chosen so that for a given signal energy and information transmission rate, the distance between two adjacent signal points is maximal [19].

To increase the equivalent energy, and, consequently, the noise immunity of the system, an effective method of signal shaping is the use of modulation in three information parameters: amplitude, phase and time.

As it is known, OFDM signals are generated by frequency multiplexing of orthogonal signals, therefore, when introducing 5G technology, it is necessary to clarify the calculation of the information transfer rate, taking into account the number of subcarriers.

In this case, the information transfer rate is calculated:

$$V = \frac{\mathrm{k\,n}}{\tau}, \tag{2}$$

where k – modulation rate (this number determines the number of signal options, which is 2k), n– number of subcarriers; τ – duration of the elementary signaling period.

Thus, forming a multidimensional signal, in which the distance between adjacent signal points increases with the same signal energy, a doubled information transfer rate is obtained.

Figure 2 shows a multidimensional signal formed in a sphere in a geometric representation. The information parameter - time - is determined by two boundaries of the integration interval D1, D2 and t3, t4. We have no right to change either the duration of the integration interval or the duration of the message, since the duration of the message determines the speed of information transmission (formula 1), and the duration of the integration interval determines the property of orthogonality of signals $T = 8,33\text{MC}$ However, we can change the boundaries of the integration interval within the message (Fig. 3). Calculations show (Fig. 4) that changes in the boundaries of the integration intervals within the duration of the message does not lead to a significant increase in the inter-channel interference power, which is characteristic of the OFDM signal [20, 21].

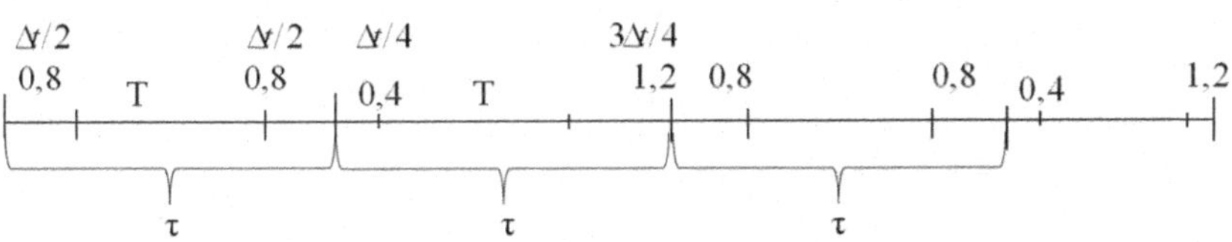

Fig. 3. The location of the boundaries of the integration interval T within the duration of the signaling period

τ is the signaling period,
T is the duration,
Δt is the duration of protection by time,
$\Delta t = \tau - T$,
where, for example:
$\tau - 10\,\text{ms}$,
$T = 8,33\,\text{ms}$
$\Delta t = 1,66\,\text{ms}$
$\Delta t/2 = 0,8\,\text{ms}$,
$\Delta t/4 = 0,4\,\text{ms}$,
$3\Delta t/4 = 1,2\,\text{ms}$.

Let us present the calculation of the dependence of the interchannel interference power (ICIP, %) on the values of the guard intervals in time $\tau - t_0$, determined by the beginning of the integration interval t_0.

The boundaries of the integration interval are shown for two cases when the guard intervals are the same $\Delta t_1 = 0,8$, and $\Delta t_2 = 0,8$ when $\Delta t_1 = 0,4$, and $\Delta t_2 = 1,2$.

Having three information parameters (amplitude, phase and time), we place the signal points in the sphere in such a way that the distances between the neighboring ones are equal in the phase plane. This doubles the equivalent signal energy.

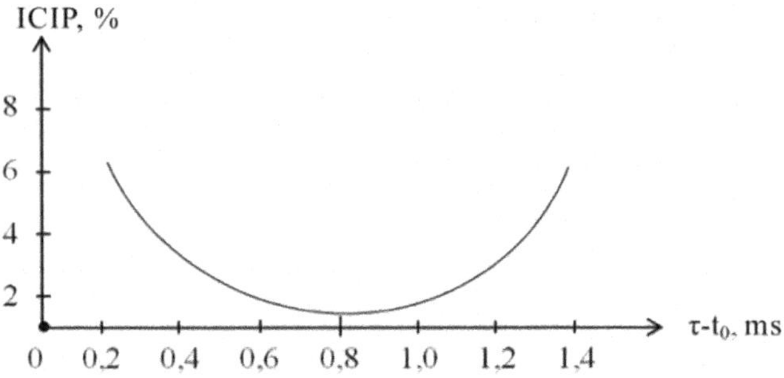

Fig. 4. Calculation of the dependence of the interchannel interference power of the ICIP, % of the values of the protection intervals over time $\tau - t_0$

In this case, the algorithm for optimal reception of the multidimensional OFDM signal, taking into account the additional parameter, will be as follows.

It should be considered the general case of digital transmission using an m-position signal with arbitrary amplitudes $a_1, a_2 ... a_n$ and initial phases $\phi_1, \varphi_2, ..., \phi_m$, without making a difference between the amplitude-phase and amplitude-phase difference modulation. With this formulation of the problem, the variant of the transmitted signal can be represented in the following form:

$$S_j(t) = a_j \sin(\omega t + \phi_j),\ j = 1, 2, ..., m. \tag{3}$$

In a channel with Gaussian uncorrelated noise, the optimal signal reception algorithm (3) can be formulated as follows: the transmitted i-th variant of the signal is fixed if for all $j \neq i$ the inequality is:

$$\int_0^T [x(t) - S_j(t)]^2 dt < \int_0^T [x(t) - S_j(t)]^2 dt, \tag{4}$$

where x (t) is the received signal, T is the duration of a parcel.

$$j = \arg\min \int_0^T [x(t) - S_j(t)]^2 dt. \tag{5}$$

In digital processing, it is convenient to switch from a high-frequency signal (4) to its reflection through coordinates in two-dimensional space, which in practice corresponds, for example, to the operations of transfer or the separation spectrum of orthogonal channel signals in a multichannel system [22].

So, let the projections of the received signal x(t) and signals onto the reference oscillations with an arbitrary phase, calculated on the interval of one message, are known:

$$\left.\begin{aligned} x_0 &= \int_0^T x(t)\cos(\omega t + \phi_0)dt, \\ y_0 &= \int_0^T x(t)\sin(\omega t + \phi_0)dt, \\ L_0 &= \int_0^T x(t)\cos(\omega t + \phi_0)dt, \end{aligned}\right\} \quad (6)$$

$$\left.\begin{aligned} x_j &= \int_0^T s_j(t)\cos(\omega t + \phi_j)dt, \\ y_j &= \int_0^T s_j(t)\sin(\omega t + \phi_j)dt, \\ L_j &= \int_0^T s_j(t)\cos(\omega t + \phi_j)dt, \end{aligned}\right\} \quad (7)$$

where $j = 1, 2, \ldots, m$.

The optimal algorithm (5) can be represented as follows:

$$j = \arg\min \int_0^T \left[(x_0 - x_j)^2 + (y_0 - y_j)^2 + (L_0 - L_j)^2\right]^2 dt, \quad (8)$$

The input values x_0 and y_0 are determined, as can be seen from (6), as a result of processing the current received signal transmission, and the values x_j, y_j and L_j, the number of which is equal to 2 m, must be known a priori or calculated (estimated) in the process of receiving the preliminary signal transmissions.

To calculate the projection estimates by the variant of the signal x_j and y_j the method of bringing and averaging the projection of the received signal is used. For the role of the averaged values, we will choose, for definiteness, the projection of the first option (3), that is, the values x_1 and y_1(7), we will also bring other versions of the received signal to them in the process of adjusting according to the information signal. If the received signal $x(t)$ contains on the interval N signal parcels $S_1(t)$ mixed with Gaussian noise, then, as is known, the plausible estimates of these quantities $\tilde{x}_1$ and $\tilde{y}_1$ are equal to:

$$\left.\begin{aligned} \tilde{x}_1 &= \frac{1}{N}\sum_{n=1}^{N} x_{0n}, \\ \tilde{y}_1 &= \frac{1}{N}\sum_{n=1}^{N} y_{0n}, \\ \tilde{L}_1 &= \frac{1}{N}\sum_{n=1}^{N} L_{0n}, \end{aligned}\right\} \quad (9)$$

where x_{0n} and y_{0n} are the values of projection (7) on the interval of the n-th premise, and the wavy line above x_1 and y_1 marks that these are estimates. Estimates (9) are unbiased and efficient. They can be formed into unbiased and effective estimates of the projection of all other signal variants included in the optimal algorithm (8). To do this, we introduce the notation $\phi_j = \phi_1 + \Delta\phi_j$, create a projection as follows:

$$x_j = \int_0^T a_j \sin(\omega t + \phi_j)\, a_0 \cdot (\omega t + \phi_0)\, dt = \frac{a_j}{a_1} \int_0^T a_1 \sin(\omega t + \phi_1 + \Delta\phi_j)$$

$$\cdot\ a_0 \sin(\omega t + \phi_0)\, dt = \frac{a_j}{a_1} [\cos\Delta\phi_j \int_0^T a_1 \sin(\omega t + \phi_1)\, Q_0 \sin(\omega t + \phi_0)\, dt$$

$$+ \sin\Delta\phi_j \int_0^T Q_1 \sin(\omega t + \phi_1)\, a_0 \sin(\omega t + \phi_0)\, dt] = \frac{a_j}{a_1} [\cos\Delta\phi_j \int_0^T a_1 \sin(\omega t + \phi_1)$$

$$\cdot\ a_0 \sin(\omega t + \phi_0)\, dt - \sin\Delta\phi_j \int_0^T a_1 \sin(\omega t + \phi_1)\, a_0 \cos(\omega t + \phi_0)\, dt]$$

$$= \frac{a_j}{a_1} \left(x_1 \cos\Delta\phi_j - y_1 \sin\Delta\phi_j\right). \tag{10}$$

Similarly, we obtain the projection y_j and L_j. Replacing now the quantities x_1 and y_1 their estimates, we get:

$$\tilde{x}_j = \frac{a_j}{a_1} \left(\tilde{x}_1 \cos\Delta\phi_j - \tilde{y}_1 \sin\Delta\phi_j + L_j \cos\Delta\phi_j\right),$$
$$\tilde{y}_j = \frac{a_j}{a_1} \left(\tilde{x}_1 \sin\Delta\phi_j + \tilde{y}_1 \cos\Delta\phi_j - L_j \sin\Delta\phi_j\right),$$
$$\tilde{L}_j = \frac{a_j}{a_1} \left(\tilde{x}_1 \cos\Delta\phi_j - \tilde{y}_1 \sin\Delta\phi_j + L_j \cos\Delta\phi_j\right), \tag{11}$$

where $\Delta\phi_j$ is the known phase difference between signals $S_j(t)$ and $S_1(t)$.

Note that when calculating estimates according to (11), there is no need to have information about the amplitude of the signal variants a_j and a_1, but it is enough to know the ratio of these amplitudes a_j/a_1.

The obtained algorithms solve the problem of coherent reception of a multi-position AFM signal in the presence of a special sync signal, which, for example, precedes the transmission of information messages: according to (7), the projections of the sync signal onto the reference oscillations with an arbitrary initial phase are calculated, then estimates of the projections of the first version of the signal $\tilde{x}_1$ and $\tilde{y}_1$, according to (10), estimates of the projections of all m variants of the signal are calculated and, finally, the estimates obtained $\tilde{x}_1$ and $\tilde{y}_1$; all signal variants are substituted instead of x_j and y_j into algorithm (8), according to which a decision is made on the OFDM information message [23–25].

Note that the considered algorithm is focused on receiving a signal with absolute PM, since the presence of a clock signal eliminates the uncertainty of the initial phase, which prevents the use of absolute phase modulation.

Let us now return to the considered algorithm and adapt it to the case, which is practically the most important, when the sync signal is absent and the "adjustments" of the projection of the signal samples have to be carried out directly according to the information messages. In this case, it is necessary to average not the projections of the received signal, but the flattened projections. In this case, the erection operation consists in converting the received projections to projections, for example, the first version of the signal using the decision made about the transmitted signal sample.

Let, as before, $\Delta\tilde{\phi}_n$ be the phase difference between the signal variant in favor of which the decision was made on the n-th parcel and the first variant of the signal, $\tilde{a}_n$ is the amplitude of the signal in favor of which the decision was made on the n-th parcel. Then $\Delta\tilde{\phi}_n$ takes values from a discrete set of allowed phases defined in (3). As for $\tilde{a}_n$, this value is equal to the actual amplitude received on the n-th message of the signal-noise mixture, however, in the future it is identified with the amplitude of the signal variant in favor of which the decision was made on the n-th message. The wavy line in both cases emphasizes that these estimates may in fact be wrong. Then the reduced projections x_{1n} and y_{1n} of the received signal on the n-th message is calculated through the received projections x_{0n} and y_{0n} by the formulas:

$$
\begin{aligned}
x_{1n} &= \frac{a_1}{\tilde{a}_n}\left(x_{0n}\cos\Delta\tilde{\phi}_n + y_{0n}\sin\Delta\tilde{\phi}_n - L_{0n}\sin\Delta\phi_j\right),\\
y_{1n} &= \frac{a_1}{\tilde{a}_n}\left(y_{0n}\cos\Delta\tilde{\phi}_n - x_{0n}\sin\Delta\tilde{\phi}_n + L_{0n}\cos\Delta\phi_j\right),\\
L_{1n} &= \frac{a_1}{\tilde{a}_n}\left(y_{0n}\cos\Delta\tilde{\phi}_n + x_{0n}\sin\Delta\tilde{\phi}_n - L_{0n}\cos\Delta\phi_j\right).
\end{aligned}
\tag{12}
$$

It should be emphasized that $\tilde{a}_n$ and $\Delta\tilde{\phi}_n$ are determined by the decision about the transmitted on the n-th parcel a variant of the signal received from the results of processing the value x_{0n} and y_{0n}.

Further, as in the algorithm for receiving by a clock signal, values (13) are averaged:

$$
\begin{aligned}
\tilde{x}_1 &= \frac{1}{N}\sum_{n=1}^{N}\frac{a_1}{\tilde{a}_n}(x_{0n}\cos\Delta\tilde{\phi}_n + y_{0n}\sin\Delta\tilde{\phi}_n - L_{0n}\sin\Delta\phi_j),\\
\tilde{y}_1 &= \frac{1}{N}\sum_{n=1}^{N}\frac{a_1}{\tilde{a}_n}(y_{0n}\cos\Delta\tilde{\phi}_n - x_{0n}\sin\Delta\tilde{\phi}_n + L_{0n}\cos\Delta\phi_j),\\
\tilde{L}_1 &= \frac{1}{N}\sum_{n=1}^{N}\frac{a_1}{\tilde{a}_n}(x_{0n}\cos\Delta\tilde{\phi}_n + y_{0n}\sin\Delta\tilde{\phi}_n - L_{0n}\sin\Delta\phi_j).
\end{aligned}
\tag{13}
$$

The difference between algorithm (13) and (9) lies in the fact that in (9) averaging is performed on the sync signal interval, and in (13) - on a "varying" interval in M messages preceding the one being processed at a given moment.

Note that when calculating estimates according to (13), there is no need for a priori information about the average power of the received signal, because this algorithm includes only the ratio of the amplitudes.

Thus, the relation (12), (13) together with (8) forms the desired coherent processing algorithm.

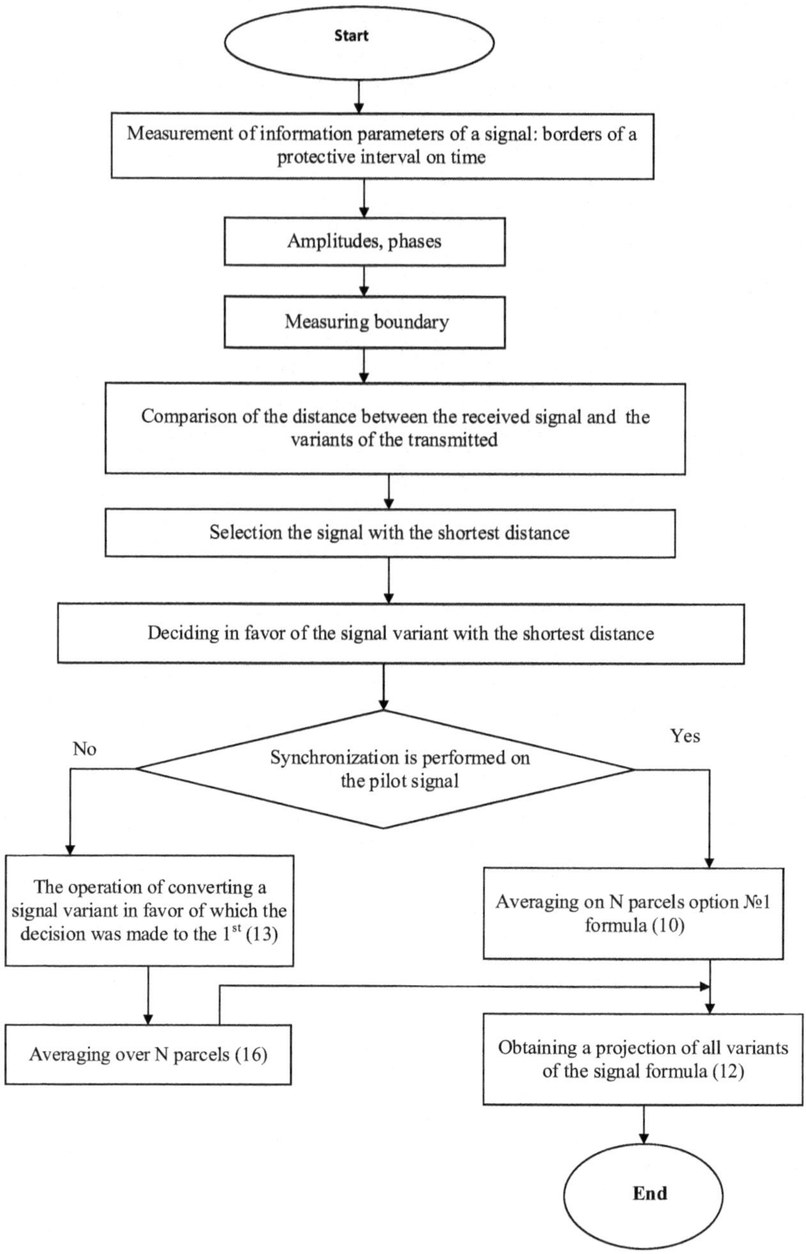

Fig. 5. Block diagram of the algorithm for optimal reception of a multidimensional signal

The proposed algorithms for coherent processing of multi-position amplitude-phase or amplitude-phase difference modulation of signals are especially convenient for multichannel (multifrequency) systems with orthogonal channel signals, since in these systems the same procedures for calculating the projections of the received signal onto two mutually orthogonal reference oscillations are used to separate orthogonal signals with an arbitrary initial phase.

In multifrequency OFDM systems, the reference oscillations of all channels are usually generated from a common master oscillator. At the same time, the initial phases of the channel signals have different shifts, and in some cases they are generally weakly coupled (for example, in radio channels with selective attenuation, they are uncorrelated). As a result, in multichannel OFDM systems, it is difficult to use the methods of reference oscillations based on adjusting the phase of the controlled oscillator.

The described methods of coherent processing do not require adjustment of the phase of the reference oscillations of channel signals and make it quite simple to implement both orthogonal separation of signals and coherent reception. Indeed, for channel separation, the same reference oscillations are used as for optimal incoherent reception. Coherent reception is performed based on the adjustment of the signal variants in each channel of the OFDM signal.

As is known [10], for each Q and a given probability β, it is possible to construct a region within which the value of the error probability (P_{er}) coincides with the experimentally found frequency value (P_{er}^{*}). The curves shown also limit the regions for different Qs at probability $\beta = 0.9$. For example, if $P_{er}^{*} = 0.1$ to ensure the confidence interval [0.09–0.11], the required sample size will be $\approx$200, while $P_{er}^{*} = 0.07$ to ensure the confidence interval [0.065–0.075], the required size will be $\approx$1000. Thus, the sample size must be at least $\frac{20}{P_{er}^{*}}$.

The graphs shown in Fig. 6 illustrate the dependence of the error probability P_{er} on the ratio of the signal energy to the spectral power density of the noise for the APM signal system shown in Fig. 1. The curves in the figure are designated by numbers 1, 2… 9. Curve 1 characterizes the potential noise immunity of the given signal system with strictly coherent reception. Curve 2 was obtained as a result of modeling with accurate reference oscillations and characterizes the potential noise immunity of the developed processing algorithm. Curves 3… 7 obtained as a result of modeling and characterize the noise immunity of OFDM systems with coherent reception of signals for different averaging intervals M, which were chosen in accordance with 1, 5, 10, 20 and 100.

- M = 100 (Curve 3),
- M = 20 (Curve 4),
- M = 10 (Curve 5),
- M = 5 (Curve 6),
- M = 1 (Curve 7).

Curve 8 characterizes the theoretical noise immunity for a 4-fold PRM with a coherent reception method. Curve 9 was obtained experimentally as a result of laboratory studies of a 48 - channel modem using a T4 channel simulator of the “channel 2” type. This curve determines the real noise immunity of OFDM systems of a 48 - channel

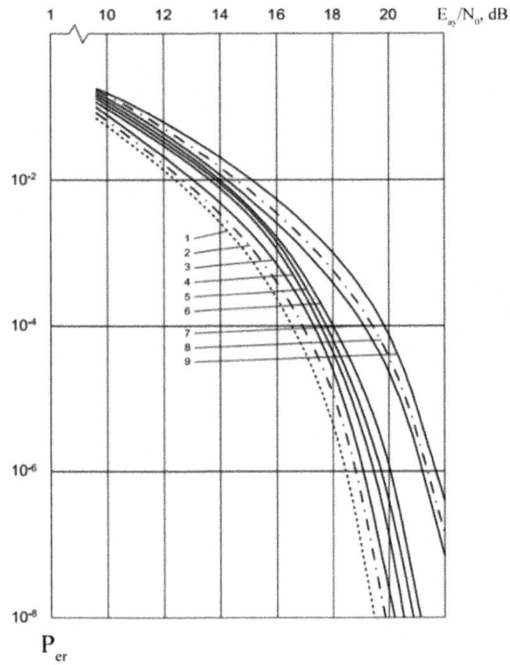

Fig. 6. Dependence of the error probability on the signal-to-noise ratio

modem with an optimal incoherent reception method, in which a 16 - position signal is used, shown in Fig. 1 (ch2 is twice as large as ch1).

Analysis of these graphs (Fig. 6) confirms the conclusion about the advantage of using coherent reception methods for multi-position signals. The energy gain can be obtained both from the use of a more efficient signal system (for example, MAPDM or APM), and from the use of the developed coherent reception algorithm.

In this case, comparing curve 3 and curve 9, it can be seen that the total gain is ≈3dB.

The minimum averaging interval, due to the number of parcels M, at which the real noise immunity is close to the potential, is ≥20 parcels.

The graph shows the quantitative dependence of the approximation to the ideal variant (the ideal is the variant in which the reference oscillations are accurate, in the graphs in Fig. 6 this is curve 2) at various averaging intervals M for the MAPDM signal systems (multidimensional signal with amplitude-phase difference modulation) and APM (signal with amplitude-phase difference modulation), and $\Delta = h_M^2 - h_{nL}^2$. For both signal systems, the number of averaged bursts at which the noise immunity is close to potential should be at least 20.

The graph in Fig. 7 shows the curves characterizing the noise immunity of two signal systems MAPDM and APM at M = 100. Comparison of these curves shows that the use of the signal system designated MAPDM gives a significant energy gain in comparison with APM by 10 dB.

It should be noted that in the process of acquiring synchronism, averaging is carried out not on a varying interval of M messages, but on an interval of received signal messages until the number of received messages reaches M. Therefore, determining that the system acquires synchronism after an interval of time parcels, it is necessary to take into account that the specified noise immunity will be provided through the time interval M parcels.

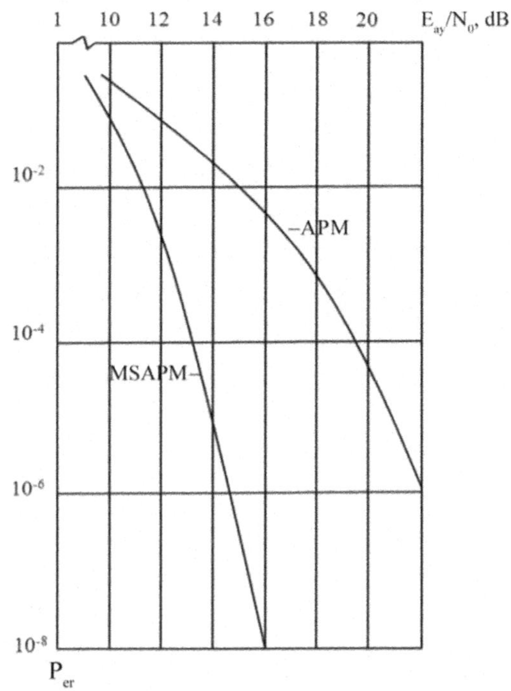

Fig. 7. Dependence of the error probability on the signal-to-noise ratio for the APM and MAPDM signal system

The graphs presented in Fig. 8 and Fig. 9 illustrate the dependence of the initial acquisition in synchronism (Fig. 8), as well as the dependence of the acquisition duration after a phase jump (Fig. 9) on the signal-to-noise ratio and the magnitude of the discrepancy of the transmitted signal on estimates of signal options in the OFDM system and the magnitude of the phase jump. The abscissa shows different angles ($\Delta\varphi$, degrees) characterizing the discrepancy between the transmitted signal and the estimates of the signal variants in the OFDM system (Fig. 9) and the magnitude of the phase jump of the transmitted signal along the ordinate axis - respectively, the duration of the initial acquisition and the duration of acquisition of synchronism after the phase jump. Each curve is obtained to determine the signal-to-noise ratio with the number of averaged bursts equal to 10 (M = 10).

It can be concluded that with a signal-to-noise ratio of at least 20 dB, as well as after a phase jump, the duration of the initial acquisition of synchronism is ≈2 parcels.

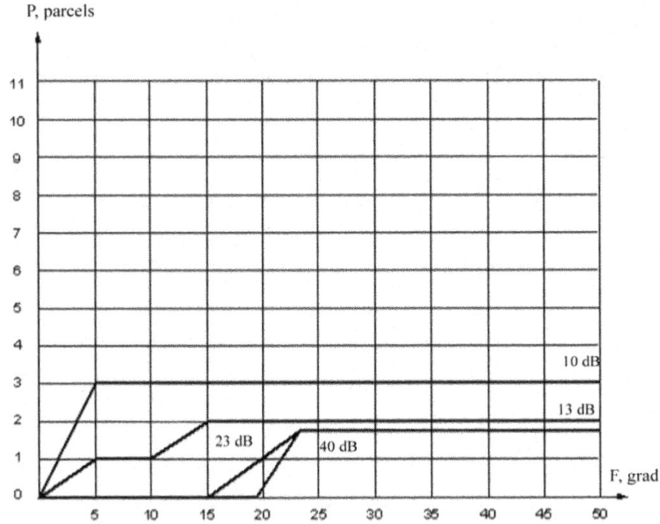

Fig. 8. Dependence of the duration of the occurrence of links on the magnitude of the phase jump at various signal-to-noise ratios

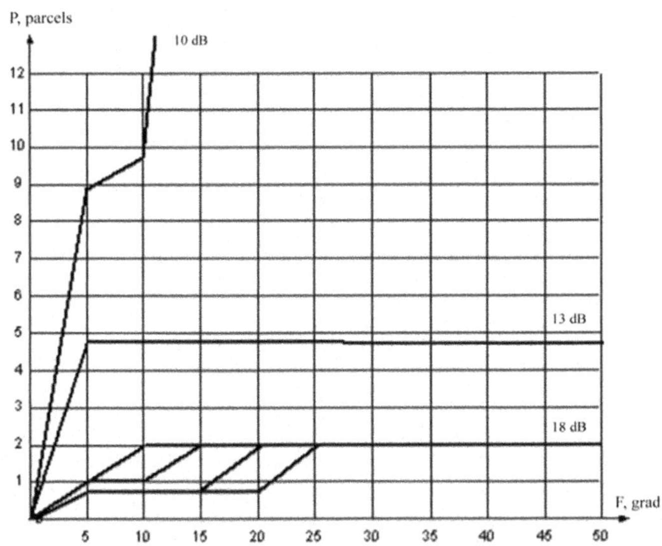

Fig. 9. Dependence of the duration of the initial entry for various signal-to-noise ratios (for M = 10)

5 Modeling a Multidimensional Information Transmission System Based on a Multidimensional Signal

The tasks solved by modeling are formulated as follows.

1. Determination of the error probability for ideal reference oscillations for various signal-to-noise ratios. This will allow determining the potential noise immunity of OFDM systems when using different groups of multi-position signals.
2. Determination of the error probability when generating signal variants for different averaging intervals. Determination of the minimum averaging interval at which the noise immunity is close to potential.
3. Comparison of the various signal systems for noise immunity with the considered processing method.
4. Determination of the duration of the initial entry into communication, as well as the duration of entry after different in magnitude jumps of the signal phase.

During the study, the method of statistical modeling was used, in which all processes were set in the form of algorithms oriented to use for calculations.

As a source of discrete information for modeling, a program for generating a pseudo-random sequence of numbers from 0 to 15 was used. The action of additive fluctuation noise modeling was based on the formation of numbers with a normal distribution law, zero mean and a given variance. These numbers, in turn, can be obtained from a sequence of pseudo-random numbers uniformly distributed over the interval (0, 1), based on the central limit theorem of probability theory.

6 Discussion of the Study Results of the Noise Immunity of 5G Mobile Networks Based on a Multidimensional Signal with Amplitude-Phase Difference Modulation

When introducing 5G networks, it is necessary to ensure the creation of ultra-dense networks with information transfer rates of 2 Gbps and higher with a minimum delay and with a given reliability. As it is known, an increase in the speed of information transmission is prevented by interference and distortions in the communication channel.

If linear signal distortions can be compensated for with the help of automatic correctors, then the effect of additive noise of the "white noise" type cannot be compensated for, and the only way to combat such noise is to form a multidimensional signal whose equivalent energy is significantly (at least twice) higher than multi-position. The methods of forming a multidimensional signal in three projections, where the amplitude, phase and time are informative have been described in the chapter. Algorithms for optimal coherent reception with synchronization of the pilot signal have been proposed.

With the help of simulation modeling, the calculation of the noise immunity of information transmission systems based on a multidimensional signal is presented. The minimum number of elementary messages required for averaging has been determined, which provides the specified noise immunity with a minimum information transmission delay.

The calculation of the noise immunity shows that the noise immunity of a system, based on a multidimensional signal, increases by more than 10 dB, which makes it possible to increase the modulation rate, and, and, accordingly, the information transfer rate by two times.

7 Conclusions

The proposed multidimensional signal provides an increase in the distance between two adjacent signal points, which makes it possible to increase the equivalent signal energy, and, consequently, its noise immunity;

The shift of the integration interval within the message for the formation of two groups of signal points does not lead to an increase in the level of inter-channel interference, characteristic of the OFDM signal;

The algorithm for processing a multidimensional signal and the found averaging intervals of the estimated parameters allow, with a minimum delay of the transmitted information, to ensure the noise immunity of the information transmission system, which is close to the potential.

References

1. Zaitsev, G.F., Steklov, V.K., Britsky, O.I.: Theory of automatic control. Engineering (2002). 688 p.
2. Tolubko, V.B., Berkman, L.N., Kozelkov, S.V., Gorokhovsky, Є.P.: Construction of ultra-dense mobile networks of the fifth generation 5G on the basis of multidimensional signals. J. Call Duty **1**, 3–7 (2017)
3. Tolubko, V.B., Berkman, L.N., Kozelkov, S.V.: Formation of 5 G technologies that are highly positioned to the signal on the basis of phase-difference modulation of high orders. Zvyazok Mag. **4**, 3–7 (2017)
4. Tikhvinsky, V.O., Barrel, G.S.: Conceptual aspects of the creation of 5G. Telecommunications **10**, 29–33 (2013)
5. Tikhvinsky, V.O., Bochechka, G.S., Minov, A.V.: Monetization of LTE networks based on M2M services. Electrosvyaz **6**, 12–17 (2014)
6. Tikhvinsky, V.O., Terentyev, S.V., Vysochin, V.P.: Mobile networks LTE/LTE Advanced: 4G technologies, applications and architecture. Media Publisher (2014). 384 p.
7. Okunev, Yu.B.: Communication systems with invariant noise immunity characteristics. Communication (1973). 79 p.
8. Tolubko, V.B., Berkman, L.N., Kozelkov, S.V., Gorokhovsky, Є.P.: Phazoriznitseva modulation of high orders for securing the intended delivery of preservation of the channels transmitted by the information. J. "Telecommun. Inf. Technol." **1**(54), 5–10 (2017)
9. Tolubko, V.B., Berkman, L.N., Otrokh, S.I., Gorokhovsky, Є.P., Yarosh, V.O.: Ratio characteristics of the performance of systems in case of vicious N-modest large-posed signals. J. "Sci. Notes UNDIZ" **2**(46), 5–11 (2017)
10. Tolubko, V.B., Berkman, L.N., Kozelkov, S.V., Gorokhovsky, Є.P.: Capturing information in one multi-channel systems from Poisson and high input flow during one hour of service. Zvyazok Mag. **3**, 3–7 (2017)
11. Banquet, V.L.: Signal-code structures in telecommunication systems. Fenix, Odessa (2009). 180 p.
12. Chin-Lin, I., Han, S., Xu, Z., Wang, S., Sun, Q., Chen, Y.: New paradigm of 5G wireless internet. IEEE J. Sel. Area Commun. **34**(3), 474–482 (2016)
13. Ma, Z., Zhang, Z., Ding, Z., Fan, P., Li, H.: Key techniques for 5G wireless communications: network architecture, physical layer, and MAC layer perspectives. Sci. China Inf. Sci. **58**(4), 1–20 (2015)

14. Bochechka, G., Tikhvinskiy, V.: Spectrum occupation and perspectives millimeter band utilization for 5G networks. In: Proceedings of ITU-T Conference «Kaleydoscope-2014», St. Petersburg, pp. 99–101 (2014)
15. Wuand, J., Fan, P.: A survey on high mobility wireless communications: challenges opportunities and solutions. IEEE Access **4**, 450–476 (2016)
16. NGMN: NGMN 5G whitepaper. http://www.ngmn.de/5g-whitepaper/5g-white-paper.html
17. Tang, Z., Cannizzaro, R.C., Leus, G., Banelli, P.: Pilot-assisted time vary in channel estimation for OFDM systems. IEEE Trans. Signal Process. **55**(5), 2226–2238 (2007)
18. Cheng, P., et al.: Channel estimation for OFDM system solver doubly selective channels: a distributed compressive sensing based approach. IEEE Trans. Commun. **61**(10), 4173–4185 (2013)
19. Andrews, J.G., et al.: Whatwill5GBe. IEEE J. Sel. Area Commun. **32**(6), 1065–1082 (2014)
20. Aboutorab, N., Hardjawana, W., Vecetic, B.: A new iterative Doppler-assisted channel estimation joint with parallel ICI cancelation for high mobility MIMO-OFDM systems. IEEE Trans. Veh. Technol. **61**(4), 1577–1589 (2012)
21. Farhang-Boroujeny, B.: OFDM versus filter bank multicarrier. IEEE Sig. Proc. Mag. **28**(3), 92–112 (2011)
22. Wunder, G., et al.: 5GNOW: non-orthogonal, asynchronous wave forms for future mobile applications. IEEE Commun. Mag. **52**(2), 97–105 (2014)
23. Dong, Z., Fan, P., Panayirci, E., Lei, X.: Power adaptation in OFDM systems based on velocity variation under rapidly time-varying channels. IEEE Commun. Lett. **19**(4), 689–692 (2015)
24. Tolubko, V., Berkman, L., Otrokh, S., Pliushch, O., Kravchenko, V.: Noise immunity calculation methodology formulti-positional signal constellations. In: 14th International Conference on Advanced Trend sin Radioelectronics, Telecommunicationsand Computer Engineering, TCSET 2018 - Proceedings (2018)
25. Berkman, L.N., Okunyev, Yu.B.: Quasi-coherent processingof signals with multiposition amplitude-phasemodulationin multichannel modems. Telecommunications and RadioEngineering (English translation of Elektrosvyaz and Radiotekhnika) (1991)
26. Tolubko, V., Berkman, L., Komarova, L., Pokhabova, I.: Miniminno and minimaksno optimum control systems of communication networks. In: 2015 2nd International Scientific-Practical Conference Problems of Infocommunications Science and Technology, PIC S and T 2015 Conference Proceedings (2015)
27. Steklov, V.K., Starodub, N.M., Berkman, L.N.: Information-entropic method for estimation invar ant controlling systems. In: CriMiCo 2001 - 11th International Conference (2001)

AI-Enabled Blockchain Framework for Dynamic Spectrum Management in Multi-operator 6G Networks

Taras Maksymyuk[1](✉), Juraj Gazda[2], Madhusanka Liyanage[3,4], Longzhe Han[5], Bohdan Shubyn[1], Bohdan Strykhaliuk[1], Oleh Yaremko[1], Minho Jo[6], and Mischa Dohler[7]

[1] Lviv Polytechnic National University, Bandery 12, Lviv 79013, Ukraine
{taras.a.maksymiuk,bohdan.p.shubyn,bohdan.m.strykhaliuk, oleh.m.yaremko}@lpnu.ua

[2] Technical University of Kosice, Letná 1/9, 04001 Košice, Slovakia
juraj.gazda@tuke.sk

[3] University College Dublin, Belfield, Dublin 4, Dublin, Ireland
madhusanka@ucd.ie

[4] University of Oulu, Pentti Kaiteran Katu 1, 90014 Oulu, Finland

[5] Nanchang Institute of Technology, Tianxiang Road 289, Nanchang 330099, Jiangxi, China
lzhan@nit.edu.cn

[6] Korea University, Sejong-ro 2511, Sejong Metropolitan City 30019, Republic of Korea
minhojo@korea.ac.kr

[7] King's College London, Strand, London WC2R 2LS, UK
mischa.dohler@kcl.ac.uk

Abstract. A smart architectural design in 5G with flexibility for various deployment scenarios and service requirements has enabled different business models for mobile network operators in both nationwide and local scales. Future 6G networks will feature even more flexible mobile network deployment driven by spectrum and infrastructure sharing among the operators. In this chapter, we propose a new multi-layer framework for 6G with decoupled operators and infrastructure planes. The proposed framework provides a flexibility of network configuration for multiple operators in condition of open spectrum and infrastructure market by using a multi-dimensional matrix representation of the data flows. In particular, the proposed model supports the dynamic switching of the operator and multi-operator service provision for the end users. As a case study, we have developed an AI-based workflow for the dynamic spectrum allocation among multiple mobile network operators. The key advantage of the proposed workflow is that it can be adjusted to the different combinations of the data flows and thus can be suitable for the spectrum allocation among multiple operators. The intelligent capabilities of the proposed workflow are provided by the deep recurrent neural network based on the long short-term memory architecture. The developed model has been trained over the custom dataset with realistic user mobility in urban area. Simulations results show that the proposed intelligent model provides a stable service quality for end users regardless of the serving operators and outperforms the static and semi-intelligent models.

M. Klymash et al. (Eds.): *Future Intent-Based Networking*, LNEE 831, pp. 322–338, 2022.
https://doi.org/10.1007/978-3-030-92435-5_19

Keywords: 6G · Blockchain · Artificial intelligence · Decentralized mobile networks · Multi-operator spectrum management

1 Introduction

With the extensive proliferation of 5G technologies we are currently on a verge of the 5th industrial revolution driven by personalized and intelligent digital ecosystems. 5G has brought together a bunch of advanced technologies and enabled an instant connectivity among them with a high throughput and reliability. Nevertheless, a further evolution to the 6G technologies is inevitable in order to facilitate a sustainable interaction among various domains and industrial verticals and provide a personalized user experience. Whereas 5G has mainly been focused on the three main pillars of quality assurance, namely eMBB (Enhanced Mobile Broadband), mMTC (Massive Machine Type Communications) and URLLC (Ultra-Reliable Low Latency Communications), in 6G we can expect much more fine-grained differentiation of the service quality [1]. In particular, 6G should take into account a ubiquitous comprehension of the user context within physical, virtual and augmented reality, while providing the instant data delivery, high service reliability and availability regardless of the serving operator [2]. Considering the challenges, which have been brought by the COVID-19 pandemic, the further development of the mobile communications is focused towards consistent improvement of the remote work and education, including the holographic telepresence, the mixed reality and the Internet-of-Skills. It is now more apparent that modern smartphones in the foreseeable future will be replaced by a set of wearable devices such as integrated display glasses, headsets, tactile and biosensors, etc. Thus, the definition of service in 6G will transform from the modern device-oriented service provision to the future user-oriented service provision, so that multiple independent data streams through different hardware means will be combined in a synchronized manner to recreate an immersive experience of the end user [3]. Such transformations are now becoming feasible due to the growing capabilities of the visualization technologies, precise sensors and specialized processing units, which open new possibilities of interaction with human senses. Another example of multi-flow service provision can be found in the area of autonomous vehicles. According to Intel's prediction, the typical self-driving car of the future will generate approximately 4 terabytes of data per hour, coming from different cameras, sensors and controllers. The main challenge here is in the difference in the data purpose. While some critical data need to be transmitted instantly with low latency, there are also useful background data, which can be collected and transmitted over longer timeframes. Therefore, a typical V2x (Vehicle-to-Everything) service in 6G era will consist of the multiple orchestrated and synchronized data flows, with different requirements for latency, throughput and packet loss [4].

However, the conventional model of the mobile networks market has been only reasonable at the very beginning of mobile networks development to ensure nationwide coverage when 2G mobile networks started to be massively deployed around the world. As a result, we observed the very inefficient duplicated infrastructure deployment by operators, because all of them have to compete and no one is willing to give up a particular coverage area [5]. Nowadays, with the rapid technological development and the continuous increasing of the traffic demand operators are forced to densify their infrastructure in order to satisfy the ever growing number of the end users. However, as mentioned above, in 6G we expect more diverse service requirements based on the cyber-physical experience of the end users. With multiple data-flows and different quality requirements, operators will need to develop and maintain even more redundant network infrastructure unless we find a new way to solve the challenge [6].

Considering the abovementioned challenges, in this chapter we propose a novel concept for the development of the future AI (Artificial Intelligence) and blockchain-enabled 6G networks. The key idea is in the economic decoupling of the network operators and network infrastructure, so that all operators will share the network infrastructure and will be able to contribute to the network development. This can democratize the mobile communications market by elimination of the current limitations and regulations for operators. In addition, we provide a case study for the decentralized spectrum management by multiple MNOs (mobile network operators) using a new AI-based workflow. The main contributions of this chapter are the following:

1. The multi-plane framework for the spectrum and infrastructure sharing among multiple MNOs is proposed by leveraging the AI and the DLT (distributed ledger technology).
2. Intelligent decentralized spectrum management workflow among multiple MNOs is proposed based on the deep recurrent neural networks and the blockchain technology.

2 Blockchain and AI-Empowered 6G Framework

Proposed framework consists of a user plane, an infrastructure plane, an operators' plane, a blockchain plane and an AI plane, as depicted in Fig. 1.

All planes are responsible for their particular parameters and functions, which are coordinated to ensure the intelligent network management in a decentralized manner. Thus, the network infrastructure can be adjusted to any intent of the MNOs and the end users in order to meet the future cyber-physical applications in 6G mobile networks.

The *user plane (U)* contains users and their corresponding data flows with different service requirements. Since user experience in 6G will be mostly cyber-physical and will depend on multiple data flows, we propose a generalized model of the user plane as a two-dimensional matrix $\underline{\mathbf{F}} \in \mathbb{R}^{I \times J}$, where each row represents a vector of the data flows of a particular end user, while each column represents a vector that indicates the users of the particular type of service:

$$\underline{\mathbf{F}} = \begin{bmatrix} f_{11}\, f_{12} \cdots f_{1J} \\ f_{21}\, f_{22} \cdots f_{2J} \\ \vdots \quad \vdots \quad \ddots \quad \vdots \\ f_{I1}\, f_{I2} \cdots f_{IJ} \end{bmatrix} \in \mathbb{R}^{I \times J}. \tag{1}$$

The smallest element of the matrix (1) represents the data flow of the user i with the service type j:

$$f_{i,j} = \underline{\mathbf{F}}(i, j). \tag{2}$$

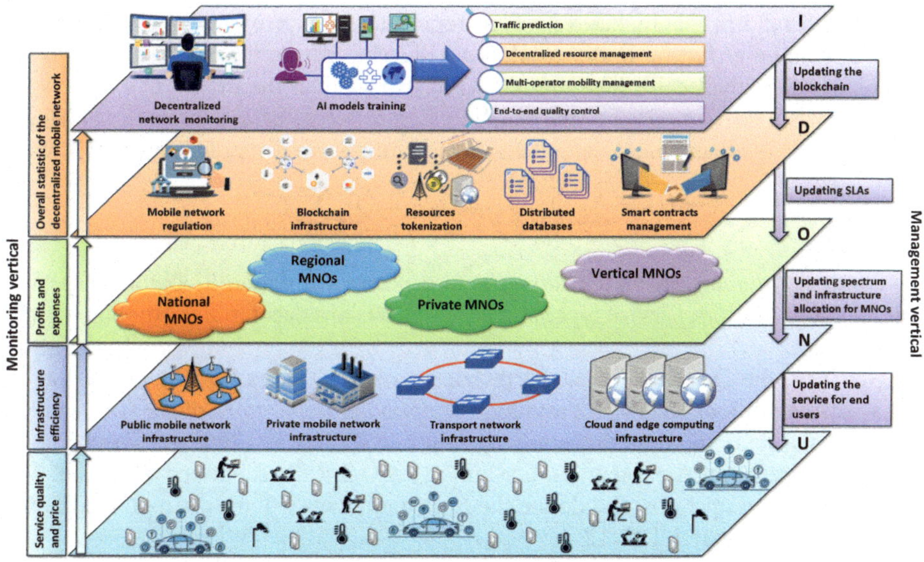

Fig. 1. The multi-plane architecture for the intelligent decentralized 6G mobile networ

Considering the fact that all users will not use all possible types of service, the matrix (1) will be sparse, because some elements $f_{i,j}$ may be equal to zeros. Note that, a row in a matrix (1) represents a person, not a device, so that all data flows that belongs to one user may not necessarily be transmitted through one device. Thus, the instantaneous total data flow of the user plane within the particular coverage area of the 6G mobile networks can be calculated as following:

$$F(t) = \sum_{i} \sum_{j} f_{i,j}(t). \tag{3}$$

By using the proposed granulation of the data flows, we can provide any combination of them by using the trivial operations of linear algebra. Therefore, such model is fully in line with the Cisco's definition of the intent-based networking that aims to configure the network upon the preferences of the end users.

The infrastructure plane (I) provides connectivity for the end users through any available wireless and wired access technology, such as macro and small cells, Wi-Fi access points, device-to-device communications, mMTC and V2x communications, Starlink, LEO (Low Earth Orbits) satellites (Starlink, etc.) and fixed optical broadband [7–9]. The data flows matrix (1) in the infrastructure plane is transformed to the three-dimensional form $\underline{\mathbf{F}} \in \mathbb{R}^{I \times J \times L}$:

$$\underline{\mathbf{F}} = \left[f_{ij1} \, f_{ij2} \cdots f_{ijL} \right] \in \mathbb{R}^{I \times J \times L}, \tag{4}$$

where the smallest element of the matrix (4) represents the data flow of the user i with the service type j through the infrastructure element l:

$$f_{i,j,l} = \underline{\mathbf{F}}(i, j, l). \tag{5}$$

Notwithstanding that by index l we can represent any type of the network infrastructure, for simplicity further in this work we will refer only to the cellular base stations like macro and small cells. The matrix (4) is sparser than the matrix (1), because all individual data flows in (1) are distributed among different base stations in (4). Hence, the instantaneous total data flow in the infrastructure plane is represented as following:

$$F(t) = \sum_i \sum_j \sum_l f_{i,j,l}(t) = \sum_i \sum_j f_{i,j}(t). \tag{6}$$

By the formalization of the infrastructure plane, we define a complete space of combinations of users, service types and base stations in a decentralized mobile network environment.

However, the last element of the puzzle is missing in the Eq. (4). Since there is not a service without a service provider, we define the *operators plane (O),* which is aimed to distribute the data flows among multiple MNOs and link the corresponding SLAs (Service Level Agreements). For brevity, we omit the economic aspects of SLA between user and operator, and define the following assumptions based on the findings of our previous research [10–12]:

- user is able to change MNO in a real-time based on his own tradeoff between service quality and service price;
- user is able to get a service from multiple MNOs simultaneously;
- MNO may not necessarily be able to provide all available types of service;
- MNO is able to transmit multiple data flows through different devices of the end user.

Thus, the main idea is to enable the spectrum and infrastructure sharing by the MNOs within the single decentralized mobile network. Such idea, can be formalized by transformation of the matrix (4) to the four-dimensional matrix $\underline{\mathbf{F}} \in \mathbb{R}^{I \times J \times K \times L}$:

$$\underline{\mathbf{F}} = \begin{bmatrix} f_{ij11} & f_{ij12} & \cdots & f_{ij1L} \\ f_{ij21} & f_{ij22} & \cdots & f_{ij2L} \\ \vdots & \vdots & \ddots & \vdots \\ f_{ijK1} & f_{ijK2} & \cdots & f_{ijKL} \end{bmatrix} \in \mathbb{R}^{I \times J \times K \times L}, \tag{7}$$

where the smallest element of the matrix (7) represents the data flow of the user i with the service type j provided by the operator k through the base station l:

$$f_{i,j,k,l} = \underline{\mathbf{F}}(i, j, k, l). \tag{8}$$

Since the matrix (7) is a transformation of the matrix (5), it is also sparse and the instantaneous total data flow can be calculated as following:

$$F(t) = \sum_i \sum_j \sum_k \sum_l f_{i,j,k,l}(t) = \sum_i \sum_j \sum_l f_{i,j,l}(t) = \sum_i \sum_j f_{i,j}(t). \tag{9}$$

In order to provide the framework for the infrastructure sharing by MNOs we introduce an additional connectivity matrix $\underline{\mathbf{O}} \in \mathbb{R}^{K \times L}$, where each element can be either 0 or 1:

$$o_{k,l} = \begin{cases} 1, & \textit{if} \text{ base station } l \text{ is used by MNO } k \\ 0, & \text{otherwise} \end{cases}. \tag{10}$$

Considering the dynamic of the 6G mobile network in decentralized deployment and infrastructure sharing conditions, the matrix $\underline{\mathbf{O}} \in \mathbb{R}^{K \times L}$ is frequently modified in time. Therefore, in order to represent the network traffic served by MNO k in a discrete time interval t, we include an instantaneous matrix state $\underline{\mathbf{O}}(t)$ as following:

$$F_k(t) = \sum_i \sum_j \sum_l \left(f_{i,j,k,l}(t) \cdot o_{k,l}(t)\right). \tag{11}$$

For simplicity, let's define that each base station can be used only by one MNO in a discrete time interval t, so that number of ones in a matrix $\underline{\mathbf{O}}$ will be always constant according to the following definition:

$$\sum_k \sum_l o_{k,l}(t) = L. \tag{12}$$

Thus, the Eq. (11) can be simplified to the following form:

$$F_k(t) = \sum_i \sum_j \sum_l f_{i,j,k,l}(t). \tag{13}$$

Hence, in an *operators' plane* we provide the flexible decentralized data flows management considering the infrastructure sharing by the MNOs and adaptive MNO selection by the end users.

In order to manage the decentralized mobile network environment there is a need for a trustable and secure mechanism, which can ensure that spectrum and infrastructure sharing by MNOs will be fair and all parties will be satisfied. Therefore, we introduce the *blockchain plane (B),* which provides a distributed ledger infrastructure on top of the conventional mobile network infrastructure. Distributed ledger (blockchain) is a decentralized system with peer-to-peer model, where there is not any single authority [14]. Blockchain is organized as an infinitely growing list of records, which are modified by transactions. Any transaction is validated and the copies of the modified ledger are shared among all nodes in the network. In order to keep the process secure, multiple transactions are assembled into a block, which is then hashed by using advanced cryptographic algorithms. The chain-like structure of the distributed ledger is achieved by linking the subsequent blocks in a way that hash value of previous block is included into next block as shown in Fig. 2. This feature ensures that any past transaction cannot be modified, because it will cause the wrong hash values of the entire subsequent chain, which will be rejected by other nodes. The process of block validation is called mining and is conducted by a consensus mechanism among all nodes in the blockchain network [15].

In this chapter, we omit detailed explanation of the different consensus mechanisms and cryptographic algorithms that are used in various types of blockchain, because they are quite widely studied in the literature [14, 15]. Our main interest within the intelligent 6G framework is in the decentralized Applications (dApp) based on the blockchain. Unlike traditional applications, which are based on centralized servers and single authority, dApp utilizes the blockchain to provide the trust between all involved parties through consensus mechanism.

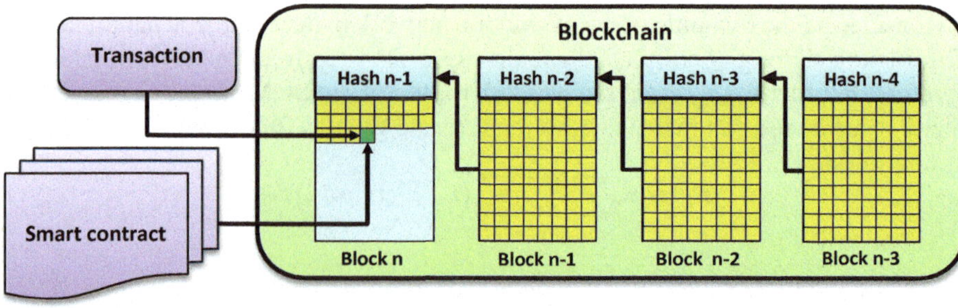

Fig. 2. An example of the blockchain operation

Since blockchain is tamper-resistant, MNOs and regulator can use dApp as a useful tool for infrastructure and spectrum management [16, 17]. The proposed structure of dApp for the intelligent 6G framework is shown in Fig. 3. We propose to use a combination of the traditional cloud-based applications, which are linked to Ethereum blockchain in order to enable spectrum and infrastructure sharing among MNOs by using smart contracts. Smart contract is a piece of code, which can be explicitly programmed to conduct several subsequent transactions in blockchain, which reflect certain financial agreements between multiple parties. For example, the MNO A can set the price of the base station rent in the smart contract, while MNO B can set a price, which he is willing to pay for the base station rent. Once conditions of both MNOs are met, the corresponding transactions will be automatically executed and MNO B will be indicated as an owner of the corresponding base station in the blockchain. Once the renting period will over, the smart contract will automatically execute the reverse transaction to return the ownership to the MNO A. Similar procedure can be applied for the spectrum sharing, as well as any other property, which can be shared among MNOs. Other role of proposed dApp is to manage SLAs between users and MNOs in a trustful and secure manner.

To bring the AI capabilities for the proposed decentralized 6G mobile network, we introduce the *AI plane (I)* to close a loop of network management in the proposed framework. The key parts of the AI plane are the data management and machine learning algorithms. The particular challenge of AI application in decentralized mobile networks is that we have the exogenous dynamic of the system, caused be the adaptive MNO switching by end user, spectrum and infrastructure sharing, etc. Hence, the conventional way of collecting and processing data by MNOs is not effective, because the external conditions are changing faster than the AI can be trained.

Thus, we introduce the model of intelligent decentralized network management (Fig. 4), which has the following features:

1. Joint network monitoring and data sharing by all MNOs.
2. Trained AI models can be either personal, which are used by MNOs to improve their services to end users, or public, which are shared among MNOs to increase the efficiency of the network resources utilization.
3. User database should be shared by all MNOs in order to maintain the proper service quality across all MNOs and network infrastructure.

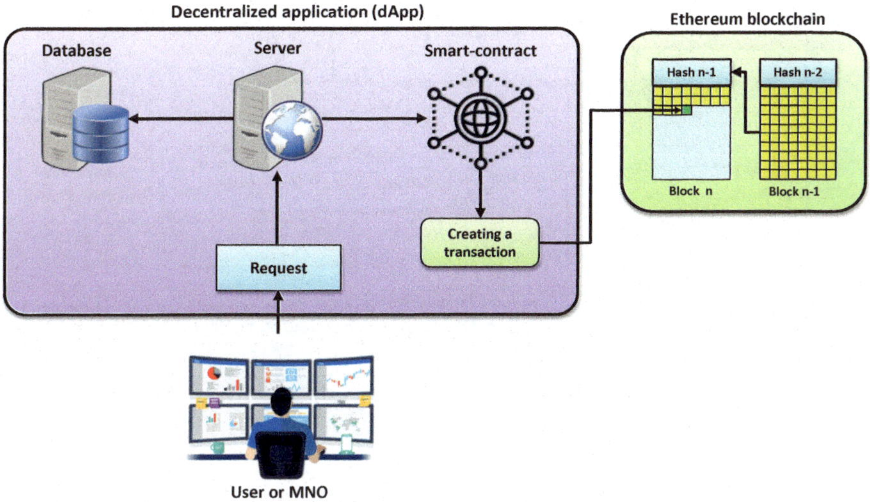

Fig. 3. The proposed structure of dApp for the decentralized 6G framework

The proposed system consists of the main core, which is responsible for the management of the AI models and interaction with shared network infrastructure and monitoring system by standardized protocols such as HTTP (HyperText Transfer Protocol), CoAP (Constrained Application Protocol) and MQTT (Message Queue Telemetry Transport). The system is integrated with the blockchain infrastructure in order to provide the secure solution for the data sharing among operators and joint decentralized network management through the dApp and the distributed ledger [18, 19]. Thus, the intelligent coordination of all planes allows to provide the target flexibility of network configuration for any given intent of the MNOs and users. Thus, we provide the new framework for the 6G development that integrates the means of artificial intelligence and the blockchain technology to leverage the advantages of both centralized and decentralized business models in the mobile network, while eliminating their corresponding drawbacks and constraints. Such a solution allows to disrupt the mobile network market by enabling a trustable spectrum and infrastructure sharing among operators, underpinned by economic and legislative mechanisms [10–13].

To further enhance the efficiency of spectrum sharing among MNOs considering their time-varying demands, in the next section we propose the intelligent spectrum management workflow based on the recurrent neural networks, which allows to predict a traffic demand of the particular network service and allocate enough spectrum in advance.

3 Deep Learning-Based Intelligent Multi-operator Spectrum Management in 6G

The typical workflow of the intelligent spectrum and infrastructure management usually defines a set of target criteria, which are used as metrics of the network efficiency.

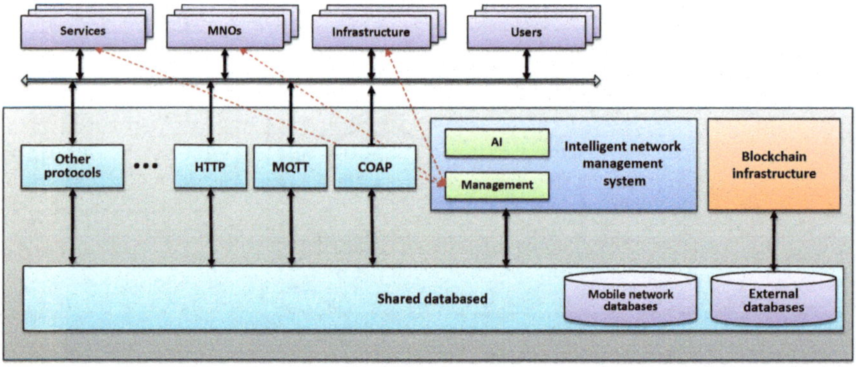

Fig. 4. The proposed functional model for the blockchain and AI-enabled decentralized 6G network management

Once, these metrics are defined, we need to find all factors, which have an impact on those parameters. Such factors can be classified into two groups: exogenous and endogenous. Exogenous factors cannot be controlled by MNOs, such as user mobility, service preferences, etc. Endogenous factors can be controlled by the MNOs, such as spectrum allocation and other configurable network parameters. Considering the impact of both groups of factors, we propose the following workflow of the intelligent network management as shown in Fig. 5.

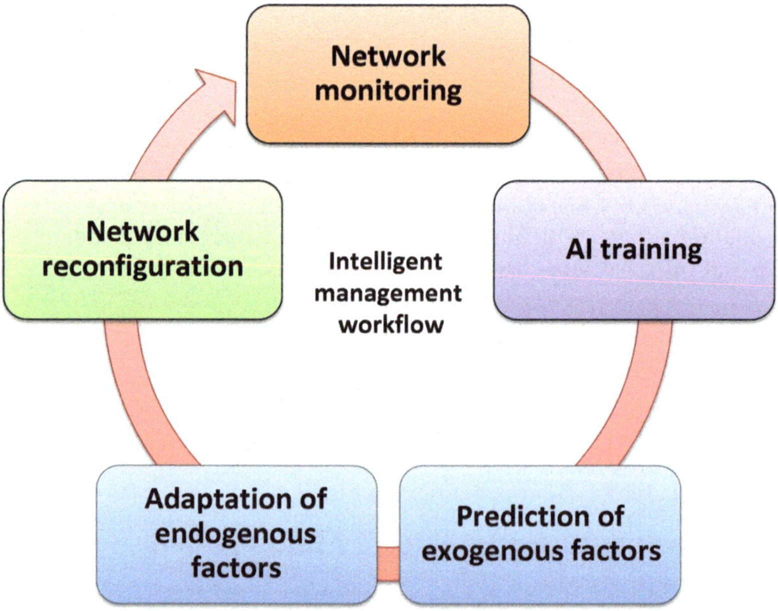

Fig. 5. The proposed workflow for the intelligent spectrum management among multiple MNOs in 6G

Initially, we have a monitoring of the network parameters, which has two important functions in our workflow. First, monitoring provides datasets for training of the deep learning models. Second, monitoring is needed to evaluate the network performance and to make a decision on the efficiency of the trained models, and determine when additional training is required. The deep learning models used in this chapter are trained explicitly for the traffic prediction of the particular base stations, i.e. for prediction of the exogenous factors caused by the behavior of the end users. Then, obtained prediction is analyzed and used to make the corresponding adjustment of the endogenous factors, i.e. spectrum allocation for each base station and MNO. Finally, the corresponding reconfigurations are made by MNOs, which can include spectrum and infrastructure sharing agreements among them. After that, the loop returns to the monitoring state to evaluate the key performance metrics of the network.

Let's consider the traffic prediction of the decentralized mobile network in details. Typically, the mobility of end users depends on many factors such as job, family, car, public transportation and many others. Therefore, each user has unique attributes and patterns, because all of them have typical favorite places and some specific service types. The important aspect here is that such patterns are self-similar and has clear periodical characteristics, such as job schedule, etc. Therefore, such periodical patterns can be easily projected to a large group of people, because a group will be a superposition of individual user patterns, so that periodic structure will be maintained.

The unique feature of the proposed intelligent workflow is in its suitability for the decentralized 6G network with multiple MNOs and many different types of service. For simplicity of the representation we consider the total traffic prediction of each base station, taking into account the part of each MNO.

Considering the resource management on the cell level, individual features of each user are not important, and the Eq. (8) can be simplified to the following form:

$$f_{k,l}(t) = \sum_{i} \sum_{j} f_{i,j,k,l}(t). \tag{14}$$

Thus, Eq. (14) generalizes the data flow of the MNO k through the base station l regardless of users and service slices. Correspondingly, the entire data flow of the particular base station l can be calculated as:

$$f_l(t) = \sum_{k} f_{k,l}(t) \equiv \sum_{i} \sum_{j} \sum_{k} f_{i,j,k,l}(t). \tag{15}$$

Equations (14) and (15) generalize the traffic prediction and spectrum allocation for each base station. In Fig. 6, we show the different types of data flows granularity, which is supported by the proposed intelligent workflow for the 6G mobile network.

As shown in Fig. 6, depends on the needs, we can either predict the individual traffic of each user with differentiation by MNO, the total traffic of each cell with differentiation by MNO or the total traffic of the each cell combined of all MNOs. As a model we use the recurrent neural network based on LSTM (Long-Short Term Memory) architecture. The model has 3 layers consisting of 256 LSTM cells, ReLU (Rectified Linea Unit) activation function and the Dropout block. The model has been chosen according to the requirements for long and short-term traffic prediction and has been validated in our

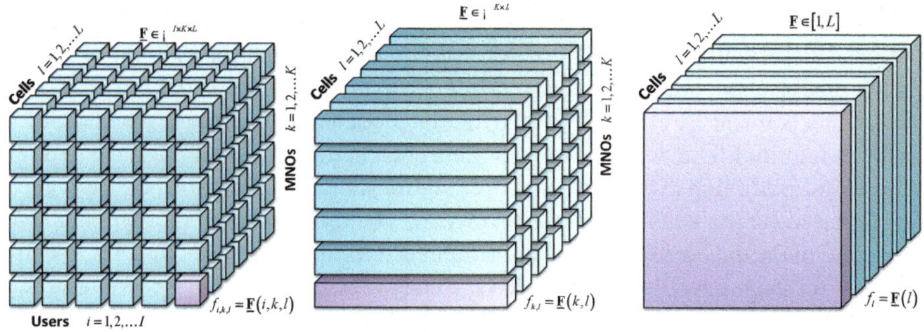

Fig. 6. The spatial representation of the data flows of multiple MNOs in a decentralized 6G mobile network

previous research works [10, 20]. In the next section, we will present the simulation results and analysis for different combinations of the data flows.

4 Simulation Results and Performance Analysis of the AI-Based Spectrum Management Workflow

The LSTM model has been trained by the custom dataset, generated in [12] for the realistic urban environment (Kosice, Slovakia) considering typical pattern of end users and real positions of base stations. The dataset consists of the 100 small cells, and 3 MNOs. Simulation results of total traffic prediction for two random cells are shown in Fig. 7.

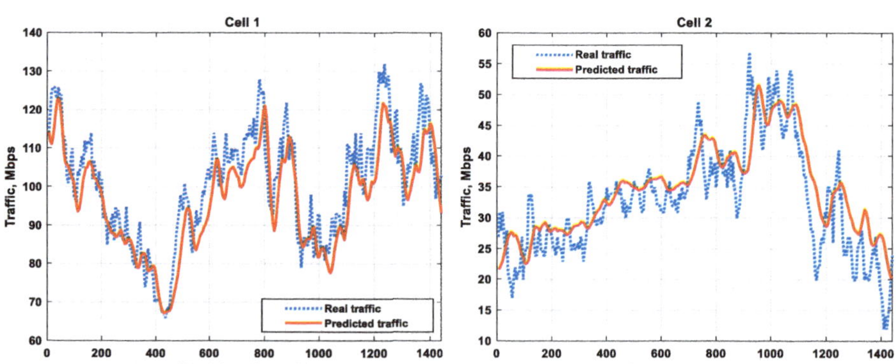

Fig. 7. Simulation results of the AI-based traffic prediction for the two arbitrary cells

The spectrum management is formalized as following. Let's define a matrix of radio resources $\underline{\mathbf{W}} \in \mathbb{R}^{K \times L}$, which contains elementary fragments:

$$w_{k,l}(t) = \underline{\mathbf{W}}(k, l)_t, \tag{16}$$

where $w_{k,l}(t)$ – is a radio resource allocated for the MNO k, at the base station l. Thus, the total radio resource is equal to:

$$W = \sum_{k} \sum_{l} w_{k,l}(t), \quad \forall t. \tag{17}$$

As a baseline let's consider a static method of uniform resource allocation among cells and MNOs. In such a case, for each MNO the bandwidth is defined independently from cells as following:

$$w_{k,l}(t) = \frac{W}{K \cdot L}, \ \forall t, \ k \in (1, K), l \in (1, L). \tag{18}$$

A more advanced method, proposed in [20] uses AI traffic prediction for each cell and static bandwidth allocation among the slices within one cell. This method requires initial traffic prediction for each cell $f_l'(t)$, and then the corresponding allocation of available radio resources for each MNO:

$$w_{k,l}(t) = \frac{f_l'(t) \cdot W}{K \sum_{b=1}^{L} f_b'(t)}, \ \forall t, \ k \in (1, K), l \in (1, L). \tag{19}$$

Unlike abovementioned methods, we propose a novel method that uses more fine-grained traffic predictions for each MNO independently. In such a case, initially we predict the traffic for each MNO within the single cell $f_{k,l}'(t)$. Thus, the equation for radio resource allocation will be modified as following:

$$w_{k,l}(t) = \frac{f_{k,l}'(t) \cdot W}{\sum_{a=1}^{K} \sum_{b=1}^{L} f_{a,b}'(t)} \approx \frac{f_{k,l}'(t) \cdot W}{\sum_{a=1}^{K} f_a'(t)} \approx \frac{f_{k,l}'(t) \cdot W}{\sum_{b=1}^{L} f_b'(t)}, \ \forall t, \ k \in (1, K), l \in (1, L). \tag{20}$$

Simulations are conducted for the different methods of resource allocation: static depicted by (18), semi-intelligent depicted by (19) and the proposed intelligent method depicted by (20). Simulation results for different cells and MNOs are shown in Figs. 8, 9. Results are shown only for 4 cells, but they reflect the overall advantage of the proposed fine-grained spectrum management in terms of the stable user experience. Note that in order to properly estimate the effect of bandwidth allocation, all users have been assigned the same channel quality indicator and the round-robin scheduling has been applied. This constraint allows to eliminate all fluctuations of the throughput caused by variable channel conditions and to focus solely on bandwidth allocation for each MNO. The important aspect, which should be mentioned, is that the total available bandwidth for each MNO is equally allocated for all active users of the particular MNO. Therefore, for static and semi-intelligent resource allocation we observe the cases, when few users of one MNO can get very high throughput, while many users of the other MNO within the same cell or in the neighbor cell may experience significant service degradation.

Considering that the percentage of users for each MNO within the cell is changing dynamically, the static and semi-intelligent bandwidth allocation experience severe

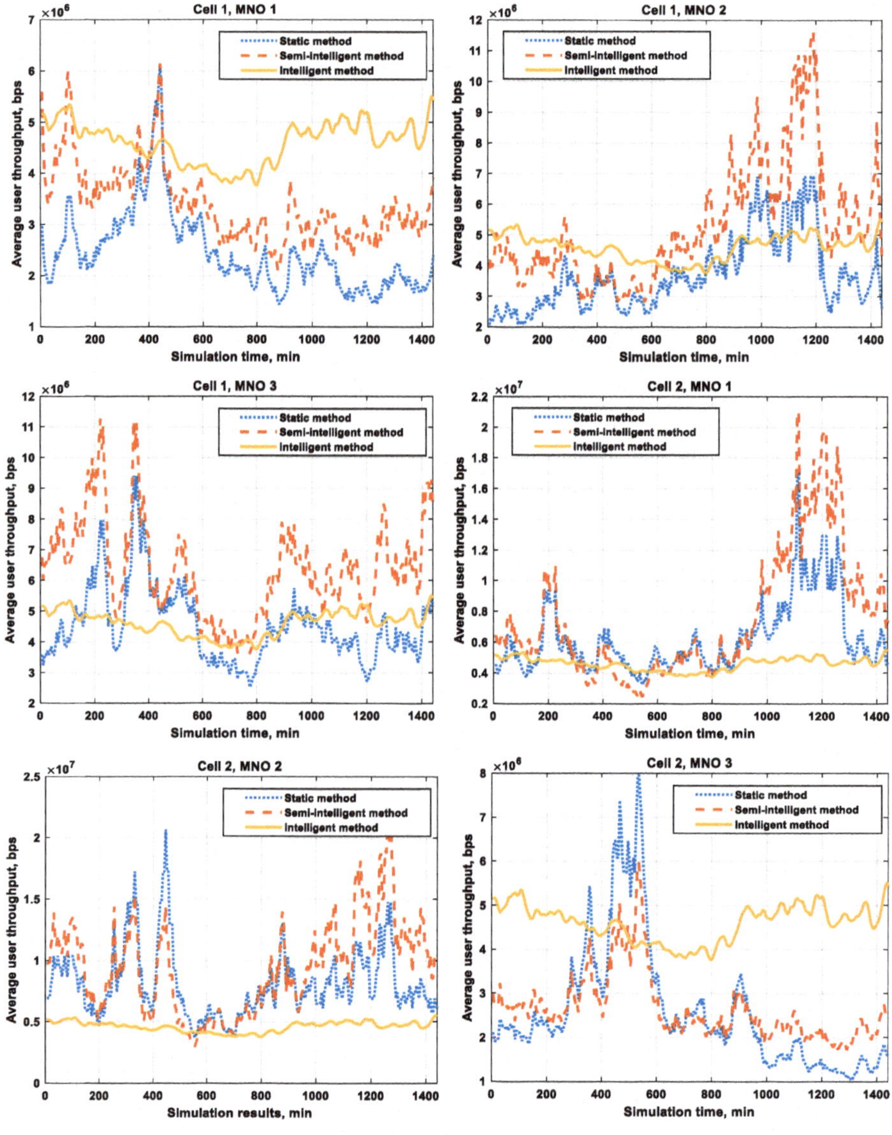

Fig. 8. Simulation results for the average user throughput of 3 MNOs for the cells 1–2

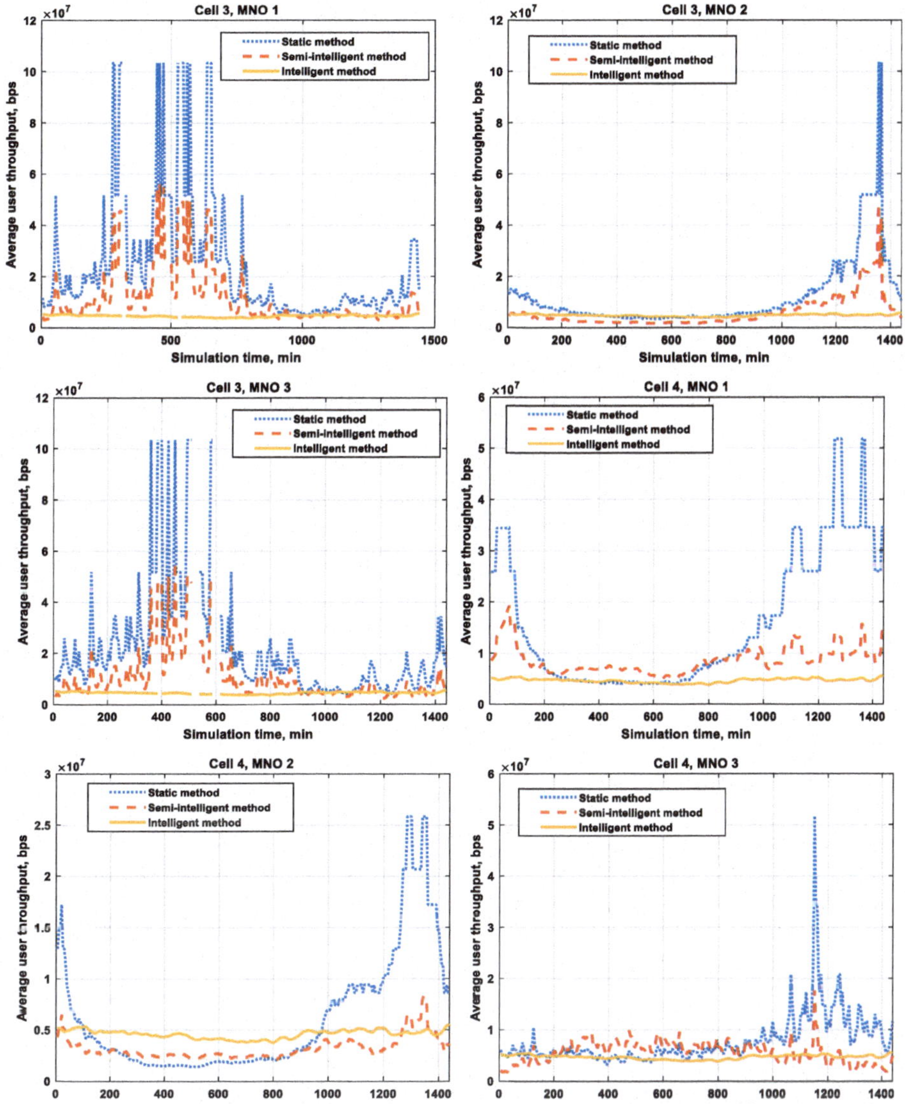

Fig. 9. Simulation results for the average user throughput of 3 MNOs for the cells 3–4

fluctuations of the service quality depends on the number of users. Thus, comparing the results for all MNOs in different cells, we observe that by using the proposed intelligent spectrum allocation among different cells, the data rate fluctuations are mitigated and all users experiencing almost the same throughput, which is close to the optimal resource allocation for the chosen simulation scenario. Hence, the proposed approach provides the fair bandwidth allocation in quasi-real time regardless of the users' mobility between different cells and MNOs.

Thus, the proposed intelligent spectrum management workflow allows MNOs in a decentralized 6G deployment to schedule their corresponding intents by using smart contracts, in order to pre-allocate the needed base stations and spectrum, considering the predicted user demand in each area and the expected number of serviced users. According to the previous findings in [10], such intelligent decentralized resource management allows MNOs to increase their profits by 19%, while maintaining the same service price for the end users.

Considering the potential of the proposed framework a further research is needed on the different types of blockchain infrastructure, which will be more suitable for the given scenario of multiple MNOs in 6G. Moreover, the number of possible AI implementation in many aspects of the 6G decentralized network management and resource allocation is uncountable, considering different service requirements and deployment scenarios. Therefore, an implementation of different AI algorithms within the proposed framework will be always welcomed to improve the performance and provide a sustainable development of the future 6G networks.

5 Conclusion

In this chapter, we have proposed the novel concept of the decentralized 6G mobile network operation in condition of open spectrum and infrastructure market by using the combination of the AI and blockchain technologies. We have proposed a high-level architecture for the interactions between MNOs in terms of spectrum and infrastructure sharing. In addition, we have developed a new mathematical model for the data flows management in the decentralized 6G scenario. Furthermore, we have developed the AI-based workflow for network resource management, which achieves better network performance for multiple MNOs in the highly variable network conditions. Simulation results show that the proposed AI workflow provides fair bandwidth allocation for all users regardless of the serving MNO by learning the short and long term patterns of the traffic demand. In this chapter, we have omitted the detailed explanation of the economic aspects and underlying AI and distributed ledger solutions, which have been described in the previous research. In our future research, a particular attention will be given to the new cutting-edge AI and distributed ledger solutions, in order to estimate the performance of the multi-operator 6G network for various scenarios with open spectrum and infrastructure market.

Acknowledgement. This research was supported by the Ukrainian government project №0120U100674 "Designing the novel decentralized mobile network based on blockchain architecture and artificial intelligence for 5G/6G development in Ukraine", by the Slovak Research and

Development Agency, project number APVV-18-0214, APVV-18-0368, by the Scientific Grant Agency of the Ministry of Education, science, research and sport of the Slovakia under the contract: 1/0268/19, by the National Natural Science Foundation of China project (No. 61962036) and by the Academy of Finland project "6Genesis Flagship" (Grant No. 318927).

References

1. Zhang, Z., et al.: 6G wireless networks: vision, requirements, architecture, and key technologies. IEEE Veh. Technol. Mag. **14**(3), 28–41 (2019)
2. Ahokangas, P., et al.: Business models for local 5G micro operators. IEEE Trans. Cogn. Commun. Netw. **5**(3), 730–740 (2019). https://doi.org/10.1109/TCCN.2019.2902547
3. Zhang, L., Liang, Y., Niyato, D.: 6G visions: mobile ultra-broadband, super Internet-of-Things, and artificial intelligence. China Commun. **16**(8), 1–14 (2019). https://doi.org/10.23919/JCC.2019.08.001
4. Chen, S., Hu, J., Shi, Y., Zhao, L., Li, W.: A vision of C-V2X: technologies, field testing, and challenges with Chinese development. IEEE Internet Things J. **7**(5), 3872–3881 (2020). https://doi.org/10.1109/JIOT.2020.2974823
5. Matinmikko-Blue, M., Yrjölä, S., Ahokangas, P.: Spectrum management in the 6G era: the role of regulation and spectrum sharing. In: 2020 2nd 6G Wireless Summit (6G SUMMIT), pp. 1–5 (2020). https://doi.org/10.1109/6GSUMMIT49458.2020.9083851
6. Bhat, J.R., Alqahtani, S.A.: 6G ecosystem: current status and future perspective. IEEE Access **9**, 43134–43167 (2021). https://doi.org/10.1109/ACCESS.2021.3054833
7. Chen, S., Sun, S., Kang, S.: System integration of terrestrial mobile communication and satellite communication —the trends, challenges and key technologies in B5G and 6G. China Commun. **17**(12), 156–171 (2020)
8. Dao, N.-N., et al.: Survey on aerial radio access networks: toward a comprehensive 6G access infrastructure. IEEE Commun. Surv. Tutor. **23**(2), 1193–1225 (2021). https://doi.org/10.1109/COMST.2021.3059644
9. Chen, S., Sun, S., Xu, G., Su, X., Cai, Y.: Beam-space multiplexing: practice, theory, and trends, from 4G TD-LTE, 5G, to 6G and beyond. IEEE Wirel. Commun. **27**(2), 162–172 (2020). https://doi.org/10.1109/MWC.001.1900307
10. Maksymyuk, T., et al.: Blockchain-empowered framework for decentralized network management in 6G. IEEE Commun. Mag. **58**(9), 86–92 (2020)
11. Khan, M., Jamali, M., Maksymyuk, T., Gazda, J.: A blockchain token-based trading model for secondary spectrum markets in future generation mobile networks. Wireless Commun. Mob. Comput., 1–12 (2020). https://doi.org/10.1155/2020/7975393. Article no. 7975393
12. Bugár, G., et al.: Techno-economic framework for dynamic operator selection in a multi-tier heterogeneous network. Ad Hoc Netw. **97**, 102007 (2020)
13. Maksymyuk, T., Gazda, J., Han, L., Jo, M.: Blockchain-based intelligent network management for 5G and beyond. In: IEEE International Conference on Advanced Information and Communications Technologies (AICT), Lviv, Ukraine, pp. 36–39 (2019)
14. Dinh, T.T.A., Liu, R., Zhang, M., Chen, G., Ooi, B.C., Wang, J.: Untangling blockchain: a data processing view of blockchain systems. IEEE Trans. Knowl. Data Eng. **30**(7), 1366–1385 (2018). https://doi.org/10.1109/TKDE.2017.2781227
15. Xiao, Y., Zhang, N., Lou, W., Hou, Y.T.: A survey of distributed consensus protocols for blockchain networks. IEEE Commun. Surv. Tutor. **22**(2), 1432–1465 (2020). https://doi.org/10.1109/COMST.2020.2969706

16. Hewa, T., Gür, G., Kalla, A., Ylianttila, M., Bracken, A., Liyanage, M.: The role of blockchain in 6G: challenges, opportunities and research directions. In: 2020 2nd 6G Wireless Summit (6G SUMMIT), Levi, Finland, pp. 1–5 (2020). https://doi.org/10.1109/6GSUMMIT49458.2020.9083784
17. Zhou, Z., Chen, X., Zhang, Y., Mumtaz, S.: Blockchain-empowered secure spectrum sharing for 5G heterogeneous networks. IEEE Netw. **34**(1), 24–31 (2020)
18. Li, W., Su, Z., Li, R., Zhang, K., Wang, Y.: Blockchain-based data security for artificial intelligence applications in 6G networks. IEEE Netw. **34**(6), 31–37 (2020). https://doi.org/10.1109/MNET.021.1900629
19. Hu, S., Liang, Y.-C., Xiong, Z., Niyato, D.: Blockchain and artificial intelligence for dynamic resource sharing in 6G and beyond. IEEE Wireless Commun. (2021). https://doi.org/10.1109/MWC.001.2000409
20. Maksymyuk, T., Han, L., Larionov, S., Shubyn, B., Luntovskyy, A., Klymash, M.: Intelligent spectrum management in 5G mobile networks based on recurrent neural networks. In: 15th IEEE International Conference The Experience of Designing and Application of CAD Systems (IEEE CADSM 2019), Polyana, Ukraine, February 2019 (2019)

Estimation of Energy Efficiency and Quality of Service in Cloud Realizations of Parallel Computing Algorithms for IBN

Igor Melnyk[1(✉)] and Andriy Luntovskyy[2]

[1] National Technical University of Ukraine "Igor Sikorsky Kiev Polytechnic Institute", Peremogy Avenue 37, Kyiv 03056, Ukraine
imelnik@phbme.kpi.ua

[2] BA Dresden University of Cooperative Education, Saxon Academy of Studies, Hans-Grundig-Street 25, 01307 Dresden, Germany
Andriy.Luntovskyy@ba-dresden.de

Abstract. The most important problems of parallel algorithm realization in the cloud computing environments are as follows: defining parallelization aspects, minimizing of energy supplying under taking into account the necessity of voluminous data transfer and supplying of powerful servers, as well as choosing the suitable methods of error-correction coding for minimizing the possible transmission errors, which can take place in a noised unsecure channel. The main approach for efficiency increasing of algorithms' parallelization, based on use of arithmetic-logic relations and recurrent matrix theory, a method of calculation of computer cooling systems, based on solving of thermodynamic balance Boltzmann equation, as well as comparative analysis of RS-codes and convolutional error-correction codes, are offered and examined in the given work. The computing examples for illustration of the discussed methods are also given. Therefore, a combined complex approach for Intent-Based Networking (IBN) is defined and proven within the chapter. The Quality of Service (QoS) parameters, such as better performance, security, data rates, and latencies are fully guaranteed herewith due to realization of the parallel computing in the cloud environments.

Keywords: IBN · QoS · Error correction · RS-coding · Convolutional coding · Algorithm parallelization · Energy efficiency

1 Introduction and Motivation

So-called Intent-Based Networking (IBN) extends the well-known Software-Defined Networking (SDN) concept under considering of "regularly filled" Quality of Service (QoS), including speed data rates, big quota of data volumes, small latencies. These virtualized networks promise a benefit for existing PHY networks like ATM, DSL, MPLS, as well as evolved 5G, LTE, Wi-Fi 6, devices for Internet of Things (IoT), LoRA, 6LoWPAN, ZigBee, EnOcean and further fog and cloud computing systems.

The main problems of parallel algorithm realization in the cloud environment are as follows (refer Fig. 1):

M. Klymash et al. (Eds.): *Future Intent-Based Networking*, LNEE 831, pp. 339–379, 2022.
https://doi.org/10.1007/978-3-030-92435-5_20

1) definition of most efficient parallelization aspects
2) minimization of energy supplying and waste heat under taking into account the necessity of voluminous data transfer and supplying of powerful servers
3) choice of a suitable method for error-correction coding for minimizing the possible transmission errors, which can take place during in the partially noised (unsecure) channels.

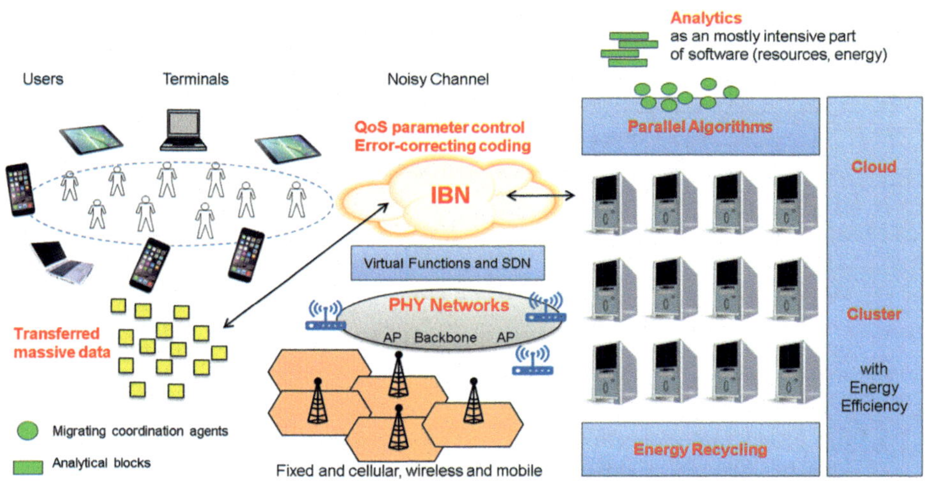

Fig. 1. IBN: state-of-the-art

Furthermore, so-called "Green IT" became the imperative of nowadays because of follows important reasons:

- IBN resources are mostly limited;
- Renewable energy is still expensive;
- Electricity costs rise daily.

Unfortunately, modern IT under use of IBN is called by the experts in general as a major cost factor. The computing hosts and IoT devices use less energy, but, generally, we have more and more devices today [1–4]. Energy inefficiency cause to the effects of higher greenhouse gas emissions. The above-mentioned important economical points of view have to be considered:

- Efficient use of the basic hardware in IBN;
- Energy efficient hardware construction;
- Energy efficient air and water-cooling;
- Recycling of waste heat.

The environmental as well as social aspects also must be taken into account. Furthermore, there are more and more ideas in the politics about the reduction of electronic waste, of CO_2 emission and for development of climate programs, based on IT environment-friendly disposal. The matured Internet community needs a raised awareness by use of IT dedicated resources.

Essential QoS parameter growth can be obtained via the parallel-operated threads and Virtual Machines (VMs). Better access to highly distributed analytical blocks and used data shorten the access time and latencies. Gained energy efficiency is proven by two parameters like PUE (Power Usage Effectiveness) and ERE (Energy Recycling Efficiency). The mentioned energy efficiency for clouds and clusters is mainly based on the following solutions [24–27]:

- waste heat removal and/or recycling (PUE and ERE values are optimized),
- optimal placement of the analytical blocks, threads, VM or mobile agents (within the clouds).

IBN obtain more QoS control and energy efficiency nowadays. The main scientific approach for efficiency increasing of algorithms' parallelization, based on use of arithmetic-logic relations and recurrent matrix theory, a method of calculation of computer cooling systems, based on solving of thermodynamic balance Boltzmann equation, as well as comparative analysis of RS-codes and convolutional error-correction codes, are presented and considered in the chapter (refer Fig. 1).

Therefore, a combined approach for Intent-Based Networking is defined and proven within the given chapter. Herewith, the Quality of Service (QoS) parameters, i.e. better performance, security, data rates, and latencies are completely guaranteed due to realization of the parallel computing in the cloud environments. Furthermore, the above-mentioned solutions are also energy-efficient.

Taken the given aspects in considering, three tasks are successfully examined and solved below: 1 – QoS by parallelization, 2 – energy-efficiency, 3 – suitable fault-tolerant and error-correction coding (Tasks 1, 2, and 3). The multilayered complex approach is here used. A diagram, which illustrated the correspondences between the above formulated Tasks 1, 2, and 3 as well as the OSI and IBN layers, is depicted in Fig. 2. The layers 2–4 are virtualized via SDN and NFV; they address Task 2. The layers 4–7 (transport and APL) represent clouds and clusters, as well as parallelized applications and mobile apps (Task 1). Task 3 corresponds mostly to layer 2 (DLL and channel coding) as well as partially to the used codecs (layer 5). Physical layer address Task 2 too. The formulated beyond Tasks 1, 2 and 3 will be considered in the next sections of the given chapter in detail. The represented chapter is organized furthermore as follows.

- In Sect. 2 – QoS control by parallelization.
- In Sect. 3 – Improvements in energy efficiency.
- In Sect. 4 – Suitable fault-tolerant and error-correction coding.
- Conclusions will crown the given chapter.

The computing examples and case studies are also provided here, which are nearly illustrating the offered approaches, methods and models [6].

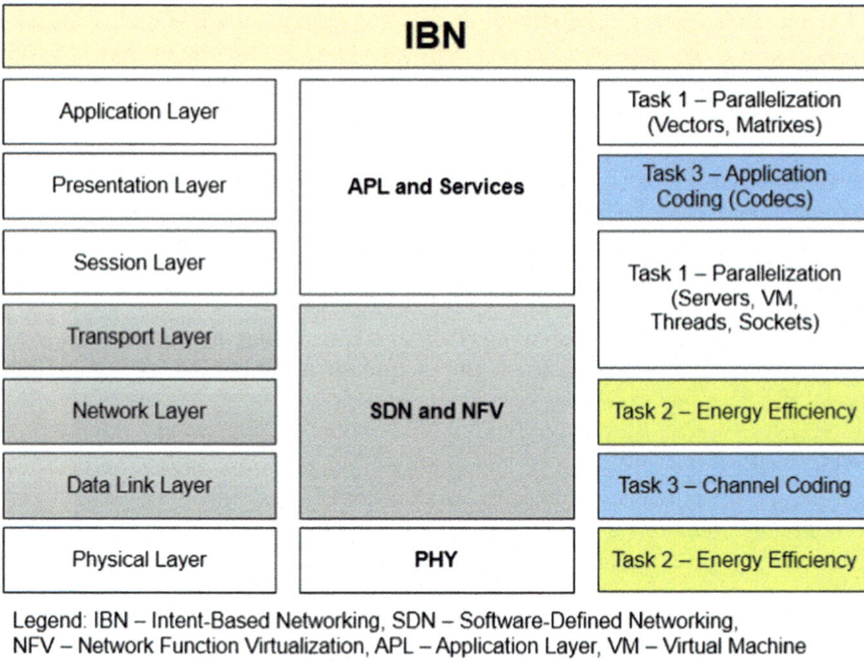

Fig. 2. Correspondence diagram for OSI and IBN layers and the formulated Tasks 1–3

2 QoS Control for IBN and Algorithm Parallelization Under Use of Matrix Approach

QoS control for IBN is provided here mainly under use of Parallel Algorithms for analytical blocks, which can be efficient placed to the clouds, to the fog (robots, sensor networks and IoT).

Parallel Computing became nowadays actual due to possibility to solve certain complex and important scientific and engineering problems. Among these problems are as follows [1–4, 24–27].

1. Genetics and pharmacology, for example, computing of protein folding. This problem is very actual today doing to find the basic structures of DNA, RNA, viruses, tumors and effective chemotherapeutical stuffs and drugs to combat and destroy it.
2. Theoretical investigation of complex spatial 3D structures, such as modelling of genome, microdevices, as well as complex mechanical instruments, aero-spatial wings of liners etc.
3. Simulation of the complex physical and technological processes in devices and installations, including the real-time calculation.

Providing the numerical calculation of biological and technical systems in global network with using the parallel algorithms allow to elaborators obtain the pervious estimation the system behavior in different conditions, therefore the time for elaboration the novel systems, equipment and technologies is significantly reduced. Today, taking into account the enormous increasing the amount of local computer devices with the medium and small computing power, the cloud and fog information exchange technologies are applied successfully for parallelization the complex calculation tasks [2–4].

However, it should be pointed out, that using the remote items in global network for computing of complex scientific and technical tasks with applying parallel algorithms is strongly connected with necessity of estimation the general factor of technic and economic efficiency of such parallelization [1–4].

The presented Sect. 2 is devoted to the parallelization approaches (matrix and vector based) as well as to the estimation models for parallelization level (among them Amdahl's, Gustafson-Barsis, Karp-Flatt, Sun-Ni etc.). For efficient implementation of parallel algorithms the standardized networking techniques like sockets, multithreading, Web Services, Micro-Services can be used as well as widespread program languages as follows: C, C++, Java or Python [1–4, 24–27].

2.1 Arithmetic-Logic Relations and Recurrent Matrixes

The method of forming the matrix approach for estimation the possibility of parallelization of numerical algorithms is based on the definition of arithmetic-logic equation. In the papers and manual books on programming arithmetic-logic relation is defined as sum of production algebraic and logic relations, which is generally written as follows [1, 7–9]:

$$\mathrm{AL}(x) = F_1(x) \cdot L_1(x) + F_2(x) \cdot L_2(x) + F_3(x) \cdot L_3(x) + \ldots + F_\mathrm{n}(x) \cdot L_n(x), \quad (1)$$

where arithmetical functions are $F_1(x) \ldots F_\mathrm{n}(x)$ and logical functions are $L_1(x) \ldots L_n(x)$, $\mathrm{AL}(x)$ is the arithmetic-logic relation.

The main requirement to logic equations $L_1(x) \ldots L_n(x)$ are follows [1, 7, 8].

1. All numerical intervals, defined by relations $L_1(x) \ldots L_n(x)$, must not crossed.
2. The region of defining the argument x of arithmetic-logic Eq. (1) must complexly and unambiguously described by the set of relations $L_1(x), \ldots, L_n(x)$.

Basic illustration of the typical location of numerical intervals, described by relations $L_1(x), \ldots, L_\mathrm{n}(x)$ on the number axis, is presented on Fig. 3.

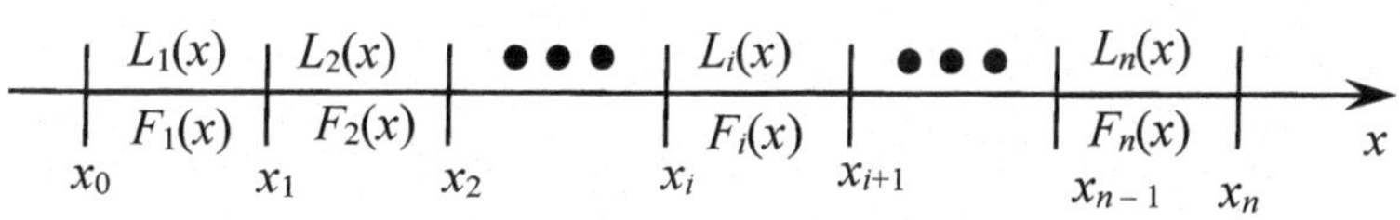

Fig. 3. Illustration the basic method of forming arithmetic-logic Eq. (1)

The examples of arithmetic-logic equations are follows [7, 8]:

$$0 \cdot (x < 0) + 0 \cdot (x \geq 0); \exp(-x) \cdot x < 0 + \exp(x) \cdot (x \geq 0);$$

$$\sin(x) \cdot (x \le 0) + \cos(x) \cdot (x > 0);\ \sin(x) \cdot (x \le 0) - \sin(x) \cdot (x > 0).$$

Considering, for example, method of forming the relation

$$\mathrm{AL}(x) = \exp(-x) \cdot (x < 0) + \exp(x) \cdot (x \ge 0).$$

It simple and generally means, that in the case $(x < 0)$ the result is $\mathrm{AL}(x) = \exp(-x)$, and in the case $(x \ge 0)$ it is $\mathrm{AL}(x) = \exp(x)$. Or, in the simple form of piecewise interval function [8]:

$$\mathrm{AL}(x) = \begin{cases} \exp(x), & \mathrm{if}(x \ge 0); \\ \exp(-x), & \mathrm{if}(x < 0). \end{cases}$$

On the base of arithmetic-logic relation definition formed the new definition of recurrent arithmetic-logic relation, where the logic relations formed in discrete from for natural numbers n. In this case relation (1) is rewritten as follows [7]:

$$\mathrm{AL}(n) = (\sum\nolimits_{i=1}^{n_r} \mathrm{AL}(i)) \cdot (i \le n_r) + (\sum\nolimits_{i=n_r+1}^{n} \mathrm{F_j}(\mathrm{AL(j)})\big|_{\mathrm{j}=i-n_r}^{i}) \cdot (i > n_r), \quad (2)$$

where n_r is the number of recursion elements, $F_j(\mathrm{AL}(j))$ are functions for calculation the elements of recurrent sequence with number j.

For example, for well-known Fibonacci row, recurrent arithmetic-logic relation is written as follows [7]:

$$\mathrm{AL}(n) = 1 \cdot (n = 1) + 1 \cdot (n = 2) + (\mathrm{AL}(n-1) + \mathrm{AL}(n-2)) \cdot (n > 2).$$

On the given beyond definition of recurrent arithmetic-logic relation the concept of recurrent matrix is formed. Usually, the recurrent matrix is defined as the set of following matrix equations [1, 7]:

$$\mathbf{M}_{<1>} = \boldsymbol{v}_1, \mathbf{M}_{<2>} = \boldsymbol{v}_2, \ldots, \mathbf{M}_{<n>} = \boldsymbol{v}_n;$$

$$\mathbf{M}_{<i>} = \boldsymbol{F}(\mathrm{i}, \mathbf{M}_{<\mathrm{i}-1>}, \mathbf{M}_{<\mathrm{i}-2>}, \ldots, \mathbf{M}_{<\mathrm{i}-n>}), \quad (3)$$

where v are vectors, $\mathrm{M}_{<i>}$ is the row of matrix with number i, F is the vector-function, which defined the function for calculation the matrix elements [1, 7]. Usually, F is defined as multidimension arithmetic-logic function with the following components F_1, F_2, …, F_n, or in the form of mathematic relations [1, 7]:

$$\mathbf{F} = \{F_1, F_2, \ldots, F_n\};\ F_1 = \mathrm{AL}(1),\ F_2 = \mathrm{AL}(2),\ F_n = \mathrm{AL}(n). \quad (4)$$

Described method of forming recurrent matrix is brightly illustrated at Fig. 4.

It is clear from the Fig. 4, that the main conception of forming the recurrent matrix is using of same set of basic functions F_1, F_2, …, F_n, for all rows of matrix.

Considering the example of applying the arithmetic-logic equation to realizing the numerical algorithm.

Example 1. Well-known, that the iterative numerical equation for calculation the potential of electric field U is based on finite-difference method and generally can be written in the following form [9]:

$$U^n(i,k) = \omega[C_a U^{n-1}(i+1,k) + C_b U^{n-1}(i,k+1) + C_c U^n(i-1,k) + C_d U^n(i,k-1) + C_\rho \frac{\rho^{n-1}(i,k)}{\varepsilon_0 \rho^n(i,k)}] + (1-\omega)U^{n-1}(i,k). \quad (5)$$

where n is the number of current iteration, i is the number of calculated item at longitudinal coordinate z, k is the number of calculated item at transversal coordinate r, ω is the relaxation parameter by the potential, which is usually used for increasing the rate of convergence of iteration process. For providing the computing for real axially-symmetric electrodes systems, the Eq. (5) can be rewritten in the form of arithmetic-logic Eq. (1) [9]:

$$m = (l > 0)\cdot(l-1) + (l = 0)\cdot 1,\ C_m = 1 + \frac{1}{2m},\ D_m = 1 - \frac{1}{2m},$$

$$U_{k,l} = \left((l > 0)\cdot \frac{\frac{U_{k-1,l}+U_{k+1,l}}{h_r^2} + \frac{D_m U_{k,m} + C_m U_{k,l+1}}{h_z^2}}{\frac{2}{h_r^2} + \frac{2}{h_z^2}} + (l = 0)\cdot \frac{\frac{U_{k-1,l}+U_{k+1,l}}{h_r^2} + \frac{4U_{k,l+1}}{h_z^2}}{\frac{2}{h_r^2} + \frac{4}{h_z^2}} \right)\cdot \left(U_p < U_{k,l} < U_{ac}\right)$$
$$+ \left(U = U_p\right)\cdot U_p + \left(U < U_p\right)\cdot\left(\frac{kh_r\left(U_p - U_a\right)}{r_p - r_a}\right) + (U \geq U_{ac})\cdot U_{ac},$$

where l is the number of the base point by coordinates z, $U_{k,l}$ is the calculated value of electric potential, U_{ac} is the acceleration voltage, U_p is the potential of near-anode electrode, U_a is the anode potential r_p is the position of near-anode electrode relatively to the cathode, r_a is the anode position relatively to the cathode, m, C_m and D_m are additional variables. In the work [9] was also proved, that arithmetic-logic Eq. (6) is iterative and can't be divided into independent thread for parallelization.

The basic relation for estimation the possibility of algorithm parallelization, obtained by analyzing the structure of recurrent matrix, will be considered in the next section of this chapter.

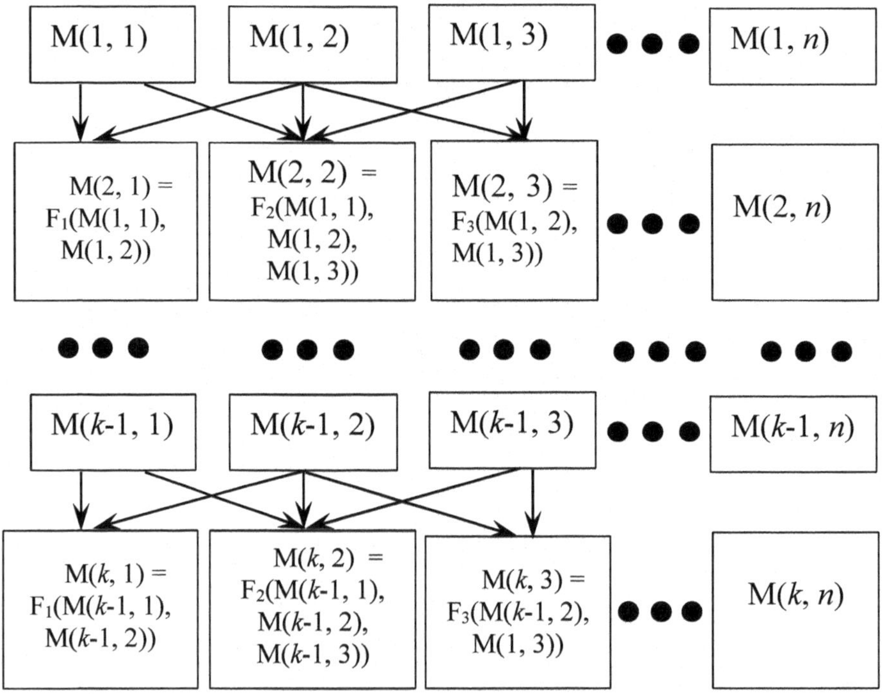

Fig. 4. Illustration the basic method of forming the recurrent matrix corresponding to relations (3, 4)

2.2 Estimation of the Level of Algorithm Parallelization by Analyzing the Structure of Recurrent Matrixes

Usually the computation time gain (speedup factor) is possible only due to higher parallelism of a math-log problem. The time estimations are as follows [1]:

$$T = s; T = s + p; T = s + \frac{p}{n}; T = s + \frac{p}{n} + k \cdot n, \tag{6}$$

where p is the potentially paralleled part of a task, T is the overall computing time, s is the sequential part of a task. The appropriate estimation of speedup factor can be done via Amdahl's law (1967), as well as Karp-Flatt metric (1990) [1]:

$$T = 1; A_n = \frac{1}{(1-p) + \frac{p}{n}} \leq \frac{1}{1-p}; A_{\max} = \frac{1}{1-p};$$

$$A_{nk} = \frac{1}{(1-p) + \frac{p}{n} + kn}, k \to 0; 1 - p = \frac{\frac{1}{A_n} - \frac{1}{n}}{1 - \frac{1}{n}}, \tag{7}$$

where A_n or A_{nk} is the speedup factor, or acceleration on n Central Processor Units (CPU) thread regarding to single one, $A_{\max}$ is the maximal speedup, k is the negative influence of communication by message passing between CPU threads.

Another approach for analyzing the speedup factor for the time of calculation of considered algorithm by it dividing into separate independent flows is based on analyzing the structure of correspondent recurrent matrix and was proposed in the paper [1]. There was pointed out, that with including the sequential connections between the elements of matrix in the row the parallelization of such algorithm is impossible. Corresponded structure of matrix is presented in Fig. 5, where the hierarchical and sequential connections between the matrix elements are marked separately.

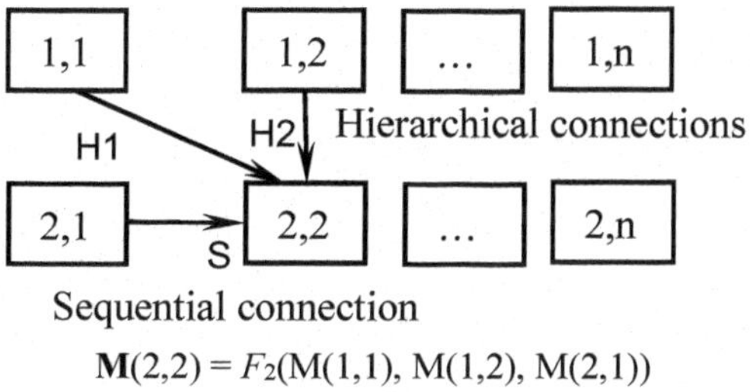

Fig. 5. Illustration of the hierarchical and sequential connections at the recurrent matrix

But for calculation schemes with only hierarchical connections, like presented in the Fig. 4, data threads T_1, T_2 ..., T_n can be considered separately. For example, thread T_2 included connections with three matrix elements of pervious row. The main idea of proposed estimation is: if we have for the thread T_n with number of connections between matrix elements N, in the simple case, when we supposed, that the time for elementary operations is the constant value, speedup factor p estimated by the following relation [1]:

$$p_1 = 1 - \frac{\max\limits_{i=1\ldots n} N(T_i)}{\sum_{i=1}^{n} N(T_i)}. \tag{8}$$

Considering the example of using the relation (8) for the recurrent matrix, which structure is presented fragmentary at Fig. 4.

Example 2. Assume, that the recurrent matrix is limited by the 3 elements in row. In such conditions all necessary connections between the elements are defined by the diagram, presented at Fig. 4. From the Fig. 4 clear also, that $N(T_1) = 2$, $N(T_2) = 3$, $N(T_3) = 2$. Corresponded threads T_1, T_2 and T_3 are shown at Fig. 6.

Using the relation (8) and providing the necessary calculations, we obtaining the follows value for speedup factor p:

$$p_1 = 1 - \frac{3}{2+3+2} = 1 - \frac{3}{7} = \frac{4}{7} = 0.5714.$$

The obtained estimation result suggests, that with using 3 CPU for 3 separated threads the time for solving corresponded task is nearly 2 times less, then for 1 CPU.

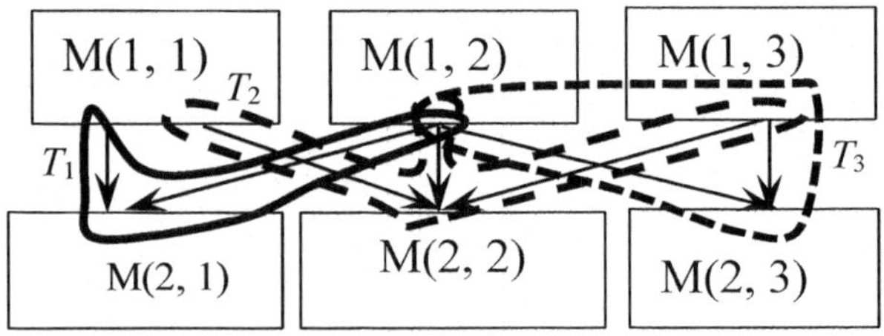

Fig. 6. The threads for parallelization the algorithm in the structure of recurrent matrix, presented in Fig. 4

It should be pointed out, that Eq. (8) can be modified for the threads with different complicity, and, corresponding, different time of treatment by CPU. In such conditions the complicity of elementary connections between elements F_j have to be analyzed and the factor of complicity of each elementary operation F_j is defined relatively to the simplest operation by factor α_j. For such condition Eq. (8) rewritten as:

$$p_2 = 1 - \frac{\max\limits_{i=1...n} \sum_{j=1}^{k_i} \alpha_j F_j}{\sum_{i=1}^{n} \sum_{j=1}^{k_i} \alpha_j F_j}. \tag{9}$$

Considering the example of using the relation (9).

Example 3. Assume, that in the recurrent matrix, which structure is presented in Fig. 6, the elementary connections have different factors of complicity, which values are noted in Fig. 7. With using relation (9) to estimate the speedup factor p.

Using the relation (9) with taking into account the value of factor α, noted for all connections between the matrix elements in Fig. 7, and providing the necessary calculations, we obtaining the follows value for speedup factor p:

$$p_2 = 1 - \frac{1 + 3,5}{(1 + 3,5) + (1 + 1 + 1) + (1 + 1)} = 1 - \frac{4.5}{4.5 + 3 + 2} = 1 - 0.474 = 0.526.$$

Therefore, for considered case, the speedup factor p, calculated with taking into account the operation complicity α, is close to pervious value, obtained in Example 2, with using more simple relation (8).

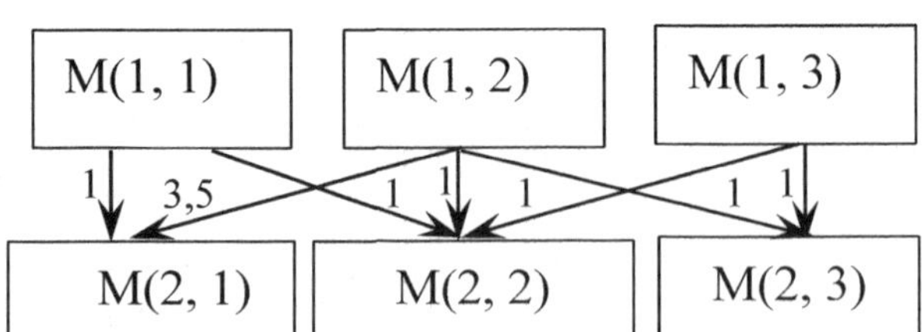

Fig. 7. Defining of complicity factor α for the connections between the elements of recurrent matrix, which structure is presented at Fig. 6

Clear, that estimation the complicity factor α of separate connections isn't practicable and usually can't be provided for complex mathematical functions. Therefore, defining the complicity factor β for the whole thread is more practicable. At such statement of task, the problem of parallelization it coming to finding the simplest threat, which complicity factor assumed as $\beta = 1$, and the factors β to another thread are defined relatively to simplest one. For example, for defining the complicity of task with knowing functional dependence between the recurrent matrix elements can be realized by counting the number of necessary elementary operations for considered CPU, which are usually estimated in TFLOPS [1]. With such statement of problem, the Eq. (8) can be rewritten throw the threads complicity factor β_i as follows:

$$p_3 = 1 - \frac{\max\limits_{i=1...n} \beta_i}{\sum_{i=1}^{n} \beta_i}. \tag{10}$$

Considering the example of using the relation (10).

Example 4. Assume, that in the recurrent matrix, which structure is presented in Fig. 6, the complicity factor β_i for the threads T_1, T_2 and T_3 was estimated as follows: $\beta_1 = 4.5$; $\beta_2 = 1$; $\beta_3 = 1$. With using relation (10) to estimate speedup factor p.

$$p_3 = 1 - \frac{4.5}{4.5 + 1 + 1} = 0.31.$$

Clear, that for this task the speedup factor is generally smaller, then for Examples 1 and 2. In such conditions the computing time for solving corresponded task at 3 CPU is only 1,44 times less, than for 1 CPU.

The standard programming means for providing the estimation with using relations (8–10) at the application layer of IBN, with taking into account such factors, as the loading of communication channels, transfer data rate, as well as the energy consumption on the servers and communication devices, will be considered in the next section of the chapter.

2.3 Efficiency of Algorithm Parallelization for IBN Under Use of Standard Programming Techniques and Functions

As it was pointed out in the Sect. 1, the estimation the efficiency of algorithm parallelization for specific formulated scientific and technical task is mostly realized on the application layer of OSI reference model (refer. Fig. 2). However, in any case, task of transferring information and finding the suitable cluster is generally corresponded to network and transfer levels. Taking into account this inconsistency, during written the programs with parallelization the calculation in IBN the specific functions for making the requests from application to transfer and network layer, have been used. Usually, the names of these functions are standard for different popular program languages, like C, C++, Java or Python, and the difference between the particularities of its using is defined only by the syntax of specific program language. For example, for program language Python these functions are follows [8, 10].

1. **socket()** – defining the socket of IBN to connection as the combination the host address and port to connection.
2. **bind()** – binding of the socket to corresponded host, located in local or global network.
3. **listen()** – the command for server to testing the network and waiting the request from any remote host.
4. **connect()** – the command for client for forming the request to connection with server.
5. **accept()** – the command for server to accept a request from client and to form a new socket for connection to it.
6. **send('Message')** – sending the message in network.
7. **recv()** – receiving the message from network.
8. **close()** – command to ending the connection.

Some examples of using functions of network programming, listed beyond, on the program language Python, are presented in manual book [10].

For realizing the effective algorithms parallelization in IBN such separate problems have to be solved [2–4].

1. Estimation the factor of parallelization p with analyzing the structure of recurrent matrix and using relation (8–10). Beyond this task was considered as Task 1.
2. Assessment of employment levels of communication channels.
3. Estimation of energy consumption on the powerful servers and communication equipment. Beyond this task was considered as Task 2.
4. Choosing the suitable protocols and methods of signals coding in noised channels. Beyond this task was considered as Task 3.

Taking into account the programming means, considered beyond, the ways of solving the listed problems are follows.

1. Forming the recurrent matrix and providing the estimation of speedup factor p with using relations (8–10).
2. Forming, with using command socket(), the list of IBN servers, in which the parallelized task can be solved. Corresponding source code on the program language Python can be written as follows [10].

Example 5.

```
import socket as soc
import time as dt
HOST1 = '234.111.20.55'
PORT1 = 5000
HOST2 = '238.124.56.1'
PORT2 = 6000
s = soc.socket(soc.AF_INET, soc.SOCK_STREAM)
s.bind((HOST1, PORT1))
```

3. Choosing the best host for solving the parallelized task by the factors of the channel's employment and energy consumption.

The energy consumption is usually independent on the channel's employment, therefore it can be defined for transferring data, depending on the amount of data treatment, as a static parameter [6]. Corresponding source code on the program language Python can be written as follows [10].

Example 6.

```
HOST1 = '234.111.20.55'
PORT1 = 5000
EC1 = 1600
HOST2 = '238.124.56.1'
PORT2 = 6000
EC2 = 1800
```

Here EC1 and EC2 are the energy consumption in W per 1 MB of transferred and treated data.

For defining the level of channel's employment can be used the method of sending of the short test to all described hosts and defining the time of reply with the same message [6]. Therefore, the source code for defining the time of receiving the testing message by the server with using the means of program language Python can be written as follows [10].

Example 7.

```
import socket as soc
import time as dt
HOST = '127.0.0.1'
PORT = 5000
s = soc.socket(soc.AF_INET, soc.SOCK_STREAM)
s.bind((HOST, PORT))
s.listen(1)
conn, addr = s.accept()
RecvDate = dt.strftime('%d:%m:%Y')
RecvTime = dt.strftime('%H:%M:%S')
RecvDT = RecvDate + ' o ' + RecvTime
TestMessg = 'Andriy Luntovskyy and Igor Melnyk. \
Estimation of Energy Efficiency and Quality of \
Service in Cloud Realizations of Parallel \
Computing Algorithms for IBN. Lector notes. \
Intend Based Network. Volume 1. 2021'
conn.send(TestMessg)
SendDate = dt.strftime('%d:%m:%Y')
SendTime = dt.strftime('%H:%M:%S')
SendDT = SendDate + ' o ' + SendTime
print('The time of receiving the test message is:', SendDT)
s.listen(1)
conn, addr = s.accept()
conn.close()
```

Since by the experiment we have defined the time t_t, necessary for transferring the testing message with amount of data V_t, and we know the amount of data V_p, should be transferring to remote host for solving the parallelized task, the estimation time t_p for transferring the data for parallelized task is simply defined as:

$$t_p = \frac{t_t V_p}{V_t}. \tag{11}$$

With knowing values of complicity factor p, defined by Eqs. (8–10), as well as the level of energy consumption E_c and the time t_p, necessary to transferring the data for parallelized task and defined by Eq. (11), the complex factor of efficiency of parallelization F_p can be estimated as follows:

$$F_p = \xi_p p + \frac{\xi_E}{E_c} + \frac{\xi_t}{t_p}, \tag{12}$$

where ξ_p, ξ_E and ξ_t are the weight coefficients for level of parallelization, energy consumption and the time of delay in communication channel correspondently. Among the few considering variants choosing one with the greater value of efficiency factor F_p.

Considering the example of using relation (12).

Example 8. Sending of testing message shown, that the time of reply from the Server 1 $t_{t1} = 10$ s, and form Server 2 – $t_{t2} = 0.1$ s. The size of testing message $V_t = 1024$ b, and the size of transferring data for solving the parallelized task is $V_p = 3$ Mb. The speedup factor of task parallelization, calculated with using Eq. (10), is $p = 0.75$. The energy consumption for Server 1 is $E_{c1} = 1300$ W/Mb, and for Server 2 – $E_{c2} = 3500$ W/Mb. The values of weight coefficients in estimated relation (12) are follows: $\xi_p = 0.95$, $\xi_E = 0.9$ and $\xi_t = 0.001$. With using relation (11), (12), define the preferable server to solving this task.

1. Defining the time for transmitting the data of parallelized tasks for both considered cases with using relation (11):

$$t_{p1} = \frac{10 \cdot 3 \cdot 2^{20}}{2^{10}} = 30 \cdot 1024 = 30720\,\text{s};$$

$$t_{p2} = \frac{0.1 \cdot 3 \cdot 2^{20}}{2^{10}} = 0.3 \cdot 1024 = 307,2\,\text{s}.$$

2. Defining the efficiency of tasks parallelization for both considered cases with using relation (12):

$$F_{p1} = 0.95 \cdot 0.75 + \frac{0.9}{1300 \cdot 3} + \frac{0.001}{30720} = 0.7125 + 2.3 \cdot 10^{-4} + 3.255 \cdot 10^{-8} = 0.7127;$$
$$F_{p2} = 0.95 \cdot 0.75 + \frac{0.9}{3500 \cdot 3} + \frac{0.001}{307.2} = 0.7125 + 8.571 \cdot 10^{-5} + 3.25 \cdot 10^{-6} = 0.7126.$$

Since $F_{p1} > F_{p2}$, conclusion is, that for solved task the first case is better.

Really, result of using estimation (12) is strongly depended on the choosing of weight coefficients ξ_p, ξ_E and ξ_t. In example 8 was considered the case, when $\xi_E >> \xi_t$, and by that reason the variant with the smaller energy consumption and greater value of calculation time is chosen. Certainly, in such conditions the QoS factor is greatly worse, than the requirement of energy economy. The estimation for weight coefficients is the separate problem and it help to solve the compromise task between improving the QoS and reduce the energy consumption.

Considering the solutions for utilizing the energy in green clouds servers will be provided in Sect. 3. It is clear also, that the volume of transferring information is strongly depend on using coding methods. Estimations of redundancy ratio for RS and convolutional error-correction codes will be considered furthermore in Sect. 4.

2.4 QoS Control for IBN and Real-Time Capability

One of the possible features is also Real-Time Capability for IBN, which can be estimated via the following expression:

$$T_r \geq (T_{tr} + T_{ex} + \Delta) \cdot (1 + a), \tag{13}$$

where T_{tr} is the transfer time, T_{ex} is the execution time, T_r is the given limit for the reaction time of the system, a is the average failure probability, Δ is the summarized process delay (latencies). By a repeated failure the given limit can even grow by the product $(1 + a + a^2)$.

The latency within modern networks can be significantly reduced down to several milliseconds: e.g. 10 ms for 4G mobile radio as well as 1 ms for the 5G, which will be deployed by years 2020–2022.

3 Energy Utilization in "Green Clouds" for IBN with Powerful Servers and Simulation of Waste Heat System

The improvements in Energy Efficiency for IBN are aimed to reach the growth of energy efficiency. There are mainly two main ways for IBN Energy Efficiency can be remarked, and they are follows.

1. Improvements in cooling processes for cloud environments.
2. Optimal workload adaptation, used data, VM and analytic blocks placement.

Under the known problem the approaches to it solving are as follows.

1. Solutions for fog computing (co-operation platforms, "green" data centers),
2. Utilization of waste heat (a "green" IT company),
3. VM optimal migration and energy efficiency of computing process.

Within this Sect. 3, we provide the innovative approaches and case studies, which can be used for betterenergy utilization in "Green Clouds" for IBN with powerful servers as well as the simulation of Waste Heat System.

3.1 PUE and ERE Criteria

A useful model task to illustrate PUE and ERE as important criteria for IBN is provided below.

Example 9. In addition to the performance parameters such as FLOPS, more and more about the energy efficiency in distributed computing is discussed, namely about the values such as PUE (Power Usage Effectiveness) and ERE (Energy Reuse Efficiency).

A powerful computer cluster obtains 1000 CPU with the integrated maximal performance of 80TFLOPS. Overall, electrical power in the cluster P_{IT} is equal 400 KW.

1. Compute the PW factor [FLOPS/W]. What does characterize this cluster parameter?
2. How much is the waste heat dissipation P_A if PUE = 1.5?
3. Due to the optimization in the cooling process and waste heat canalization in the cluster P_R became 100KW via recycling of the ambient heating and water boiling within the cluster building. Calculate, please, the ERE factor.

Solution

$$PUE = \frac{P_{IT} + P_A}{P_{IT}}; \; ERE = \frac{P_{IT} + P_A - P_R}{P_{IT}}, \tag{14}$$

$P_{host} = P_{IT}/N$, where N = 1000 hosts, therefore, it means P_{IT} = 1000 CPUs * 400 W. Each host possesses electrical power of P_{host} = 400 W.

Computing efficiency per Watt for the given cluster: PW = Performance/P_{IT} = 80T FLOPS/400 kW = 0,2 [GFLOPS/W], but it's less than by the leading computing systems from Top500 (cp. an excerpt in Table 1) like:

- Summit (USA) – 200 PFLOPS;
- Sierra (USA) – 125 PFLOPS;
- Sunway TaihuLight (China) – 125 PFLOPS;
- Tianhe-2 Milky Way (China) – 33,9 PFLOPS;
- Titan (USA) – 17,6 PFLOPS.

Furthermore, for the considered case we realize:

- Power Usage Effectiveness:

$$PUE = \frac{400\,\mathrm{KW} + P_A}{40} KW = 1.5,$$

where $P_A = 200\,\mathrm{KW}$ of waste heat.

- Energy Reuse Efficiency:

$$ERE = \frac{(400 + 200 - 100)\,\mathrm{KW}}{400} KW = 1.25,$$

where $P_A = 200\,\mathrm{KW}$ of waste heat, $P_R = 100\,\mathrm{KW}$ of recycled power.

Table 1. The performance and electrical parameters of the world-leading clusters (own representation by year 2020 based on Top 500)

Computing cluster	Location	Performance in PFLOPS	Architecture	P_{IT} in MW	PW in GFLOPS/W	Deployment
Summit	Oak Ridge National Laboratory (Tennessee, USA)	200	RAM 10PB, 4608 nodes je 2x (22-core CPU + 6 GPU)	10	20	Climate simulation, relativistic quantchemistry, bio- and plasma-physics, computational chemistry
Sierra	National Nuclear Security Administration, USA	125	1,572,480 cores	7,438	16,8	Simulation of nuclear weapons, civic craft engineering tasks simulations
Sunway TaihuLight	National Supercomputing Center in Wuxi, Jiangsu, China	125	40.960 CPU, 1,31 PB RAM	15,37	8,13	Finances and economics
Tianhe-2 Milky Way	National Supercomputing Center in Wuxi, Jiangsu, China	33,9	32.000 CPU, 1,4 PB RAM	17,8	1,9	Chemistry, physics, oil and aircraft simulation

(*continued*)

Table 1. (*continued*)

Computing cluster	Location	Performance in PFLOPS	Architecture	P_{IT} in MW	PW in GFLOPS/W	Deployment
Titan	Oak Ridge National Laboratory (Tennessee, USA)	17,6	18.688 CPU, 18.688 GPU, 693,5 TB RAM	8,2	2,146	Physical simulations

3.2 Fourier Thermodynamic Equation and Basic Methods of Its Solving

The solutions of utilizing the energy on the powerful servers is generally based on effective cooling of its heated items and on further utilizing this energy to heating of the buildings or to production needs [5]. For providing the effective cooling such systems of heat wasting should to be optimized by choosing the suitable materials and geometry parameters. Therefore, the methods of computer engineering with using numerical simulation are usually used for designing such waste heating systems [5]. Generally, the methods of simulation of cooling systems are based on the analytical or numerical solving the Fourier thermodynamic Part Differential Equation (PDE), which in the general form in cartesian coordinates is written as follows [11]:

$$\frac{\partial T(x,y,z,t)}{\partial t} - a^2\left(\frac{\partial^2 T(x,y,z,t)}{\partial x^2} + \frac{\partial^2 T(x,y,z,t)}{\partial y^2} + \frac{\partial^2 T(x,y,z,t)}{\partial z^2}\right) = f(x,y,z,t),\ a = \frac{\chi}{c_p \rho}, \quad (15)$$

where T is the temperature, t is the time, x, y, z are the spatial coordinates, a is the thermal diffusivity coefficient, $f(x, y, z, t)$ are the sources of heats, χ is the thermal conductivity, c_p is the isobaric thermal conductivity and ρ is the density of material. Since the inhomogeneous Eq. (15) is usually difficult to solving both analytically and numerically, often enough considered the homogeneous thermodynamic equation without right part, namely [11]:

$$\frac{\partial T(x,y,z,t)}{\partial t} - a^2\left(\frac{\partial^2 T(x,y,z,t)}{\partial x^2} + \frac{\partial^2 T(x,y,z,t)}{\partial y^2} + \frac{\partial^2 T(x,y,z,t)}{\partial z^2}\right) = 0. \quad (16)$$

Analytical solving of homogeneous Eq. (16) usually can be formalized by solving the Cauchy problem and considering the kernel of this equation, which in the general form is written as [11]:

$$\Phi(\mathrm{x}, \mathrm{t}) = \frac{1}{\left(2a\sqrt{\pi t}\right)^n}\exp\left(-\frac{|x|}{4a^2 t}\right). \quad (17)$$

Instead of sophisticated relation (17), another and simpler method of analytical solving the homogeneous Eq. (16) is widely used for providing the pervious estimation of heat wasting systems geometry parameters. This method id based on reduction of the Cauchy problem for Eq. (16) to the well-known Boltzmann Thermodynamic Equation (BTE), which in the general form can be given as follows [5, 11, 12]:

$$P_a = c_m m_m \frac{dT_s}{dt} + P_t,\ P_t = \frac{S_c(T_s - T_w)}{R(T)},\ R(T) = \frac{l_m(T)}{\lambda_m(T)}, \quad (18)$$

where T_s is the temperature of considered heating or cooling surface, T_w is the temperature of cooling liquid or gas, $R(T)$ is the thermal resistance of heated or cooling materials, P_a is the full power, absorbed by the system, P_t is the useless power, expended to the thermal conductivity, m_m is the mass of heated or cooling materials, c_m – its' thermal capacity by the mass, which depended on their temperature, λ_m is the thermal conductivity of material, l_m are their thickness. Taken into account, that for stationary regime of heat exchanging $\frac{dT_s}{dt} = 0$, the equation system (18) can be rewritten in the simplified for n as follows [5, 11, 12]:

$$P_a = P_t = \frac{S_c(T_s - T_w)\lambda_m(T)}{l_m(T)}. \tag{19}$$

The results of solving the Eq. (19) for cooling the multiprocessors of computing blocks by the moving liquid or gas will be presented in the next part of the chapter.

3.3 Simulation of Waste Heat System by Solving the Boltzmann Thermodynamic Equation

In the paper [5] was presented the analytical solution of Eq. (19) for waste the heat from multiprocessor block by the cool liquid or gas, transporting in the tube with rectangle cross-section. Obtained relation for Power Usage Effectiveness (PUE) is written as follows [5]:

$$PUE_\Sigma = \frac{N_p(P_c + P_{ww1} + P_{r1}) + \left(N_p - 1\right)(P_{ww2} + P_{r2})}{N_p(P_c + P_{ww1}) + \left(N_p - 1\right)P_{ww2}}, \tag{20}$$

where N_p is the number of processors, from which heat is wasted, P_c is the full electrical power of processor unit, P_{ww1} is the power, dissipated at the cooling system to the thermal conductivity between processor block and cooled liquid or gas, P_{ww2} is the power, dissipated at the cooling system to the thermal conductivity between the cooled liquid or gas and the free air and P_{r1} is the power, expended to heating the cooling liquid or gas, transferring in the tube.

The iterative relations for calculation the average operation temperature of processor block with number i $\overline{T^i_{\text{cr}}}$ and the temperature of cooling water or gas at the last processor block with number N $T_{\text{c}2}$ are written as follows:

$$\overline{T^i_{\text{cr}}} = \frac{\frac{P_{CPU}\,abh_{\text{tb}}\left(\frac{d_{\text{g}1}}{\lambda_{\text{g}1}} + \frac{d_e}{\lambda_e} + \frac{d_{\text{g}1}}{\lambda_{\text{g}1}} + \frac{d_{\text{w}}}{\lambda_{\text{w}}} + \frac{1}{\alpha_c}\right)}{v_{\text{c}}} + \overline{T^{i-1}_{\text{cr}}}}{2}, T_{\text{c}2} = \overline{T^N_{\text{cr}}} - \frac{P_{CPU}\,T_{\text{c}1}\left(\frac{d_{\text{g}1}}{\lambda_{\text{g}1}} + \frac{d_e}{\lambda_e} + \frac{d_{\text{g}1}}{\lambda_{\text{g}1}} + \frac{d_{\text{w}}}{\lambda_{\text{w}}} + \frac{1}{\alpha_c}\right)}{c_{\text{c}}\rho_c v_c^2},$$

$$A_i = \frac{c_{\text{c}}\rho_c v_c^2}{(a + h_{\text{tb}})l_i R_{c2}}, R_{c2} = \frac{d_w}{\lambda_w} + \frac{1}{\alpha_{\text{c}}}, T_c^{i-1} = T_c^i(1 + A_i) - A_i T_0, \tag{21}$$

$$\alpha_c = k_1 + k_2\sqrt{v_c}.$$

where P_{CPU} is the power, dissipated on the CPU blocks, a, and b are the geometry sizes of CPU blocks, h_{tb} is the highness of the tube's hollow, d_e is the thickness of CPU enclosure, d_{g1} is thickness of the gap between crystal and enclosure, d_{g2} – thickness of the gap between enclosure and tube, d_w is thickness of the tube's wall, T_{c1} is the start temperature of cooling liquid or gas at the inlet of cooling system, α_c is the heat transfer coefficient, l_i is the length of part of tube between CPU blocks, c_c ρ_c and v_c, is the heat capacity, density and velocity of cooling liquid or gas correspondently, k_1 and k_2 is the semiempirical coefficients for choosing liquid or gas and tube material [11, 12].

The cross-section of considered heat wasting system with noting the corresponded geometry parameters, have been used in Eqs. (18, 19), is presented at Fig. 8. The considered heat flow at Fig. 8 is signed as Q. The results of simulation of heat wasting system, obtained with using Eqs. (18, 19), are presented at Fig. 9. The simulation was provided for such geometry and thermodynamic parameters: material of the first gap – quartz glass $\left(\lambda_{g1} = 1,36\frac{\text{W}}{\text{m}\cdot\text{K}}\right)$, material of the second gap – air $\left(\lambda_{g2} = 0,034\frac{\text{W}}{\text{m}\cdot\text{K}}\right)$, material of enclose – polyvinyl chloride $\left(\lambda_e = 0,17\frac{\text{W}}{\text{m}\cdot\text{K}}\right)$, material of tube – copper $\left(\lambda_w = 384\frac{\text{W}}{\text{m}\cdot\text{K}}\right)$, $k_1 = 350$, $k_2 = 2100$, $d_{g1} = 10^{-4}$ m, $d_e = 5\cdot10^{-4}$ m, $d_{g2} = 10^{-4}$ m, $d_w = 1\cdot10^{-3}$ m, $a = 0.05$ m, $b = 0.05$ m, $h_{tb} = 0{,}05$ m, $T_{c1} = 10$ °C, $N_p = 10$, $l_i = 0.1$ m [11–13].

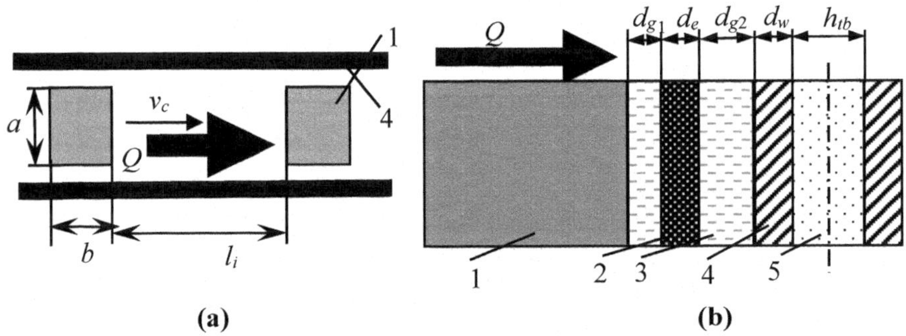

Fig. 8. Horizontal (a) and vertical (b) cross-section of simulated heating waste system 1 – crystal of cooling processor; 2 – processor's enclosure; 3 – the gap between the processor and the tube with the cooling liquid or gas; 4 – the wall of the tube; 5 – the hollow of the tube

The provided calculations for waste heat system are very important for forming practice recommendations about CPU cooling. The results of simulation shown, that for dissipated CPU power $P = 500$ W for system with 1 CPU the water velocity $v_c = 15$ l/min is enough, but for 10 CPU and the same dissipated power in its the water velocity $v_c = 30$ l/min is necessary. The estimations of water velocity is based on the presumption, that the crystal temperature $T_{cr} > 50$ °C for most CPU is critical. But, corresponding to obtained results, increasing of water velocity v_c isn't so critical. Furthermore, as the number of processors increases several times, the water supply speed increases significantly slightly. This conclusion is simply explained by the fact, that with location of CPU at the required distance, which must be greater than critical distance l_{cr}, the water in wasting systems has a time to cooling. In such conditions the number of CPU in computing block isn't so significant to providing the effective heat wasting. The heated water can be used to heating of the buildings or production needs.

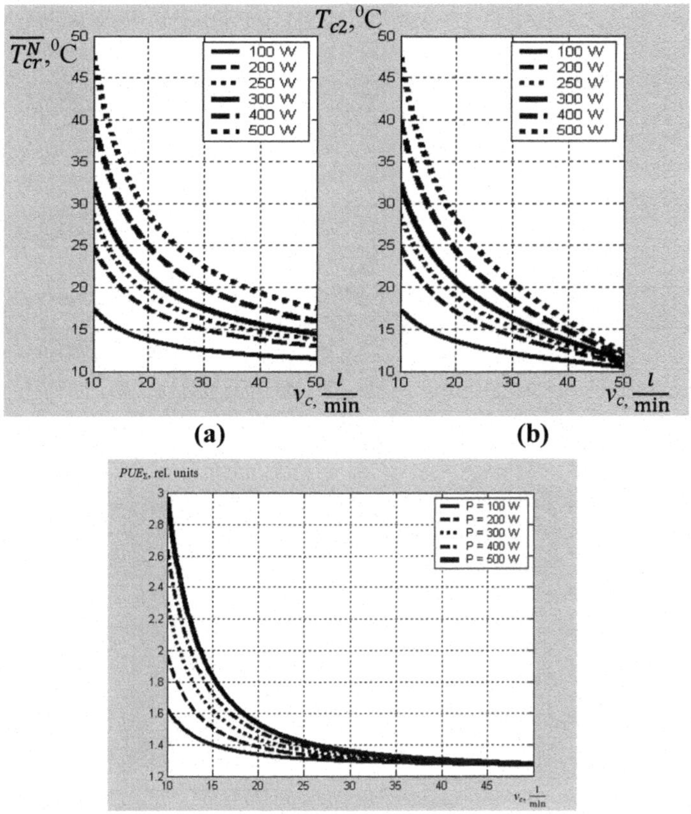

Fig. 9. Dependences of the temperature of last CPU block (a), temperature of cooling water at the last block (b) and PUE (c) on the water consumption vc and the power P, dissipated on the CPU blocks, calculated by solving the set of Eqs. (19–21)

4 Advanced Error-Coding Methods for IBN

4.1 Estimation of Redundancy of RS-Code

As was pointed out in the first and third sections, the using of the best error-correction coding is also necessary to providing QoS in realizing the parallel computing in IBN [14–19]. Analyzing the redundancy of two well-known and widely used error-correction codes, such as RS-code and convolutional code, is the subject of this part of chapter.

It is well-known, that the RS-codes are generally based Galois Field (*GF*) theory. The order of *GF* m is corresponded to the number of bites in the 1 symbol of RS-code [14–20]. Therefore, corresponding to the well-known law of probability theory, it can be assumed, that the probability of distortion 1 symbol of message, coded by the RS-code, is defined as follows [21]:

$$p_m = 1 - (1 - p_b)^m, \tag{22}$$

where p_b is the probability of distortion 1 bit in symbol, p_m is the probability of distortion the symbol.

With such assumptions, taking into account well-known Bernoulli's law of binomial distribution [21], the probability of distortion of θ bits in the massage with n symbols is defined as follows [16, 20]:

$$(\tau = \theta) = C_n^{\theta} \cdot \left(1 - (1 - p_b)^m\right)^{\theta} \cdot \left((1 - p_b)^m\right)^{n-\theta};\ C_n^{\theta} = \frac{n!}{\theta! \cdot (n - \theta)!}. \tag{23}$$

where C_n^{θ} is the number of combinations from n elements by θ [21, 22].

Now taking into account, that in RS-code with n symbols t errors can be corrected. It is knowing from the coding theory, that for original message, which consist of k information symbols, the length of coded message n is defined as [14–20]:

$$n = k + 2t. \tag{24}$$

With such assumptions and with taking into account the well-known law about adding the probabilities of independent events [21], corresponding to written beyond Eqs. (23, 24), the probability of receiving the correct message, coded by RS-code, is defined as follows [16, 19, 20]:

$$P_c = P(\tau \le \theta) = \sum\nolimits_{\theta=0}^{t} C_n^{\theta} \cdot (1 - (1 - p_b)^m)^{\theta} \cdot ((1 - p_b)^m)^{n-\theta}, t \le n. \tag{25}$$

Corresponding to obtained relation (25), the probability P_e, corresponded to possible error in receiving of message, coded by RS-code, is defined as follows [16, 19, 20]:

$$P_e = 1 - P_c = P(\tau > \theta) = \sum\nolimits_{e=t+1}^{n} C_n^{\theta}(1 - (1 - p_b)^m)^{\theta}((1 - p_b)^m)^{n-\theta}, t < n, \tag{26}$$

The maximal number of corrected symbols $t_{\max}$ is defined as follow [16, 19, 20]:

$$t_{\max} = \begin{cases} 2^{m-1} - \frac{k}{2} - 1, & \text{when the value } k \text{ is even;} \\ 2^{m-1} - \frac{k+1}{2}, & \text{when the value } k \text{ is odd.} \end{cases} \tag{27}$$

The dependences of probability of error in receiving coded message on probability of bit error $P_e(p_b)$, obtained with using relation (26), for different values of m and t code parameters, are presented at Fig. 10 [20].

Analyzing now the graphical dependencies, given at Fig. 10. Firstly, was taken the RS-code with parameters $k = 21$ and $t = 5$ at Galois Field *GF* (5). But it is clear from relations (27), that for $m = 5$ and $k = 21$ the value $t = 5$ is corresponded to limit value $t = t_{\max}$. In such conditions only one way to increasing the corrective possibility of RS-code is existed, and it can be realized by increasing the order m of Galois Field *GF* from 5 to 6. Therefore, the second dependence is also obtained for $k = 21$ and $t = 5$, but with another value of *GF* order $m = 6$. As clear form the Fig. 10, in such condition the corrective ability of RS-code is lose, and this fact can be simply explained. Really, the number of errors, which can be corrected, is the same ($t = 5$), but the length of code n is 2 times increased, from $n = 31$ to $n = 62$. But the advanced of such transforming is also

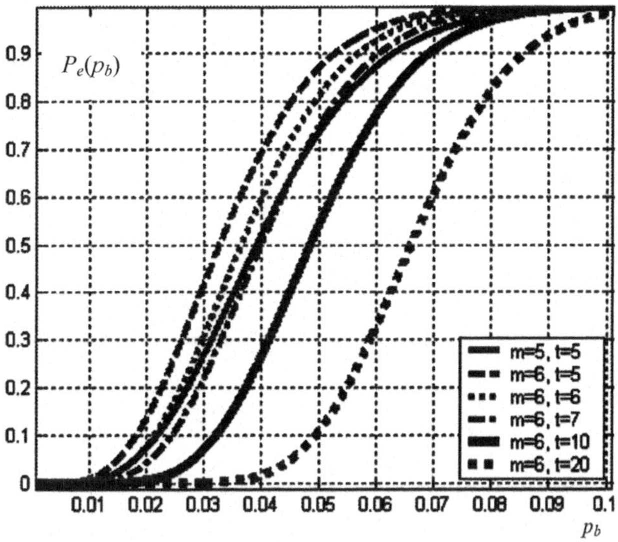

Fig. 10. Dependences of $P_e(p_b)$ for RS-codes, obtained with using the relation (26)

clear, because for $m = 6$ the number of corrected errors t can be significantly greater. Really, in such conditions, corresponding to relations (27) the value of $t_{\max}$ is:

$$t_{\max} = 2^{m-1} - \frac{k+1}{2} = 2^5 - \frac{21+1}{2} = 32 - 11 = 21.$$

It is clear from Fig. 10, that in $GF(6)$ the corrective ability of RS-code $P_e(p_b) = 1 - P_c(p_b)$ is losing only for the values $t = 5$ and $t = 6$. For the number of corrected errors $t = 7$ value of $P_e(p_b)$ is generally same to the corresponded value for RS-code with parameter $t = 5$ in $GF(5)$, and with further increasing the value of t in $GF(6)$, corrective ability of code is significantly increased. But the greatest considered value in the $GF(6)$ $t = 20$ is not much different from the calculated limit value $t_{\max} = 21$.

In any case, it is clear from dependences, presented at Fig. 10, that for $p_b > 0.1$ corrective ability of RS-code is significantly decrease to the negligible value. Therefore, using of RS-codes in the noised communicated channels, where probability of bit error is usually significantly greater, than $p_b = 0.1$, generally is not recommended [14–19].

4.2 Estimation of Redundancy of Convolutional Codes

4.2.1 The General Principle of Formation of Convolutional Codes Constructions and Its Basic Parameters

In the previous section the corrective ability of RS-codes is considered. It should be pointed out, that systematic linear and block codes, like RS-code, usually have the standard description by such parameters, as number of information symbols k and number of control symbols r [14–20]. For example, in considered beyond RS-code, $r = 2t$. The theory of error correction for linear and block codes is built on such simple assumption about code parameters [14–19] and on the base of combinatory analyze [21, 22]. For example, by using this approach the relations (22–27) have been obtained [16].

The structure of convolutional codes is generally different form the linear and block ones. One of the important parameters of convolutional codes is constraint length, which usually signed by K. And the main particularity of realization the hardware and software for forming the convolutional codes sequences is changing the size of using memory depending on number of symbols, which have been received. As the result, the current symbol, which have to be formed, is depend both on input digital symbol and on the pervious $K - 1$ symbols. In such conditions the structure of convolutional code can be described correctly by the corresponded structure scheme of Finite-Sequence Machine (FSM) [14–20]. The simplified digital scheme for forming the convolution code sequence is presented in Fig. 11.

It is clear from Fig. 11, that the scheme for forming of convolution code consist on the XOR operation blocks, which number is n, and included the input shift register R with number of bits kK.

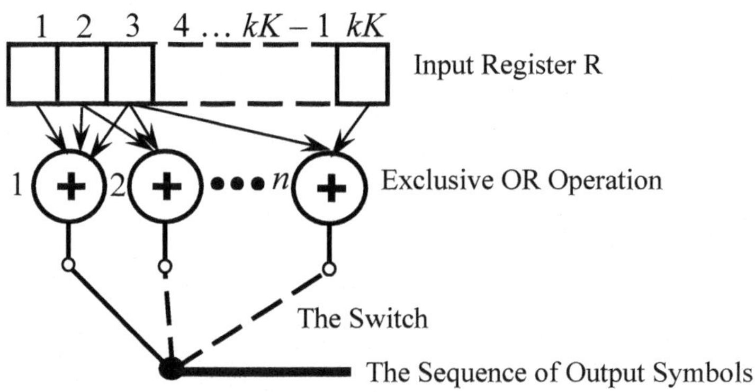

Fig. 11. The simple general digital scheme for forming the convolutional code

The encoder scheme, presented on Fig. 11, is operated as follows. At the correspondent time the register R receives k bits of the input sequence m and all bits in this register are shifted to k positions to right. Simultaneously all kK bits are fed to n XOR schemes and n bits of the convolutional code are formed as the result of this operation. The coding digital signal is transmitted to the communication channel.

On the general scheme, presented at Fig. 11, possibility of using the multibit symbols with number of bits k is assumed, but for making of error-correction estimations considering of simple binary symbols is enough. In such condition, the number of bits in the shift register R is K [14, 20]. Another parameter of convolutional code is its redundancy factor, which, correspondently to the scheme, presented at Fig. 11, is correspond to the number of XOR schemes n and usually signed at the literature as $\frac{1}{n}$ [14–20]. Therefore, the constraint length K and redundancy factor $\frac{1}{n}$ are considered usually as the main parameters of convolutional codes [14–20].

4.2.2 Basic Principles of Defining the Corrective Ability of Convolutional Codes

The advantage of using convolutional codes in communication systems with the noised channels relatively to linear and block ones is the limited number of allows symbols, taking into account, that the sequence of $K-1$ symbols, have been received early, is also analyzed. Usually, for the binary symbols, convolution encoding systems are analyzed by the FSM-model. In such analyze from any FSM state only 2 transmission is allowed: one of its corresponded to the input signal 0, and the second – to the input signal 1. All another sequences of $N = 2^{K-1}$ symbols of convolution code are always considered as errors, and finding the correct sequence by the receiver is usually provided with using the maximum likelihood principle [14–20]. For example, in modern communication systems Viterbi algorithm is widely used to defining the most likelihood symbols sequences [14–20]. Therefore, the systems, based on convolution codes, are usually operated with prediction of the most likelihood symbol, and the main principle of its operation is comparing of the most likelihood and received sequences [14–20].

But, causing described beyond sophisticated principle of its operation, finding the corrective ability of convolutional codes is usually more complicated problem, than for linear and block ones. Solving of this generalized problem is lead to considering such interdependent tasks [14–20].

1. On the base of structure scheme of encoder, the FSM scheme is formed.
2. By considering the FSM scheme formed the set of linear relations, which described the transmission between the FSM states.
3. By solving the obtained system of linear relations defined the transfer function of considered FSM $T(D, L, N)$ in polynomial presentation, where D – Hemming distance between the zero and waiting codes sequences, L – the counter of transmissions between the start and current FSM state, and N – the mark to transmissions, which corresponded to the input signal 1.
4. Defining the clearance parameter of the convolution code d_f as the minimal power of variable D at the polynomial presentation of function $T(D, L, N)$. Really, parameter d_f characterized the minimal possible difference between the bits in the allows sequences of convolution code and it corresponded to the minimal code distance d in the linear and block codes [14–20].
5. Defining the dependence of maximal value of the probability of bit error in the sequence of convolutional code P_B on the probability of single bit error p by finding the partial derivations for the transfer function $T(D, L, N)$. It should be pointed out, that for different methods of forming the digital signals on the physical layer of OSI reference model (refer. Fig. 2), the relations for bit error in the convolutional code $P_B(p)$ are also different [14–20].

   Considering the relations $P_B(p)$ for 2 important cases, which are widely used for forming the binary signals on the physical layer of OSI reference model.

   1. For standard potential coding, like Alternative Mark Inversion (AMI) [6, 23], the relation for finding $P_B(p)$ is written as follows [14, 20]:

$$P_B(p) \leq \left. \frac{dT(D, N, L)}{dN} \right|_{N=1, L=1, D=2\sqrt{p(1-p)}}. \quad (28)$$

2. For the Phase Shift Coding (PSC) binary signal and its transferring in the Gaussian Noise Channel (GNC) [6, 23], the analytical relation for defining P_B is written as follows [14, 20]:

$$P_B\left(\frac{E_c}{N_0}\right) \leq \mathrm{Q}\left(\sqrt{2d_f\frac{E_c}{N_0}}\right)\exp\left(d_f\frac{E_c}{N_0}\right)\frac{dT(D,N)}{dN}\bigg|_{N=1,L=1,D=\exp\left(-\frac{E_c}{N_0}\right)}, \tag{29}$$

where $\frac{E_b}{N_0}$ is the relation of the energy of information signal E_b to the density of noise power spectrum N_0, $\frac{E_c}{N_0} = \frac{rE_b}{N_0}$ is the relation of full energy of coded message E_b to the density of noise power spectrum N_0, $r = \frac{k}{n}$ is the redundancy factor of convolutional code, d_f is the clearance parameter of convolution code, considered beyond, Q(x) is the Gaussian error function, or integral error function, which is well known from the basic principles and laws of probability theory and written as follows [21, 22]:

$$\mathrm{Q}(x) = \frac{1}{\sqrt{2\pi}}\int_x^{\infty} \exp\left(-\frac{u^2}{2}\right)du. \tag{30}$$

Considering now some examples of defining the transfer functions $T(D, L, N)$ and the clearance parameter of convolution codes d_f, as well as calculation the probability of bit errors in these codes by using the relations (28–30). It is clear, that the possible bit error in the communication channel is lead to reducing the QoS parameter for solving the task of algorithm parallelization in IBN.

Examples of calculation the transfer functions $T(D, L, N)$ and the clearance parameter of convolution codes d_f will be presented below.

4.2.3 Some Examples of Defining the Transfer Function and Corrective Ability of Convolutional Codes

Example 10. The scheme of convolution code encoder is presented at Fig. 12. Defining the constraint length of this code K, the redundancy factor $\frac{1}{n}$ and clearance parameter of the code d_f.

From the scheme, presented at Fig. 12, clear, that the number of bits in input register is 3 and number of XOR blocks is 2. Therefore, it is clear from describing beyond the parameters of convolutional code, that constraint length of code $K = 3$ and redundancy factor $\frac{1}{n} = \frac{1}{2}$. The convolutional code with such parameters is usually described in literature as $\left(3, \frac{1}{2}\right)$.

The scheme of FSM for encoder, presented at Fig. 12, is presented at Fig. 13.

The singularities of FSM scheme, presented at Fig. 13 and described the structure of convolution code $\left(3, \frac{1}{2}\right)$, are follows [14, 19].

1. For forming the states of FSM only 2 right bits of input register R are used, because the last bit, have been received, is corresponded to input signal.
2. The transmissions, corresponded to input signal 0, are marked by the solid lines, and the transmissions, corresponded to input signal 1, are marked by dash lines.

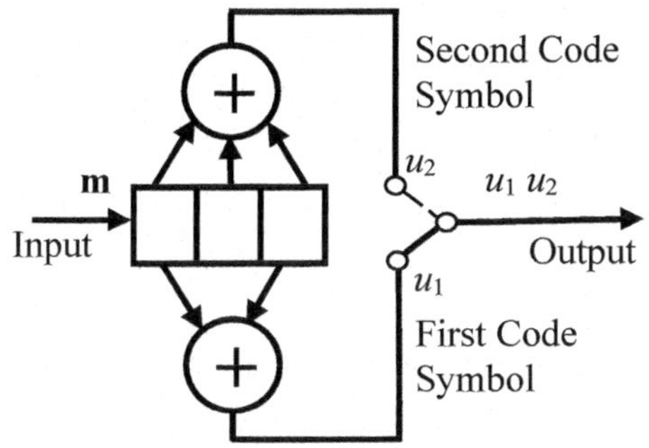

Fig. 12. Scheme of encoding device for example 10

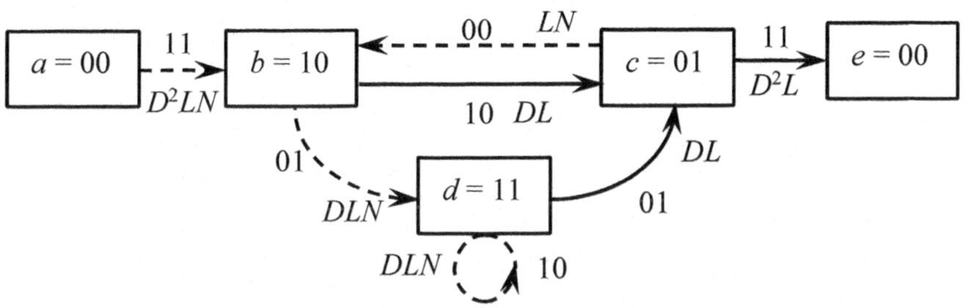

Fig. 13. The structure of FSM for the encoding scheme, presented in Fig. 12

3. The output signals of encoder are marked at the transmission's lines.
4. At the transmission's lines the code parameters *D*, *L* and *N*, described beyond, are also marked.
5. The FSM zero state 00 is consider at the scheme 2 times, and its's corresponded to the input and output states of FSM. By this transforming the loop connection on the zero state is excluded.

Such presentation of FSM is allowed to exclude from FSM scheme all loopback connections, therefore, the set of linear equations, formed on the base of this scheme, with using such approach is generally simplified [14, 20]. Such set of linear equation for the FSM scheme, presented at Fig. 13, formed on the base of weight of connections between the states and can be written as follows [14, 20]:

$$\begin{cases} X_b = D^2LNX_a + LNX_c; \\ X_c = DLX_b + DLX_d; \\ X_d = DLNX_b + DLNX_d; \\ X_e = D^2LX_c, \end{cases} \tag{31}$$

where X_a, X_b, X_c, X_d and X_e are the variables, which described to the state *a*, *b*, *c*, *d* and *e* correspondently. Solving of the system (31) given the such result [14, 20]:

$$T(D, L, N) = \frac{D^5L^3N}{1 - DLN(1 + L)}. \tag{32}$$

But disadvantage of presentation of transferring function in the from (32) is that it is the fractional rational function, but not polynomial one. For transferring the fractional rational functions, like (32), to polynomial functions, well-known basic relation of Taylor row theory can be used [14, 21, 22]:

$$f(x) = \frac{1}{1 - ax} = 1 + ax + a^2x^2 + a^3x^3 + \ldots + a^nx^n + \ldots. \tag{33}$$

Taking into account the relation (33), fractional rational function (32) can be rewritten in the polynomial form as follows [14, 20]:

$$\begin{gathered}T(D, L, N) = \frac{D^5L^3N}{1 - DLN(1 + L)} = D^5L^3N + D^6L^4(1 + L)N^2 \\ + D^7L^5\left(1 + L^2\right)N^3 + \ldots + D^{l+5}L^{l+3}N^{l+1} + \ldots\end{gathered} \tag{34}$$

Since the term with least degree of the variable D in Eq. (34) is D^5L^3N, the clearance parameter of considered convolution code $\left(3, \frac{1}{2}\right)$ is $d_f = 5$.

Example 11. The scheme of convolution code encoder is presented at Fig. 14. Defining the constraint length of code K, the redundancy factor $\frac{1}{n}$ and clearance parameter of the code d_f.

Form the scheme of encoder, presented at Fig. 14, clear, that in this case the code parameters are constraint length of code $K = 4$ and redundancy factor $\frac{1}{n} = \frac{1}{3}$, therefore in formalized description such code can be considered as $\left(4, \frac{1}{3}\right)$.

On the base on the encoding scheme, presented at Fig. 14, created the FSM scheme, presented at Fig. 15 [20]. The basic principles of forming this FSM scheme are generally similar, as was used beyond in example 10. Only one difference is, that in the FSM scheme, presented at Fig. 15, loop connection for the state h is existed. But, as in the previous case for example 10, the zero state of FSM is divided into input state a and output state i, and by this transferring loopback connection from output to input of FSM is excluded.

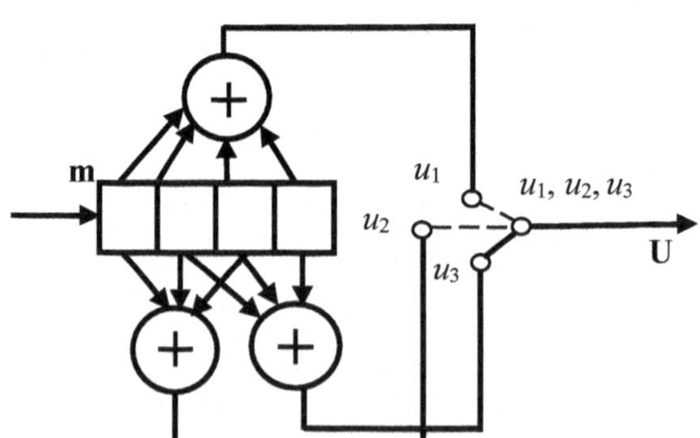

Fig. 14. Scheme of encoding device for example 11

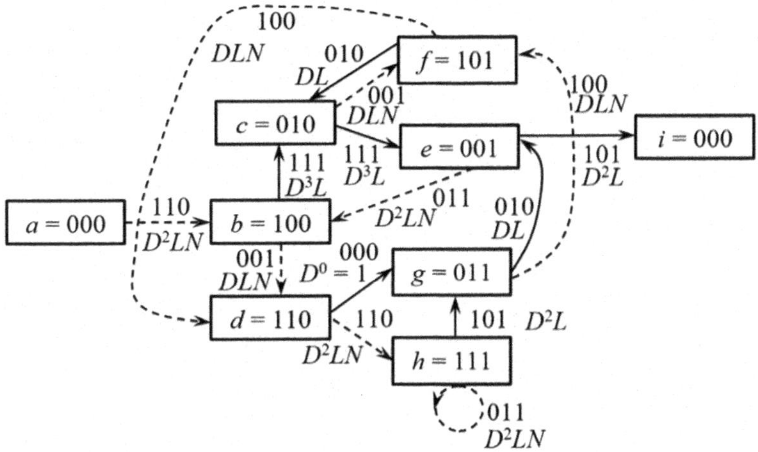

Fig. 15. The structure of FSM for the encoding scheme, presented in Fig. 14

The set of equations, described the structure of FSM, presented at Fig. 15, is written as follows [20]:

$$\begin{cases} X_b = D^2LN(X_a + X_e); \\ X_c = D^3LX_b + DLX_f; \\ X_d = DLN(X_b + X_f); \\ X_e = D^3LX_c + DLX_g; \\ X_f = DLN(X_g + X_c); \\ X_g = X_d + D^2LX_h; \\ X_h = D^2LN(X_d + X_h); \\ X_i = D^2LX_e. \end{cases} \tag{35}$$

Solving of the set of Eq. (35) given such sophisticated fractional rational relations for the denominator of transfer function $T_d(X_h)$ and its nominator $T_n(X_h)$ [20]:

$$T_d(X_h) = \frac{D^4L^3N^2(L-N-1) - D^3L^3N^2 - D^2LN(1+LN) + 1}{D^5L^3N^3(1-D^2L^2N)(1-D^2L^2N+D^3L)}X_h.$$

$$T_n(X_h) = \frac{-D^{20}L^{12}N^5 + D^{19}L^{10}N^5(L-1) + 2D^{18}L^{10}N^5(LN+1)}{D^5L^3N^3(1-D^2L^2N)(1-D^2L^2N+D^3L)}X_h$$

$$+\frac{D^{17}L^9N^4(1-N(L+1)) + D^{16}L^8N^4(1+(N+1)L-L^2N(N-1))}{D^5L^3N^3(1-D^2L^2N)(1-D^2L^2N+D^3L)}$$

$$+\frac{2D^{15}L^8N^4(L-1) + D^{14}L^7N^2(L^3N^3 + L(L+1)N^3 - N^2 + LN + 1)}{D^3L^2N^2(1-D^2L^2N)(1-D^2L^2N+D^3L)}X_h$$

$$+\frac{2D^{13}L^7N^3(1-L) + D^{12}L^5N^5(2L^3N^2 - L^2N^2 - L(N+1) - 1)}{D^3L^2N^2(1-D^2L^2N)(1-D^2L^2N+D^3L)}X_h$$

$$+\frac{D^{11}L^6N^5(LN+1)+D^{10}L^5N\left(LN^2+1\right)-D^9L^5N^2(L+1)+D^7L^3N(L+1)+D^7+D^6L^4N^2}{D^3L^2N^2(1-D^2L^2N)(1-D^2L^2N+D^3L)}X_h. \quad (36)$$

But, taking into account, that in relations (36) only the terms with the lower power of variable D has to be considered to defining the clearance parameter of convolution code d_f, simplified transferring function for FSM, presented at Fig. 15, in the fractional rational form, with taking into account (36), can be rewritten as follows [20]:

$$T(D,L,N)=\frac{D^6L^4N^2}{(1-DLN)(1+DLN)}=\frac{D^5L^3N}{2}\left(\frac{1}{1-DLN}-\frac{1}{1+DLN}\right). \quad (37)$$

By using the relations (33), the obtained fractional rational relation (37) can be transformed to the polynomial form as follows [20]:

$$T(D)=\frac{D^5L^3N}{2}\left(\frac{1}{1-DLN}-\frac{1}{1+DLN}\right)$$

$$=\frac{D^5L^3N}{2}\left(2DLN+2D^3L^3N^3+2D^5L^5N^5+\ldots\right)=D^6L^4N^2+\ldots \quad (38)$$

Therefore, it is clear from relation (38), that the clearance parameter of considered convolution code $\left(4,\frac{1}{3}\right)$ is $d_f=6$.

Example 12. The scheme of convolution code encoder is presented at Fig. 16. Defining the constraint length of code K, the redundancy factor $\frac{1}{n}$ and clearance parameter of the code d_f.

Form the scheme of encoder, presented at Fig. 16, clear, that in this case the code parameters are follows: constraint length of code $K=5$ and redundancy factor $\frac{1}{n}=\frac{1}{3}$. Therefore, in formalized description such code can be considered as $\left(5,\frac{1}{3}\right)$.

On the base on the encoding scheme, presented at Fig. 16, created the structure of FSM, which is presented at Fig. 17 [20].

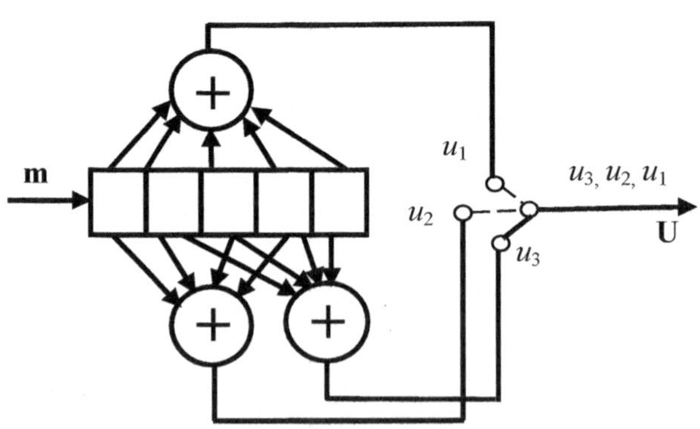

Fig. 16. Scheme of encoding device for example 12

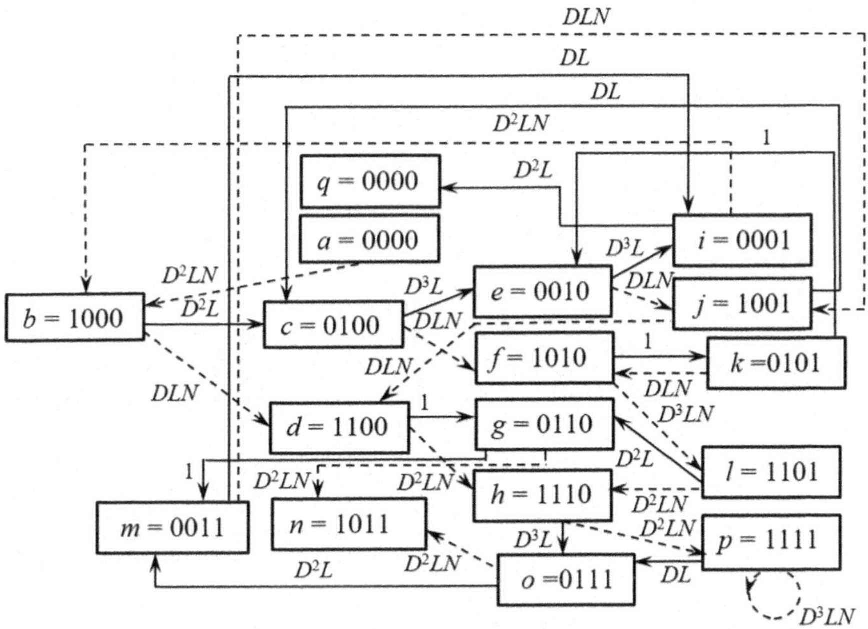

Fig. 17. The structure of FSM for the encoding scheme, presented in Fig. 16

The set of equations, described the structure of FSM, presented at Fig. 17, is written as follows [20]:

$$\begin{cases} X_b = D^2LN(X_a + X_i); \\ X_c = D^2LX_b + DLX_j; \\ X_d = DLN(X_b + X_j) \\ X_e = D^3LX_c + X_k; \\ X_f = DLN(X_c + X_k) \\ X_g = X_d + D^2LX_l; \\ X_h = D^2LN(X_d + X_l); \\ X_i = D^3LX_e + DLX_m; \\ X_j = DLN(X_e + X_m) \\ X_k = X_f + D^2LX_n; \\ X_l = D^2LN(X_f + X_n); \\ X_m = X_g + D^2LX_o; \\ X_n = D^2LN(X_g + X_o); \\ X_o = D^3LX_h + DLX_p; \\ X_p = D^2LNX_h + D^3LNX_p; \\ X_q = D^2LX_i. \end{cases} \tag{39}$$

Solving of the sophisticated set of Eqs. (39) was complexly described in the tutorial book [20]. It was obtained with using the MATLAB symbolic processor [7] for providing the necessary cumbersome polynomial operations and given such result for FSM transfer function in polynomial presentation [20]:

$$T(D, L, N) > \frac{L^{28}N^{24}D^{57}}{5L^{23}N^{20}D^{51} - L^{22}N^{20}D^{50}} = \frac{L^6N^4D^7}{5LND - 1}$$
$$= -L^6N^4D^7\left(1 + 5LND + 25L^2N^2D^2 + 125L^3N^3D^3 + \ldots\right) \quad (40)$$
$$= -L^6N^4D^7 - 5L^7N^5D^8 - 25L^8N^6D^9 - 125L^9N^7D^{10} - \ldots$$

Therefore, it is clear, that the clearance parameter of considered convolution code $\left(5, \frac{1}{3}\right)$ is $d_f = 7$.

4.2.4 Comparison of Manual and Computer Methods of Defining the Clearance Parameter of Convolution Codes and Analyzing the Possibility of Realizing This Calculation in the IBN Cloud Computing

At the previous section was analyzed the structures of FSMs and the clearance parameters d_f for convolution codes $\left(3, \frac{1}{2}\right)$, $\left(4, \frac{1}{3}\right)$ and $\left(5, \frac{1}{3}\right)$ were defined with using the algebraic operations with the fractional rational relations, as well as polynomial operations. Clear, that if for the simple codes $\left(3, \frac{1}{2}\right)$ and $\left(4, \frac{1}{3}\right)$, for which the schemes of FSMs are presented at Fig. 13 and Fig. 15 correspondently, the manual calculation is possible, in contrary, for the more sophisticated code $\left(5, \frac{1}{3}\right)$, for which the scheme of FSM is presented at Fig. 17, using of symbolic processor and computer calculations for providing necessary analytical transformers is inevitability [20].

Generally, it is clear, that using of manual calculation methods is possible only for the simple coding systems, where the number of FSM states is smaller than 10. In such condition the maximal polynomial degree is smaller, than 20, and the number of elementary operations, corresponded to summing the polynomial terms, is smaller, than 100. Therefore, although such analytical transformations are quite cumbersome, it's really can be performed manually. On another hand, a significant advantage of the manual method of providing the polynomial operations is the ability to find a most compact record of the corresponding coefficients with their grouping by the several variables. For example, the analytic relation $D^{14}(L^{10}N^3 + L^8N^2 - L^3N - 3)$ can be rewritten as the complex polynomial function of variables D, L and N as follows:

$$D^{14}\left(L^{10}N^3 + L^8N^2 - L^3N - 3\right) = D^{14}\left(L^3N\left(L^7N^2 + L^5N - 1\right) - 3\right)$$
$$= D^{14}\left(L^3N\left(L^5N\left(L^2N + 1\right) - 1\right) - 3\right). \quad (41)$$

Sometimes the relations with the great number of brackets, like the final polynomial relation in (41), are difficult to further analyzing, but it is clear, that its' also have certain advantages, among which the following can be noted [20].

1. The relations with the grouping of polynomial terms by the few variables are always more compact.
2. The powers of variables in such relations are usually smaller.
3. In most cases using of such presentation of polynomials allows to set the important additional regularities, which described the general particularities of operation for considered FSM.

4. The possibility of introducing auxiliary variables in case of excessive complication of analytical expressions.

In a contrary, the MATLAB symbolic processor always presents the results of algebraic polynomials operations in a such way, that all polynomial coefficients are also polynomials with respect to other variables [7]. By this reason the analytical expressions, derived from computer calculations, are usually disproportionately cumbersome [7, 20]. The same disadvantage is inherent to the symbolic processor of ANACONDA computer system, which is based on the means of Python program language [10].

Analyzing the particularities of applying the polynomial transforming for defining the interconnections between the states of FSM, lead to such important conclusions [14–20].

1. It is desirable to use manual calculations at the first stage on general statement of the task of forming the convolutional code and finding of most important interrelations between FSM states.
2. The manual calculations are usually very effective for providing the further transforming, when the forming and analyzing of simplified analytical relations are realized and finding of important polynomial dependences in the general form is provided.
3. For providing the standard cumbersome polynomial operations with the polynomials, which order is higher, than 20, using of computer software with symbolic processors, like MATLAB or ANACONDA, is more convenient and effective and strongly recommended [7, 24].

Really, all convolutional codes with the constraint length $K > 10$ are formed today on the base of computer analyze with applying suitable combinator algorithms [14–19].

But it should be pointed out, that for the polynomials with power of collected variables grater, than 200, the time of calculation with using symbolic processor is enormous enlarges. Generally, it caused by the enlarging the number of elementary operations with polynomial terms, which, corresponded to combinatory law, is enlarges as $n!$, where n is the polynomial power [20, 21]. It also can be explained by the FSM theory, Really, although from any FSM state only 2 transmission is possible, with increasing the number of states possible ways from zero state to last one became prohibitively large [20]. For example, for obtaining the transfer function $T(D, L, N)$ for convolutional code with parameters $\left(5, \frac{1}{3}\right)$ (Example 12) on the PC with 4-core CPU Intel Xenon, frequency 2.83 MHz and RAM 16 GB, the appropriative time of solving such task was more than few days. Certainly, that such formalizing of solved problem generally isn't correct for any computer system. Really, such problem can be formalized to providing calculations in cloud, since the sets of polynomials terms can be treated independently, therefore the factor of parallelization p, defined by the Eqs. (12–14), is $p \approx 1$. But, in any case, for enormous number of elementary operations the time, necessary to providing the parallel calculation, as well as the time t_p, necessary to transferring the data for parallelized task, usually isn't suitable to reach the high level of QoS in IBN. In any case, estimation with using Eqs. (12–14) is necessary.

Another, more effective way of significant reducing the time of providing polynomial operations for defining the clearance parameter of convolution code d_f is significant reduction the number of polynomial terms. As was shown beyond, in example 11, only the polynomial terms with lowest power are necessary to defining clearance parameter d_f. Insofar as the information about the terms with the highest polynomial power is also usually necessary, considering only the terms with lowest and highest power is possible, and the terms with medium value of power is usually negligible and its can be excluded from consideration. For example, in the manual book [20] such relations for reducing the number of polynomial terms were proposed:

$$X_{nom}^{short}(D, L, N, X_h) = \left(\sum_{i=n_{\max}^{snom}-k_{snom}}^{n_{\max}^{snom}} K_{inom}(L, N)D^i + \sum_{j=n_{\min}^{snom}}^{n_{\min}^{snom}+l_{snom}} K_{jnom}(L, N)D^j\right)X_h, \quad (42)$$

$$X_{den}^{short}(D, L, N, X_h) = \left(\sum_{i=n_{\max}^{den}-k_{sden}}^{n_{\max}^{den}} K_{iden}(L, N)D^i + \sum_{j=n_{\min}^{den}}^{n_{\min}^{den}+l_{sden}} K_{jden}(L, N)D^j\right)X_h,$$

where $n_{\max}^{snom}$ is the value of maximal power of variable D in the nominator polynomial, $n_{\max}^{sden}$ is the value of maximal power of variable D in the denominator polynomial, $n_{\min}^{snom}$ is the value of minimal power of variable D in the nominator polynomial, $n_{\min}^{sden}$ is the value of minimal power of variable D in the denominator polynomial, k_{snom} is the number of terms of nominator polynomial by the highest power, l_{snom} is the number of terms of nominator polynomial by the lowest power, k_{sden} is the number of terms of denominator polynomial by the highest power, l_{sden} is the number of terms of denominator polynomial by the lowest power, K_{inom}, K_{jnom}, K_{iden}, and K_{jden} are the polynomial coefficients for the terms of nominator and denominator correspondently, X_h is the output stage of FSM.

The parameters of polynomials reducing l_{enom}, l_{eden}, k_{enom}, k_{eden}, l_{inom}, l_{iden}, k_{inom} and k_{iden} are chose as the compromise factor between the time of calculations and requirement for correct defining the terms with lower and highest power in the transfer function. During providing the calculations for convolutional code $\left(5, \frac{1}{3}\right)$, considered in example 12, such parameters have been choosed: $l_{enom} = 10$, $k_{enom} = 13$, $l_{eden} = 9$, $k_{enom} = 13$; $l_{inom} = 11$, $k_{inom} = 13$, $l_{iden} = 19$, $n_{\max}^{enom} = 239$, $n_{\min}^{enom} = 14$, $n_{\max}^{eden} = 234$, $n_{\min}^{eden} = 42$ and $n_{\max}^{inom} = 343$, $n_{\min}^{inom} = 35$, $n_{\max}^{iden} = 335$, $n_{\min}^{iden} = 60$ [20].

In any case, computer simulation in IBN clouds is very effective for creating the new effective constructions of convolutional codes with improved correction parameters, but for simplifying the calculation task and its effective realizing in IBN the reducing of number of polynomial terms with using relations, similar to (42), is also necessary [20].

The examples of defining the corrective ability of convolutional codes with using relations (28–30) will be considered below.

4.2.5 Defining the Corrective Ability of Convolutional Codes

Example 13. With using relations (28–30), finding the dependences of error in convolutional codes verse bit error for the code's constructions $\left(3, \frac{1}{2}\right)$, $\left(4, \frac{1}{3}\right)$ and $\left(5, \frac{1}{3}\right)$, considered in examples 10, 11 and 12.

The transfer function of the convolutional code $\left(3, \frac{1}{2}\right)$ in the fractional rational form is defined by Eq. (32). Considering relation (32) and finding its particle derivation led to follows result [14, 20]:

$$\left.\frac{dT(D,N)}{dN}\right|_{N=1,L=1} = \left.\frac{D^5(1-2DN)+2DD^5}{(1-2DN)^2}\right|_{N=1} == \frac{D^5-2D^6+2D^6}{(1-2D)^2} = \frac{D^5}{(1-2D)^2}. \tag{43}$$

Taking into account, that in relation (28) $D = 2\sqrt{p(1-p)}$, as well as obtained relation (43) for the derivation $\left.\frac{dT(D,N)}{dN}\right|_{N=1,L=1}$, the maximal value of bit error $P_B(p)$, correspondently to relation (28), for the code construction $\left(3, \frac{1}{2}\right)$ is calculated as follows [14, 20]:

$$P_B(p) \le \left.\frac{dT(D,N)}{dN}\right|_{N=1,D=2\sqrt{p(1-p)}} = \left.\frac{D^5}{(1-2D)^2}\right|_{D=2\sqrt{p(1-p)}} = \frac{\left(2\sqrt{p(1-p)}\right)^5}{\left(1-4\sqrt{p(1-p)}\right)^2}. \tag{44}$$

For the convolutional code with parameters $\left(4, \frac{1}{3}\right)$ the transfer function in the fractional rational form is defined by Eq. (37). By considering the relation (37) and finding its particle derivation, we obtained the following result [20]:

$$T(D,N)|_{L=1} = \left.\frac{D^6L^4N^2}{1-D^2L^2N^2}\right|_{L=1} = \frac{D^6N^2}{1-D^2N^2}. \tag{45}$$

Correspondently to (45), the partial derivation $\left.\frac{dT(D,N)}{dN}\right|_{N=1}$ is written as follows [20]:

$$\left.\frac{dT(D,N)}{dN}\right|_{N=1} = \left.\frac{2D^6N\left(1-D^2N^2\right)+2D^2ND^6N^2}{(1-D^2N^2)^2}\right|_{N=1} = \left.\frac{2D^6N-2D^8N^3+2D^8N^3}{(1-D^2N^2)^2}\right|_{N=1} = \frac{2D^6}{(1-D^2)^2}. \tag{46}$$

Substituting the obtained expression for the partial derivative $\left.\frac{dT(D,N)}{dN}\right|_{N=1}$ (46) into the given above relation for the bit error $P_B(p)$ (28), for considered convolutional code with parameters $\left(4, \frac{1}{3}\right)$ the following result has been obtained [20]:

$$P_B(p) \le \left.\frac{dT(D,N)}{dN}\right|_{N=1,D=2\sqrt{p(1-p)}} = \frac{64(p(1-p))^3}{(1-4p(1-p))^2}. \tag{47}$$

For the convolutional code with parameters $\left(5, \frac{1}{3}\right)$ the transfer function in the fractional rational form is defined by Eq. (40). For this code construction, taking into account (40) and finding the particle derivation $\left.\frac{dT(D,N)}{dN}\right|_{N=1}$ for the function $T(D,L,N)$, written in the fractional rational form, the follows result have been obtained [20]:

$$\left.\frac{dT(D,N)}{dN}\right|_{N=1} = \left.\frac{3N^3D^7(5ND-1)-5DN^4D^7}{(5ND-1)^2}\right|_{N=1} = \left.\frac{15N^4D^8-3N^3D^7-5N^4D^8}{(5ND-1)^2}\right|_{N=1}$$

$$= \left.\frac{10N^4D^8 - 3N^3D^7}{(5ND-1)^2}\right|_{N=1} = D^7\frac{10D-3}{(5D-1)^2}. \qquad (48)$$

Taking into account (48) and substituting it in (26), for considered convolutional code with parameters $\left(5, \frac{1}{3}\right)$ the relation $P_B(p)$ is written as follows [20]:

$$P_B(p) \le \left.\frac{dT(D,N)}{dN}\right|_{N=1, D=2\sqrt{p(1-p)}} = D^7\left.\frac{10D-3}{(5D-1)^2}\right|_{D=2\sqrt{p(1-p)}}$$

$$= \left(2\sqrt{p(1-p)}\right)^7 \frac{20\sqrt{p(1-p)}-3}{\left(10\sqrt{p(1-p)}-1\right)^2}$$

$$= 128\left(\sqrt{p(1-p)}\right)^7 \frac{20\sqrt{p(1-p)}-3}{\left(10\sqrt{p(1-p)}-1\right)^2}$$

$$= \frac{2560(p(1-p))^4 - 384(p(1-p))^3\sqrt{p(1-p)}}{\left(10\sqrt{p(1-p)}-1\right)^2}. \qquad (49)$$

The graphic dependences, obtained with using relations (44), (47) and (49), are presented at Fig. 18.

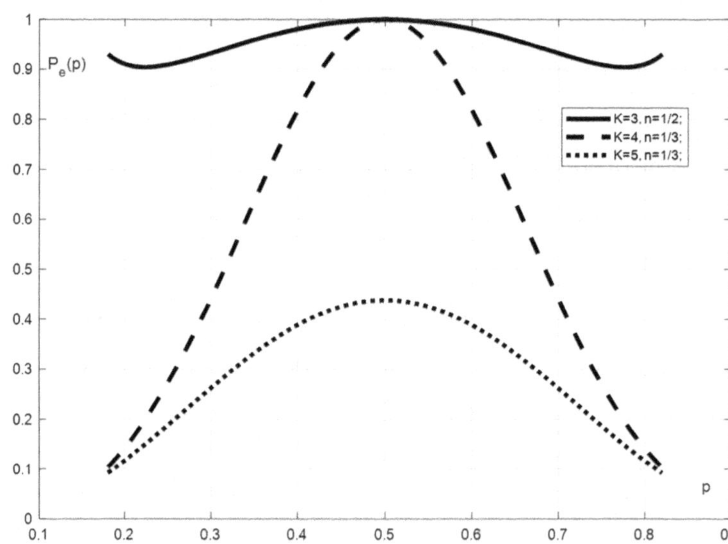

Fig. 18. The graphic dependences of maximal value of error probability in the convolutional codes $\left(3, \frac{1}{2}\right)$, $\left(4, \frac{1}{3}\right)$ and $\left(5, \frac{1}{3}\right)$ verse the bit error $P_e(p)$, obtained with using relations (44), (46) and (49)

It is clear, that for convolutional code $\left(3, \frac{1}{2}\right)$ the minimal value of maximal error probability is $P_B = 0{,}905$, and it take place in 2 cases: namely for the bit errors $p = 0{,}215$ and $p = 0{,}785$. In another side, the maximum value of $P_B = 1$ is take place for

$p = 0{,}5$. The function $P_B(p)$ is symmetric relatively to the line $p = 0{,}5$, therefore, to any value of p the condition $P_B(p) = P_B(1 - p)$ is correct. The value of P_B, taking from the graphic dependences, presented at the Fig. 18, are generally great, but it should be taking into account, that it's corresponded to the probability of maximal bit error in the convolutional code $\left(3, \frac{1}{2}\right)$. The similar dependence for the convolutional code $\left(4, \frac{1}{3}\right)$, presented at Fig. 18, as well as dependence for the convolutional code $\left(5, \frac{1}{3}\right)$, are also symmetric relatively to the line $p = 0{,}5$. This symmetry can be simply explained by the correction properties of convolutional codes, which are operation with presumptions. Really, if the probability of bit error is high, the code construction is usually known by the pervious symbol about this error and it corrected automatically. But in contrary, when the probability of bit error is $p = 0.5$, defining the correct symbol by the prediction is impossible, because the probabilities both correct and wrong received symbols are the same. Therefore, for the convolutional code $\left(4, \frac{1}{3}\right)$ maximal value of error probability in the case $p = 0.5$ is also $P_B(0.5) = 1$, but for other values of bit error probability p the maximal values of code error probability $P_B(p)$ for this code are significantly smaller, than for code $\left(3, \frac{1}{2}\right)$. And for the convolutional code $\left(5, \frac{1}{3}\right)$ the maximal value of code error P_B also corresponded to the value of bit error $p = 0.5$, but for this case $P_B(0.5) = 0{,}4375$. It is clear also, that for all values of bit error p the values of code error $P_B(p)$ is the smallest for the most sophisticated convolutional code construction $\left(5, \frac{1}{3}\right)$.

Analytical transforming of relation (27), with taking into account (28) and the value of clearance parameter of convolution code d_f, give the following results for dependence of error in the code combination on the relation of bit signal power E_b to the noise power spectrum N_0 $P_B\left(\frac{E_b}{N_0}\right)$ for PSC signal in the GNC [20].

1. For the convolutional code $\left(3, \frac{1}{2}\right)$, $d_f = 5$:

$$P_B\left(\frac{E_b}{N_0}\right) \leq Q\left(\sqrt{\frac{5E_b}{N_0}}\right)\exp\left(\frac{5E_b}{2N_0}\right)\frac{\exp\left(-\frac{5E_b}{2N_0}\right)}{\left(1 - 2\exp\left(-\frac{E_b}{2N_0}\right)\right)^2} - \frac{Q\left(\sqrt{\frac{5E_b}{N_0}}\right)}{\left(1 - 2\exp\left(-\frac{E_b}{2N_0}\right)\right)^2}. \tag{50}$$

2. For the convolutional code $\left(4, \frac{1}{3}\right)$, $d_f = 6$:

$$P_B\left(\frac{E_b}{N_0}\right) \leq Q\left(\sqrt{\frac{4E_b}{N_0}}\right)\exp\left(\frac{2E_b}{N_0}\right)\frac{2\exp\left(-\frac{2E_b}{N_0}\right)}{\left(1 - \exp\left(-\frac{2E_b}{3N_0}\right)\right)^2} == \frac{Q\left(\sqrt{\frac{4E_b}{N_0}}\right)}{\left(1 - \exp\left(-\frac{2E_b}{3N_0}\right)\right)^2}. \tag{51}$$

3. For the convolutional code $\left(5, \frac{1}{3}\right)$, $d_f = 7$:

$$P_B\left(\frac{E_b}{N_0}\right) \leq Q\left(\sqrt{\frac{14E_b}{3N_0}}\right)\exp\left(\frac{7E_b}{3N_0}\right)\exp\left(-\frac{7E_b}{3N_0}\right)\frac{10\exp\left(-\frac{E_b}{3N_0}\right) - 3}{\left(5\exp\left(-\frac{E_b}{3N_0}\right) - 1\right)^2}$$

$$= Q\left(\sqrt{\frac{14E_b}{3N_0}}\right)\frac{10\exp\left(-\frac{E_b}{3N_0}\right)-3}{\left(5\exp\left(-\frac{E_b}{3N_0}\right)-1\right)^2}. \tag{52}$$

The estimated graphic dependences for the probability of error in convolutional code combination P_B on the relation of bit signal power E_b to the noise power spectrum N_0, obtained with using Eqs. (50–52), are presented at Fig. 19.

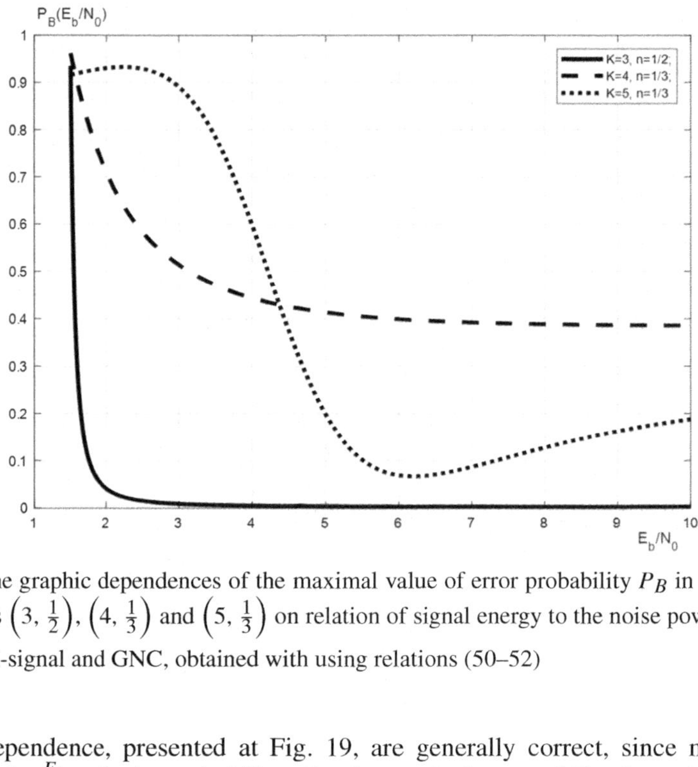

Fig. 19. The graphic dependences of the maximal value of error probability P_B in the convolutional codes $\left(3, \frac{1}{2}\right)$, $\left(4, \frac{1}{3}\right)$ and $\left(5, \frac{1}{3}\right)$ on relation of signal energy to the noise power spectrum $\frac{E_b}{N_0}$ for PSK-signal and GNC, obtained with using relations (50–52)

The dependence, presented at Fig. 19, are generally correct, since mostly with increasing the $\frac{E_b}{N_0}$ ratio the probability of code error is decreased. For the code $\left(5, \frac{1}{3}\right)$ this regularity is also generally satisfied, but not for all values of ratio $\frac{E_b}{N_0}$. Namely, it is clear from the Fig. 19, that for the value $\frac{E_b}{N_0} = 2.5$ the calculated dependence $P_B\left(\frac{E_b}{N_0}\right)$ has the local maximum, and for the value $\frac{E_b}{N_0} = 6$ it has the local minimum. Furthermore, the regularity of reduction of the probability of error in the convolutional code P_B in the case of improving the corrective ability of the code in the dependences, presented at Fig. 19, generally isn't observed. This is primarily due to the fact that the estimates with using the relations (50–52) are approximative and usually slightly inflated. Namely, with using these relations estimated only the maximal value of error probability in the convolutional code P_E, therefore its real value is usually smaller. For example, the minimal value of error probability for the convolutional code $\left(4, \frac{1}{3}\right)$, even with the great value of $\frac{E_b}{N_0}$ relation, corresponding to the Fig. 19, is $P_B \approx 0.45$. Clear, that for the real convolutional encoding systems $\lim_{\frac{E_b}{N_0} \to \infty} P_B\left(\frac{E_b}{N_0}\right) = 0$ [14–19]. Similarly, from Fig. 19 is clear, that for the convolutional code $\left(5, \frac{1}{3}\right)$ with increasing the ratio $\frac{E_b}{N_0}$ from 6 to 10 the value of P_B is also increased from 0.09 to 0.2.

In any case, obtained estimations of convolutional codes parameters with using relations (28, 32, 37, 45), are generally correct [14–20]. As for approximative relations (29, 30) for estimation the probability of code error in the GNS, forming of more precision models for GNC and complex digital signals is impossible today by the reason of very sophisticated task [14–19].

5 Conclusions

A combined approach for Intent-Based Networking (IBN) is defined and proven within the given work. The Quality of Service (QoS) parameters, such as better performance, security, data rates, and latencies are herewith fully guaranteed due to realization of the parallel computing in the cloud environments.

Using of complex factor of efficiency of parallelization F_p, defined by Eq. (11–14), is allow defining the real possibility of parallelizing the computational task in IBN. With using Eq. (11–14) the amount of transmitted data V_p should be defined with taking into account the redundant factor, caused by applying the convolutional code with the defined constraint length K. The singularities of forming convolutional codes and estimation of its redundancy and error correction parameters are described in the Sect. 4. For estimation the clearance parameter of convolution code the FSM theory have been used. Clear, that, corresponding to Fig. 2, applying of convolutional code is corresponded to the channel layer of OSI reference model. For estimation of the level of parallelization with using relations (11–14) is recommended, and, as it pointed out in the third section and corresponded to Fig. 2, it realized on the application layer of OSI model. The novel approach to estimation the level of algorithm parallelization is considered in the Sect. 2.3 and it based of forming and analyzing the connections between the elements of recurrent matrixes. Programming means of program language Python for realizing algorithm parallelization in network computing are considered in the Sect. 2.3.

Estimations of energy consumptions with using Eqs. (20, 21), considered in the Sect. 3.3, are also necessary for realizing "Green IT", but, as it was pointed out in the Sect. 2.3, such estimations for knowing amount of data V_p are generally static and can be give as constant value at application layer at the user's program, which forming the data threads for parallelized task. In the Sect. 4 separately, as the possible task for parallelization, were considered the polynomial operations, necessary for defining the clearance parameter d_f of convolutional codes. However, it was also pointed out, that, in any case, reducing the number of polynomial terms during forming this task for parallelization in IBN, is generally necessary. For providing this operation using of Eq. (42) is recommended.

The investigation of correction ability of convolutional codes, provided in the forth part of this chapter, allows make a main conclusion, that probability of code error P_B is generally decreased with increasing of constraint length K and redundancy factor $\frac{1}{n}$. Mostly the correction ability of the convolution code is defined by the clearance parameter d_f, which calculated by analyzing of the FSM states. For defining the error probability using of Eqs. (28–30) is possible, correspondent calculations were considered in the Example 13. Also, it was proof by calculations of code error, that with high probability of bit error corrective ability of convolutional codes is generally better, than the same parameter to RS-codes, considered in the Sect. 4.1. Choosing of the constraint length K and redundancy factor $\frac{1}{n}$ for convolutional codes is usually the compromise solution between complication of the encoding and decoding electronic devices and corresponded software applications and increasing the code corrective ability. The redundant information, appears with the using of convolutional codes, can be simply estimated as $I_c = I_i \cdot n$, where I_i – initial information, I_c – coded information. Since the correction ability of convolutional codes is very high value and it increased with reducing the probability of code error P_B, QoS factor of IBN with using of convolutional codes in communication systems is generally increased. Nowadays, the convolutional codes are widely used for certain well-known IEEE communication standards, as well as for Wi-Fi, LTE [6, 23].

It is clear from the contents of this chapter, that such important factors, as level of parallelization, energy consumption and method of coding are very significant for providing of QoS in realizing parallel computing in IBN. The main methods for estimation these factors are complexly described. Correspondent estimative relations, as well as examples of its' using, are also given.

The chapter may be interesting for the experts on creating of software applications for realizing of parallel computing in networks, including the cloud realization in IBN.

References

1. Luntovskyy, A.O., Melnyk, I.V.: Simulation of technological electron sources with use of parallel computing methods. In: XXXV IEEE International Scientific Conference "Electronic and Nanotechnology (ELNANO)", Conference Proceedings, Kyiv, Ukraine, pp. 454–460 (2015). https://ieeexplore.ieee.org/document/7146929
2. Luntovskyy, A.O.: Technologii Rozpodilenyh Programnyh Dodatkiv. Monography. Published in Ukrainian Language. Kyiv, State University of Information and Communication Technologies DUIKT (2010). 452 p. ISBN 978-966-2970-51-7
3. Luntovskyy, A.O., Klymash, M.M., Semenko, A.I.: Rozpodileni Servisy Telekomunikatsiinyh Merezh ta Povsiakdennyy Computing i Cloud-techno;ogii. Monography. Published in Ukrainian Language. Lviv, National University 'Lvivs'ka Politechnika' (2012). 368 p. ISBN 978-966-2405-87-3
4. Luntovskyy, A.O., Klymash M.M.: Informatsiina Bezpeka Rozpodilenyh System. Monography. Published in Ukrainian Language. Lviv, National University 'Lvivs'ka Politechnika' (2014). 444 p. ISBN 978-966-322-397-1
5. Globa, L.S., Melnyk, I.V., Luntovskyy, A.O.: Waste heat transport models for green clouds. In: 2016 IEEE International Black Sea Conference on Communications and Networking, BlackSeaCom (2016). https://ieeexplore.ieee.org/document/7901554
6. Tanenbaum, A., Wetherall, D.: Computer Networks, 5th edn. Pearson Prentice Hall, Hoboken (2011)
7. Melnyk, I.V.: Systema Naukovo-Technichnyh Rozrahunkiv MatLab ta ii Vykorystannia dlia Rozviazannia Zadach z Electroniky. Tom 2. Osnovy Programuvannia ta Rozviazuvannia Prykladnyh Zadach. Published in Ukrainian Language. Kyiv, University 'Ukraina' (2009). 327 p. ISBN 978-966-388-288-8

8. Melnyk, I.V.: Osnovy Programuvannia na Movi Python. Tom 1. Bazovi Pryncypy pobudovy Movy Programuvannia Python ta ii Golovni Syntaksychni Konstrukcii. Published in Ukrainian Language. Kyiv, Kafedra (2020). 372 p. ISBN 987-617-7301-73-0
9. Melnyk, I., Tyhai, S., Pochynok, A.: Universal complex model for estimation the beam current density of high voltage glow discharge electron guns. In: Ilchenko, M., Uryvsky, L., Globa, L. (eds.) MCT 2019. LNNS, vol. 152, pp. 319–341. Springer, Cham (2021). https://doi.org/10.1007/978-3-030-58359-0_18
10. Melnyk, I.V.: Osnovy Programuvannia na Movi Python. Tom 2. Rozvyneni Zasoby Movy Programuvannia Python. Published in Ukrainian Language. Kyiv, Kafedra (2020). 492 p. ISBN 987-617-7301-73-7
11. Isachenko, V.P., Osipova, V.A., Sukhomel, A.S.: Teploprovodnost. Published in Russian Language. Moscow, Energoatomizdat (1975). 488 p. https://www.c-o-k.ru/images/library/watermarked/cok/334/33429/961f467d6c5fc3c6a02d5a5fbff490af.pdf
12. Horst, K.: Taschenbuch der Physik. Published in German Language. Hanser Verlag. 21 Edition (2014). 711 p.
13. Espe, W.: Materials of High Vacuum Technology. Pergamon Press, Oxford (1966). 912 p.
14. Sklar, B.: Digital Communications: Fundamentals and Applications, 2nd edn. University of California, Los Angeles (2001)
15. Berlekamp, E.R.: Algebraic Coding Theory. McGraw-Hill Book Company, New-York (1968). 478 p.
16. Lathi, B.P.: Modern Digital and Analog Communication Systems. Holt. Rinehart and Whinston Inc. (1989). 720 p.
17. Birkhoff, G., Bartee, T.C.: Modern Applied Algebra. McGraw-Hill Book Company, New-York (1970). 471 p.
18. Peterson, W.W., Weldon, E.J.: Error-Correcting Codes, 2nd edn. MIT Press, Cambridge (1972). 560 p.
19. Blahut, R.: Theory and Practice of Error Control Codes. Addison-Wesley Press, Boston (1983)
20. Melnyk, I.V.: Koduvannia Sygnaliv v Elektronnyh Systemah. Chastyna 3. Sposoby Koduvannia Sygnaliv. Tom 2. Grupovi, Iteratyvni ta Zgortkovi Kody. Published in Ukrainian Language. Kyiv, Kafedra (2021). 633 p.
21. Gubner, J.A.: Probability and Random Processes for Electrical and Computer Engineers. Cambridge University Press, Cambridge (2006)
22. Anderson, J.A.: Discrete Mathematics with Combinatorics, 2nd edn. University of South Carolina, Spartanburg (2004)
23. Irvine, J., Harle, D.: Data Communication and Networks: An Engineering Approach. Wiley, Hoboken (2001)
24. Luntovskyy, A., Guetter, D., Melnyk, I.: Planung und Optimierung von Rechnernetzen: Methoden, Modelle, Tools für Entwurf, Diagnose und Management im Lebenszyklus von drahtgebundenen und drahtlosen Rechnernetzen. Springer/Vieweg + Teubner Verlag Wiesbaden (2011). 411 p. ISBN: 978-3-8348-1458-6, (in German)
25. Luntovskyy, A., Spillner, J.: Architectural Transformations in Network Services and Distributed Systems: Service Vision. Springer, Heidelberg (2017). Case Studies, XXIV, 344 pages and 238 color pictures, ISBN: 978-3-658-14840-9
26. Luntovskyy, A., Gütter, D.: Moderne Rechnernetze-Übungsbuch: Aufgaben und Musterlösungen zu Protokollen, Standards und Apps in kombinierten Netzwerken. Springer Fachmedien Wiesbaden GmbH (2020). ISBN: 978-3-658-25619-7 (in German)
27. Luntovskyy, A., Gütter, D.: Moderne Rechnernetze-Theoriebuch: Protokolle, Standards und Apps in kombinierten drahtgebundenen, mobilen und drahtlosen Netzwerken. Springer Fachmedien Wiesbaden GmbH (2020). ISBN: 978-3-658-25617-3 (in German, foreword Alexander Schill, in German)

Modeling of 5G Energy Efficiency on Example of Germany as Technological Basis for Intent-Based Networking

Daniel Wasiutinski and Volodymyr Vasyutynskyy(✉)

BA Dresden University of Cooperative Education (Saxon Study Academy), Hans-Grundig-Str. 25, 01307 Dresden, Germany

Abstract. Two topics dominate the public debates today: environmental protection and digitalization. Both are very closely related: for example, digital technology in Germany consumes about as much energy per year as a large German city. At the same time, the total amount of data transmitted over Internet is increasing, which requires new methods for transmission of higher data volumes. As consequence, we must answer the question of how technical progress of digitalization could go along with environmental protection. This chapter addresses the question whether the new 5G network, which should be not only fast but also energy efficient, would be a reasonable solution for this. In this respect, we compare the current 3G and 4G mobile networks with the new 5G network regarding workload and energy efficiency based on the different variants of the development of transmitted data volumes in mobile networks in Germany. We estimate the energy consumption, the future data consumption and especially the energy efficiency of current networks and of the fifth generation of mobile communications for the coming years in Germany by statistical analysis using different models and extrapolation. The result shows that 5G is indeed more energy efficient than 4G, but only above a certain data volume per mobile cell. Still, with the increasing mobile traffic in the coming years, the 5G will clearly lead to better energy efficiency compared to previous technologies.

Keywords: 5G · 4G · Energy efficiency

1 Motivation

Climate change and environmental protection are of central importance in our society and are often the subject of public discussion. That is why technical innovations are not only received with enthusiasm, but also checked with skepticism for environmental friendliness and efficiency. It is noticeable that the focus is no longer on the highest possible profit, but rather the profit is weighed against the consumption of the resources, especially the energy consumption. The resources like energy or resources for energy generation (e.g. crude oil, natural gas, and coal) are available on Earth to a limited extent and are usually very polluting during energy generation. Renewable energies are therefore part of environment-friendly solutions, but these in their turn are limited by

M. Klymash et al. (Eds.): *Future Intent-Based Networking*, LNEE 831, pp. 380–391, 2022.
https://doi.org/10.1007/978-3-030-92435-5_21

environmental factors, whereby the potential of reducing the consumed energy is far from being exhausted nowadays. For this reason, there are a lot of activities to create new, environmentally friendly technologies, whether directly in energy production (e.g. renewable energies) or on the consumption side by curbing energy consumption and increasing efficiency. If you look at the energy consumption in Germany, it is noticeable that Internet with approximately 43.2 TWh energy consumption per year [1] also has a significant share in the total electrical energy consumption of approximately 500 TWh per year. This proportion increases due to the process of digitization and thus the expansion of the Internet infrastructure. End devices, servers and communication networks are included in the energy consumption. The focus is primarily on servers and communication networks, because while end devices can be further optimized, mobile networks are facing a rapidly increasing data traffic. The reason for the ever-increasing demand for data is mainly video streaming, but also the general increase in the use of the Internet among the population. This means that the need to transmit more data in a shorter period is growing, which is why the energy consumption in the mobile communications sector is also growing.

But the increasing energy demand would contradict to the environmental protection needs. The 5G technology is intended to remedy this by exchanging larger amounts of data to the user faster, and at the same time more energy efficient than the fourth generation (4G).

The expansion of 5G standard has started in Germany in year 2020. The topic is being controversially discussed: on the one hand it's seen as the starting signal for complete digitization, on the other hand 5G is also described as inefficient and sometimes even as harmful to health. That is the topic of the possible energy efficiency and consumption in the 5G networks after their broader adoption is so appealing.

In this chapter we compare the 5G energy consumption of the whole mobile network infrastructure with that of current standards 3G and 4G based on different scenarios of future data consumption in Germany in the next 8 years. The mobile network itself is considered, not the energy requirements of the data centers. We want to see if the promise of the 5G to improve the data rates along with the energy efficiency could be fulfilled when the 5G will be broader adopted.

2 State-of-the-Art

2.1 Mobile Networks and 5G

Currently, the mobile market in Germany is dominated by 3G and 4G networks. Below, we shortly overview these technologies based on [2, 3].

3G technology, introduced in 2000 as UMTS standard, is still widespread and supports the data rates of up to 42 MBit/s in the frequency band of 1.920 to 2.170 MHz.

The LTE (Long Term Evolution) cellular standard is the successor to the third generation (3G) and is in use since 2010 [4]. At that time, it was supposed to replace 3G to achieve much higher data rates. In some cases, all LTE development stages are incorrectly named as 4G. However, only technologies that exceed the 3G standard are considered 4G cellular standard (therefore LTE is not the same as 4G); so, they must achieve speeds of up to 100 MBit/s for a driving user and 1 GBit/s for a stationary/standing user. Another feature of 4G is the lack of so-called circuit switching. With packet switching, which is now used in 4G, resources are only claimed when there is information to be transmitted. This technology is used to "squeeze" more data into the same bandwidth and thus increases efficiency with improved data rate and capacity. This means that download speeds of up to 300 MBit/s are possible, and even more with further developments. The 4G network uses the frequency ranges of 0.7–2.6 GHz, with 700MHz frequency primarily used for rural areas due to required large ranges. With this increased bandwidth, up to 90% of the corresponding broadband gaps could be closed in Germany and for 2019 there should have been network coverage of 4G networks of 82.2% in metropolitan areas and 73.5% outside of them [5]. Still, the transfer from 3G and 4G in Germany is not completed and these standards are used in parallel, with some mobile operators like O2 starting to stop 3G only by middle of 2021.

5G standard was introduced in 2016, while mobile network companies started deploying it worldwide in 2019. It was developed meet future requirements on data volume, latencies, and energy efficiency. 5G should offer the users the improvements in three areas:

1) Smooth access available for the user always.
2) 5G should enable improved networking of intelligent systems, e.g. in transport or in industry.
3) 5G offers scalability, i.e. the flexible adaptation of the infrastructure to requirements, for example in industry, by varying such factors like data rate, latency and frequency and adapting them to the usage purpose. Not only new technologies will be used and applied, but also old applications on the QoE and QoS principle. The QoE principle (= Quality of Experience) describes subjective empirical values in which this quality of a product/service is expressed. The QoS principle (= Quality of Service) in turn describes the objective, rational quality of a product/service, i.e. which one can also measure and understand. In other words, 5G offers application-specific better data rates, latency, and energy efficiency.

This shall be enabled by several new technologies sketched below.

Higher frequency ranges are introduced in 5G: mid-band with 2.5–3.7 GHz, allowing speeds of 100–900 Mbit/s, and high-band with 25–39 GHz and potential data rates of up to several Gbit/s. However, this comes with smaller spatial ranges and more impediments by walls and windows, as well as with higher costs for additional antennas and transceivers.

Beamforming is a method where the covered area of a radio cell is no longer covered isotopically with signal to build up a weaker, but extensive network, where electromagnetic waves are sent specifically to a point (here: a user). This bundling of the transmission energy on the active users leads to a better signal-to-noise ratio and increases the throughput. This is achieved by dynamic massive MIMO (MIMO = Multiple Input Multiple Output) antennas which consist of 64 * 64 small antennas.

According to [6] and [7], massive MIMOs reduce the radiated energy by a factor proportional to the square root of the number of antennas, while the data rate remains unchanged (applies to an ideal, single-cell MIMO system). Hardware performance was considered in [8], with the result that network energy efficiency is maximized, but for a limited number of antennas.

According to [9], the key to the energy efficiency of networks is to condense them. This can be achieved by using many base stations or many antennas per base station. Analyzing the individual optimization processes in [9] showed that a reduction in cell size has a positive effect on energy efficiency. This effect occurs when the consumption of the cell's circuit exceeds its transmission power. A further positive increase in energy efficiency arises from the increase in the number of base station antennas, i.e. in principle the structure of MIMOs. These energy gains are made by reducing interferences in the cell, since more cells lead to partially smaller data volumes and thus there are fewer disruptive factors.

Spatial density of the networks is essential to increase efficiency and the number of possible devices. In 4G, the radio cells are divided into a small number of smaller areas that are assigned to less-powered base stations. Dense heterogeneous networks increase the number of base stations per area and form a larger number of heterogeneous infrastructures, which are activated as required, as can be seen in Figure, for example. The cells are divided unevenly to optimize the network for the respective use, instead of using uniform cells that are only scalable to a limited extent (Fig. 1).

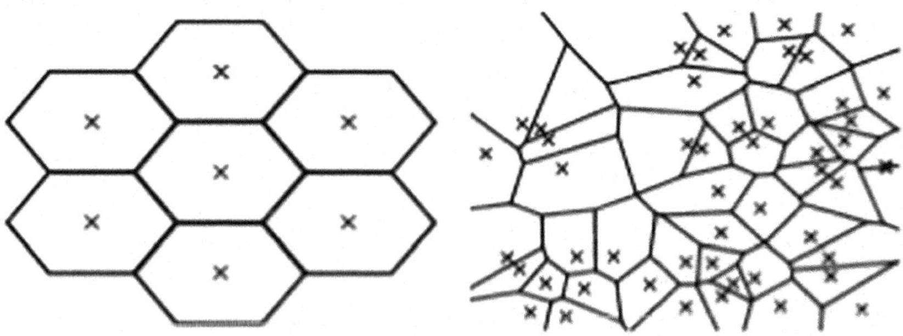

Fig. 1. Mobile cells of traditional structure (left) and heterogeneous networks (right) [10]

The problem is finding of nodes through which the cell shapes are formed. It is relatively difficult to plan such nodes, which is why stochastic means must be used. Nevertheless, this concentration of nodes leads to a reduction in the distances between the end devices. This in its turn leads to higher data rates at lower transmission power, but this can result in additional interference [11]. Analyzes of the optimal density of cells and disruptive factors for increasing energy efficiency are not yet particularly precise.

To further improve the energy efficiency, the mobile networks can be combined with further radio access technologies like device-to-device communication (D2D), visible light communication (VLC), local caching or millimeter wave mobile communication with frequency of up to 28 GHz. However, these technics have rather limited scenarios and thus will probably have only a small influence on the overall consumption.

2.2 Internet Consumption

Germany as industry country with a mix of urban and rural areas is quite representative for analysis of mobile data consumption and its energy efficiency. With estimated 43.2 TWh in 2020, the consumption of digital end devices and infrastructure makes up to 8% of the overall energy consumption, according to the German Agency for renewable energy [1].

Although the energy consumption of telecommunications is expected to further decrease in the coming years, it is more explained by the optimization of digital end devices (computers, TV, smartphones). However, the consumption of the telecommunication infrastructure, including the mobile network infrastructure, data centers and telecommunication networks, are expected to grow by around 60% from 2007 to 2025. A fundamental factor for this growth is the increase of the data volume that is expected in Germany and worldwide.

According to the forecasts, the total volume of data generated in the world will increase by factor of 5 from 33 ZB in 2018 up to 175 ZB in 2025 [12]. There are many reasons for such a rapid increase, such as increased number of users and duration of use, digitalization, and video streaming. So, streaming platforms such as YouTube, Netflix and Twitch already account for almost 60% of the data volume [13] and this share of the data volume is growing. They also contribute a large part to the increasing volume of data, and thus also to the energy consumption by the Internet. This development exposes the engineers to two fundamental problems. On the one hand, we must transmit larger volumes of data faster. On the other hand, the focus is also on energy consumption, because if we want the energy consumption to keep pace with the increasing amounts of data, energy efficiency must be increased.

3 General Assumptions

In this chapter, we evaluate the overall consumption of the mobile network infrastructure in Germany in the coming years depending on such factors as the possible variants of the mix of the mobile technologies (4G, 5G) and different scenarios of data volumes.

3.1 Energy Consumption of Cells

The information about the energy consumption of the base stations is still quite scarce due less spread of this technology. For this work, we will use the data from Huawei, see Fig. 2. It shows that today (4G) consumption is 6 kW on average per base station and 8 kW during maximum utilization. In 3 years, with partial introduction of 5G, this maximum output should be 14 kW and the average output 11 kW. In 5 years, the values of the maximum power increase to 19 kW and on average to 14 kW.

Furthermore, the maximum throughput for 4G is assumed to be 300 Mbit/s and the typical throughput of 100 Mbit/s. For 5G, the assumed typical throughput is 1 Gbit/s and the maximum value is 10 Gbit/s [14]. The data throughputs are assigned to the associated performance values of the base stations. In our modeling, we will use the exponential interpolation for data throughputs between these values.

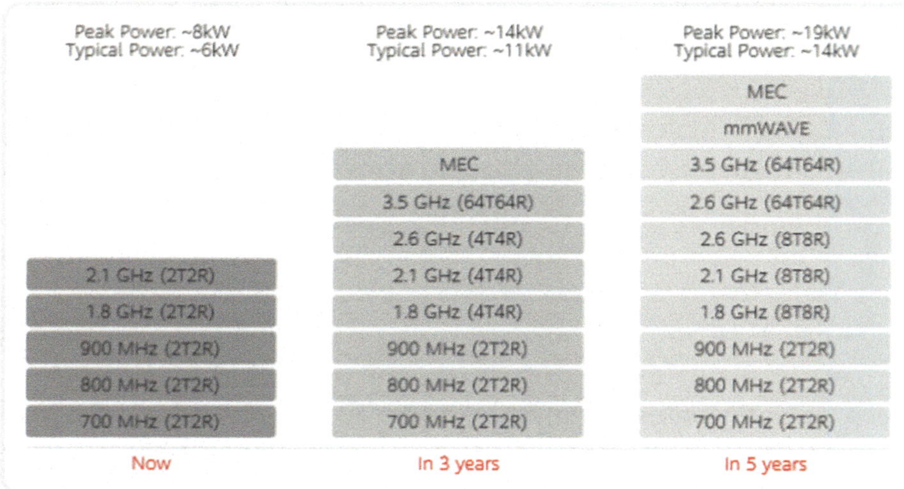

Fig. 2. Maximum and average power of the respective base station of the antennas (from Huawei [15])

A very high value for the specific energy requirement for 5G and a lower value for the 4G for the data rates under 10 Mbit/s can be explained by the higher basic consumption of the 5G base stations, which broadcast in a broader spectrum and therefore require more hardware modules and antennas. Both lines intersect at around 600 Mbit/s and the 5G graph continues to increase slower than 4G consumption at higher data rates. This makes 5G work more energy-efficient than 4G over the data rates of 600 Mbit/s per mobile cell (Fig. 3).

These figures show the ambiguity of 5G regarding the energy efficiency. On the one side, 5G has larger basic consumption compared to 4G, on the other side the energy efficiency of 5G excels that of 4G at higher data rates per base station. Thus, to evaluate the overall effect on the energy consumption, the analysis of the overall consumption scenarios is needed, which is in scope of this chapter.

3.2 Data Traffic

The data volume in mobile networks of Germany in last years is presented in the Figure below. Its analysis shows that it can follows the exponential model, which allows to forecast the overall data volume for the coming years. The exponential growth of data volume goes in line also with the data volume forecasts for the world until 2025. The possible drivers for further growth of data volumes would be high-resolution video streaming, digitalization, as well as broader usage of sensors and IoT technologies.

For our estimation of energy consumption, the number of the base stations and the transmitted data rates per cell play an important role. The number of the base stations in Germany continuously increased in the past, cp. Fig. 5. We can assume that with 5G technology and their need for denser mobile networks, the number of the base stations will further increase.

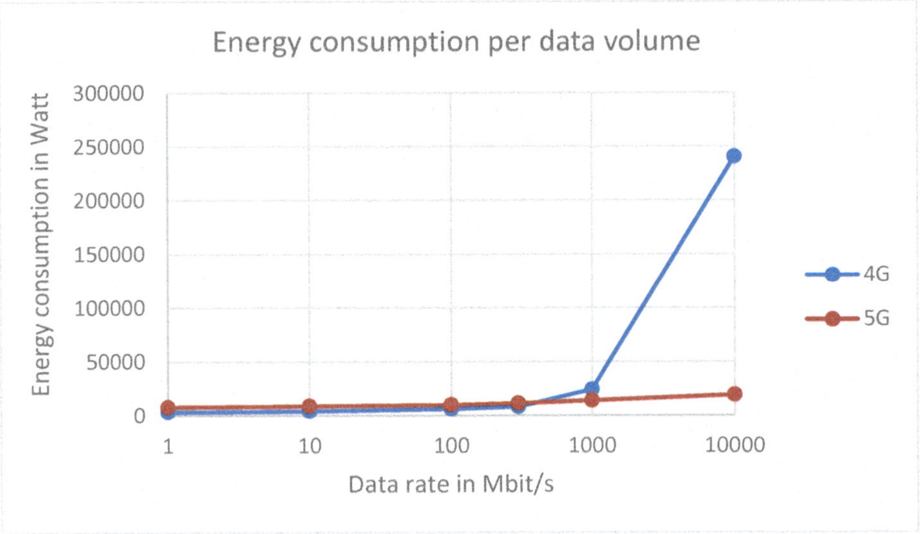

Fig. 3. Energy consumption per data volume, interpolated based on data from [15]

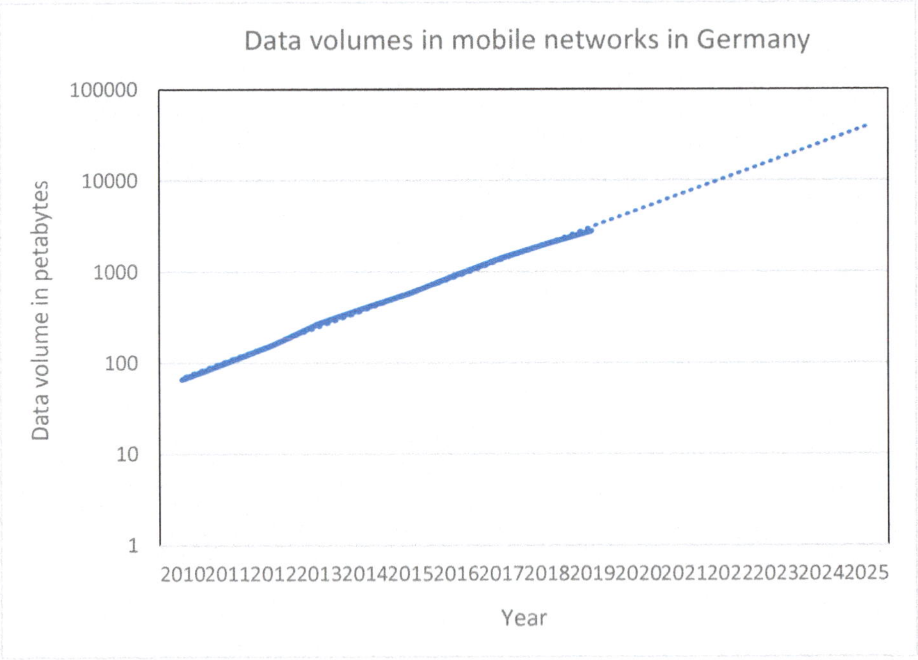

Fig. 4. Data volume in mobile networks in Germany in years 2010–2020 and forecast until year 2025 (dotted line, extrapolated with exponential model), based on [16]

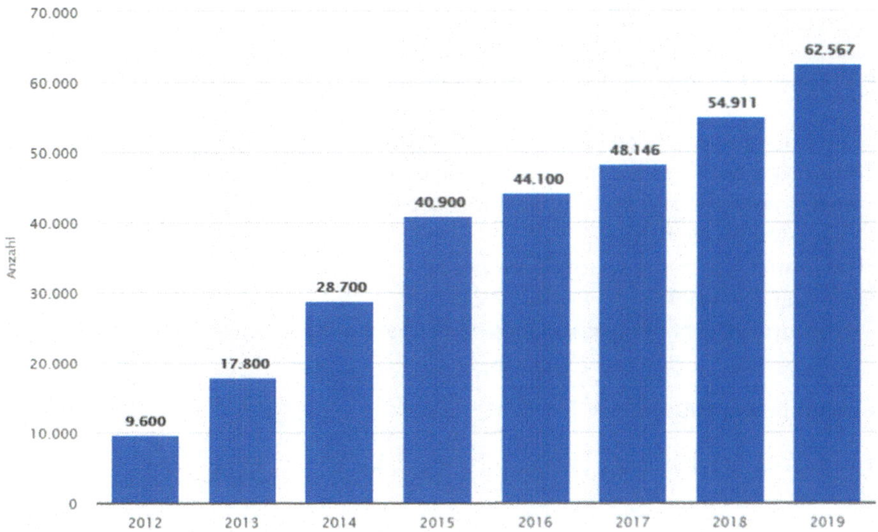

Fig. 5. Number of LTE base stations in Germany in years 2012–2019 [17]

4 Energy Consumption Simulation

4.1 Scenario and General Assumptions

We shall simulate the energy consumption of 4G and 5G technologies in Germany for years 2020 to 2028 based on following potential scenarios:

- Scenario 1: Constant number of base stations. It is assumed that the number of mobile cells for 4G and 5G remains the same, i.e. 5G replaces 4G one-to-one in year 2020.
- Scenario 2: Increasing number of base stations. It is assumed that the number of mobile cells for 5G further increases to leverage the advantages of the 5G regarding more dense mobile cells and better usage of higher frequency bandwidths.
- Scenario 3: This is a combined scenario for scenarios 1 and 2 with changing mix of 4G and 5G technologies. It is assumed that the ratio of 5G base cells will linearly grow from 2020 to 2028, starting with 0% of 5G cells and 100% of 4G cells in 2020, and ending with 100% of 5G cells and 0% of 4G cells in 2028. This scenario reflects the realistic assumption that the mobile networking companies will gradually retrofit the current antennas to 5G or build new 4G and 5G antennas due to significant necessary investments.

For all scenarios, we make further assumptions:

- The data volumes will increase in coming years according to the model introduced in Fig. 4.
- The daily traffic per cell consists of 0.5h peak traffic and 23.5h of average traffic.

4.2 Simulation Methodology

The simulation consists of the following steps:

1. Determination of the average and peak mobile traffic per base station based on the forecasted numbers for number mobile cells and mobile data traffic. The forecast models are validated by comparison of the numbers for last years with the numbers forecasted by the model.
2. Determination of the energy consumption per cell based on the mobile traffic for different scenarios.
3. Projection of the energy consumption per cell to the German-wide consumption.

4.3 Simulation Results

The results of simulation are presented in the figures below (Figs. 6 and 7).

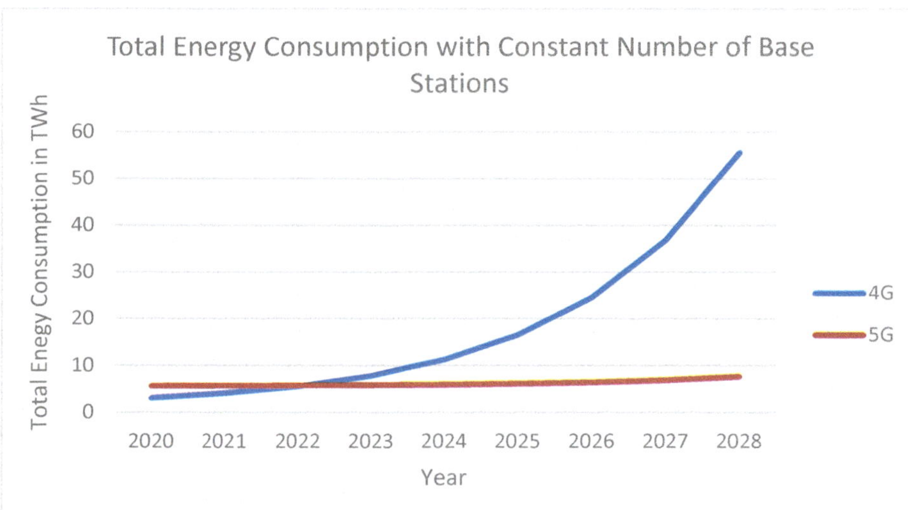

Fig. 6. Simulation of total energy consumption for mobile networks in years 2020–2028 in Germany for Scenario 1 (constant number of base stations)

The simulations clearly demonstrate the advantages of 5G regarding energy consumption compared to 4G with the increasing data rates. The breakeven point for energy consumption for both technologies will be achieved already in year 2022 for Scenario 1 (constant number of base stages) or in year 2023 for Scenario 2 (increasing number of base stations). In year 2028, the 4G energy consumption would increase with 55 TWh by factor 17.7, making up to 11% of the total electricity consumption of Germany in 2020 (approximately 500 TWh [18]), compared to 0,8% in year 2020. Compared to that, the energy consumption for the 5G technology increases by factor of 2.4 and thus quite moderate, increasing the share of the mobile network energy consumption in the total

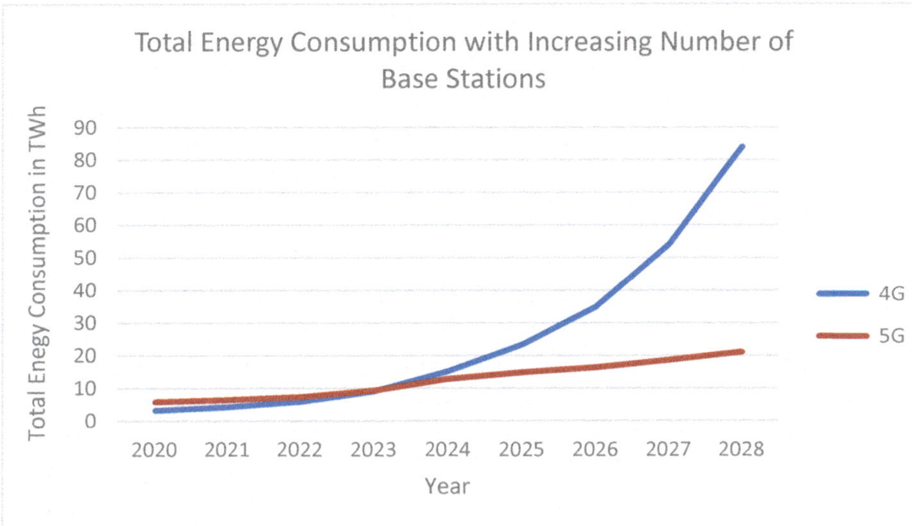

Fig. 7. Simulation of total energy consumption for mobile networks in years 2020–2028 in Germany for Scenario 2 (increasing number of base stations)

energy consumption to 1.5%, which is still acceptable. Thus, 5G would be indeed the possible way to cover the increasing demand of mobile data traffic from the energy point of view.

5G scales also well with the increasing number of base stations. When according to our forecast the number of base stations increases from 62.6 thousand in 2019 to 240,8 thousand in 2028, the total energy consumption would increase linearly from 6.31 TWh to 21.0 TWh. However, in contrast to [9], the denser mobile networks with higher number of base stations do not lead to the increase of the energy efficiency on the large scale, but rather increase the overall energy consumption. Thus, we rather expect the usage of denser mobile network only in selected areas like high populated city centers or production facilities with higher number of sensor data exchanged over 5G.

Probably, the most realistic scenario 3 shown in Fig. 8 still demonstrates the advantages of 5G. Even if the increased mobile traffic in years 2024–2007 will lead to a significant increase of the total consumption due to a big ratio of "old" 4G base stations, after a complete transition to the 5G the total energy consumption would decrease to a level between 7.6 TWh (for constant number of mobile cells) and 21.0 TWh (for increased number of mobile cells).

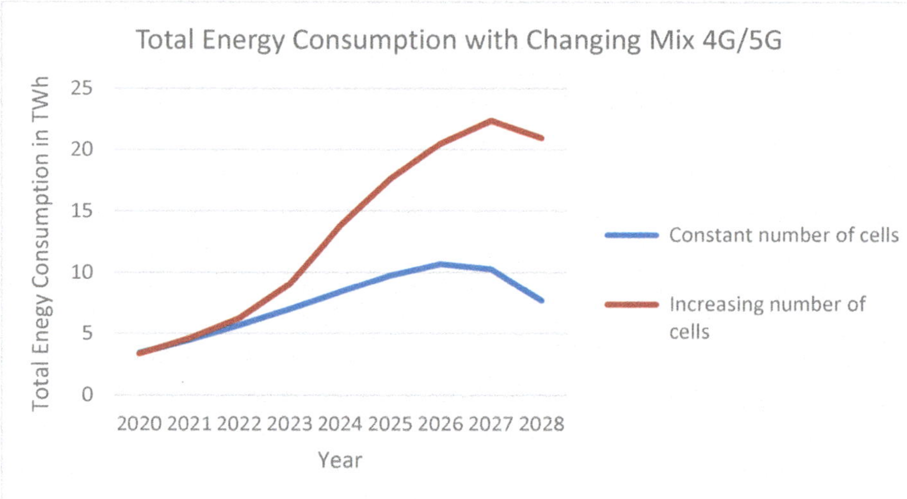

Fig. 8. Simulation of total energy consumption for mobile networks in years 2020–2028 in Germany for Scenario 3 (changing mix of 4G and 5G base stations)

5 Conclusions and Outlook

The simulation results clearly demonstrate that with the increasing mobile traffic in the coming years, the transition to 5G will significantly improve the energy efficiency of the mobile networks. Further using of 4G technology would lead to a devastating energy demand for mobile network infrastructure with up to 11% of the overall electrical energy consumption of Germany in 2028. But the complete transition to 5G until 2028 would allow to keep the share of the mobile network energy consumption at 1.5%, which is still acceptable. Further benefits of the 5G like larger data rates, lower latencies and enabling further scenarios like IoT underline the advantage of this mobile technology.

The simulation method of this work neglected certain details like spatial distribution of the mobile networks, different traffic types and economical cost/benefit analysis. However, on such a large scale, these details would contribute less to the overall estimation of the energy consumption in a country. Along with that, considering these factors would be important for example during the planning of base station placement.

Acknowledgement. Authors acknowledgements to Dresden University of Cooperative Education (Berufsakademie Dresden) and in special to Professor Daniel Gembris for support and inspiration. This work was elaborated as part of the Special Learning Activity at Martin-Andersen-Nexö Gymnasium in Dresden, Germany.

References

1. German Agency for Renewable Energies (2018). https://www.unendlich-viel-energie.de/newsletter/?nl=79:0

2. Arasan, E.: A review on mobile technologies: 3G, 4G and 5G. In: Second International Conference on Recent Trends and Challenges in Computational Models-2017 (2017). https://www.researchgate.net/profile/Ezhil_Arasan9/publication/324182620_A_Review_on_mobile_technologies_3G_4G_and_5G/links/5d380698299bf1995b453f8d/A-Review-on-mobile-technologies-3G-4G-and-5G.pdf.
3. Fagbohun, O.O.: Comparative studies on 3G,4G and 5G wireless technology. IOSR J. Electron. Commun. Eng. (2014). https://pdfs.semanticscholar.org/92cf/1e3eb16a2f18db2766897164e084c995e36e.pdf
4. Huang, J., Qian, F., Gerber, A., Mao, M., Sen, S., Spatscheck, O.: A close examination of performance and power characteristics of 4G LTE networks. In: MobiSys 2012: Proceedings of the 10th international conference on Mobile systems, applications, and services, June 2012, pp. 225–238 (2012). https://doi.org/10.1145/2307636.2307658
5. t-online.de. Investigation to LTE coverage of German cities (in German). see www.t-online.de/digital/handy/id_85926356/studie-zu-lte-abdeckung-diese-staedte-haben-das-beste-4g-mobilfunknetz.html
6. Marzetta, T.L.: Noncooperative cellular wireless with unlimited numbers of base station antennas. IEEE Trans. Wirel. Commun. 9(11) (2010). http://iwct.sjtu.edu.cn/personal/zychen/Massive%20MIMO.pdf
7. Ngo, H., Larsson, E., Marzetta, T.: Energy and spectral efficiency of very large multiuser MIMO systems. IEEE Trans. Commun. (2012). https://arxiv.org/pdf/1112.3810.pdf
8. Björnson, E., Sanguinetti, L., Hoydis, J., Debbah, M.: Optimal design of energy-efficient multi-user MIMO systems: is massive MIMO the answer? IEEE Trans. Wirel. Commun. **14**(6) (2015). https://arxiv.org/pdf/1403.6150.pdf
9. Björnson, E., Sanguinetti, L., Kountouris, M.: Deploying dense networks for maximal energy efficiency: small cells meet massive MIMO (2016). https://arxiv.org/pdf/1505.01181.pdf
10. Buzzi, S., Lin, C., Klein, T., Poor, V., Zappone, A.: A survey of energy-efficient techniques for 5G networks and challenges ahead (2016). https://arxiv.org/pdf/1604.00786.pdf
11. Björnson, E.: Pilot Contamination in a Nutshell. ellintech.se (2017). https://ma-mimo.ellintech.se/2017/01/14/pilot-contamination-in-a-nutshell/
12. statista.com. Forecast of the world wide digital data consumption (2018). https://de.statista.com/statistik/daten/studie/267974/umfrage/prognose-zum-weltweit-generierten-datenvolumen/
13. eon.de. Why is the power consumption of interent as much bad for the environment as the air traffic (in German). https://www.eon.de/de/eonerleben/warum-der-stromverbrauch-im-internet-die-umwelt-genauso-belastet-wie-der-weltweite-flugverkehr.html
14. telekom.com. 5G Speed is Telecommunication in Real Time (in German): https://www.telekom.com/de/konzern/details/5g-geschwindigkeit-ist-datenkommunikation-in-echtzeit-544496
15. Huawei 5G Telecom Energy Target Network White Paper. https://carrier.huawei.com/~/media/CNBGV2/download/products/network-energy/5G-Telecom-Energy-Target-Network-White-Paper.pdf
16. statista.com. Transmitted Data Volume in German Mobile Networks from 2005 to 2019. https://de.statista.com/statistik/daten/studie/172798/umfrage/datenvolumen-im-deutschen-mobilfunkmarkt-seit-2005/
17. statista.com. Number of LTE base stations in Germany from 2012 to 2019. https://de.statista.com/statistik/daten/studie/793776/umfrage/anzahl-der-lte-basisstationen-in-deutschland/
18. statista.com Netto consumption of electrical power in Germany from 1991 to 2019. https://de.statista.com/statistik/daten/studie/164149/umfrage/netto-stromverbrauch-in-deutschland-seit-1999/)

Methods of Signal Detection and Recognition to Perform Frequency Resource Sharing in Cognitive Radio Networks

Valeriy Bezruk[1], Stanislav Ivanenko[1], Oleksii Fedorov[1](✉), Zdeněk Němec[2], and Jan Pidanič[2]

[1] National University of Radio Electronics, Kharkiv, Ukraine
{valerii.bezruk,stanislav.ivanenko,oleksii.fedorov}@nure.ua
[2] University of Pardubice, Pardubice, Czech Republic
{zdenek.nemec,jan.pidanic}@upce.cz

Abstract. The paper addresses the problem of spectrum sensing in cognitive radio networks and aims at improving efficiency of signal detection procedures by employing unconventional methods of signal detection and recognition. Such the methods allow us to reference unknown signals to a special class of signals for which no prior information is provided.

Conventional methods for signal sensing are as follows: (i) constructing adaptive decision rules based on fitting observed signals with typical stochastic models (like Gaussian and Rician). The main drawback of these methods is the lack of robustness to possible violations of initial model assumptions; (ii) testing signals for cyclostationarity. Despite their effectiveness such the methods rely on prior knowledge about spectral width of signal to detect and are computationally expensive; (iii) energy level detection, based on comparing signal's energy with a threshold, is easy to implement however it grants poor accuracy for low SNRs.

As of the moment the lack of efficient facilities of blind signal detection and radio emission type recognition slows down further progress in development and use of cognitive radio techniques. Thus the problem is urgent. The paper focuses on researching into the algorithms based on methods to detect changes in probabilistic properties of signals. Research is performed with respect to samples of real signals, typical for both VHF/UHF and the IEEE 802.22 frequency bands.

Keywords: Frequency · Sensing · Channel · Signal · Noise · Detection · Recognition · Resource · Sharing · Cognitive radio

1 Introduction

Nowadays, wireless radio technologies are an intensively developing branch of the telecommunications industry. The reason for their high popularity is the convenience of their use for civil and specialized applications. Currently there

M. Klymash et al. (Eds.): *Future Intent-Based Networking*, LNEE 831, pp. 392–409, 2022.
https://doi.org/10.1007/978-3-030-92435-5_22

are lots of radio means sharing the air, and their number grows rapidly every day. Naturally, the frequency resource is already quite jammed [23]. Empty frequency bands are required to introduce new radio means; the lack of frequency bands at hand leads to electromagnetic compatibility issues.

An alternative solution to the problem is an efficient (in some sense) utilization of the available frequency resource. The described above circumstances have become prerequisites for the emergence of the IEEE 802.22 standard. This standard incorporates the cognitive radio (CR) technologies [3,9,16,21,27] with the aim to provide electromagnetic compatibility of broadband radio access systems and ground TV broadcasts operating in the 54 – 862 MHz frequency range.

A CR network is a self-organizing radio system with dynamic access to the radio frequency resource. Such a system is capable of learning its operational and locational environment to adapt its functional parameters and protocols to them. Also the CR network is capable of employing the knowledge gained in the process of operation to change its operational environment, subject to the established regulatory policies and the CR network's functional state [13,16].

The CR technology adaption is intended to an increase in the efficiency of the radio frequency resource utilization through identification of those frequency bands that are not used by primary (licensed) users and providing them for a short-term use by secondary users.

Considering the above, we extract these three operational tasks of the CR:

(i) frequency resource acquisition, that is identification of temporarily unoccupied frequency channels in the given range for the secondary use;
(ii) frequency resource access prioritization, that is primary users go first, while secondary users get prioritized, depending on the capacity requirements and delay sensitivity;
(iii) frequency resource utilization optimization, that is optimal algorithms selection to achieve secondary channel capacity maximization and arrange frequency resource access.

Among the listed so far tasks the most crucial one is related to the ability to observe dynamics of radio emission (RE) changes within a specified frequency range. Information on the frequency resource utilization may be obtained in two ways [3,19,21,27,28]:

(i) from the CR system database;
(ii) by automated spectrum sensing of a given frequency range.

Spectrum sensing [3,6,28] is a special case of radio monitoring, which is a complex task of space-spectral-time processing of REs observed in a wide range of frequencies [18,19,25]. With the simplicity in mind this task is decomposed into a number of relatively independent processing tasks, namely, (i) detection of specified REs, (ii) selection and recognition of specified RE types, (iii) detection of unknown REs, (iv) recognition and estimation of modulation parameters of unknown REs.

Conventionally these tasks can be accomplished by processing signals taken from the output of scanning radio systems [11,17,25,28]. At the first stage, for every frequency channel a decision is made on the presence of either a mixture of signal and noise or noise only, i.e. the problem of signal detection at presence of noise is actually solved. As a result of such an analysis of observations, unoccupied frequency channels are determined. At the second stage, once a signal has been successfully spotted, it undergoes a recognition procedure done to determine whether this signal is known to the CR system. In other words, every detected signal is to be assigned to one of two classes, namely, signals peculiar to primary (licensed) or secondary users. It is also possible that the detected signal breaks the pattern, i.e. has nothing to do with the signals the CR system was trained to recognize. To the best of our knowledge, such a variation of the statement of the problem of detecting unoccupied frequency channels along with classifying signals found in the occupied frequency channels when conducting spectrum sensing in the CR networks was not considered by other researchers.

This paper presents results of research into the problem of detection and recognition of specified known signals in the presence of unknown signals for the purpose of applying these techniques to determine occupancy of frequency channels during spectrum sensing in cognitive radio networks.

2 Frequency Resource Acquisition

2.1 Problem Analysis and a Proposed Approach

The bare bones solution to the problem of signal detection in the presence of noise is normally considered within a parameter estimation context [7], namely, we judge on the value of the parameter $\theta \in \{0, 1\}$ given the output signal:

$$x(t) = \theta \cdot s(t) + n(t), \tag{1}$$

where $s(t)$ is for the known signal and $n(t)$ is for the additive channel noise. Usually, the noise samples are treated as i.i.d. normal with 0 mean and unknown variance. To move any further, we need to put forward a hypothesis on θ. There is an uncertainty about the null hypothesis choice, which should be done carefully, keeping in mind that [4]: a null hypothesis should never be accepted; it is either rejected, or not rejected, based on the fact whether the data are consistent or inconsistent with the null hypothesis.

A conventional [7,15,16] choice is $H_0 : \theta = 0; \quad H_1 : \theta = 1$. This choice is in a good agreement with the above. Indeed, once the type I error (i.e. probability of a false alarm) α is fixed, then the probability of confusing noise with the signal is α and in terms of the CR, making a mistake leads us to non-occupying a currently free channel which is not beneficial from the perspective of the secondary channel capacity utilization, but guards us against possible collisions with primary users, which is of the most priority for us. Imagine we do a vice versa and say H_0' :

$\theta = 1$; $H_1' : \theta = 0$, then the type I error is to declare noise only presence when there is a mixture of the signal and noise. This is a much worse scenario.

The case (1) allows a relatively easy generalization for a set of multiple signals to detect [22]. This situation is quite typical for the cognitive radio, as it is not very unusual that primary users have a bunch of different signals to transmit within a prefixed range of frequencies. The main drawback of this M-hypothesis [22] approach is that we are forced to make a decision. That is, if we are looking to discriminate noise from a bunch of peculiar to some primary users signals then multiple signal misdetections are possible and we will never be sure it is safe to occupy the frequency band of interest by transmitting a secondary user signals. It means a completely different approach is needed. This approach must account for these main facts:

(i) observations that are consistent with the null hypothesis H_0 do not imply the alternative H_1 is false [4];
(ii) the search for unoccupied frequency channels is done within prefixed sub-bands of the given frequency range. Hence there should be awareness of typical signals peculiar to primary users which utilize this frequency range;
(iii) primary users may transmit signals of a fresh type that is not in the CR data base;
(iv) signals from another secondary user may be present in the frequency band of interest, that the CR is not aware of;
(v) cognitive radio system should be capable of discovering previously unregistered signals emitted by primary or secondary users.

The first item in the list implies that only a rejected hypothesis is capable of providing us with statistically significant information about the observed sample. That is, a new strategy is necessarily two-stage one:

1. At the first stage, the input signal is tested for membership of the class of known signals, i.e. those for which training samples are at hand. If so, optimal or quasi optimal decision rules may be used to achieve low rates of recognition error. In turn, previously unknown signals (training samples for which are missing in the cognitive radio database) can appear at the input. In this case, in order to avoid the error of false assignment of an unknown signal to the class of known signals, emitted by primary users, it is proposed to enroll such an unknown signal into a special class, say $(M + 1)$.
2. The signal that has been referenced to the $(M + 1)$-th class is unknown to the system, but it may well also turn out to be noise that indicates the channel non-occupancy. Thus at the second stage it is necessary to check whether an unknown signal is noise. To achieve this we may resort to the use of such decision rules, which rely solely on statistical properties of the noise. Alternatively, once the null hypothesis on the noise presence has been rejected, the input signal must be treated as completely new to the CR system and additional inquiry should be undertaken to clarify the signal's nature and reference it to the base of primary or secondary signals.

Note, to keep the probability of false alarm tolerable, we need to resort to the theory of multiple comparisons [12]. More details on this will be provided in the subsequent sections.

2.2 $M + 1$ Hypothesis Problem

We assume that signal being recognized is represented by a finite dimensional random vector $\vec{x}$ of equidistantly spaced samples of the signal. Decision on signal's membership is made with respect to observed realizations. We put forward $(M+1)$ hypotheses about the received signals, namely, H_i, $i = \overline{1, M}$ are for statistically defined signals, H_0 is for signals gathered into the $(M+1)$-th class and possessing unknown probabilistic characteristics [2]. Probability densities $W(\vec{x}|\vec{\alpha}_i)$, $i = \overline{1, M}$ of the statistically defined signals are specified accurate within vector parameters $\vec{\alpha}_i$, $i = \overline{1, M}$ and the probability density is unknown for the $(M+1)$-th class. Prior probabilities of hypotheses $\mathsf{P}(H_i) = P_i$ are also given and $\sum_{i=0}^{M} P_i = 1$. It is assumed that learning samples for M defined signals $\{\vec{x}_{ir},\ r = \overline{1, n_i};\ i = \overline{1, M}\}$ are given and a learning sample for the $(M+1)$-th class of unknown signals ($i = 0$) is either absent or unrepresentative.

Now we proceed with the average risk analysis [22,24]. Here we are, the average risk is

$$\mathcal{R} = \sum_{l=0}^{M} \sum_{\substack{i=1 \\ l \neq i}}^{M} C_{li}\, P_i\, \mathsf{P}(y_l|H_i) + \sum_{i=1}^{M} C_{0i}\, P_i\, \mathsf{P}(y_0|H_i) + P_0 \sum_{l=0}^{M} C_{l0}\, \mathsf{P}(y_l|H_0), \quad (2)$$

where C_{li} is for the cost of each course of action to be taken, the first subscript indicates the hypothesis chosen and the second, the hypothesis that was true; $\mathsf{P}(y_l|H_i)$ is for the probability of deciding in favor of the l-th signal when the i-th signal is present.

A non-randomized decision rule of recognition does the observation space partitioning into M non-overlapping domains. Allowing for that, the first term in (2) is the component of the average risk caused by a misclassification within the class of defined signals. The second term of the average risk is due to referencing a defined signal to the $(M+1)$-th class of unknown signals. The third term is the component of the average risk due to referencing an unknown signal from the $(M+1)$-th class to one of M defined signals.

According to the available information it is possible to find within the stated problem of recognition, estimates of the first two components in (2). It does not seem possible to estimate the third component. With the aim to take into account the third term we offer to introduce a scalar parameter that is equal to the volume of the rejection region $Y = \cup_{l=1}^{M} y_l$ for the hypothesis H_0 on the presence of a signal from the $(M+1)$-th class. This region has meaning of the acceptance region of M defined signals. From the intensional point of view, the recognition problem we have under consideration consists in making a decision on presence of one of M defined signals and referencing unknown signals to the $(M+1)$-th class. In connection with the above, this problem of recognition can be treated as the problem of selection and recognition of defined random signals.

2.3 Decision Rule for Selection and Recognition of Defined Signals

Solution to the formulated above unconventional problem of signal selection and recognition gives the following decision rule of signal recognition [2]:

$$H_0 : \max_{l=\overline{1,M}} \{P_l W(\vec{x}|\vec{\alpha}_l)\} < \Delta \tag{3a}$$

we conclude that the data do not contradict the hypothesis H_0 on the presence of a signal from the $(M+1)$-th class of unknown signals;

$$H_i : \max_{l=\overline{1,M}} \{P_l W(\vec{x}|\vec{\alpha}_l)\} \geq \Delta, \tag{3b}$$

$$\begin{gathered} P_i W(\vec{x}|\vec{\alpha}_i) \geq P_l W(\vec{x}|\vec{\alpha}_l), \\ l = \overline{1, M}, \qquad l \neq i \end{gathered} \tag{3c}$$

we conclude that the hypothesis H_i is consistent with the claim that the specified i-th signal is present.

Here parameters $\vec{\alpha_i}$, $i = \overline{1, M}$ are estimated with respect to learning samples for M specified signals; the threshold Δ is defined from the condition of providing the given probability of correct recognition of the specified signals. Note, we do not use any information on probability distribution density of the $(M+1)$-th class signal as well as its learning samples.

Geometrical meaning of the decision rule (3) is explained in Fig. 1, where $\hat{\vec{G}}_p$ is the estimate of the mathematical expectation found for the p-th signal; y_p is the acceptance region of the p-th signal; ρ_{pz} is the Euclidean distance

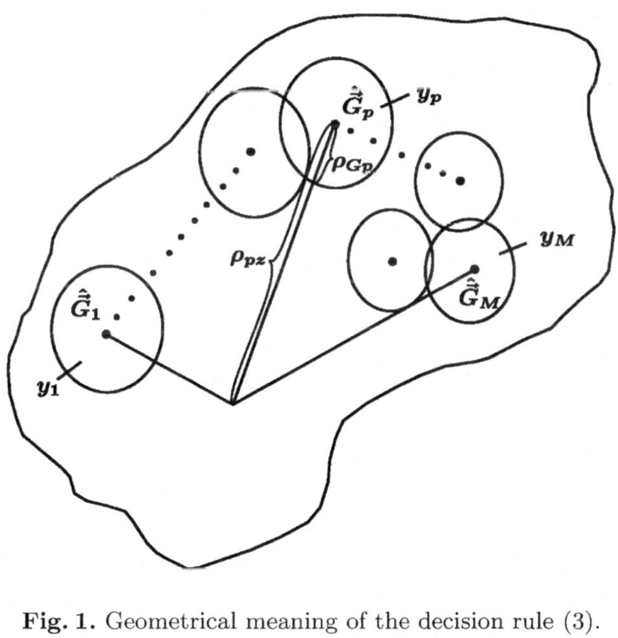

Fig. 1. Geometrical meaning of the decision rule (3).

measured from the observed realization of a signal to the centre of the p-th signal acceptance region; ρ_{G_p} is the value that defines the volume of the p-th signal acceptance region.

2.4 Decision Rules for Detecting Unknown Signals in the Presence of Noise

Moving back to the case (1) we put forward the following two hypothesis: H_0, the hypothesis, which is valid if there is no signal in the frequency band to analyze and the alternative hypothesis H_1, which is valid if H_0 fails, i.e. the hypothesis about the presence of a signal mixed with noise. Employing the introduced above mathematical notation we write [11]

$$W(\vec{x}|\vec{\alpha}_0) \underset{H_1}{\overset{H_0}{\gtrless}} \Delta_0, \tag{4}$$

where $W(\vec{x}|\vec{\alpha}_0)$ is for the multivariate density of the noise represented by the L-dimensional vector $\vec{x}$ of samples; $\vec{\alpha}_0$ is the distribution density parameter; Δ_0 is a certain threshold value selected to provide the specified probability of false alarm (Type I error).

Let $N < L$ and $\mathbf{I}_{NL} = \|a_{i,j}\|_{i,j=1}^{L}$ be a block matrix, where

$$a_{ij} = \begin{cases} 1, & i = j;\ i, j \leq N, \\ 0, & \text{otherwise.} \end{cases}$$

If $\vec{x}$ comes from a multivariate Gaussian distribution, decision rule for detecting unknown signals in the presence of noise is [2,8,26]

$$(\vec{x} - \vec{\mu}_0)^T \mathbf{I}_{NL} \mathbf{R}_0^{-1} \mathbf{I}_{NL} (\vec{x} - \vec{\mu}_0) \underset{H_1}{\overset{H_0}{\lesseqgtr}} \Delta_0, \tag{5}$$

where $\vec{\mu}_0$ and $\mathbf{R}_0$ are respectively mean vector and covariance matrix of the noise; T stands for matrix transposition and $\mathbf{I}_{NL}$ allows for taking into account not all of the components of the vector $\vec{x}$, but the first N out of L. Estimates for $\vec{\mu}_0$, $\mathbf{R}_0$ are determined from a learning sample of noise $\vec{x}_0$.

If components of $\vec{x}_0$ are uncorrelated, covariance matrix $\mathbf{R}_0$ becomes diagonal and the decision rule (5) reduces to comparing to the threshold Δ_0 Euclidian distances between the noise reference and signals. Namely,

$$\sum_{j=1}^{N} \frac{(x_j - \mu_{j0})^2}{\sigma_{j0}^2} \underset{H_1}{\overset{H_0}{\lesseqgtr}} \Delta_0, \tag{6}$$

where x_j and μ_{j0} are components of $\vec{x}$ and $\vec{\mu}_0$ respectively; σ_{j0}^2 are for diagonal elements of $\mathbf{R}_0$.

In turn, when for decision-making a set of observation vectors $\{\vec{x}_q\}_{q=1}^{v}$ is used, the decision rule (5) takes the form:

$$\sum_{q=1}^{v}(\vec{x_q}-\vec{\mu}_0)^T\mathbf{I}_{NL}\mathbf{R}_0^{-1}\mathbf{I}_{NL}(\vec{x_q}-\vec{\mu}_0)\underset{H_1}{\overset{H_0}{\lessgtr}}\Delta_{0v} \tag{7}$$

or

$$\sum_{q=1}^{v}\sum_{j=1}^{N}\frac{(x_j^{(q)}-\mu_{j0})^2}{\sigma_{j0}^2}\underset{H_1}{\overset{H_0}{\lessgtr}}\Delta_{0v} \tag{8}$$

for the case of diagonal covariance matrix $\mathbf{R_0}$. Here $x_j^{(q)}$ stands for the j-th component of $\vec{x}_q$.

Another variation of the rule (5) is

$$\mathrm{Tr}\mathbf{I}_{NL}(\mathbf{R}-\mathbf{R}_0)\mathbf{I}_{NL}(\mathbf{R}-\mathbf{R}_0)^T\mathbf{I}_{NL}\underset{H_1}{\overset{H_0}{\lessgtr}}\Delta_{0\mathbf{R}}, \tag{9}$$

where $\mathbf{R}_0=\frac{1}{n_0}\sum_{q=1}^{n_0}(\vec{x}_{0q}-\vec{\mu}_{0q})(\vec{x}_{0q}-\vec{\mu}_{0q})^T$, $\mathbf{R}=\frac{1}{v}\sum_{q=1}^{v}(\vec{x}_q-\vec{\mu}_{0q})(\vec{x}_q-\vec{\mu}_{0q})^T$, n_0 is for the size of a training sample and $v \ll n_0$ is the size of a control sample; $Tr(\cdot)$ designates matrix trace operator. If matrices $\mathbf{R}_0$ and $\mathbf{R}$ happen to be diagonal, expression (9) becomes

$$\sum_{j=1}^{N}(r_j-r_{j0})^2\underset{H_1}{\overset{H_0}{\lessgtr}}\Delta_{0\mathbf{R}}, \tag{10}$$

where for $j=\overline{1,N}$ $r_{j0}=\frac{1}{n_0}\sum_{q=1}^{n_0}(x_{j0}-\mu_{j0})^2$, $r_j=\frac{1}{v}\sum_{q=1}^{v}(x_{jq}-\mu_{j0})^2$ and $v \ll n_0$.

For the sake of comparison, a conventional energy detector

$$\vec{x}^T\vec{x}\underset{H_1}{\overset{H_0}{\lessgtr}}\Delta_0 \tag{11}$$

was used.

The above decision rules (5)–(11) define some possible algorithms for detecting unknown signals in the presence of noise, and can be used for acquisition of unoccupied frequency channels in CR networks at the second stage of the proposed technique based on $M+1$ hypotheses approach.

Spectrum sensing is mostly done in the frequency domain. That is, in what follows to monitor changes in signal and noise environment, we process bins of magnitude and energy spectra of observed signals; to get the spectra, the DFT is used.

2.5 Multiple Comparisons: Post Hoc Hypotheses Testing

A technique of frequency resource acquisition presented so far in Subsects. 2.1 through 2.3 is necessarily two stage one. And every stage requires testing hypotheses about observed signals. To keep tolerable the false alarm probability we resort to the theory of multiple comparisons [12]. A brief how-to guide on the subject one may find in [10].

Since at the very beginning of our testing activities we are unable to predict the number of stages to be passed on, we deal with the post hoc comparisons, thus we need to be more conservative than with the planned comparisons. Indeed, once the input signal is observed, we former test if the signal is known to the CR sensing system and if it is not we determine whether this signal features statistical properties of noise. If all the tests are negative we treat the observed signal as unknown one and update the CR data base. Here we stick with the Bonferroni procedure [1,10] and set the Type I error rate for both of our tests equal $\alpha_{\text{familiwise}}/2$, where $\alpha_{\text{familywise}}$ is for the significance level for the whole sensing procedure.

Figure 2 depicts a flow chart describing the proposed approach to detecting unoccupied frequency channels and sharing them with secondary CR users.

3 Results of Research into Methods for Detecting Unknown Signals

This section reports results of comparative efficacy analysis of operation of decision rules for detecting signals in the presence of noise. The ability of efficient detection of unoccupied frequency channels is crucial for a smooth operation of the CR system. Decision rules listed in Subsect. 2.4 were implemented in MATLAB; the research was done with respect to peculiar to radio networks samples of signals and noise. In particular, we dealt with a wide-band WFM signal and a narrow-band AFS FSK 130 Bd signal. Digital records were obtained with the aid of an SDR receiver, capable of scanning and analyzing a given frequency range, as well as making digital records of signals present in a particular frequency band.

Figures 3 and 7 depict magnitude spectra of the WFM and AFS FSK signals. These signals were treated as unknown ones, which appeared in the presence of noise in the frequency band of interest. A Monte Carlo experiment was used to estimate performance of the signal detection rules. In our experiment we employed both control samples of the signals and training sample of the noise having size 1000 realizations composed of $L = 64$ time bins each. Realizations of noise served the goal of determining threshold values to plug them into detection procedures, while realizations of the signals were used to get estimates for probabilities of the signal correct detection, $\hat{\mathsf{P}}(1|1)$, in the presence of noise in a given frequency channel. Note, that with regard to the stated in Subsect. 2.4 hypothesis (4), $\hat{\mathsf{P}}(1|1)$ is actually a correct rejection rate. In what follows, we use as mutually replaceable both of the terms, namely, probability of signal correct detection and probability of noise hypothesis correct rejection.

The detection performance was studied in the form of dependencies of the probability of correct detection, $\mathsf{P}(1|1)$, on the signal-to-noise ratio (SNR) with a fixed false alarm probability (Type I error) $\mathsf{P}(1|0) = 0.04$. Estimates $\hat{\mathsf{P}}(1|1)$

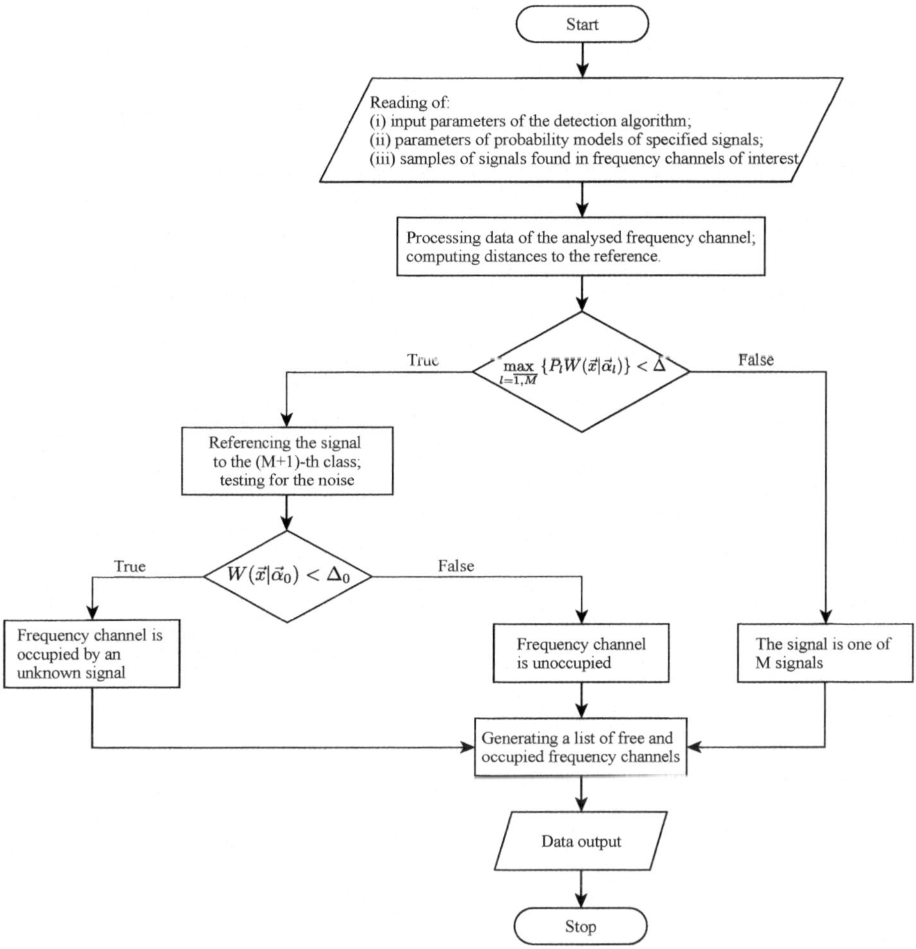

Fig. 2. A flow chart describing operation of the system to detect unoccupied frequency channels in the CR.

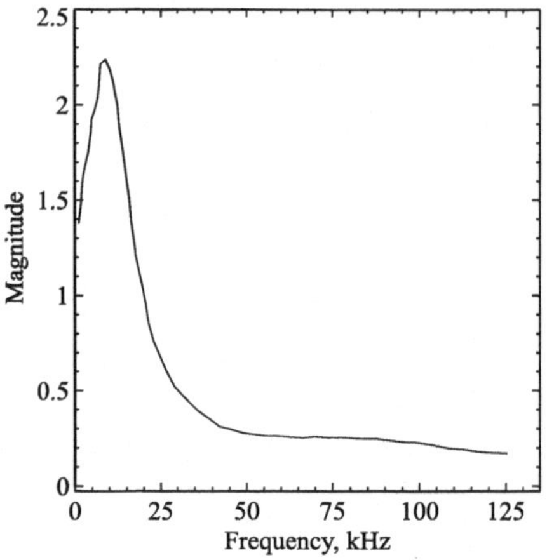

Fig. 3. Magnitude spectrum of a wideband FM (WFM) signal.

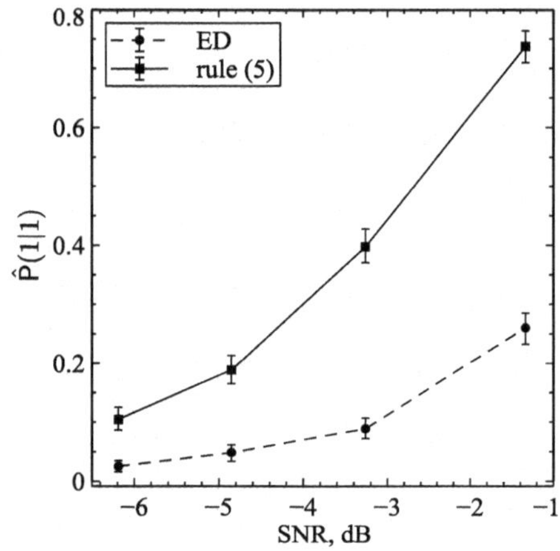

Fig. 4. $\hat{\mathsf{P}}(1|1)$ vs SNR; WFM signal; decision rules (5) and (11).

were obtained as proportions $\hat{\mathsf{P}}(1|1) = n_1/n$, where n_1 is for the number of experiments in which the correct decisions were made on the detection of the signal, $n = 1000$ is the total number of experiments.

Figures 4 through 6 and 8 through 10 show the dependences obtained for decision rules (5), (8) and (10) which were implemented in the frequency domain with $N = L/2$. Also, for the sake of reference, the above figures contain dependency $\hat{\mathsf{P}}(1|1)$ vs SNR for the energy detector (ED) represented by (11). Note, decision rule (8) deals with bins of the magnitude spectrum, while decision rule (10) has bins of the power spectrum as its input. The figures analysis suggests there

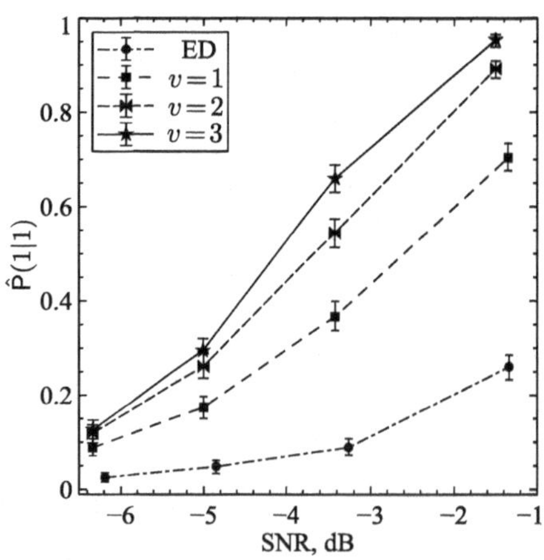

Fig. 5. $\hat{\mathsf{P}}(1|1)$ vs SNR; WFM signal; decision rules: (8) for $v=\overline{1,3}$ and (11).

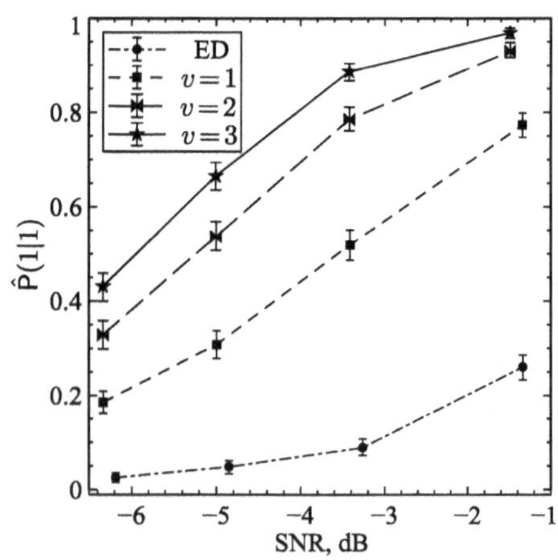

Fig. 6. $\hat{\mathsf{P}}(1|1)$ vs SNR; WFM signal; decision rules: (10) for $v=\overline{1,3}$ and (11).

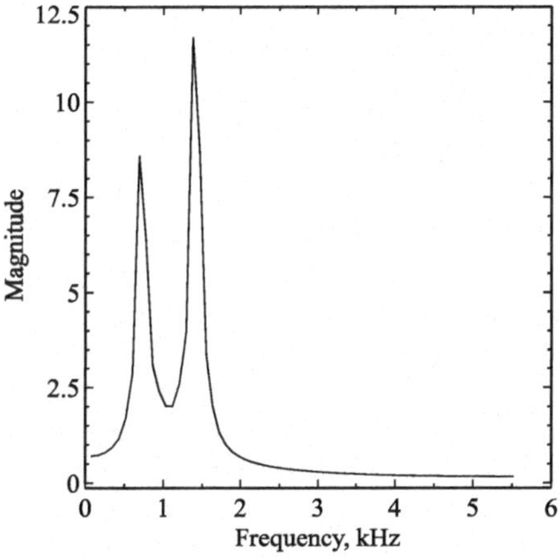

Fig. 7. Magnitude spectrum of a narrow-band AFS FSK 130 Bd signal.

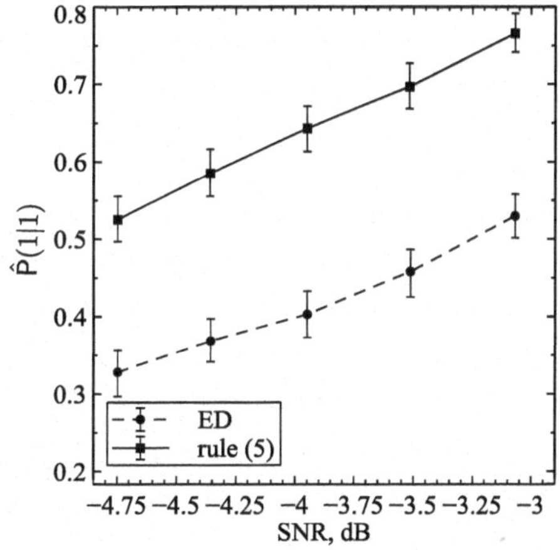

Fig. 8. $\hat{\mathsf{P}}(1|1)$ vs SNR; AFS signal; decision rules (5) and (11).

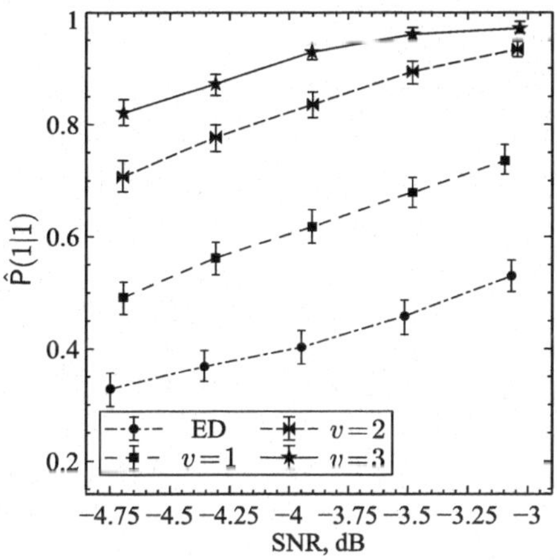

Fig. 9. $\hat{\mathsf{P}}(1|1)$ vs SNR; AFS signal; decision rules: (8) for $v=\overline{1,3}$ and (11).

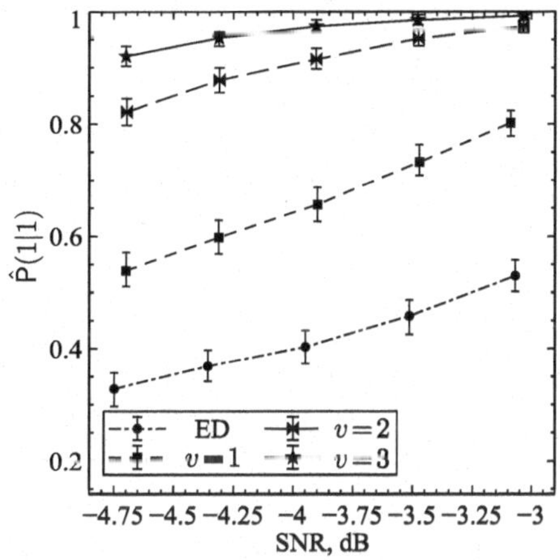

Fig. 10. $\hat{\mathsf{P}}(1|1)$ vs SNR; AFS signal; decision rules: (10) for $v=\overline{1,3}$ and (11).

are two ways to gain the probability of the unknown signal correct detection, namely, by increasing either the SNR or the number of realizations v employed at the moment of working out the decision (Figs. 4, 5, 6 and 8, 9, 10).

4 Employing the Concept of Pareto Optimality to Meet Trade-Off Requirements for Algorithms to Detect Unknown Signals

The previous section did a comparison of decision rules for detecting unknown signals with respect to the correct rejection rate $\hat{\mathsf{P}}(1|1)$ (i.e. power) which was determined under two types of alternatives, namely, wide- and narrow-band signals (see Figs. 3 and 7). Despite being informative this comparison payed no attention to computational complexity of the rules. In turn, such a complexity is closely related to the amount of data to be collected in order to make a decision.

Here we come up with two objectives: (i) miss rate $\hat{\mathsf{P}}(0|1)$ or the Type II error, i.e. the probability of declearing only noise presence where there is a mixture of signal and noise at the input and (ii) computational complexity index, which measures performance of the detection algorithms in terms of time required for a detection algorithm to accumulate the necessary sample size and make a decision.

When it comes to practical implementation of detection algorithms with the aid of high-performance computing means the required observation duration does the main contribution to a resulting value of the computing complexity index. Such a duration is computed with the equation $t = v\,T$, where T is for the duration of one realization and v is the number of realizations the algorithm employs to make a decision. Sampling frequency F_s determines a particular value of T, $T = L/F_s$. Once the sufficient length of observations has been accumulated, the DFT is computed to get the spectral representation of the observed realizations. Only unique spectral bins are retained, thus $N = L/2$.

As with the correct rejection rate $\hat{\mathsf{P}}(1|1)$, the miss rate $\hat{\mathsf{P}}(0|1)$ values, corresponding to different decision rules, were obtained by Monte Carlo experiments, $\hat{\mathsf{P}}(0|1) = n_0/n$, where n_0 is for the number of experiments in which wrong decisions on only noise presence were made with respect to a mixture of signal and noise, $n = 1000$ is the total number of experiments.

Values of t peculiar to different variants of the decision rules from Subsect. 2.4 are listed in Table 1; column $k_1 = \frac{t}{t_{max}}$ contains the normalized values. In turn, column $k_2 = \hat{\mathsf{P}}(0|1)$ reports the miss rate values. One may see easily we deal with conflicting objectives. That is, when choosing in favor of a particular version of detection algorithm, we improve one of the objectives and deteriorate another one. Therefore to perform a comparative analysis and select a preferred version of the algorithm for detecting unknown signals we resort to the theory of multi-objective optimization [5,14,20].

Let $S \subset \mathbb{R}^3$ be a feasible region associated with possible variations of the decision rules and $\vec{z} \in S$, where $\vec{z} = (z_1, z_2, z_3)^T$; z_1 is for the rule number, i.e. (5), (8), (10) and (11); $z_2 = L$ and $z_3 = v$ (see Table 1). Say Y is the image of S and $\vec{k}(\vec{z}) \in Y$, $\vec{k}(\vec{z}) = (k_1(\vec{z}), k_2(\vec{z}))^T$. Here Y is the feasible objective region and $\vec{k}(\cdot)$ is the vector-valued function, values of which coincide with those listed in columns k_1 and k_2 of Table 1. We are looking to extract the set of Pareto optimal decision vectors from the feasible region S. For the case when the feasible region S is discrete and finite, the inclusion of $\vec{z}^* \in S$ into the Pareto set $\mathscr{P}(S)$ of optimal decision vectors is done iff there is no other $\vec{z} \in S$ such that $k_i(\vec{z}) \leq k_i(\vec{z}^*)$ for all $i \in \{1, 2\}$ and $k_j(\vec{z}) < k_j(\vec{z}^*)$ for at least one index j [14].

When constructing the Pareto subset in accordance with the above approach the worst decision vectors from the feasible region S are rejected as well as the corresponding to them variants and variations of the decision rules. In particular,

Table 1. Pareto efficient algorithms to detect unknown signals

Rule	#	L	v	t, μs	$\hat{\mathsf{P}}(1\|1)$	$k_1 = \frac{t}{t_{max}}$	$k_2 = \hat{\mathsf{P}}(0\|1)$	$\mathscr{P}(S)$
(5)	1	64	1	256	0.740	0.083	0.260	–
	2	128	1	512	0.902	0.167	0.098	–
	3	256	1	1024	0.977	0.333	0.023	+
(8)	4	64	1	256	0.734	0.083	0.266	–
	5		2	512	0.927	0.167	0.073	–
	6		3	768	0.965	0.250	0.035	–
	7	128	1	512	0.894	0.167	0.106	–
	8		2	1024	0.964	0.333	0.036	–
	9		3	1536	0.990	0.500	0.010	+
	10	256	1	1024	0.954	0.333	0.046	–
	11		2	2048	0.991	0.667	0.009	–
	12		3	3072	0.998	1.000	0.002	–
(10)	13	64	1	256	0.784	0.083	0.216	+
	14		2	512	0.935	0.167	0.065	+
	15		3	768	0.970	0.250	0.030	+
	16	128	1	512	0.907	0.167	0.093	–
	17		2	1024	0.967	0.333	0.033	–
	18		3	1536	0.989	0.500	0.011	–
	19	256	1	1024	0.953	0.333	0.047	–
	20		2	2048	0.995	0.667	0.005	+
	21		3	3072	0.999	1.000	0.001	+
(11)	22	64	1	256	0.262	0.083	0.738	–
	23	128	1	512	0.330	0.167	0.670	–
	24	256	1	1024	0.481	0.333	0.519	–
Rule	#	L	v	t, μs	$\hat{\mathsf{P}}(1\|1)$	$k_1 = \frac{t}{t_{max}}$	$k_2 = \hat{\mathsf{P}}(0\|1)$	$\mathscr{P}(S)$

there were identified 7 Pareto optimal decision algorithms out of 24 feasible ones. The selected algorithms have a "+" sign in $\mathscr{P}(S)$ column of Table 1. Note, that found so far Pareto optimal variants $\vec{z}^* \in \mathscr{P}(S)$ remain incomparable among themselves in terms of the used objectives. That is it, any version of the algorithm $\vec{z}^*$ from the Pareto subset can be treated as the optimal one for solving the problem of detecting unknown signals. To narrow down the Pareto subset to a unique preferred variant of the detection rule, one may employ a certain conditional preference criterion, which utilizes some extra information on preferability among the Pareto optimal detection algorithms, for example

$$k_{\mathscr{P}}(\vec{z}) = w_1\, k_1(\vec{z}) + w_2\, k_2(\vec{z}), \qquad (12)$$

where $w_{1,2}$ are the weights to specify the relative importance of every objective. Setting $w_1 = w_2 = 0.5$ leaves us with one preferred variant, namely, number 14 in Table 1. This decision rule is the minimizer of $k_{\mathscr{P}}(\vec{z})$, i.e. $\vec{z}^* = \arg\min_{\vec{z}\in\mathscr{P}(S)} k_{\mathscr{P}}(\vec{z})$, where $\vec{z}^*$ stands for rule (10) with $L = 64$ and $v = 2$. The choice in favor of such a rule ensures the correct rejection rate $\hat{\mathsf{P}}(1|1) = 0.935$ when detecting unknown signals in the presence of statistically defined noise; the false alarm rate was set to $\hat{\mathsf{P}}(1|0) = 0.04$. In this case the required to make a decision duration of observations equals $t = 512\,\mu$s.

5 Results of Research into the Procedure for Selection and Recognition of Defined Signals

As it was described previously, we treat differently signals that the CR system is aware of and signals that are new to the system (see flow chart in Fig. 2 for details). We assume Gaussianity of the input vectors which represent signals and employ the Mahalanobis distance [8] between an input signal $\vec{x}$ and the reference (i.e. one of M known signals, $\vec{\mu}_l$) D_l, $l = \overline{1, M}$ to rewrite the decision rule (3):

$$\begin{aligned} H_0 : &\quad \max_{l=\overline{1,M}} D_l > \Delta; \\ H_i : &\begin{cases} \max\limits_{l=\overline{1,M}} D_l \leq \Delta, \\ D_i \leq D_l; \quad i, l = \overline{1, M};\ i \neq l, \end{cases} \end{aligned} \tag{13}$$

where $D_l = (\vec{x} - \vec{\mu}_l)^T \mathbf{I}_{NL}\, \mathbf{R}_l^{-1}\, \mathbf{I}_{NL}\, (\vec{x} - \vec{\mu}_l)$; $\vec{\mu}_l$ and $\mathbf{R}_l$ are respectively mean vector and covariance matrix of the l-th known signal; $\mathbf{I}_{NL}$ is the block matrix introduced in Subsect. 2.4 and Δ is some threshold value selected to keep tolerable the false alarm probability, $\alpha = 0.05/2 = 0.025$. Such a value of α was selected to meet the Bonferroni procedure requirements. Additionally, we supposed that components of the input vectors were uncorrelated, which resulted in the following expression for the Mahalanobis distance

$$D_l = \sum_{j=1}^{N} \frac{(x_j - \mu_{jl})^2}{\sigma_{jl}^2}, \quad l = \overline{1, M},$$

where $N \leq L$ and $\vec{x} = [x_1, x_2, \ldots, x_L]^T$, $\vec{\mu}_l = [\mu_{1l}, \mu_{2l}, \ldots, \mu_{Ll}]^T$, $\mathbf{R}_l = \mathrm{diag}(\sigma_{1l}^2, \sigma_{2l}^2, \ldots, \sigma_{Ll}^2)$.

Figure 11 plots normalized averaged magnitude spectra of signals which were treated respectively as known signals (1 through 4) and unknown signals (5 through 8). Among the unknown ones there was a noise signal, which had sequence number 5. The signals are 6 kHz ones [18]. Accumulated learning and testing samples included 100 realizations each. Every realization was composed of 128 time bins. Results of recognition are summarized in Table 2. Analyzing Table 2 data, we need to keep in mind that signals 1–4 were introduced to the system by their learning samples, while signals 5–8 were not. This explains misclassification of signal 7 due to the lack of corresponding to it learning samples. The rest of unknown signals (namely #5, #6 and #8) have been referenced properly to the $(M+1)$-th class. From the perspective of cognitive radio smooth operation the previously observed confusion of the 7-th signal with the 4-th one influences nothing as it is crucial for us not to occupy a currently engaged by the primary user frequency band and making a decision in accordance with the results of our classification does it for us.

Table 2. Results of signal selection and recognition

Probability of referencing Signal j to the given class	Signal j							
	1	2	3	4	5	6	7	8
$P(1\|j)$	0.97	0.00	0.00	0.00	0.00	0.00	0.00	0.00
$P(2\|j)$	0.00	0.66	0.00	0.00	0.00	0.00	0.00	0.00
$P(3\|j)$	0.00	0.34	0.96	0.00	0.00	0.00	0.00	0.00
$P(4\|j)$	0.00	0.00	0.00	0.90	0.00	0.00	0.65	0.00
$P(M+1\|j)$	0.03	0.00	0.04	0.10	1.00	1.00	0.35	1.00

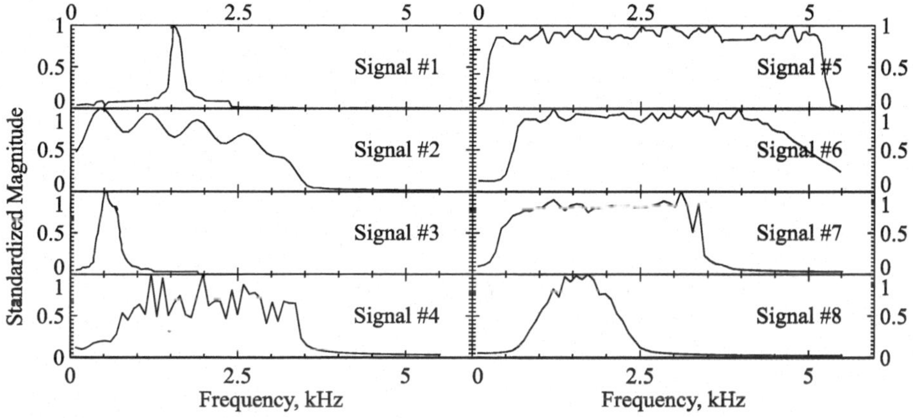

Fig. 11. Standardized averaged magnitude spectra of radio signals.

6 Conclusion

Analysis of the provided results of research into methods of signal detection and recognition to perform frequency resource sharing in cognitive radio networks shows efficiency of the proposed two stage decision-making procedure. At the first stage of this procedure the input signal is tested for membership of the class of known signals (i.e. those for which training samples are at hand); while at the second stage, all the signals which have been referenced to the class of unknown ones, are tested for exhibiting features of the channel noise.

Employing such a procedure in contrast to conventional techniques of signal recognition allows us to reduce substantially chances for occupying a previously engaged by the primary user channel, which is good from the perspective of preserving integrity of data transmitted by primary users. Keeping such an integrity is the main objective of the cognitive radio system. A native drawback of the proposed approach is that our two stage procedure being conservative leads to a decrease in capacity of the secondary channel. One of possible ways to address this problem is to increase the operability of the cognitive radio system when it comes to spotting new empty frequency sub-bands, which was done in Sect. 4.

References

1. Abdi, H.: Bonferroni test. In: Salkind, N. (ed.) Encyclopedia of Measurement and Statistics, vol. I, pp. 103–107. SAGE Publications, Thousand Oaks, Calif (2006)
2. Bezruk, V.M., Pevtsov, G.V.: Theoretical Foundations of Signal Recognition Systems Design for Automated Radio Monitoring. Collegium, Kharkov (2007). (in Russian)
3. Cabric, D., Mishra, S.M., Brodersen, R.W.: Implementation issues in spectrum sensing for cognitive radios. In: Conference Record of the Thirty-Eighth Asilomar Conference on Signals, Systems and Computers, 2004, vol. 1, pp. 772–776 (2004). https://doi.org/10.1109/ACSSC.2004.1399240
4. Christensen, R.: Analysis of Variance, Design, and Regression: Applied Statistical Methods. CRC Press, Boca Raton (1996)
5. Deb, K.: Multi-Objective Optimization Using Evolutionary Algorithms. Wiley-Interscience Series in Systems and Optimization. John Wiley & Sons, Inc., Hoboken (2001)
6. Diaz, L.M.G., Marrero, L.M., Gómez, J.T.: Performance comparison of spectrum sensing techniques for cognitive radio networks. In: 2016 16th International Convention and Fair Informática, Havana, Cuba, pp. 1–7 (2016)
7. Franks, L.E.: Signal Theory. Dowden & Cluver Inc, Stroudsburg (1981)
8. Gallego, G., Cuevas, C., Mohedano, R., García, N.: On the mahalanobis distance classification criterion for multidimensional normal distributions. IEEE Trans. Signal Process. **61**(17), 4387–4396 (2013). https://doi.org/10.1109/TSP.2013.2269047
9. Haykin, S.: Cognitive radio: brain-empowered wireless communications. IEEE J. Sel. Areas Commun. **23**(2), 201–220 (2005). https://doi.org/10.1109/JSAC.2004.839380
10. Helberg, C.: Multiple comparisons. In: Salkind, N. (ed.) Encyclopedia of Measurement and Statistics, vol. I, pp. 644–648. SAGE Publications, Thousand Oaks, Calif (2006)
11. Helstrom, C.W.: Statistical theory of signal detection: International Series of Monographs in Electronics and Instrumentation. Pergamon (1968)
12. Hsu, J.: Multiple Comparisons: Theory and Methods. CRC Press, Boca Raton (1996)
13. Khattab, A., Perkins, D., Bayoumi, M.: Cognitive Radio Networks. Springer, New York (2013). https://doi.org/10.1007/978-1-4614-4033-8

14. Miettinen, K.: Nonlinear Multiobjective Optimization, International Series in Operations Research & Management Science, vol. 12. Kluwer Academic Publishers, Boston (1999)
15. Minkoff, J.: Signal Processing Fundamentals and Applications for Communications and Sensing Systems. Artech House, Norwood (2002)
16. Mishra, S.M., Sahai, A., Brodersen, R.W.: Cooperative sensing among cognitive radios. In: 2006 IEEE International Conference on Communications, vol. 4, pp. 1658–1663 (2006). https://doi.org/10.1109/ICC.2006.254957
17. Poor, H.V.: An Introduction to Signal Detection and Estimation. Springer, Heidelberg (1994). https://doi.org/10.1007/978-1-4757-2341-0
18. Proesch, R., Daskalaki-Proesch, A.: Technical handbook for radio monitoring VHF/UHF. Books on Demand (2017)
19. Rembovsky, A.M., Ashikhmin, A.V., Kozmin, V.A., Smolskiy, S.M.: Radio Monitoring. SCT, Springer, Cham (2018). https://doi.org/10.1007/978-3-319-74277-9
20. Sawaragi, Y., Nakayama, H., Tanino, T.: Theory of Multiobjective Optimization, Mathematics in Science and Engineering, vol. 176. Academic Press, Inc., Cambridge (1985)
21. Taboković, Ž.: A survey of cognitive radio systems (2008). https://www.researchgate.net/publication/228450166_A_Survey_of_Cognitive_Radio_Systems
22. Trees, H.L.V.: Detection, Estimation, and Modulation Theory, Part I. John Wiley & Sons, Inc., Hpbpken (2001). https://doi.org/10.1002/0471221082
23. UCRF: Register of radio frequency assignments (centralized assignments)
24. Webb, A.R., Copsey, K.D.: Statistical Pattern Recognition. Wiley, Hoboken (2011)
25. Weber, C., Peter, M., Felhauer, T.: Automatic modulation classification technique for radio monitoring. Electron. Lett. **51**(10), 794–796 (2015). https://doi.org/10.1049/el.2015.0610
26. Wickens, T.D.: Elementary Signal Detection Theory. Oxford University Press, Oxford (2002)
27. Wild, B., Ramchandran, K.: Detecting primary receivers for cognitive radio applications. In: First IEEE International Symposium on New Frontiers in Dynamic Spectrum Access Networks, 2005. DySPAN 2005, pp. 124–130 (2005). https://doi.org/10.1109/DYSPAN.2005.1542626
28. Yücek, T., Arslan, H.: A survey of spectrum sensing algorithms for cognitive radio applications. IEEE Commun. Surv. Tutor. **11**(1), 116–130 (2009). https://doi.org/10.1109/SURV.2009.090109

Model of Increase of Spectral Efficiency of Use of Frequency Resource of Low-Orbit System with Architecture of the Distributed Satellite

Volodymyr Saiko(✉), Serhii Toliupa, Volodymyr Nakonechnyi, Mykola Brailovskyi, and Volodymyr Domrachev

Taras Shevchenko National University of Kyiv, Kiev, Ukraine
{tolupa,Nvc2006}@i.ua, {bk1972,mipt}@ukr.net

Abstract. The chapter deals with an overview analysis of known ways to increase the efficiency of OFDMA mobile communication systems. In order to increase the spectral efficiency of the frequency resource use of a low-orbit satellite network with a distributed satellite architecture, a model of cognitive multiuser access with OFDMA is proposed, which includes a unit for determining the required number of frequency subchannels as well as a new algorithm for determination of distorted carriers and algorithm for selecting the operating frequency section OFDMA (RU) for the respective frequency channels. With a view to evaluating the effectiveness of the developed model, simulations were performed for channels with 20 MHz bandwidth when exposed to interference with 1 MHz bandwidth, which simulate Bluetooth signals. It follows that in comparison with the original ITU algorithm, the proposed method of suppressing RU gives a significant gain in modeling, namely the efficiency of the use of subcarriers has increased significantly compared to the original algorithm.

Keywords: Low-orbit satellite network LEO-system · OFDMA technology · Preamble puncturing mechanism · Resource frequency units (RU) · Mobile telecommunication systems

1 Introduction

Given the development of fundamentally new functionalities of space systems and the need to increase their intellectualization in order to ensure a high level of efficiency and reduce time for operational decisions, there is a need to create a new type of space object called cluster or in English terminology «swarm». The use of a cluster of small spacecraft (SMA) allows you to get a number of known benefits. In particular, the group flight of the ICA increases not only the ability of such a system to perform the target functions under specified conditions and modes of application, but also its ability to reconfigure in the event of various outrageous factors. The ballistic construction of an ICA group can be both global in nature with a uniform distribution of ICA in several orbital planes and in each of these planes, and local in which the group retains its ballistic structure at small intersatellite distances. In this case, one of the ICA groups is assigned

M. Klymash et al. (Eds.): *Future Intent-Based Networking*, LNEE 831, pp. 410–423, 2022.
https://doi.org/10.1007/978-3-030-92435-5_23

the role of "leader", which can be transferred to another ICA of the same type. The main function of the ICA-leader is to coordinate the actions of the subordinate group of ICA, including the transfer of control commands and the provision of autonomous navigation information of the subordinate ICA.

From this point of view, domestic research are aimed at developing a low-orbit satellite communication system, which is a group of small spacecraft (LEO-system) with a distributed satellite architecture [1, 2, 12–14]. The development of innovative methods and devices for technical implementation of multi-user transmission in a low-orbit system to increase the spectral efficiency of the satellite network frequency resource with a distributed satellite architecture is among the currently unresolved pressing issues for such a system.

2 The State of Development of the Issue Under Study

The desire to minimize the composition of the active equipment of the ICA satellite system led to attempts to maximize the use of the included systems and equipment of the spacecraft. Therefore, communication between the root and repeater satellites in a distributed satellite is carried out using a new generation wireless broadband access network, such as LTE, WiMAX [1], which use OFDMA mode. An important feature of OFDMA technology is that data transmission can be performed on those subcarriers that are least prone to this user to frequency-selective interference. To select such subcarriers, each radio sends transmission quality reports using different subcarriers. According to experts, the choice of subcarriers, taking into account the feedback on the quality of transmission, can increase the bandwidth by up to 50% compared to the random choice of subcarriers.

There are ways to increase the efficiency of OFDMA radio communication systems, which are based on this principle of operation: [3] - a method of adaptive disconnection of carriers; [4] - a method of adaptive distribution of OFDM carriers of radio communication systems; [5] a way to increase the spectral efficiency of OFDM radio communication systems in channels with random parameters.

The disadvantage of these methods is that they do not allow to technically implement multi-user transmission in a low-orbit system "distributed satellite", i.e. simultaneous transmission of different data by the root satellite to several satellites-repeaters of the microgroup (DL MU MIMO technology) and (multi-user transmission, which is performed synchronously with several satellites-repeaters of the micro-group, and the root satellite acts as a coordinator and determines the start of the transmission - UL MU MIMO technology).

There is a method of maximum fairness (FA, Fairness algorithm), which aims to allocate users a frequency resource in order to increase the minimum data rate [6]. In essence, this corresponds to the equalization of data rates of all users, hence the name of this method.

The main purpose of FA is to maximize the minimum data rate in the system. There are problems with its implementation due to the complexity of optimizing the method of maximum justice. The general approach of this algorithm is that firstly each subcarrier is allocated equal power, and then step by step available subcarriers are assigned to users with the lowest speed, but the best characteristics in the channel.

The disadvantage of the method of maximum fairness is that the distribution of speeds between users is not flexible. In addition, the total bandwidth is largely limited by the user with the lowest signal-to-noise ratio (SINR), because most of the resources are allocated to him, which is a very suboptimal approach.

There is a method of proportional rate limitation (PRC, Proportional Rate Constraints), which aims to maximize the total bandwidth with an additional limit so that the data rate of each user is proportional to a set of predefined system parameters [7].

The disadvantage of this method is that it is particularly acute problem of optimization and, accordingly, its ability to technically implement.

There is a method of proportional fair distribution (PF, Proportional Fairness), the main purpose of which is to compromise between two priority requirements in the network: an attempt to maximize the overall bandwidth of the wireless network, and the level of service should not be less than the minimum [8]. The PF method is a method that allows you to maximize bandwidth in the long run of the user depending on the average transmission conditions, while satisfying the conditions of objectivity for each of them.

The disadvantage of this method is that it cannot support real-time applications such as voice services and video streaming services.

There are known systems and methods, which are described in the patent [9], for the implementation of a multi-user scheme that allows multiple users, groups of users or carriers to share one or more channels. In [9], the available bandwidth is divided into several subchannels with equal bandwidth according to standard OFDM practice. The transmitter is informed by the application that it needs to transmit data at a specific transmission rate. The transmitter determines the minimum number of subchannels and the maximum power (or noise) limit for each subchannel that are required to achieve this data rate, and selects a set of subchannels that meet these requirements. It is not necessary that the subchannels be contiguous in the spectrum or belong to the same channel. Once the transmitter has selected the required number of subchannels, it starts transmitting simultaneously on these subchannels across the entire bandwidth used by these subchannels.

The disadvantage of these systems and methods is that they do not allow to technically implement multi-user transmission in a low-orbit system with a distributed satellite architecture, ie simultaneous transmission of different data by the root satellite to multiple satellites. uplink (multi-user transmission, which is performed synchronously with several satellites-repeaters of the microgroup, and the root satellite acts as a coordinator and determines the start of the transmission - UL MU MIMO technology).

3 Statement of the Task

The closest to the proposed model is a known method of frequency control in the wireless network standard IEEE 802.11ah, which is being developed by ITU [15]. Its feature is that it uses the known mode of increasing the spectral efficiency of the OFDMA channel and the preamble puncturing mechanism, as a new way to adapt to interference from other devices, through more flexible control of traffic multiplexing in the channels.

In [16], each transfer may occupy one or more resource blocks. In the frequency domain, each resource block may consist of 26, 52, 106, 242, 484 or 996 subcarriers (including a number of service subcarriers). In the time domain, the duration of the resource block is equal to the duration of the multi-user transmission and each time is determined by the access point. Thus, for this multi-user transmission, the frequency bands 20 MHz, 40 MHz, 80 MHz and 160 MHz are one resource block with 242 subcarriers, two resource blocks with 242 subcarriers, two resource blocks with 484 subcarriers and two resource blocks with 996 subcarriers, in accordance. Each such resource block, in turn, can be divided into two narrower resource blocks, etc. However, there are a number of exceptions. For example, in a channel with a width of 20 MHz, a resource block with 242 subcarriers may be replaced by two resource blocks, each of which consists of 106 subcarriers, and in a channel with a width of $\geq$ 40 MHz, a resource block with 242 subcarriers may be replaced by two resource blocks, each of which consists of 106 subcarriers, and one resource block of 26 subcarriers.

Although OFDMA technology can be used for both uplink and downlink transmission, this solution does not allow uplink and downlink transmission at the same time. In the case of transmission in the downlink, the frame contains a common preamble for all recipients, which indicates information about the assignment of specific resource blocks to each of the recipients. Implementing multi-user transmission in the uplink is a more complex task. Multi-user transmission is performed synchronously by several stations, and the access point acts as a coordinator and determines the start time of the transmission. First, it must obtain information about the data available at the stations for transmission, and secondly, assign resource blocks for transmission to those stations that have data. It is also worth noting that the access point does not know whether there will be a free channel in terms of the station at the time of the scheduled transmission.

The preamble puncturing mechanism for 802.11ax networks as well as OFDMA mode are new access algorithms, alternatives and analogues of which did not exist in previous generations of the standard, therefore the study of preamble puncturing efficiency, unlike OFDMA (in LTE network research) is little studied. Therefore, the authors proposed a method for assessing the quality of the algorithm and its modeling based primarily not on data rates, but on the aspect of efficient use of frequency resources of the channel, because bandwidth does not occupy the entire bandwidth, 20 MHz subchannel.

From the simulation results shown in Fig. 1, it is seen that when the interference band does not overlap the signal band, it does not affect the operation of the preamble puncturing mechanism, ie does not initiate the suppression of one or two subchannels.

The depressions on the graphs correspond to the moments when the interference band overlaps two adjacent subchannels, respectively, in these cases, the preamble of both adjacent subchannels is suppressed. In general, for band channels, the efficiency of subcarriers is quite low, as only a certain part of the spectrum is affected by interference, but the entire 20 MHz subchannel is suppressed, and in some cases, when the interference is located within two subchannels - two. In general, the original preamble puncture mechanism [10] is not well adapted to adapt to band interference, as it suppresses one or two 20 MHz subchannels, thus failing to provide OFDMA subcarriers for the transmission of unaffected such interference.

Thus, the disadvantage of this method is that the implemented mechanism of preamble puncturing in IEEE 802.11ax networks does not provide both the rational use of

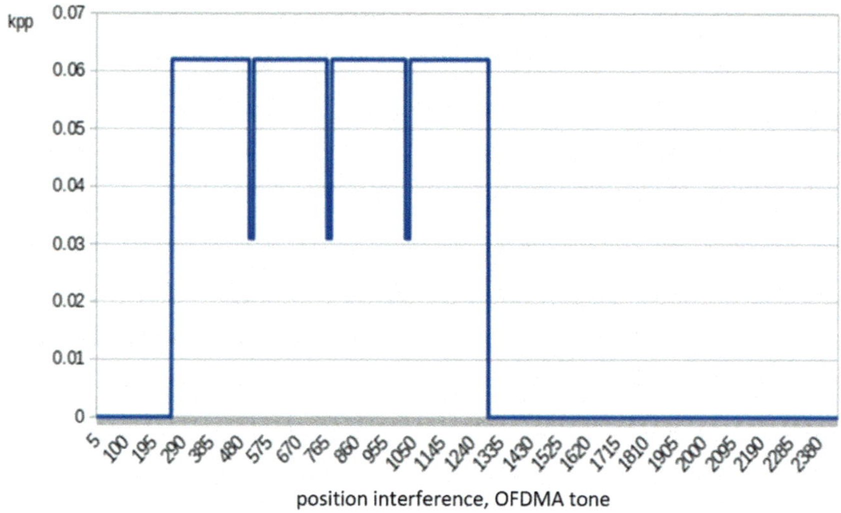

Fig. 1. Dependence of the efficiency factor of the use of subcarriers for the 80 MHz channel on the interference of bluetooth

frequency resources for the system and, accordingly, high spectral efficiency of the system.

In the proposed solution, aspects of the OFDMA method and the preamble puncturing mechanism are combined in a new way to create a cognitive multiuser access system with more efficient use of OFDMA subchannels for wireless communication in a low-orbit system with a distributed satellite architecture.

The objective of the proposed innovative solution is to improve the low-orbit satellite communication system by developing a method of cognitive multi-user access from OFDMA satellite network with a distributed satellite architecture to increase the spectral efficiency of its frequency resource.

The problem is solved by the fact that in the method of cognitive multiuser access with OFDMA, which uses a multiplexing protocol with frequency division multiplexing, containing a unit for determining the required number of frequency subchannels, based on the relevant performance of the data transmission system, additionally introduced an algorithm for the assignment of non-carrier and a vibration algorithm of the OFDMA (RU) operating frequency band for the respective frequency channels, which use the granularities of those frequency spectrum bands that can be suppressed, by improving the operation of the preamble puncturing mechanism, the introduction of cognitive radio techniques and the use of OFDMA mode.

The use of OFDMA in conjunction with the preamble puncturing mechanism provides suppression not of the entire channel, but of a single frequency section of OFDMA RU, which allows one to use unaffected subcarrier interference for data transmission.

The essence of the proposed scientific and technical solution can be explained as follows. Borrowed from LTE networks, OFDMA mode allows the distribution of one or more sub-channels (carrier set) with 20 MHz bands between several stations in the DL channel (downlink channel), and to transmit in the UL channel (uplink channel) data from several stations simultaneously. In this case, the 20 MHz subchannel is divided into so-called resource frequency units (RU), which have different bandwidth depending on the requirements and the amount of data required for transmission. Depending on this in the channels of 20 MHz can be allocated RU, which are divided into the following types:

- 26-tone;
- 52-tone;
- 106-tone;
- 242-tone.

Therefore, depending on the availability of workstation resources (STA), other parts of the spectrum may be provided, in addition to standard channels with bandwidths of 20, 40, 80 or 160 MHz. It is obvious that such a solution organically complements the mechanism of preamble puncturing in terms of increasing the granularity of parts of the spectrum that can be suppressed under the influence of band noise.

From this point of view, it is advisable to apply subchannel suppression for different types of RU so as to cover only 20 MHz subchannel with different RU allocation schemes so as to overlap the maximum of all sections of the 20 MHz subchannel in order to suppress these areas. By reducing parts of the spectrum, it is possible to increase the efficiency of the preamble suppression mechanism by suppressing not only 20 MHz subchannel, but only a certain part of it using the OFDMA scheme with different RU allocations. To do this, do the following:

- determine the set of distorted carriers of the used channel;
- select RU allocation schemes that will be involved in signal transmission, one of the RUs of which will be suppressed;
- select a suppressed RU that will not participate in the transfer.

Obviously, it does not make sense to apply all schemes where all RUs are involved in the preamble suppression mechanism, as many of them are repeated, and combinations of RU use simply change between stations. The main goal when choosing unique distribution schemes is the maximum efficiency of data transmission in comparison with other RU allocation schemes, ie it is necessary to choose such schemes in which tones are not suppressed, RU which will be used for data transmission should have the highest bandwidth. In the Table 1 shows the optimal distribution schemes RU developed by the authors.

Next, based on the combination of OFDMA mode and preamble puncturing mechanism, it is necessary to develop an algorithm that will take into account which RUs to use from the distribution scheme mentioned above in order to suppress bandwidth by disabling single RUs.

Here it is necessary to make clarifications about the 20 MHz subchannel. In the original algorithm [10], according to the recommendations of the standard, it was recommended not to use preamble puncturing, for channels 20 and 40 MHz. This applied primarily to the feasibility of suppressing one or two subchannels, if the total bandwidth was small in relation to the number of these subchannels. However, in the proposed

Table 1. Sample data

Year	2003	2004	Average
Efficacy I type	70.00%	80.00%	75.00%
Error I type	30.00%	20.00%	25.00%
Efficacy II type	70.00%	60.00%	65.00%
Error II type	30.00%	40.00%	35.00%

Table 2. Optimal RU distribution schemes of different types

Type RU	Index suppression of RU	Distribution scheme RU								
26- tonal	0–4	26	26	26	26	26	106			
26- tonal	4–8	106				26	26	26	26	26
52- tonal	0–4	52		52		26	106			
52- tonal	5–8	106				26	52		52	
106- tonal	0–4	106				26	106			
106- tonal	5–8	106				26	106			

algorithm, what is new is that the suppression of tones is carried out not only for the entire subchannel at once, but also for certain parts of it, thereby using undisturbed tones for data transmission.

Thus, our solution uses a tone suppression mechanism, both in channels with 20 MHz bandwidth and for channels with 40 MHz bandwidth, which is respectively new.

Therefore, the proposed method operates as subchannels to localize changes in RU schemes, and specific RUs, the indices of which are transmitted during the sending of OFDMA HE MU PPDU (multi-user data packet format). Thus, the sequence of the proposed method is as follows:

- determine the distorted carriers of the channel used (the algorithm of this operation is shown in Fig. 2);
- determine the transmission scheme according to the channel used;
- determine the subchannels within which the suppression of RU will be carried out;
- for each subchannel in the RU allocation scheme, the corresponding RU channel is suppressed using digital signal processing techniques (for example, using a digital notch filter);
- on the receiving side OFDMA HE MU PPDU is interpreted as OFDMA PPDU, according to the presence of suppressed subchannels in the information header of the frame;
- according to the OFDMA RU scheme, which is also transmitted in the information header, the signal is ignored within the suppressed RU.

The claimed method can be implemented as follows. The technical result of the proposed model is to increase the spectral efficiency of the frequency resource of the satellite network with a distributed satellite architecture. To do this, the multi-user scheme allows the grouping of low-orbit spacecraft (LEO-system) with a distributed satellite architecture, which includes groups of root (leading) satellites and repeater satellites (slave), to share one or more channels. In the proposed model, the available bandwidth of the channel is divided into several subchannels with the corresponding bandwidth in accordance with standard OFDMA practice. The transceiver of the root satellite is informed by the application program that it needs to provide data transmission with a specific transmission rate on the uplink (to the root satellite) or downlink channels (to the satellite repeaters). The device determines the distorted carriers and selects the operating frequency sections OFDMA (RU) for the respective frequency channels, taking into account the granularity of those sections of the frequency spectrum that can be suppressed. The use of OFDMA in conjunction with the preamble puncturing mechanism provides suppression not of the entire channel, but of a single frequency section of OFDMA RU, which allows to use unaffected interference subcarriers for data transmission required to achieve this data rate. Once the device has selected the required number of subchannels, it transmits the control channel to the appropriate receivers of the satellite repeaters identifiers of the selected subchannels. It then starts transmitting data simultaneously on said subchannels in the corresponding bandwidth used by said subchannels between several repeater satellites in the DL channel or transmits data in the UL channel from several repeater satellites simultaneously. In the UL channel, data transmission is performed synchronously by several repeater satellites, and the root satellite acts as a coordinator and determines the start time of transmission. To inform the root satellite about the availability of data on the repeater satellites for transmission from time to time, they send him reports.

A key aspect of the described system is the method of selecting subchannels for use. This method applies the principles of cognitive radio to the OFDMA method and the preamble puncturing mechanism. This process can be implemented in hardware or software. First, the application requests the data rate for transmission. This data rate will mainly depend on the type of data transmitted. The root satellite transceiver then begins an iterative subchannel selection process to meet the required criteria.

For channels with appropriate bandwidths (eg 20 MHz), the following RU distribution schemes are used, which were shown in Table 2. According to them, the algorithm selects the allocation scheme RU, which is suitable for suppressing this interference. The algorithm performs the following (see Fig. 3):

- determines the distorted carriers of the used channel (the algorithm is shown in Fig. 2);
- preserves the vector of tones that have been subjected to interference, all other tones, including information, are equated to zero;
- in the main cycle, a search is made for a suitable RU allocation scheme according to whether the stored tone vector is completely overlapped by suppressing a certain RU in a certain distribution scheme.

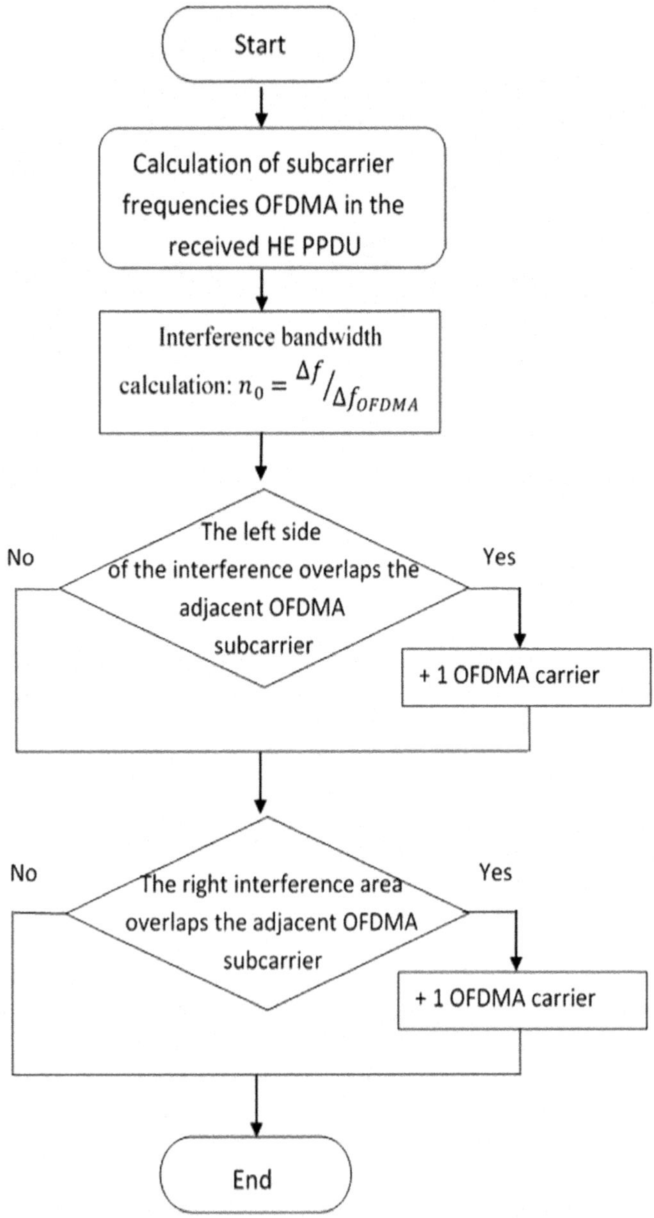

Fig. 2. Block diagram of the algorithm for determining the distortion of the carriers

This estimate is determined by the following expression:

$$k_{INT} = \sum_{i=0}^{n_{20\text{МГц}}} tone_i, \tag{1}$$

where k_{INT} is the interference factor in the channel; $tone_i$ is the element of the tone vector; $n_{20\text{МГц}}$is the number of tones in the 20 MHz channel.

The estimate is determined based on whether the interference factor in the channel is equal to or not equal to zero. The loop searches for the RU allocation scheme and the specific RU to be suppressed. In this case, in accordance with the frequency distribution of RU in the channel 20 MHz carry out the imposition of the mask on the vector of tones and zeroing those tones of the position that coincide with the position of a particular RU in a particular distribution scheme. Accordingly, after each iteration, evaluate whether the suppressed RU completely covers the interference band or not, and then perform an overall assessment according to the following criteria:

- if the RU distribution and suppression scheme has completely eliminated the interference, this RU suppression scheme and index are selected as working to send OFDMA HE PPDU, where RU suppression is not involved in the transmission;
- if the distribution scheme and the proposed RU have not completely removed or have not eliminated the interference, then search for the next RU in the current distribution scheme RU or select the next distribution scheme.

Since the types of RU depending on the distribution schemes contain different bandwidth, it is advisable to first start searching for RU with a lower bandwidth, and only then try allocation schemes where RU have a wider bandwidth. Therefore, for channels with 20 MHz bandwidth, the RU selection algorithm and the RU allocation scheme is a RU selection scheme, where narrowband RUs are used first, and then broader RUs.

To evaluate the effectiveness of the developed method, the above algorithm was simulated for channels with bandwidth of 20 MHz under the influence of interference with a bandwidth of 1 MHz, which simulate Bluetooth signals. During each simulation series, about 2300 model starts were performed according to the different position of the interference relative to the signal band.

For a channel with a bandwidth of 20 MHz, the obtained simulation results are shown in Fig. 4.

From the results of the study it follows that in comparison with the original algorithm [10] the proposed method of suppression of RU gives a significant gain in modeling, namely the efficiency of subcarriers has increased significantly compared to the original algorithm. The increase along the edges of the graph is explained by the fact that the obstacle when moving gradually takes the position of RU, respectively, the number of subcarriers that are exposed to the noise gradually increases. The local minima and maxima on the graph correspond to the moments when there is a change in the distribution scheme RU, ie when there is a change in the bandwidth RU. The minimum coefficient in the central part of the graph is explained according to the RU distribution scheme in the 20 MHz channel, namely that in the central zone there are 106 - tone and 26 - tone RU, respectively.

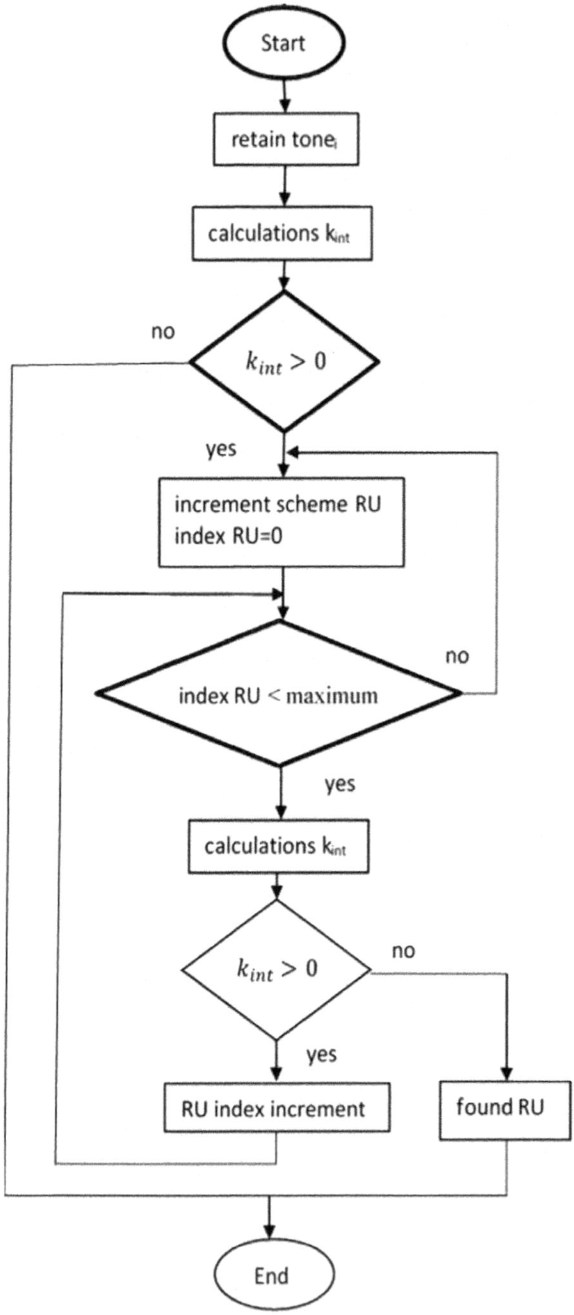

Fig. 3. Selection algorithm for 20 MHz channel

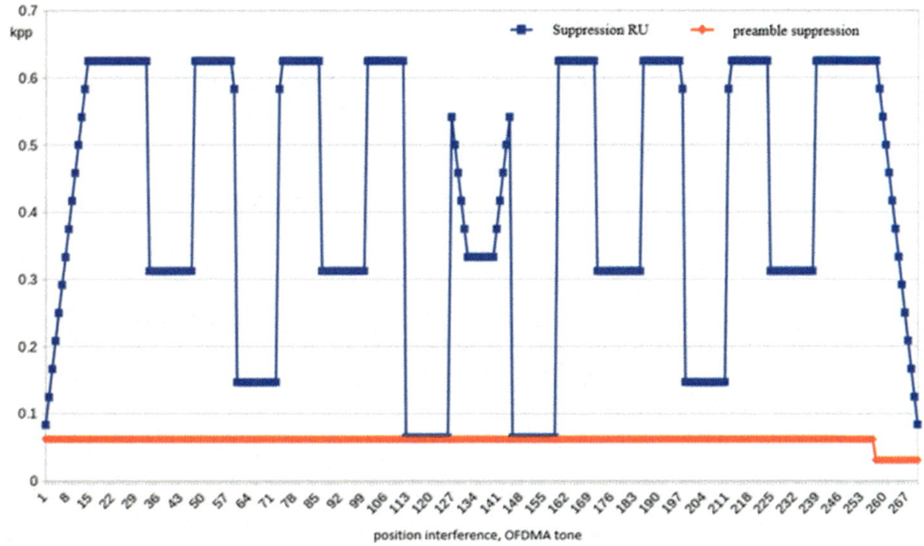

Fig. 4. Dependence of the efficiency factor of the use of subcarriers for the 20 MHz channel on the interference of bluetooth

4 Conclusion

1. The original preamble puncture mechanism known today is poorly adapted to adapt to narrowband interference, as it suppresses one or two 20 MHz subchannels, thus not using OFDMA subcarriers to transmit unaffected transmissions.
2. The proposed method of improving the operation of the preamble puncturing mechanism is associated with the need to increase the granularity of those parts of the spectrum that can be suppressed, by implementing the principles of cognitive radio techniques or using OFDMA technique.
3. With a view to increasing the spectral efficiency of the use of the frequency resource of a low-orbit satellite network with a distributed satellite architecture in the method of cognitive multiuser access with OFDMA, a new algorithm for determining distorted carriers and algorithm for selecting the working frequency band OFDMA (RU) spectra that can be suppressed by improving the operation of the preamble puncturing mechanism, the introduction of cognitive radio techniques and the use of OFDMA. The use of OFDMA in conjunction with the mechanism of preamble puncturing provides suppression not of the entire channel, but of a single frequency section of OFDMA RU, which allows you to use unaffected interference subcarriers for data transmission.
4. The results of the research allow to recommend the joint use of OFDMA and preamble puncturing techniques in the scenarios of the impact of various narrowband interference in the study of the features of the operation of 5G /IoT wireless networks.

References

1. Saiko, V.G., Domrachev, V.M., Narytnyk, T.M., Sivkova, N.M.: Patent of Ukraine for a utility model №141528 Ukraine. Low-orbit satellite communication system with FC-architecture. Application, Bull. No. 7. 10/24/2019; publ. 10.04.2020
2. Saiko, V.G., Domrachev, V.M., Narytnyk, T.M., Sivkova, N.M.: Patent of Ukraine for a utility model №142478 Ukraine. System of low-orbit satellite communication with inter-satellite communication channels of terahertz range. Application, Bull. No. 11. 11/21/2019; publ. 10.06.2020
3. Maltsev, A.A., Rubtsov, A.E.: Investigation of the characteristics of OFDM radio communication systems with adequate subcarrier deviation. Herald of the N. I. Lobachevsky State University of Nizhny Novgorod, no. 5, pp. 43–49 (2007)
4. Saiko, V.G., Dikarev, O.V., Grishchenko, L.M., Lysenko, D.O., Dakova, L.V.: Patent for utility model 114470 Ukraine, H04B 7/00, H04B 7/165. The method of determining the optimal values of the parameters of signals with high noise immunity in FH-OFDMA radio networks/declared, Bull. No. 5. 08.09. 2016; publ. 10.03. 2017
5. Andrianov, I.M.: Increasing the spectral efficiency of communication systems with orthogonal frequency compaction in channels with random parameters. Herald of the Bauman Moscow State Technical University. Chapter. Instrument making, no. 7. pp. 71–77 (2010)
6. Rhee, W., Cioffi, J.M.: Increase in capacity of multiuser OFDM system using dynamic sub channel allocation. In: Vehicular Technology Conference Proceedings (May 2000)
7. Ibnkahla, M.: Adaptation and Cross Layer Design in Wireless Networks. CRC Press, Boca Raton (August 2008)
8. Viswanath, P., Tse, D., Laroia, R.: Opportunistic beam forming using dumb antennas. In: IEEE Transactions on Information Theory, vol. 48, no. 6 (June 2002)
9. Hassan Amer, A., Hutema, C.: Russian Patent No. 2461996. Cognitive multi-user access with orthogonal frequency division multiplexing. Bull. No. 6. publ. 20.09.2012
10. Stefan, B., Patrice, N.: Russian Patent No. 2718958. Extended control of the AU in the transmission mode of multi-user EDCA in a wireless network. Bull. No. 11. publ. 15.04.2020
11. Tikhvinsky, V.O., Terentyev, S.V., Visochin, V.N.: LTE/LTE Advanced Mobile Networks: 4G Technologies, Applications and Architecture. Publishing house Media Publisher, Moscow, 384 p. (2014)
12. Saiko, V., Nakonechnyi, V., Narytnyk, T., Brailovskyi, M., Toliupa, S.: Increasing noise immunity between LEO satellite radio channels. In: Proceedings - 15th International Conference on Advanced Trends in Radioelectronics, Telecommunications and Computer Engineering, TCSET 2020, стр. pp. 442–446 (2020). 9088630
13. Saiko, V., Nakonechnyi, V., Narytnyk, T., Brailovskyi, M., Lukova-Chuiko, N.: Terahertz range interconnecting line for LEO-system. In: Proceedings - 15th International Conference on Advanced Trends in Radioelectronics, Telecommunications and Computer Engineering, TCSET 2020, стр. pp. 425–429 (2020). 9088550
14. Saiko, V., Domrachev, V., Gololobov, D. Improving the noise immunity of the inter-satellite communication line of the LEO-system with the architecture of the distributed satellite. In: 2019 IEEE International Conference on Advanced Trends in Information Theory, ATIT 2019 - Proceedings, стр. pp. 33–136 (2019). 9030501

15. Beshley, H., Klymash, M., Beshley, M., Kahalo, I.: Improving the efficiency of LTE spectral resources use by introducing the new of M2M/IoT multi-service gateway. In: 2019 IEEE 15th International Conference on the Experience of Designing and Application of CAD Systems (CADSM), pp. 114–117 (2019). https://doi.org/10.1109/CADSM.2019.8779270
16. Maksymyuk, T., Beshley, M., Klymash, M., Petrenko, O., Matsevityi, Y.: Eavesdropping-resilient wireless communication system based on modified OFDM/QAM air interface. In: 2018 14th International Conference on Advanced Trends in Radioelecrtronics, Telecommunications and Computer Engineering (TCSET), pp. 1127–1130 (2018). https://doi.org/10.1109/TCSET.2018.8336392

Optical Signal Decay and Information Data Loss in Wireless Atmospheric Communication Links with Fading

Irina Bronfman[1,2](✉), Irit Juwiler[1], Nathan Blaunstein[2], and Anatolii Semenko[3]

[1] Department of Electrical and Electronics Engineering, Sami Shamoon College, 77245 Ashdod, Israel
irinamo@post.bgu.ac.il, irinamo@ac.sce.ac.il, iritj@sce.ac.il

[2] School of Electrical and Computer Engineering, Ben-Gurion University, 74105 Beer-Sheva, Israel
nathan.blaunstein@hotmail.com

[3] University of Communication and Information Sciences, Kiev, Ukraine
setel@ukr.net

Abstract. This chapter is based on recent research work in two fields: a) optical signal decay in wireless atmospheric communication links, and b) performance and quality of service (QoS) in such links that accounts for the destructive effects of fading phenomena on signal data streams transmitted through such channels. The total signal decay was formulated based on predictions of the main losses occurring in the atmospheric communication link, accounting for the effects of attenuation and absorption by gaseous structures and hydrometeors (rain, snow and clouds), and turbulence-induced fast fading of radio and optical signals. Signal data parameters, including capacity, spectral efficiency and bit-error-rate (BER), were analyzed for the prediction and increase of QoS, taking into account all features occurring in atmospheric communication links. An optimal algorithm for the prediction of the total signal decay was found, considering various meteorological conditions occurring in the real atmosphere at different heights and various frequencies of radiated signals. Lastly, a method was proposed to evaluate the data stream parameters: capacity, spectral efficiency and BER, accounting for effects of fast fading atmospheric turbulence that corrupt information data signals transmitted through such channels.

Keywords: Atmosphere · BER · Capacity · Clouds · Hydrometeors · K-parameter of fading · Optical waves · Rain · Scintillation index · Spectral efficiency · Turbulence

1 Introduction

There are three main transmission channels for wireless communications: terrestrial, atmospheric and ionospheric. In this work, we will discuss the atmospheric channel,

M. Klymash et al. (Eds.): *Future Intent-Based Networking*, LNEE 831, pp. 424–443, 2022.
https://doi.org/10.1007/978-3-030-92435-5_24

and, in particular, the troposphere, which ranges over the altitudes of 10–20 km from the ground surface. During recent decades the most widely researched wireless communication channels were between ground-based subscribers and air-based subscribers (airplanes, helicopters, drones, and more) [1–7]. The content of the troposphere is too wide: from gaseous particles - aerosols, to hydrometeors, such as rain, clouds, fog, hail and snow [1–14]. All these features give a huge impact on signal decay and signal intensity loss caused by attenuation and absorption of radio and optical signals by all kinds of these hydrometeors. More dangerous for signal data corruption, regularly observed in the troposphere, called *atmospheric turbulences* (Fig. 1). These gaseous structures, having stochastic nature, result sporadic air streams and motions in space, accompanied by sporadically varied wind direction and speed, the wind intensity, by temporal variations of air humidity, moisture, and temperature. Due to irregular weather conditions in the troposphere, when optical signals propagate through tropospheric channels, their amplitude or intensity verities sporadically. This phenomenon is called *fading*, fast and slow - in the time domain, or small-scale and large-scale - in the space domain, respectively.

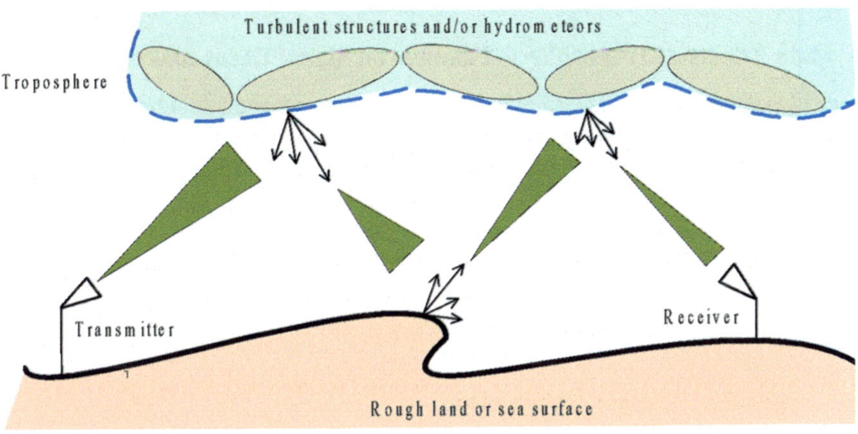

Fig. 1. The «Earth-Atmosphere-Earth» propagation channel

The aim of the presented work reporting our recent investigations is to analyze the total optical signal decay, that is, to evaluate the total signal loss inside the atmospheric channel based on predictions of tropospheric parameters related to hydrometeors, gaseous structures, and turbulences. A second aim is to derive a quality of service (QoS) metric for the link by estimating parameters of the information signals, called *baseband* signals [7], as it passes through such atmospheric wireless channels with fading caused by multiple scattering, reflection and diffraction of optical waves. Finally, the aim of this work is to provide designers of wireless atmospheric communication links with a stable algorithm that can predict the total signal loss and evaluate the capacity of the data stream, or the spectral efficiency and bit-error-rate (BER) in such channels, affected by attenuation, absorption and fading phenomena.

The rest of the chapter is organized as follows. In Sect. 2, we present the atmospheric content and the different types of hydrometeors, as well as specific features of atmospheric turbulence. Then, in Sect. 3, we describe the effects of these atmospheric features on optical signal propagation through such channels with fading. Section 4 presents the main parameters of the signal data stream, such as capacity, spectral efficiency and BER, and outlines their computation, accounting for effects of fading. Sample computations of these parameters for various propagation conditions occurring in wireless atmospheric channels with fading are presented.

2 Effects of Tropospheric Features on Signal Propagation

For a homogeneous gaseous atmosphere in conditions of line-of-sight between a source of radiation and the receiver, as an optical detector, in conditions of absence of hydrometeors fading phenomena of optical waves can prevent the availability of 99.999% for the paths of 5 km and more, with the fade margin of 28 dB. However, the existence of hydrometeors and turbulent structures significantly affects conditions of optical signals propagation and can decrease the efficiency of atmospheric communication links due to scattering, absorption and diffraction by hydrometeors and turbulences. Cumulative effect of these natural features causes strong fast fading of signal data.

2.1 Main Effects of Tropospheric Features on Signal Decay and Fading

We will briefly examine separately the main effects of each feature on signals with information data (called *band path* signals [7]) passing the irregular troposphere, caused by rain, and clouds on the signal attenuation, caused by turbulences on the signal, as the main source of fast fading. As was shown in [1–14], there are three main features that cause signal attenuations: molecular absorption, effects of rain and effect of clouds. The effect of turbulence on the scattering of signals is presented separately. We will show the main parameters, the corresponding formulas, and will compute and plot their characteristics in this paragraph.

2.1.1 Molecular - Gaseous Absorption

Gaseous molecules and *atoms* are among many types of atmospheric molecules and atoms, and at the middle-latitude troposphere mostly include O_2, O, CO_2, NO, N_2, etc. [1–5].

Aerosols are usually organized in form of liquid or solid interacting particles spatially distributed in the atmosphere. Aerosol particles play an important role in the processes of condensation and freezing forming molecules of water or ice. They also, via particles ionization processes, influence the electrical properties of the atmosphere. For aerosols stability, their actual size should be between a few nanometers and a few micrometers; aerosols with sizes larger than 50 μm are unstable. An amount of aerosol particles depends on the altitude of their location z, and on the number of aerosols near the ground surface, $N(0)$, that is:

$$N(z) = N(0) expexp\left(\frac{z}{z_s}\right), \tag{1}$$

where $N\ (z)$ is the number of aerosols at altitude z (in meters), and z_s is a scaled height given by $1\,\text{km} < z_s < 1.4\,\text{km}$ [1–5].

Generally speaking, all gaseous molecules, including aerosols, forming atmosphere absorb energy from optical waves passing through them, and, therefore, causing attenuation [3–5]. Signal strength/power loss depends on the frequency of radiation, deviations of temperature, pressure, and water vapor concentration, and, finally, it increases with the increase of frequency and all meteorological parameters of the atmospheric gases [2].

The total absorption signal passing atmospheric link of the path length r is given by:

$$A = \int \gamma(r)dr, \tag{2}$$

where $\gamma(r)$ is the normalized attenuation per length of link (in dB/km). It consists of the sum of two components: $\gamma_o(r)$ and $\gamma_w(r)$, which are defined the attenuation of oxygen and water vapor, respectively:

$$\gamma(r) = \gamma_o(r) + \gamma_w(r), \tag{3}$$

where $\gamma_o(r)$ and $\gamma_w(r)$ are oxygen and water vapor specific attenuation characteristics at ground level, where pressure is 1013 mbar and temperature are 15 °C were defined experimentally and given empirically in [3–5]:

$$\gamma_o = \left[7.19 \cdot 10^{-3} + \frac{6.09}{f^2 + 0.227} + \frac{4.81}{(f - 57)^2 + 1.5}\right] f^2 \times 10^{-3} \left[\frac{\text{dB}}{\text{Km}}\right], \tag{4}$$

$$\gamma_w = \left[0.05 + 0.0021\rho + \frac{6.09}{(f - 22.2)^2 + 8.5} + \frac{10.6}{(f - 18.3)^2 + 9}\right] f^2 \rho \times 10^{-4} \left[\frac{\text{dB}}{\text{Km}}\right]. \tag{5}$$

Here f is the frequency (in GHz) and ρ is the water vapor density (in g/m^3). In [3–5], the temperature parameters were considered by using the correction factors: for dry air - 1.0% per 1 °C from 15 °C, and for water vapor - 0.6% per 1 °C from 15 °C. If so, the absorption specific factors in the tropospheric link with the path length L can be given by [3–5]:

$$A_o = \gamma_o L_o; \qquad A_w = \gamma_w L_w[\text{dB}]. \tag{6}$$

The total atmospheric attenuation L_a (in dB) for a tested link can then be found by integrating the specific attenuation characteristic γ_a (*in* dB/m) over the total path length in the atmospheric link:

$$L_a = \int_0^{\pi} \gamma_a(l)dl = \int_0^{\pi} \left[\gamma_o(l) + \gamma_w(l)\right]dl, \tag{7}$$

by assuming an exponential decrease in gas density with altitude. In [6] the total signal attenuation in the gaseous quasi-homogeneous atmospheric was defined as the zenith attenuation and denoted by L_z. Then, attenuation for an inclined path with an elevation angle $\theta > 10^{\circ}$ can be found from the zenith attenuation L_z as [6]:

$$L_a = \frac{L_z}{sin\theta}, \tag{8}$$

and was computed from Eqs. (3)–(8) following the standard ITU-676 approach [1].

2.1.2 Effects of Rain

Rain is globally distributed in space and along the height of liquid water drops with diameters greater than 0.5 mm. When the drops' sizes are smaller, they are usually called drizzle droplets or drizzles. The concentration of drops typically ranges from 100 to 1000 m^{-3} [6–11]. Raindrops have diameters larger than 4 mm because the concentration generally decreases as the diameter increases, except heavy rain. Drizzles reduce the visibility of the path much higher than raindrops. Usually, in meteorology rain is classified according to its rate of fall.

In meteorology usually consider three types of rain: light, moderate and heavy, which correspond to dimensions less than 2.5 mm, between 2.8 mm and 7.6 mm, and more than 7.6 mm, respectively. Rainfall with a rate less than 250 mm per year or more than 1500 mm per year, approximately has been found as extreme limits on all the world continents [6–11].

As was mentioned by numerous experimental observations carried out and described in [7–11], the attenuation of optical waves caused by rain increases:

- with the number of raindrops along the transmission path,
- the size of the drops, and
- the length of the path through the rain.

There are several models for finding the attenuation caused by rain: empirical [7, 8], semi-empirical and statistical-analytical models [7–11], close to that created by Saunders [6]. The Saunders' model, which coincides with the ITU models (see [7–11]), does not depend on any particular place, not frequency dependent, has a satisfactory processing time and can be easily implemented. In our work, we follow the Saunders model as more attractive to experimental observations of rainfalls worldwide. However, the Saunders model is correct only when the density and shape of the raindrops are constant. According to [6] the received power P_r at a given detector is found to decrease exponentially with optical path r through the rain and with α as the parameter of attenuation along with the distance r along with the link. It shows receiving power $Pr(r)$ decrease to e^{-1} of its initial value $Pr(0)$:

$$Pr(r) = Pr(0)e^{-\alpha r}. \tag{9}$$

As for the value of α, it is given by the integral of a one-dimensional (1-D) distribution of the drops' diameter D, denoted by $N(D)$, and by the effective cross-section $C(D)$ of scattering of signal power by raindrops distributed over the drops' diameter D:

$$\alpha = \int_{D=0}^{\infty} N(D) \cdot C(D)dD. \tag{10}$$

In real tropospheric optical traces, the drop diameter distribution $N(D)$ is not a constant value and can be found from the following expression [6]:

$$N(D) = N_0 exp\left(-\frac{D}{D_m}\right), \tag{11}$$

where N_0 and D_m are the parameters that depend on the rainfall rate premeasured above the ground surface in millimeters per hour and equals [6]:

$$N_0 = 8 \times 103 \text{ m}^{-2}\text{mm}^{-1}, \quad D_m = 0.122R^{0.21} \text{ mm}. \tag{12}$$

As for the effective cross-section C(D), it can be found using the Rayleigh approximation that is valid for lower frequencies, when the average drop size is small compared to the incident wavelength [6, 7]. In this case, can be used a very simple expression for $C(D)$:

$$C(D) \propto \frac{D^3}{\lambda}. \tag{13}$$

As was mentioned in [7], when *N(D)* is not constant (as we described earlier in (11) following [6]), we can now take the value of the specific attenuation at a given point on the trace, $\gamma(r)$, and integrate it over the whole path length r_R to find the total path loss:

$$L = \int_0^{r_R} \gamma(r)dr, \tag{14}$$

and find the relation of $\gamma(r)$ introduced in [7] with the specific attenuation α (as shown in [6]):

$$\gamma = \frac{L}{r} = 4.34\alpha. \tag{15}$$

Then, expressing (14) on a logarithmic scale can be rearranged as:

$$L = 10\, log\, log\left(\frac{P_T}{P_R}\right) = 4.34\alpha r. \tag{16}$$

In practice, usually used an empirical model, which combines all effects of rainfall, where γ is assumed to depend only on *R*, the rainfall rate measured on the ground in millimeters per hour. According to [6–10]:

$$\gamma(f, R) = a(f)Rb(f). \tag{17}$$

The attenuation coefficients, $a(f)$ and $b(f)$, can be calculated from [6, 8, 10]. In order to calculate the attenuation for a given path where the elevation angle θ is less than 90° it is necessary to account for the variation in the rain in the horizontal direction.

Sometimes, a reduction factor *s* of the path loss is introduced for total rain attenuation [6, 8–10]:

$$L = \gamma s r_R = a(f)Rb(f)sr_R. \tag{18}$$

Here, r_R and s can be obtained from [6]. It should be noticed that rain varies in time over various scales: seasonal, annual and diurnal. To take into account all of the temporal variations, it was estimated expression (18) for rain attenuation and was found that the effect of rainfalls occurrence does not exceed 0.01% of the time. If so, was propose another formula than (18):

$$L_{0.01} = a(f)R_{0.01}b(f)s_{0.01}r_R. \tag{19}$$

where values of $s_{0.01}$ can be found in [6–11]. For a temporal percentage, another than 0.01%, the attenuation can be corrected by introducing the relevant time percentage P, which changes over a broad range from 0.001% to 1%, that is:

$$L_p = L_{0.01} \times 0.12P^{-(0.546+0.043 logP)}. \tag{20}$$

However, rain precipitation is defined by variations in both horizontal and vertical directions, which makes it very hard to describe the spatial distribution of rain. The correction factor (i.e., the term $s_{0.01}r_R$, we used in the effective path loss evaluation according to Eq. (19)).

2.1.3 Effects of Clouds

All main characteristics of clouds, the dimensions, their shape and structure, are defined by air movements, which change their formation and growth, and by the properties of the cloud particles. Usually, sky cover is the observer's view, whereas cloud cover is areas that are smaller or larger than the overall space of the sky dome [7]. There are several well-known models that are used for the probability distribution of the sky cover [12–14]. For the prediction of the cloud attenuation, we will follow the ITU-R model given in [12].

Specific Attenuation for Clouds. The specific attenuation of optical signal caused by a cloud can be determined by [12]:

$$\gamma_c = K_1 M \left[\frac{\text{dB}}{\text{Km}}\right], \tag{21}$$

where γ_c is the specific attenuation of the clouds, in dB/km, K_1 is the specific attenuation coefficient [in (dB/km)(g/m^3)], and M is liquid water density (in g/m^3).

As was mentioned above, for small-scale cloud drops, the Rayleigh approximation can be used for the calculation of specific signal loss factor [12]. A mathematical model based on Rayleigh scattering, which uses a double-Debye model for the dielectric permittivity $\epsilon(f)$ of water, can be used to calculate the value of K_1:

$$K_1 = \frac{0.8197f}{\epsilon''(1+\eta^2)}\left(\frac{\text{dB}}{\text{km}}\right)\left(\frac{\text{g}}{\text{m}^3}\right), \tag{22}$$

where f is the frequency in GHz and η is defined as:

$$\eta = \frac{2+\epsilon'}{\epsilon''}, \tag{23}$$

where ϵ' and ϵ'' are the real and imaginary components of the complex dielectric permittivity of water, respectively. The complex dielectric permittivity of water can be computed by knowledge on the primary and secondary frequencies of the double Debye model [12]:

$$f_{prim} = \{20.09 - 142(\phi - 1) + 294(\phi - 1)^2, (\epsilon' > \epsilon'')590 - 1500(\phi - 1), (\epsilon' < \epsilon''), \quad (24)$$

where $\phi = 300/T$, and T is the temperature in Kelvin. If so, the complex dielectric permittivity of water can be found by using the following expressions:

$$\epsilon'(f) = \frac{(\epsilon_0 - \varepsilon_1)}{1 + \left(\frac{f}{f_p}\right)^2} + \frac{(\epsilon_1 - \epsilon_2)}{1 + \left(\frac{f}{f_s}\right)^2} + \epsilon_2, \quad (25)$$

$$\epsilon''(f) = \frac{f(\epsilon_0 - \varepsilon_1)}{f_p\left(1 + \left(\frac{f}{f_p}\right)^2\right)} + \frac{f(\epsilon_1 - \epsilon_2)}{f_s\left(1 + \left(\frac{f}{f_s}\right)^2\right)}, \quad (26)$$

where $\epsilon_0 = 77.6 + 103.3(\phi - 1)$, $\epsilon_1 = 5.48$, $\epsilon_2 = 3.51$.

Total Cloud Attenuation. The total cloud attenuation can be found as:

$$A = \frac{LK_1}{sinsin(\theta)}, \quad (27)$$

where θ is the elevation angle ($5° \leq \theta \leq 90°$), K_1 is the specific attenuation coefficient defined by (22), and L is the total content of liquid water (in kg/m^2).

Table 1 shows total cloud attenuation versus radiated frequency varying from 10 to 1000 THz and for elevation angles varying from 5 to 30°. In our computations, a water density of 0.29 g/m^3 was used, according to [19].

Table 1. Total cloud normalized attenuation loss, A [dB/m], vs. inclination angle θ[°] and temperature T[°C] for radiation frequencies f varying from 10 to 1000 THz

θ°/T [°C]	3°	10°	20°	30°
−10°	183.7236	55.3726	28.1134	19.2307
0°	234.3845	70.6413	35.8655	24.5335
10°	288.7611	87.0299	44.1863	30.2252
30°	399.3016	120.3458	61.1012	41.7957

It is clearly seen that an increase of frequency from 10 to 1000 THz has a weak impact on total cloud attenuation (the standard deviations does not exceed 1% at 1000 THz that obtained for 10 THz). As seen from Table 1, the total loss of optical waves caused by clouds is too strong and, in practice, decreases efficiency of optical communication links in cloudy environments.

3 Effects of Turbulence

Atmospheric turbulence is a gaseous chaotic structure (called eddy) generated in the troposphere by a sporadic and random distribution of the overall gas temperature, overall wind magnitude and its direction variations along the signal propagation paths [15]. This chaotic behavior results in fluctuations of index of refraction and causes Doppler shift, i.e., fast fading phenomena. It was postulated by numerous investigations, experimental and theoretical [15–20] that exists only one common way to describe atmospheric turbulence - by use turbulence power spectra divided into three separate regions by introducing two turbulence scales: L_0, as the outer scale of the turbulence, varied between 10 m and 100 m, and l_0, the inner scale varied between 1 mm and 30 mm [15–19]. According to such a decision, the spatial and temporal evolution of turbulence, small-scale, moderate-scale, and large-scale, either weak or strong, can be described by the so-called *cascade process* – from large structures with scale L_0 to small structures with scale l_0, due to the process of turbulence energy dissipation, introduced independently by Richardson and then by Kolmogorov and Obukhov [15–17]. The process of random variations of optical signal intensity passing tropospheric link filled by such turbulent structures has been defined as scintillations of signal intensity via the so-called scintillation index, as the normalized variance of signal intensity fluctuations [15–19].

Scintillation Index. The scintillation index, denoted as σ_I^2, describes fluctuations in optical power or intensity as measured by a point detector. According to [15–19] the scintillation index is defined by:

$$\sigma_I^2 = \frac{\langle I \rangle^2 - \langle I \rangle^2}{\langle I \rangle^2} = \frac{\langle I \rangle^2}{\langle I \rangle^2} - 1. \tag{28}$$

Since the signal intensity deviations caused by turbulence is distributed by log-normal PDF and CDF (or Gaussian PDF and CDF) with a zero-mean value [15–19], we can present the normalized signal intensity scintillations (called the index of intensity deviations) in the following manner [15–19]:

$$\sigma_I^2 = 0.12 \cdot C_\epsilon^2 \cdot k^{\frac{7}{6}} \cdot d^{\frac{11}{6}}, \tag{29}$$

where C_ϵ^2 is the structure parameter of dielectric permittivity of the turbulence averaged over the path length, d is the distance in km, and k is the wave number, $k = 2\pi/\lambda$.

It should be noticed that below we investigate cases of both weak and strong turbulence occurring in the atmosphere. Indeed, as was observed by numerous experiments, both in line-of-sight (LOS) conditions with weak fading and non-line-of-sight (NLOS) conditions with strong fading occur in different atmospheric regions over the world [15–20]. Therefore, it is interesting to examine "good-good" and "bad-bad" situations defined in [6]). We start our discussion by analyzing weak and moderate turbulence occurring in the irregular atmosphere. Thus, in Fig. 2a to Fig. 2c, the index of signal intensity scintillations, σ_I^2,(computed, according to (29)) versus the structure constant of the turbulence C_ϵ^2 computed for $C_\epsilon^2 = 10^{-12} - 5 \cdot 10^{-12}$ is presented. Here, we accounted for weak and moderate turbulent structures occurring in the atmospheric optical channel. Parameter

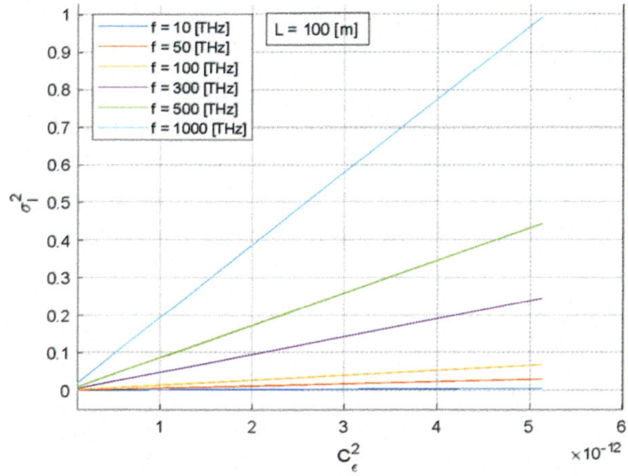

(a)

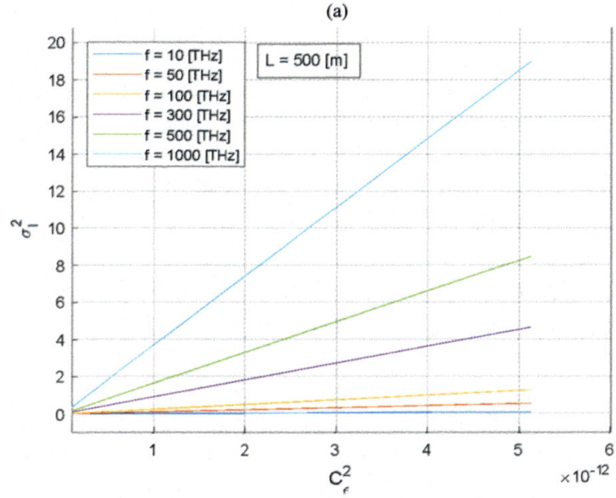

(b)

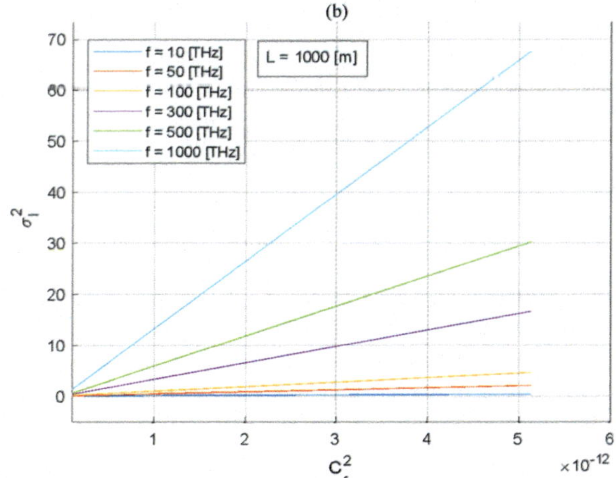

(c)

Fig. 2. Index of signal intensity scintillations vs. the structure constant of the turbulence $C_\epsilon^2 = [10^{-12},\ 5 \cdot 10^{-12}]$, averaged over the path for frequencies from 10 THz to 1000 THz, for links: (a) $L = 100$ m, (b) $L = 500$ m, and (c) $L = 1000$ m.

σ_I^2 was averaged over the optical paths $L = 100$, 500 and 1000 m, respectively, for different frequencies from 10 THz to 1000 THz.

It can be seen that above $C_\epsilon^2 = 2 \cdot 10^{-12}$ the dimensionless scintillation index σ_I^2 does not exceed unity (for all studied frequencies) over short optical atmospheric links ($L = 100$ m), exceeds ten over a link of $L = 500$ m, reaching several tens for long optical atmospheric links ($L = 1000$ m). In other words, weak and moderate turbulence affects the optical signal in the visual and IR ranges only for long optical atmospheric links and does not lead to corruption of signal data propagating through channels characterized by weak and moderate turbulence.

Now we consider an atmospheric optical channel characterized by strong turbulence structures, that is, a channel with a structure constant of the turbulence, C_ϵ^2, varying within the range of $C_\epsilon^2 = [10^{-9}, 5 \cdot 10^{-9}]$.

In Fig. 3a to Fig. 3c, we present the computed index of signal intensity scintillations, σ_I^2, versus the structure constant of the turbulence C_ϵ^2 computed for $C_\epsilon^2 = 10^{-9} - 5 \cdot 10^{-9}$. That is, we analyze strong turbulent structures occurring in the atmospheric channel, and averaged over the optical paths $L = 100$, 500 and 1000 m, for different frequencies from 10 THz to 1000 THz. It can be seen that above around $C_\epsilon^2 = 2 \cdot 10^{-9}$ the dimensionless scintillation index σ_I^2 exceeds values of several hundred (for all frequencies of observation and $L = 100$ m); that is, the optical data-carrying signal (called the band path signal [17]) in the visual and IR ranges is fully annihilated. Moreover, as the frequency increases from 10 to 1000 THz and for ranges $L > 500$ m, the tendency of the index of signal intensity scintillations, σ_I^2, to increase becomes much stronger, reaching from several hundred to tens of thousands. This result is very important for designers of arbitrary wireless atmospheric or land-atmospheric links, because it makes it possible to predict the real fast fading of a signal passing through the turbulent troposphere for optical networks operating at frequencies of $f > 10$ THz.

Fast fading of a signal over open optical links passing into the gaseous quasi-homogeneous troposphere with the absence of hydrometeors (called LOS conditions) is caused mainly by multiway propagation due to turbulences and, therefore, stochastic variations of the refractive index. The fluctuations of the signal intensity due to turbulence are distributed log-normally with the normalized standard deviation described by (29) [15, 16]. For moderate fluctuations of an optical signal, according to Rytov's approach [15, 16], the effect of turbulence corresponds to the turbulence attenuation γ_R, following Rytov's theory of regular turbulence, and can be written as:

$$\gamma_R = 2 \cdot \sqrt{23.17 \cdot C_\epsilon^2 \cdot k^{\frac{7}{6}} \cdot d^{\frac{11}{6}}}. \tag{30}$$

Here we should notice that Rytov's approach is based on Kolmogorov's spectral theory of weak-to-strong turbulence behavior, which, as was shown in [16, 18, 19], is similar to the modified spectral theory proposed by Andrew and Phillips in [16]. Therefore, we follow in our analysis both formulas (29) and (30), which are followed by both these theories.

Relation Between Scintillation Index and K-parameter of Fast Fading. Another way to calculate the total signal pathloss caused fast fading effects is to use the relations between the K-parameter of fading usually used in land-land communication links [7, 17] and the

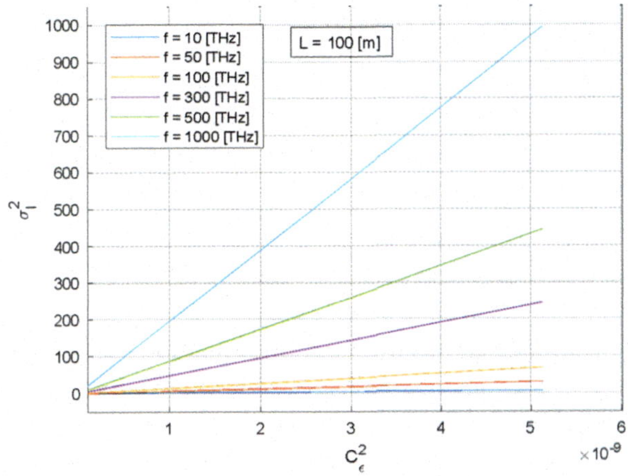

(a)

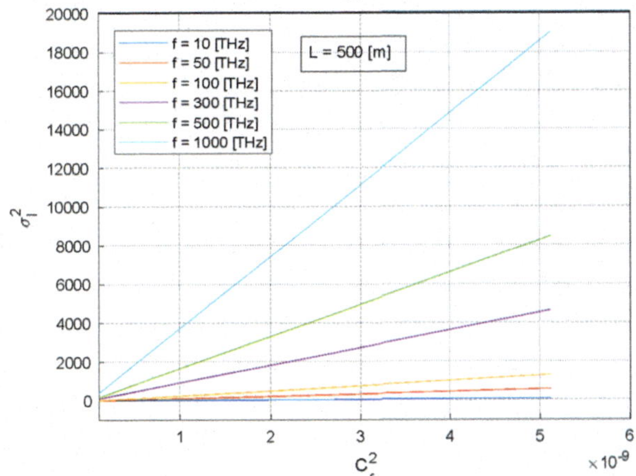

(b)

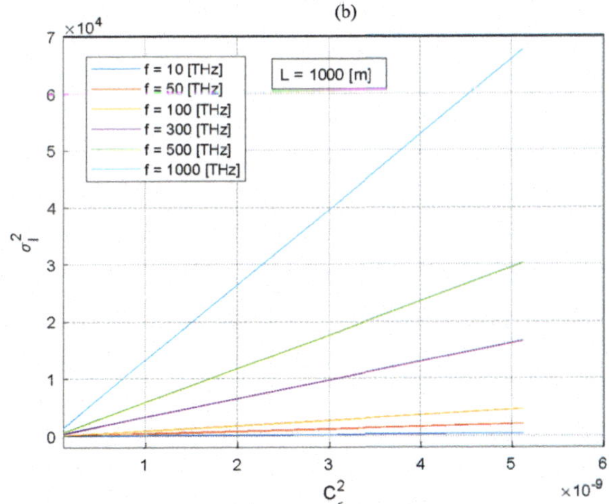

(c)

Fig. 3. Index of signal intensity scintillations, σ_I^2, vs. the structure constant of the turbulence $C_\epsilon^2 = [10^{-9}, 5 \cdot 10^{-9}]$, for frequencies from 10 THz to 1000 THz and averaged over the paths: (a) $L = 100$ m, (b) $L = 500$ m, and (c) $L = 1000$ m

scintillation index, σ_I^2. Usually, in land wireless communication the Ricean parameter of fading $K = I_{co}/I_{inc}$ is used [7, 17], but in atmospheric communication links always used the normalized scintillation index σ_I^2 [18–20]. For a zero-mean Gaussian distribution usually observed experimentally in the turbulent atmosphere, in [19] was found the relation between the parameter of fading, K, introduced in [7, 17], and the mean square of the scintillation index $\langle \sigma_I^2 \rangle$:

$$\langle \sigma_I^2 \rangle = \frac{\langle [I - \langle I \rangle]^2 \rangle}{\langle I \rangle^2} = \frac{I_{inc}^2}{I_{co}^2} \equiv K^{-2}, \tag{31}$$

where I_{co} and I_{inc} are the coherent and incoherent components of the total signal intensity.

Taking $C_\epsilon^2 \sim 10^{-9}$ for computations of the average value of the structure parameter of the refractive index, and computing the dimensionless square root of the average intensity scintillation index, $[\langle \sigma_I^2 \rangle]^{1/2}$, according to (31) for the trace of 500 m, we defined the K-fading factor. The results of these computations are plotted in Fig. 4.

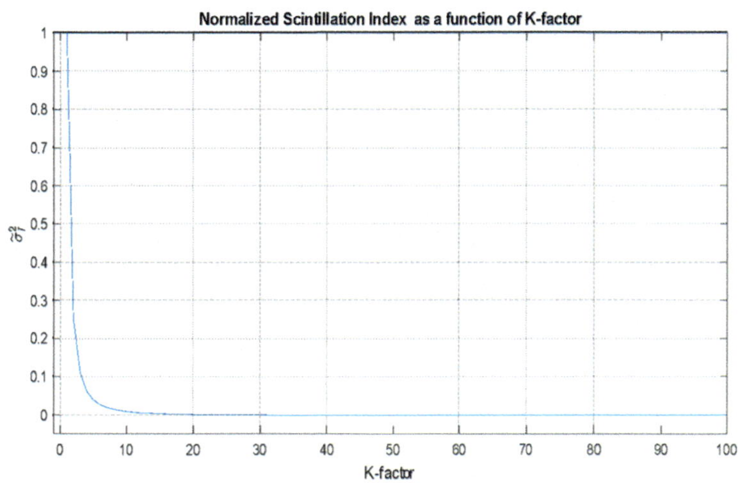

Fig. 4. Mean square of signal scintillation index, $\left[\left\langle \sigma_I^2 \right\rangle\right]^{1/2}$, vs. K-factor

We note that from numerous experiments where relation between $\langle \sigma_I^2 \rangle$ and the refractivity of the turbulence in the irregular troposphere were taken into account [17–19], it was found that the range of the latter parameter ranged from ~$C_\epsilon^2 \approx 10^{-13}$ to ~$C_\epsilon^2 \approx 10^{-10}$, for nocturnal and diurnal atmospheres at the heights of 1–2 km, respectively. In other words, the researchers have mostly investigated weak and moderate turbulence occurring in the atmosphere, and, therefore, their results are too optimistic, giving conditions of LOS with weak distortion of the carrier signal and information within it for most geographic places. The above obtained results make it possible to discuss situations occurring in channels with strong turbulence, i.e., with strong fading (for $K < 1$).

Thus, as follows from the non-linear relation between the mean square of the signal intensity scintillation index and K, illustrated in Fig. 4, when K is high the mean square of the signal intensity scintillation index, $\left[\langle\sigma_I^2\rangle\right]^{1/2}$, is low, and vice versa; when $\left[\langle\sigma_I^2\rangle\right]^{1/2}$ approaches its maximum value of unity, the parameter of fading, K, becomes lowest ($K < 2$). We are considering the worst (or "bad-bad") Rayleigh distribution describing strong fast frequency selective fading causing total obstruction (detonation) of the signal with information. At the same time, as the scintillation index decreases, saturation of the fading parameter K is observed (see Fig. 4). This is clear to understand, because with an increase of $K > 2$ (i.e., the signal is more than double the multiplicative noise), the situation becomes closer to line-of-sight (LOS) propagation conditions, where the incoherent component of the signal becomes weaker with respect to the coherent one. In this case, only flat slow fading can be observed without any sporadic scintillations of optical signal intensity.

Thus, by obtaining information on the average scintillation index, $< \sigma_I^2 >$, using measured data of the average refractive factor $< C_\epsilon^2 >$ according to (38), the K-parameter of fading can be used to determine the capacity, spectral efficiency and BER of a data stream sent via a wireless atmospheric optical communication channel.

4 Effects of Turbulence on Signal Data Propagating through Atmospheric Communication Links

4.1 Characteristics of Signal Data in Atmospheric Communication Links

According to the classical approach, the Shannon-Hartley formula, which defines the relationship between the maximum data rate via any channel, defines the capacity of a channel with additive white Gaussian noise (AWGN) of bandwidth B_ω, which is linearly proportional to the bandwidth B_ω [in Hz] and logarithmically - to the white or additive signal-to-noise ratio ($SNR \equiv N_{add}$).

$$\langle\sigma_I^2\rangle = \frac{\langle[I - \langle I\rangle]^2\rangle}{\langle I\rangle^2} = \frac{I_{inc}^2}{I_{co}^2} \equiv K^{-2}, \tag{32}$$

where the power of additive noise in the AWGN channel is $N_{add} = N_0 B_\omega$, S is the signal power (in Watts, W), and N_0 is the signal power spectrum (in W/Hz).

Usually, in optical communication, wireless and wired, another characteristic called the *spectral efficiency* of the channel [7, 17] is used:

$$\tilde{C} = \frac{C}{B_\omega}\left(1 + \frac{S}{N_0 B_\omega}\right). \tag{33}$$

Based on the *approximate* approach proposed in [7, 17], which accounts for the flat or frequency-selective or time-selective fading, we will introduce in (33) the multiplicative noise in terms of the Ricean *K-factor* of fading defined by relation (31). In [7, 17], the K-factor of fading was defined as a ratio between the coherent and the multipath (incoherent) components of the signal intensity, $K = I_{co}/I_{inc}$, or $K = S/N_{mult}$. Using these notations, the capacity as a function of the K-factor and the signal to additive noise ratio (SNR_{add}) can be introduced instead of (33), i.e.:

$$C = B_{\omega}\left(1 + \left(SNR_{add}^{-1} + K^{-1}\right)^{-1}\right) = B_{\omega}\left(1 + \frac{K \cdot SNR_{add}}{K + SNR_{add}}\right). \tag{34}$$

If so, one can easily obtain from (34) the spectral efficiency of the channel with fading as a source of multiplicative noise:

$$\tilde{C} = \frac{C}{B_{\omega}}\left(1 + \frac{K \cdot SNR_{add}}{K + SNR_{add}}\right), \tag{35}$$

where the bandwidth B_{ω} corresponds to the optical link/system under investigation. Comparison, made in [7, 17] between the two approaches, classical Shannon and approximate, showed that (33) with $SNR_{add} = N_0 B_{\omega}$ only, and (34) with K and $SNR_{add} = N_0 B_{\omega}$, are equivalently described of the channel/system capacity only in the cases when the K-factor is larger than SNR_{add}.

As can be seen in Fig. 4, the fading parameter K indicates whether or not strong direct visibility exists between the source and the detector, accompanied by the weak additional effects of multipath phenomena caused by the multiple scattering of signals by the turbulent structures filled the perturbed atmospheric regions [7, 17–20].

Using information regarding the K-factor, one can now predict deviations of the signal data parameters of the atmospheric multipath channels during their pass through the moderate and strong turbulences-filled gaseous quasi-homogeneous troposphere with an absence of hydrometeors.

Thus, the capacity and spectral efficiency, described in terms of the *K-factor* by (34) and (35), respectively, can easily be estimated for various scenarios occurring in the atmospheric channel and for different conditions of the internal noise of the source and the detector at both sides of the link under consideration. One of numerous computed examples is presented in Fig. 5 for different additive SNRs and for a "point" receiver (with respect to the range of 1 km between the terminals).

In Fig. 5, the K-parameter ranged from 0.1 to 30, that is, it covers the "worst" case, corresponding $K < 1$, describing by Rayleigh's law in the NLOS case [7, 17], through $K \approx 1$ (quasi-LOS case), and reaching $K > 1$, describing in the LOS case of signal propagation by a delta-shaped Gaussian law [7, 17].

As can be seen, in the strongly perturbed irregular atmosphere (with strong turbulences), when $1 < K < 10$ the spectral efficiency is around of 0.7–1.1 bit/s/Hz for SNR = 1 dB, while when $10 < K < 30$, it is around 3.4 and 4.6 bit/s/Hz for SNR = 20 dB.

Since the total band path optical signal consists of a carrier signal with carrier frequency fc combined with baseband signal as a carrier of digital information, i.e., a sequence of bits, effects of multipath fading occurring in any optical communication channel leads to errors in the detected bits, which is characterized by the bit-error-rate (BER). Using the Rayleigh distribution for the "worst" case of strong fading, one can determine the probability of bit error defined BER, occurring in the multipath optical channel using the following formula [7, 17]:

$$P_r(e) = \frac{1}{\sigma^2} \int_{r_T}^{\infty} r e^{-\left(\frac{r^2}{2\sigma_N^2}\right)} dr = e^{-\left(\frac{r_T^2}{2\sigma_N^2}\right)}, \tag{36}$$

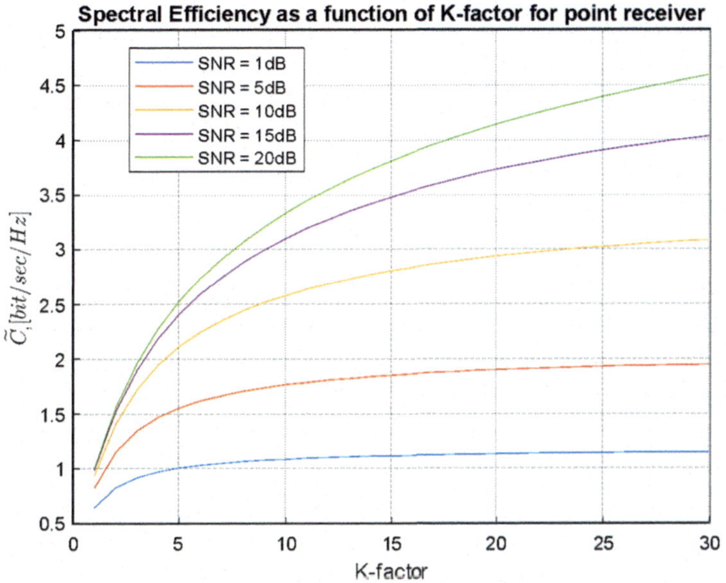

Fig. 5. Spectral efficiency vs. K-factor for $SNR = 1, 5, 10, 15$, and 20 dB for point receiver

where $P_r(e)$ represents the evaluated probability of a bit error, σ_N^2 is the intensity of interference at the optical receiver (usually defined as the multiplicative noise [7, 17]), r_T determines the threshold between detection without multiplicative noise (i.e., the "good case" [6, 7]), and with multiplicative noise (i.e., the "worst case" [6, 7]).

In our investigations, we present BER as a function of the Ricean *K-factor* of fading, following results obtained in [7] based on a Ricean probability density function (PDF), which is more general than the Rayleigh PDF [7, 17]. Thus, following [17] and using a classical formula for BER [15, 16], we finally get:

$$BER = \frac{1}{2} \int_0^\infty p(x) erfc\left(\frac{SNR}{2\sqrt{2}} x\right) dx, \tag{37}$$

where $p(x)$ is the PDF (in our case Ricean), and $erfc(\cdot)$ is the well-known error function [7]. Using the BER definition (37), where $p(x)$ is the Ricean PDF and the SNR includes the multiplicative noise, we finally get the following expression for the *BER*:

$$BER\left(K, \frac{S}{N_{add}}, \sigma\right) = \frac{1}{2} \int_0^\infty \frac{x}{\sigma^2} \cdot e^{-\frac{x^2}{2\sigma^2}} \cdot e^{-K} \cdot I_0\left(\frac{x}{\sigma}\sqrt{2K}\right) \cdot erfc\left(\frac{K \cdot \frac{S}{N_{add}}}{2\sqrt{2}\left(K + \frac{S}{N_{add}}\right)} x\right) dx. \tag{38}$$

This is an important formula, which gives the relation between the BER and the additive SNR, the Ricean parameter K, describing the multipath fading phenomena occurring within the land-atmospheric links between the ground-based subscribers and moving and flying vehicles. Finally, for each specific channel one can obtain the BER and loss of bits inside the information data stream passing a channel.

4.2 Relations Between Signal Data Parameters in Atmospheric Links and Fading

We can present a detailed analysis of the key parameters of the wireless atmospheric optical channel, and of the data stream parameters, based on the approximate approach outlined above. The analysis is divided into the following steps:

Firstly, the refractive index structure parameter, C_n^2, that characterizes the strength of atmospheric turbulence, was measured experimentally [7, 17, 18, 20].

Secondly, the intensity of the incoherent component was estimated from the measured beam scintillation/fluctuation energy, $< \sigma_I^2 >$, according to (29). As shown in [18–20], the effects of fading become stronger/weaker depending on the kind of turbulence: strong (with $\langle C_\epsilon^2 \rangle = 5 \cdot 10^{-12}$) occurring at altitudes up to 100–200 m/weak (with $\langle C_\epsilon^2 \rangle = 4 \cdot 10^{-14}$) occurring at altitudes of 1–2 km.

Thirdly, the parameter K, which represents the ratio between the coherent and incoherent components of the optical signal within the communication link, can be estimated either via (29) or via (31) by the corresponding fading measurements. It was shown that, for horizontal atmospheric channels, the fading factor K exceeds unity.

($K > 1$) at higher atmospheric altitudes where the LOS component exceeds the multipath non-line-of sight (NLOS) component.

In the fourth step of our algorithm, the effects of fading (e.g., the changes of the K-parameter) on the BER in a tropospheric wireless communication link featuring weak and strong turbulences can now be studied. Results of the BER in moderate ($\sigma = 2$ dB) and strong ($\sigma = 5$ dB) atmospheric turbulent media as a function of the K-factor of fading for different SNR values, are given in Fig. 6a and Fig. 6b, respectively, based on the experimental data described in [17–20].

As seen in Fig. 6a the BER decreases from $2.3 \cdot 10^{-1}$ for $K \approx 1$ to 10^{-5} for $K \approx 5$, for all SNRs, that is, as the LOS component becomes predominant with respect to NLOS multipath components (i.e., for atmospheric links at altitudes of 100–500 m containing turbulent structures [14, 17]). From Fig. 6b we see a decrease in the BER from 10^{-1} to 10^{-5} for $K \approx 1$ and $K \approx 5$, respectively, for all SNRs. These results imply a huge decrease in bit errors in the data flows passing through the optical atmospheric channel. This effect depends on the SNR in the optical channel, and additionally decreases with increasing SNR.

Further, comparing Fig. 5 and Fig. 6, one can understand the dependence of the BER on $\tilde{C} = C/B_\omega$. As seen in Fig. 5, the spectral efficiency increases with an increasing fading K-factor, and for a small value ($K = 1$–3), increases from 0.7 to 0.9 bit/s/Hz even for a low SNR (SNR = 1 dB). The spectral efficiency reaches 2.2 bit/s/Hz for the same fading K-factor, but with SNR > 5 dB. Considering the results presented in Fig. 6a and Fig. 6b for the same small range of $K = 1$–3, we observe a sharp decrease in the BER from 0.23 to 0.001 for low SNRs (SNR = 1–5 dB) and from 0.1 to 10^{-5} for high SNRs (SNR = 10–20 dB). In summary, the BER decreases significantly with an increase of spectral efficiency of the wireless optical communication link as the K-factor increases for a constant SNR, or for a constant K-factor, but with increasing SNR values.

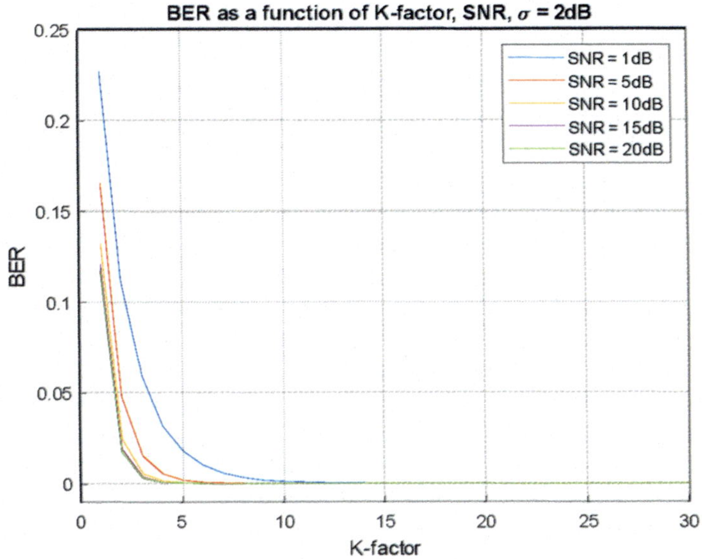

(a)

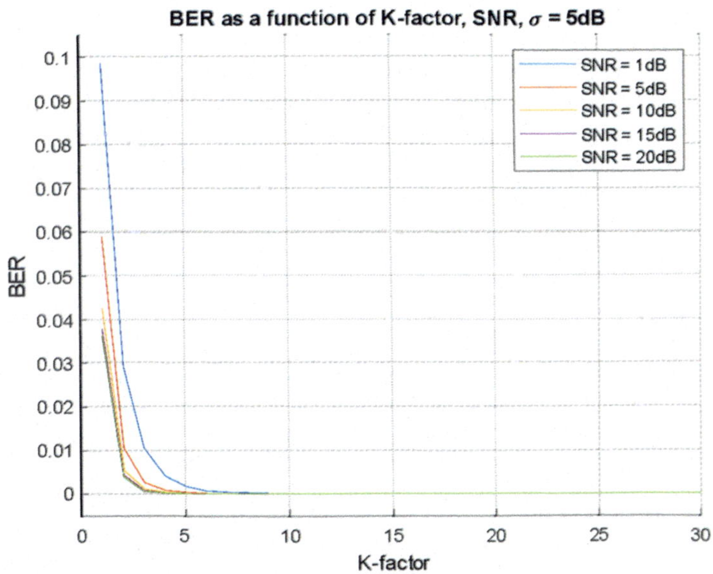

(b)

Fig. 6. BER vs. K-factor for SNR = 1 dB to 20 dB for a) moderate (with $\sigma = 2$ dB) and b) strong (with $\sigma = 5$ dB) atmospheric turbulence

5 Conclusions

This chapter is based on the recent research work in two fields of research: (1) optical signal decay in wireless atmospheric communication links, and (2) quality-of-service (QoS) in such links accounting for the destructive effects of fading phenomena on signal data streams passing through such channels.

The optical signal decay was studied based on the attenuation and scattering effects of gaseous structures and hydrometeors (rain, snow and clouds), as well as on turbulent structures, which have a predominant impact on the fast fading of optical signals propagating through atmospheric channels with fading. Data stream parameters (capacity, spectral efficiency and BER) were analyzed, taking into account all relevant atmospheric communication link features for the prediction of, and increase in, QoS.

The impact of each of the above atmospheric features on the total loss of signal passing such fading channels was analyzed for optical signal decay prediction.

Then, the influence of fading phenomenon, defined by the K-factor (instead of the more commonly used scintillation index), on the degradation of data stream parameters, such as spectral efficiency and BER, was investigated both theoretically and numerically. Such an approach affords a unified examination of various optical channels based on both the scintillation index and the K-factor of fading, enabling a unified approach to the analysis of optical and radio channels simultaneously.

Finally, an optimal prediction algorithm was found for the optical signal decay in various meteorological situations occurring in the real atmosphere at different heights and for various frequencies of radiated signals. Further, a method was proposed to evaluate the data stream parameters, spectral efficiency/capacity and BER, that accounts for the effects of atmospheric turbulence on fast fading, which corrupt information transmission in these channels.

References

1. Jaenicke, R.: Aerosol physics and chemistry. In: Fisher, G. (ed.) Physical Chemical Properties of the Air, Geophysics and Space Research, vol. 4(b). Springer-Verlag, Berlin (1988)
2. d'Almeida, G.A., Koepke, P., Shettle, E.P.: Atmospheric Aerosols. Global Climatology and Radiative Characteristics, Deepak Publishing, Hampton (1991)
3. Rosen, J.M., Hofmann, D.J.: Optical modeling of stratospheric aerosols: present status. Appl. Opt. **25**(3), 410–419 (1986)
4. ITU-R International Telecommunication Union, ITU-R Recommendation: Attenuation by atmospheric gases, pp. 676–683 (1997)
5. Seinfeld, J.H.: Atmospheric Chemistry and Physics of Air Pollution. Wiley, New York (1986)
6. Saunders, S.R.: Antennas and Propagation for Wireless Communication Systems, 2nd edn. WileySons, New York (1999)
7. Blaunstein, N., Arnon, S., Zilberman, A., Kopeika, N.: Applied Aspects of Optical Communication and LIDAR. CRC Press, Taylor & Francis Group, New York (2010)
8. Crane, R.K.: Prediction of attenuation by rain. IEEE Trans. Commun. **28**, 1717–1733 (1980)
9. International Telecommunication Union, ITU-R Recommendation, P.838: Specific attenuation model for rain for use in prediction methods, Geneva (1992)
10. ITU-R Recommendation International Telecommunication Union: Specific attenuation model for rain for use in prediction methods, p. 838 (1992)

11. ITU-R Recommendation International Telecommunication Union: Propagation data and prediction methods required for the design of terrestrial line-of-sight systems, pp. 530–537 (1997)
12. ITU-R Recommendation International Telecommunication Union: Attenuation due to clouds and fog, pp. 840–842 (1992)
13. Chou, M.D.: Parametrizations for cloud overlapping and shortwave single scattering properties for use in general circulation and cloud ensemble models. J. Clim. **11**, 202–214 (1998)
14. ITU-R Recommendation International Telecommunication Union: Characteristics of precipitation for propagating modeling, p. 837 (1992)
15. Ishimaru, A.: Wave Propagation and Scattering in Random Media. Academic Press, New York (1978)
16. Andrews, L.C., Phillips, R.L.: Laser Beam Propagation through Random Media, 2nd edn. SPIE Press, Bellingham (2005)
17. Tiker, A., Yarkoni, N., Blaunstein, N., Zilberman, A., Kopeika, N.: Prediction of data stream parameters in atmospheric turbulent wireless communication links. Appl. Opt. **46**(2), 100–199 (2007)
18. Bendersky, S., Kopeika, N., Blaustein, N.: Atmospheric optical turbulence over land in middle-east coastal environments: prediction, modeling and measurements. J. Appl. Opt. **43**, 4070–4079 (2004)
19. Blaunstein, N., Kopeika, N.: Optical Waves and Laser Beams in the Irregular Atmosphere. CRC Press, Taylor&Frances Group, Boca Raton (2018)
20. Belov, V., et al.: NLOS communication: theory and experiments in the atmosphere and underwater. Atmosphere **11**(10), 1122–1129 (2020)

Control Methods Research of Indicators for Intelligent Adaptive Flying Information-Telecommunication Platforms in Mobile Wireless Sensor Networks

Leonid Uryvsky, Oleksandr Lysenko, Valeriy Novikov, and Serhii Osypchuk(✉)

Igor Sikorsky Kyiv Polytechnic Institute, Industrialnyi Lane2 (Campus 30), Kyiv 03056, Ukraine
leonid_uic@ukr.net

Abstract. This chapter is devoted to the problem statement of managing the intelligent adaptive flying information-telecommunication platforms (FITP) network. Construction and operation peculiarities of mobile wireless sensor networks (MWSN) with FITP are outlined in this study. The approach to creating new architectural, algorithmic, and technical solutions for intelligent control systems construction based on MWSN with FITP features is proposed. The chapter analyzes construction methods and operation protocols for intelligent adaptive FITP. The intelligent adaptive control concept of FITP is developed for applying in emergency or critical infrastructure protection zones. The MWSN with FITP throughput increase methods are investigated and math model is introduced. Methods of noise protection increase in channels for MWSN with FITP are analyzed. The math model creation general problem statement of noise protection research for MWSN channels with FITP is outlined. Noise immunity quality indicators of MWSN channels with FITP are given. The math model of research and comparison noise immunity for wireless communication systems in Gaussian and Rayleigh channels is proposed. Suggestions for improving MWSN with FITP noise immunity based on the developed wireless communication channels math model description are given.

Keywords: Flying Information-Telecommunication Platforms (FITP) · Mobile Wireless Sensor Networks (MWSN) · Math model · Control method

1 Introduction

The key global tendencies of infocommunication systems development are shown in [1–3]. One of the current areas is intelligent adaptive flying information and telecommunication platforms (FITP) in mobile wireless sensor networks (MWSN), and control methods for their indicators [4, 5].

The object of study is a high-performance sensor network management system based on the robotic flying objects and computing FOG-infrastructure use.

M. Klymash et al. (Eds.): *Future Intent-Based Networking*, LNEE 831, pp. 444–467, 2022.
https://doi.org/10.1007/978-3-030-92435-5_25

The subject of research is a set of intellectual control methods for high-performance sensor networks based on the robotic objects and computing FOG-infrastructure use for terrestrial sensor network and flying information and telecommunication robots group in the emergency zone or area for tactical tasks.

The work aims to develop the systems intellectual control methods control for high-performance sensor networks, based on the use of robotic objects and computing FOG-infrastructure, to increase reliability, timeliness, accuracy and reliability for critical infrastructure information systems in emergency zone search and rescue operations, to use flying information and telecommunication robots with mobile sensors and telecommunication ground and air platforms and their effective management, while ensuring their structural and functional connectivity in conditions of rapid and unpredictable objects movement in dual-use systems.

2 Concept Development of Intellectual Adaptive Control in the Emergency Zone for Flying Information and Telecommunications Robots

2.1 Management Tasks for Flying Information and Telecommunication Robots Network

Network management tasks for FITP are divided into stages: planning, deployment, and operational management [6].

Planning stage is carried out by the MWSN control center with FITP.

Based on the projected situation and available resources, the planning scope is:

1. Topology planning for the FITP network (finding the required number of FITP, determining their location or movement in space), which implements a specific management goal, based on the requirements for the parameters of the MWSN and the requirements for the messages transmission.
2. Resources allocation (hardware, frequency, energy, spatial) for the FITP network; methods, control algorithms, parameters, and modes definition for technical means operation.

Deployment stage is to run a given number of FITP and control their flight to some barrage specified areas. At the same time, some deployment phase tasks (for example, topology re-planning) of the FITP network can be performed during operational management with significant changes in the network topology. Control over the FITP flight and its onboard systems operation is carried out from the network control center (CC).

MWSN with FITP state ***at the operational management stage*** is assessed according to the accepted efficiency criteria and, in accordance with the plan and the real situation, measures are taken to maintain the MWSN with FITP performance within the specified limits or their optimization [7]. In contrast to planning tasks, operational management tasks are solved in real time in a mixed way (centralized / decentralized), and their content can be repeated many times.

FITP network management tasks are divided into the following two groups ***by functions***:

Special management tasks – determining the FITP flight routes and coordinating their movement [8].

2. Universal control tasks – typical for any MWSN with FITP [9]: topology management; routing management; load management; radio resource management, energy resource management, security management. However, specific implementation of management methods for each functional subsystem must take into account the purpose of the FITP network and the features of its architecture (dimension, mobility, performance, etc.).

The functional model of the FITP network management system is presented on Fig. 1.

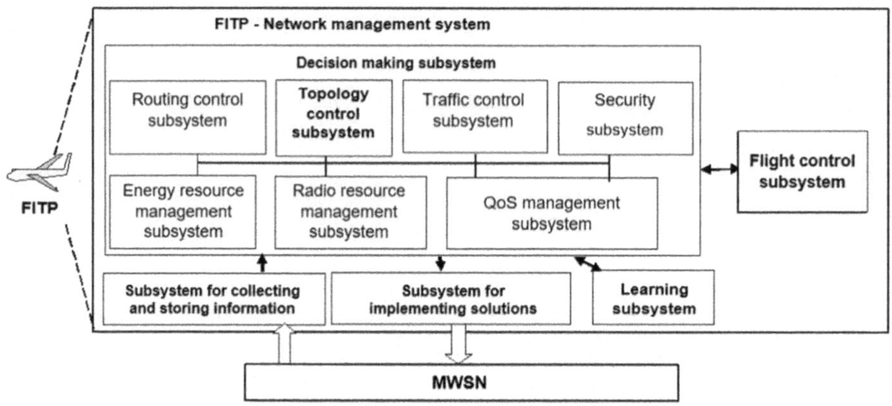

Fig. 1. Functional model of the FITP network operational management system

The FITP network management system efficiency in general case is evaluated by the vector criterion $K = [K_1, K_2, \ldots, K_n]$:

- $K_1 = [K_1^{\,1}\ K_1^{\,2}\ K_1^{\,3}]$ are the routes quality for all MWSN from FITP or its zone ($K_1^{\,1}$ is the length of transmission routes in network zones, $K_1^{\,2}$ is the average delivery time, $K_1^{\,3}$ is the throughput, etc.);
- K_2 is the bandwidth of all MWSN from FITP or its zone;
- K_3 are the FITP zones coverage degree by fixed amount, or by selected and prioritized mobile sensor nodes (MSN);
- K_4 is the structural reliability (connectivity);
- K_5 is the FITP number.

The efficiency criteria set presence determines the management tasks multicriteria nature and significantly complicates the formal methods development. To find a compromise management, it is proposed to use the leading criterion method: to determine the main efficiency criterion (based on the current situation), which is subject to optimization, and others to consider limitations. To determine the leading criterion, it is proposed to use the hierarchical target dynamic alternatives evaluation method [10].

One of the FITP network management main tasks is the FITP network topology (location) management task. The FITP network topology management problems classification is shown on Fig. 2.

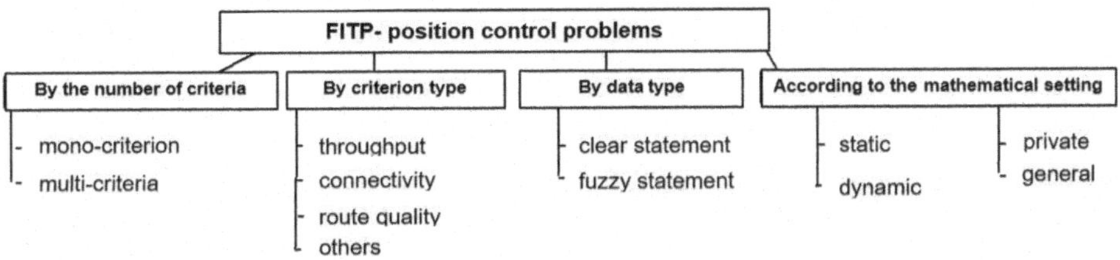

Fig. 2. FITP network topology (location) management tasks classification

2.2 Stages of Planning FITP Network Management Cycles

According to [11], the topology (location) management primary goal for the FITP network is to increase the MWSN bandwidth while ensuring MSN structural connectivity and quality of service (QoS). Therefore, as the FITP location management effectiveness main criterion, it is proposed to choose the MWSN bandwidth, and others to be classified.

Then the FITP location management will be performed according to the following principles (stages): planning (re-planning), deployment, operational management.

At the ***planning stage (re-planning)*** the control center (CU) is carried out:

- collecting information about the initial topology $((x_i, y_i), v_i, X_{0k}, V_{0k})$ and network operation (Π_i) and input the original data $(s_0, t_3^0(l^0), d^0, D^0, N, K, \Gamma)$;
- structural connectivity and duration parameters calculation, as well as route quality parameters, using advanced and existing mathematical models discussed below;
- compliance analysis with the structural connectivity and routes quality;
- network S bandwidth calculation, according to the mathematical model considered in Subsect. 4.2;
- search for the initial (next) placement of the next FITP, which maximizes network bandwidth, in the possible solutions presence, using the proposed algorithm, which are also discussed in Sect. 4.2.

At the ***deployment stage*** is carried out:

- FITP output (movement) to the initial (next) placement point.
- message transmission routes adjustment and channel loading.

At the ***operational*** management stage is carried out:

- FITP operation adaptation to real operating conditions (traffic level generated by each MSN).
- periodic check of necessity for the FITP position change need (while all MWSN MSNs with FITP are considered fixed at the set moment of time).

These principles (stages) are implemented sequentially and quickly for each FITP according to the method discussed below, depending on how much the network topology has changed relative to its previous state (i.e. almost depending on the network topology change rate).

3 Mobile Wireless Sensor Networks with FITP Infrastructure Construction Peculiarities

An example of building an MWSN architecture with FITP is shown on Fig. 3.

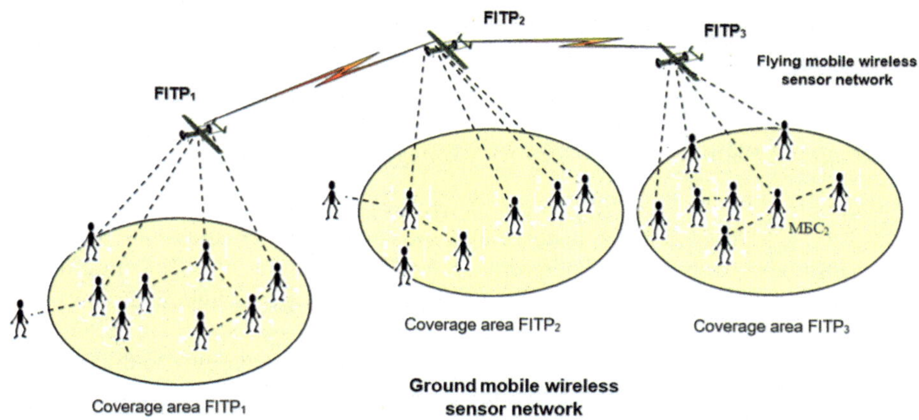

Fig. 3. Two-level MWSN with FITP architecture example

The MWSN with FITP first level (primary information sensors level) consists of MSN clusters (groups).

The second level (air) consists of the core network FITPs that are designed to ensure MSNs ($FITP_i$, i = 1, …, m) remote clusters connectivity, and the FITPs for "problematic" MSN clusters increasing connectivity $j, j = 1, \ldots, p$ (for example, between clusters Ω_1 and Ω_2 or Ω_4, Ω_5 and Ω_6, as shown on Fig. 4).

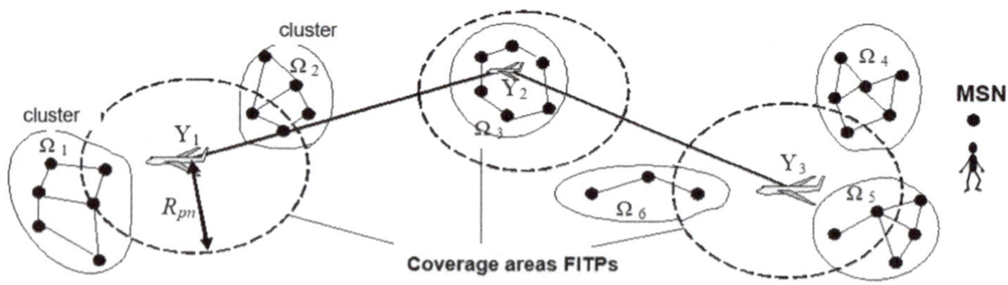

Fig. 4. FITP network scheme with PSNs

Each FITP is equipped with two sets of equipment: antenna system and transceiver (for communication with MSNs and for communication between FITP), network router, GPS-navigator, buffer storage device. There are 3 different functionalities of FITP:

- FITP bridge;
- FITP router;
- FITP gateway.

The FITP gateway has the most complex structure. It provides a wide range of message processing operations and can act as a device for pairing with an external network. FITP-bridges provide relay, and FITP-router – messages routing within one MWSN with FITP. The most multifunctional and rational FITP is such when each FITP will combine all functionalities stated above.

The end devices of MWSN with FITP are MSN (Fig. 1). Those must be equipped with:

- two sets of receiving and transmitting equipment, – for communication between MSN and for communication with FITP. The use of two separate radio interfaces allows the messages transmission both through the terrestrial MWSN and through the FITP network, which creates more routes between a pair of MSNs. The second advantage of using separate radio interfaces is the different frequency bands use on each of the interfaces, which provides better quality of service and load balancing in the network;
- network bridge-router, for retransmission and messages routing assigned to other MSNs;
- functionality for converting information into a user-friendly form (data, voice, video).

The following information transfer methods between MSNs are used in MWSN with FITP:

1) messages transmission between MSNs (without FITP use);
2) messages retransmission to recipient through FITP in the event that both the message source and the source recipient are within the one FITP coverage area;
3) communication through intermediate MSNs that are called base nodes. In this case, each message received by the FITP must be transmitted to the base node, which determines the further message route;
4) retransmission with transfer to the FITP storage device. In this case, the message is received by the FITP from the message source, stored in its storage device, and transmitted to the consumer during the flight of the FITP over the consumer;
5) communication with the interstation channels use between FITP.

The most useful is methods combination #1, #2 and #5. The following two options for organizing channels are possible. In the first option, the message between two MSNs is transmitted over a network with multiple re-emission of FITP on radio channels created only at the time of message transmission. In the second option, trunk channels are created between all MSNs that observe each other, through which messages are transmitted if necessary. Each of these two options has its advantages and disadvantages.

The first option advantages are ease of implementation, ability to use one set of receiving and transmitting equipment FITP for communication with the MSN and with neighboring FITP. However, the economized use of channel resources complicates the communication process and is accompanied by difficulties in multiple access organization. This option is quite effective at low and medium inbound traffic (inbound network load) and, as will be shown below, ineffective at high inbound traffic.

In the case of large inbound traffic, it is advisable to use trunk communication channels. The increase in bandwidth is achieved by a five to six time increase in the transceivers number on the FITP board, which necessitates the mutual influence elimination of trunk channels among themselves, as well as mutual messages interference in the middle of trunk channels.

The *functioning feature* of MWSN with FITP is its *topology dynamic change* (both due to MSN and FITP movements). Therefore, there are scientific tasks when designing MWSN with FITP, – to increase packet messaging efficiency, MSN to MWSN multiple access organization, routes determination for messages retransmission through intermediate network nodes, FITP topology (location) management.

Each message is divided into information packets when packet messages are transmitted. The packet message transmission method allows to increase the network bandwidth by simultaneously servicing MSNs large number.

The rules called *multiple access protocols* enable the *network use* by a large MSN number and FITP interaction with them and with each other. These protocols can be divided into types: random, deterministic and hybrid [12].

Deterministic protocols streamline the MSN and FITP work in such a way as to completely eliminate conflicts in which two or more MSNs (FITPs) simultaneously transmit messages to the same MSN. In such protocols, conflict resolution is performed by static or dynamic channel resources fixation for MSN and FITP: time (TDMA), frequency (FDMA), spatial (SDMA), code (CDMA), or hybrid (TDMA / CDMA, STDMA), and requires high level of network management organization.

Random access protocols allow conflicts to occur, i.e. packet collisions in communication channels. Historically, the first method with random access is the ALOHA method. It works effectively in low network load cases, when network bandwidth partial loss due to packet collisions is smaller than losses associated with network resource downtime. With a large number of packets, there is a situation where the network node is unable to receive or transmit the packet. This problem is partially solved by random access clocking, in which all network nodes begin to transmit packets with the onset of the next clock (S-ALOHA method). The time interval between cycles covers the packet duration and its maximum propagation time in space.

The carrier-controlled access methods (e.g., CDMA) are used to reduce the conflicts likelihood using radio channel state preliminary check. In this case, the node monitors channel status (the carrier presence or packet transmission), before transmitting the packet. If the channel is busy, the node postpones the packet transmission to a later time. When the channel is released, packet transmission can begin in the following ways: immediately, using "hard" CDMA, or at random intervals, using "soft" CDMA, with transmission randomization time into segments or with probability p (p-persistent) [13].

To solve the "open" and "hidden" terminal problems, a number of methods (protocols) with carrier control and conflict prevention (CSMA/CA) have been proposed. The best known of these are FAMA, MACA, MACA-BI, MACAW, DBTMA and IEEE 802.11 DCF [13]. When using these methods, the communication channel is temporarily reserved for the message transmission period by exchanging between the sender and the recipient short service packets: the sender's request for transmission (RTS) and the recipient consent (CTS).

The analysis [14] shows that in the MSN high mobility conditions, which is typical for MWSN with FITP, the IEEE 802.11 protocols set has an advantage, although it is far from perfect. Also, studies in [14] indicate the IEEE 802.11 protocols using possibility not only inside the building but also outside. With a cell radius of up to 6 km, the MAC sublayer protocol complies with all 802.11 standards.

In addition, the MAC access layer should address such important tasks as authentication, synchronization, encryption, power management, roaming, and more. Particular attention should be paid to roaming, i.e. the procedure of MSN entry into the FITP service area, MSN transition from one FITP zone to another and switching of MSN between FITP zones when MSN is in the service areas of several FITPs simultaneously.

Route definition for each package is a complex operation and can be carried out by both on-board FITP and MSNs. Note that the MWSN and FITP routers have the same functional properties, which allows the existing routing protocols use and allows to abandon the auxiliary protocol use, which coordinates the FITP arrival and departure with the routing protocol [15].

In general, MWSN with FITP in comparison with existing information transmission systems are characterized by a much higher complexity and organization level.

So, the approach is proposed for new architectural, algorithmic and technical solutions level for the intelligent control systems construction based on MWSN with FITP features:

- The network structure complexity: the radio channels between the MSN message source and the FITP, between the FITPs, between the FITP and the MSN message recipient, – are complex information transmission systems.
- The MWSN topology changes due to MSN movement causes connectivity instability, the routing complexity, the Doppler shifts carrier frequencies emergence.
- The MWSN with FITP operation requires an effective management system, an integral part of which the FITP network subsystem management is.

4 Methods Analysis for Mobile Wireless Sensor Networks with FITP Capacity Increasing

4.1 Methods Review of MWSN with FITP Capacity Increase

Today, the most promising for use in the disaster area are MWSN with FITP class of unmanned aerial vehicles (UAVs) [16]. Recently, attention is growing to small UAVs (UAVs), which are more cheap, easy to operate and do not require a runway or special launch pad installations [17]. MWSN-type UAVs that are equipped with communication means, – can better explore inaccessible areas (natural disasters areas, man-made disasters, etc.), and ground-based MWSN to increase the MWSN efficiency.

The UAV example is the model R-100 LLC "Iuavia" (Kyiv), which is shown on Fig. 5.

Fig. 5. Miniature UAV (R-100) manufactured by Iuavia

To date, the problem is insufficiently solved for optimal operational FITP set placement to increase the bandwidth of MWSN while ensuring the structural and functional MSN connectivity.

Significant attention has recently been paid to ensuring network connectivity, including the MWSN context. It is shown in [18] that base stations distributed network use can significantly increase network connectivity. In [19] the authors determine the critical power from which a node should transmit messages in the network to ensure network connectivity with probability 1, in the case when the network nodes number goes to infinity.

Miller in [20] calculated the probability that two network nodes can be connected in a two-way better than directly. In work [21] the authors studied what should be the transmission range that provides network connectivity with a higher probability.

In [22] the authors shown that in a network with N nodes placed randomly, if each node is connected to less than 0.074 log$N$ neighboring nodes, the network is *asymptotically disconnected* with a probability 1 with increasing N. In the case where the node is connected with more than 5.1774 log$N$ neighbors, then the network is *asymptotically connected* with a probability 1 with increasing N.

In [23] the authors investigated how to place *additional nodes* in the network so the extended network will be *connected*. Huller in [24] studied the *increasing connectivity problem* and determined the *edges set of the minimum weight* that must be added in order to form a k-connected graph. The works [25] and [26] present analytical expressions that allow to determine the required nodes set, which is formed at a given density almost without a doubt k-connected network. In work [27] the methods proposed for managing MWSN connectivity based on the message routing and node capacity management. In

[15] the MWSN connectivity management is offered on the basis of network nodes antennas pattern control.

The problem of MWSN capacity increasing with the FITP help is also studied by domestic and foreign scientists. Ways to increase the MWSN bandwidth with FITP based on FITP coverage for the maximum number of ground nodes are presented in [15, 16, 17, 18, 19, 20, 21, 22, 23, 24, 25, 26, 27, 28, 29]. Increase network bandwidth using a single UAV is investigated in [10]. Increasing network bandwidth by load control at network nodes was studied in [30, 31, 32, 33].

So, currently available methods embedded in the topology (location) FITP management system, **solve only partial problems of MWSN disconnected components geometric connectivity, and do not take into account the channel resources limited capacity, load distribution and packet maintenance in MSN**. Also, most methods solve only static problems, and do not take into account MSN mobility and FITP maneuverability, and therefore are subject to improvement. Existing planning methods for MWSN with stationary sensors are also not effective, because they have a high complexity and computation time, which does not allow FITP to work out the obtained solutions in real time.

Thus, there is an urgent scientific task – method development for increasing the MWSN capacity with FITP and its location control in case of rapid and unpredictable LSM movement.

4.2 Problem Statement for Creating the Mathematical Model of System Bandwidth Research

To state a problem of creating the mathematical model for studying the system bandwidth, let's consider the MWSN with FITP functional model (Fig. 6).

The MWSN with FITP 1st level is a MSNs network that can move arbitrarily in some area r. In the direct visibility presence, MSNs communicate with each other through a common broadcast channel on the frequency f_1, and in the direct visibility absence, the messages transmission is carried out through intermediate nodes. MSNs are multifunctional devices that combine transceiver, modem, codec, router, and storage device, operating in single-frequency half-duplex mode on a store-and-forward basis. For FITP messaging, each MSN also has a second set of equipment operating in duplex dual frequency mode at frequencies f_2–f_3.

The MWSN with FITP 2nd level is a FITP network that barrage at a height h on a minimum radius circle around its optimal location point (x_{0k}, y_{0k}), $k = \overline{1, K}$, where K is the FITP number in the network, forming cells with radius R. The FITP on-board equipment is also a complex multifunctional device with separate radio interfaces (for communication with MSN and FITP with each other), capable for retransmitting messages in the middle of the cell or beyond. Between cellular connections (FITP-FITP) operate in duplex mode with frequency compaction, using a distributed carrier frequencies set on a cellular principle, with a separate demodulator at each frequency. With a single transmitter help, messages are sent to the neighboring FITP according to the available requests in the time division mode. It is considered that each FITP has information onboard on its location and frequencies distribution on cells, what allows to define at FITP position change which frequencies should be used at present.

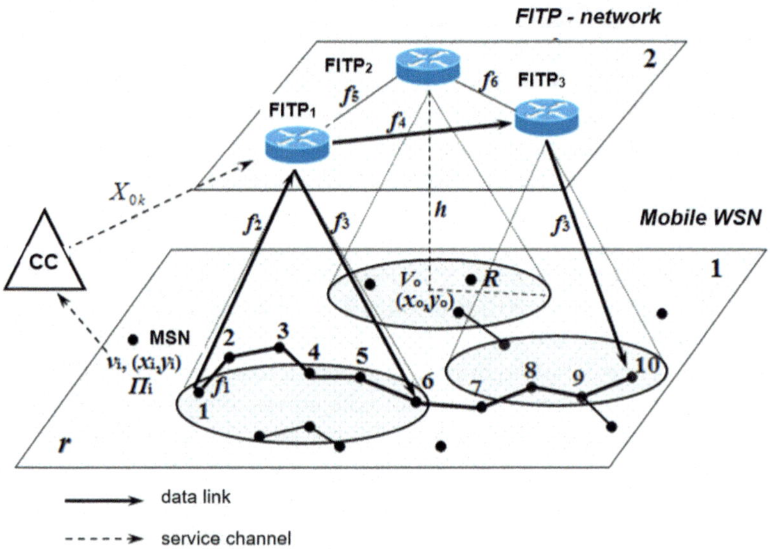

Fig. 6. MWSN with FITP functional model

Therefore, the following message transmission routes options between MSN (for example, between MSN_1 and MSN_{10}) are possible [34], which are indicated by solid bold arrows on Fig. 6:

1. through the MSN network (MSN_1 – MSN_2 – … – MSN_{10});
2. through the FITP network (MSN_1 – $FITP_1$ – $FITP_3$ – MSN_{10});
3. mixed way (MSN_1 – $FITP_1$ – MSN_6 – … – MSN_{10}).

The following *requirements* apply to the routes between the specified pair *sender a – recipient b*:

1) S_{mab} is the route bandwidth m_{ab}: $m_{ab} \geq s^0, a, b = \overline{1, N}, m = \overline{1, M}$, where N is the MSNs number in the network, M is the routes number in the network, s^0 – the minimum allowable route bandwidth level;
2) t_{3ab} are the transmission delays (or relays number) on the route: $t_{3ab} \leq t_3^0 \left(l(m_{ab}) \leq l^0\right)$;
3) $d_{ij}(D_{ik})$ is the structural connectivity on all route sections: $d_{ij} \leq d^0 \left(D_{ik} \leq D^0\right) \forall ij || ik \in m_{ab}, i, j, a, b = \overline{1, N}, k = \overline{1, K}$, where d_{ij}, d^0 is the distance between the MSN and the corresponding restriction from above, and D_{ij}, D^0 is the distance between MSN and FITP, and the corresponding restriction from above;
4) T_{3ij} is the each route section ij connectivity duration: $T_{3ij} \leq T_3^0$, where T_{3ij} is the minimum time during which the FITP can work out the specified location, set the route and transmit the minimum information amount.

Route selection is based on one of the known routing methods. For the FITP topology (location) management convenience, it is better to use tabular-oriented methods (for example, OLSR), then each MSN has its own shortest paths route table Π_i to all other network nodes.

The purpose of FITP position management is to increase the MWSN with FITP bandwidth while ensuring the MSN structural connectivity and the data transmission routes quality between them.

In this case, the network bandwidth will be understood as the maximum traffic value γ, which the network can process per unit time with a constant matrix of traffic distribution G.

So, developed mathematical model for intelligent adaptive flying information and telecommunication system optimal functioning:

determine the FITP group X location to maximize the bandwidth of MWSN with FITPs S, i.e.:

$$S = \sum_{a=1}^{N} \sum_{b=1}^{N} s_{mab}(X) \to \max_{X \in \Omega}, \tag{1}$$

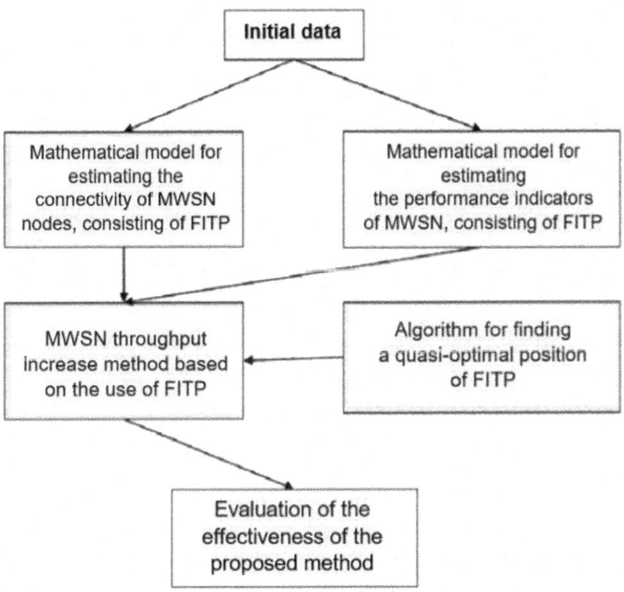

Fig. 7. Mathematical model structure

where Ω is the permissible values range, which is determined by the requirements for MWSN with FITP connectivity and performance indicators;

$$X = \begin{bmatrix} X_{01} \\ \dots \\ X_{0k} \end{bmatrix}, \text{ where } X_{01} = \begin{bmatrix} x_{01} \\ y_{01} \\ z_{01} \end{bmatrix}, \dots X_{0k} = \begin{bmatrix} x_{0k} \\ y_{0k} \\ z_{0k} \end{bmatrix}, k = \overline{1, K} \quad (2)$$

The mathematical model structure and its components interaction is shown on Fig. 7.

Thus, the use of increasing bandwidth methods by MWSN based FITP analysis shown that the existing methods, which are embedded in the FITP topology (location) management system, solve only partial problems that ensure MWSN disconnected components geometric connectivity, do not take into account channel resources capacity and MWSN nodes load distribution. Most methods solve only static problems, and do not consider the MSN and FITP maneuverability and mobility nature, and therefore are subject to improvement. Existing planning ground MWSN methods are also not effective, as they have a high complexity and computational time, which does not allow FITP to work out the obtained solutions in real time.

5 Increasing Noise Immunity Methods in Channels of MWSN with FITP

5.1 General Problem Statement for Mathematical Model Research of MWSN with FITP Channels Noise Immunity

The starting point is the main characteristics description in the classical (ideal) channel with additive white Gaussian noise (Additive White Gaussian Noise, AWGN) with statistically independent Gaussian noise samples that distort information samples in the wireless communication systems characteristics analysis, and without any presence of inter-symbol interference (ISI). The main performance degradation sources are divided into internal (thermal noise generated by the receiver) and external (natural and artificial noise and interference sources). In mobile communication systems, external noise and interference are often more significant than the receiver thermal noise. Such noise includes interference from other radio channels, as well as extraterrestrial noise and atmospheric noise.

If the channel characteristics are not specified, it is usually assumed that the signal is attenuated with a distance similar to the propagation in an ideal free space, where the area between the transmitter and receiver antennas is considered free from signal absorbing and reflecting objects. With such an ideal propagation, the received signal strength is quite predictable. For most real channels, such a free space propagation model inadequately describes the channel behavior and does not allow to predict the wireless information transmission systems characteristics [35].

5.2 MWSN with FITP Channels Noise Immunity Quality Indicators Comparative Analysis

The most significant are the following MWSN with FITP channels noise immunity quality indicators.

Delay τ is the signal expansion consequence in time caused by suboptimal impulse response in the channel with fading.

Transmission time (observation) t is associated with antenna movement or spatial changes in the propagation paths that determine the non-stationary channel behavior.

Channel coherence band $\Delta F_{\kappa}$is the frequency range statistical measure in which the signal spectrum components are considered correlated, and the channel passes them with approximately the same attenuation coefficient and linear phase change.

Coherence time (correlations) $T_{\kappa}$is the time period during which the channel characteristics do not change significantly.

Maximum signal delay τ_{max} is the time between the first and last component reception, after which the multi-beam signal scattered components power is lower than the threshold level set relative to the most powerful component. The threshold level is usually chosen 10 or 20 dB below the most powerful beam level. Sizes τ_{max} and ΔF_{κ}related by an approximate ratio $\Delta F_{\kappa} \approx \frac{1}{\tau_{max}}$..

A more accurate parameter that characterizes the signal delay is the delays scatter, which is often described by the root mean square value σ_{τ}. Sizes $\Delta F_{\kappa}$and σ_{τ} are closely related as follows: $\Delta F_{\kappa} \approx \frac{1}{50\sigma_{\tau}}$..

Rapid fading is channels characteristic with $T_{K} < T_{S}$where the correlation interval T_{κ}does not exceed the character duration T_S. During time T_s the fading nature changes many times, which leads to signal distortion.

Slow fading occurs when $T_{K} > T_{S}$. The symbols are not distorted in shape, but the decrease in transmission quality is caused by a decrease in the average signal-to-noise ratio $\overline{\gamma}_b$ (Signal-to-Noise Ratio, SNR) due to signal components incoherent addition, as well as amplitude fading.

In the channel with fading the relationship between τ_{max} and T_s (similarly, between the channel ΔF_{κ}coherence band and the signal bandwidth ΔF_S) considered from two different standpoint categories of transmission quality deterioration:

– frequency-selective fading (frequency-selective fading) when τs_{max}, or $\Delta F_{\kappa} > \Delta F_{S}$;;

– frequency-nonselective, or amplitude fading (flat fading) when τs_{max}, or $\Delta F_{\kappa} < \frac{1}{T_S} \approx \Delta F_{S}$..

The subscriber devices mobility is crucial in modern telecommunications, and it causes problems number that are directly related to terminal movement.

The Doppler effect essence is the following. In case when moving the transmitter and/or signal receiver relative to each other (or/and the propagation medium), there is a change in received signal wavelength λ. Change λ generates a change in received signal carrier frequency f_0. Such a change f_0 on the Doppler frequency shift magnitude $f_{Д} = \frac{V}{\lambda}$, where V is the receiver and / or transmitter relative speed, leads to the so-called Doppler expansion (parasitic frequency deviation) and carrier frequency spectrum scattering in the band $\Delta f = f_0 \pm f_{Д}$.. Because Doppler spread and channel coherence time $T_{K}$are inversely proportional ($T_{K} \approx \frac{1}{f_{Д}}$), value f_{∂} can be considered as channel fading rate.

5.3 Wireless Communication Systems Noise Immunity Comparison in Gaussian and Rayleigh Channels

The approximate analytical expressions and graphs of bit error probability (BER) P_B from SNR $\frac{E_b}{N_0}$ dependency for modulations BFSK, BPSK, and differential BPSK (DBPSK) in the channel with AWGN [35] are shown on Fig. 8 and Table 1.

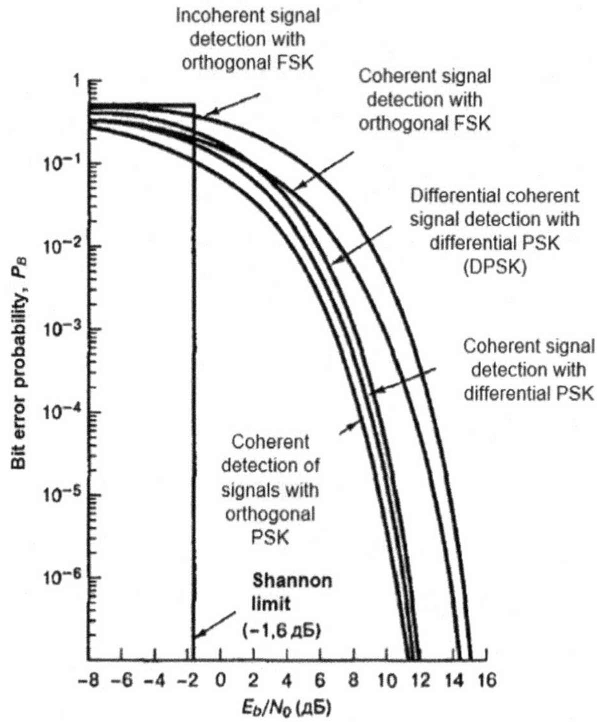

Fig. 8. BER for several binary modulation types in channel with AWGN [35]

Table 1. BER calculation for different binary modulations types and detection

Modulation	$\hat{l}_3$
BPSK (coherent detection)	$Q\sqrt{2\gamma_b}$
DBPSK (differential coherent detection)	$\left(\frac{1}{2}\right)\exp(-\gamma_b)$
BFSK (coherent detection)	$Q\sqrt{\gamma_b}$
BFSK (incoherent detection)	$\left(\frac{1}{2}\right)\exp\left(-\left(\frac{1}{2}\right)\gamma_b\right)$

In Table 1, the SNR values are $\gamma_b = \frac{E_b}{N_0}$, where E_b is the energy required to transmit one information bit, N_0 is the white noise spectral density power in the channel, $Q(x) = \frac{1}{\sqrt{2\pi}} \int_x^{\infty} e^{-\frac{t^2}{2}} dt$ is the error integral.

All dependency graphs P_B from $\frac{E_b}{N_0}$ (Fig. 8) show the classical exponential dependence is a waterfall-like species associated with the Gaussian channel. However, in the conditions of multibeam propagation [36], the analytical expressions for the mentioned above modulation types, have a different form (Table 2). In the Table 2 formulas, the average SNR $\overline{\gamma}_b = \left(\frac{E_b}{N_0}\right)\boldsymbol{E}\left(\alpha^2\right)$, where $\boldsymbol{E}(\cdot)$ is a mathematical expectation (average value α^2), α is a fading factor (random variable with Rayleigh distribution), α^2 is described by the probability density χ^2 with two freedom degrees. Figure 9 shows noise immunity graphs for such Rayleigh fading.

Table 2. BER Rayleigh limit probability at $\overline{\gamma}_b >> 1$

Modulation	P_B
BPSK (coherent detection)	$\frac{1}{4\overline{\gamma}_b}$
DBPSK (differential coherent detection)	$\frac{1}{2\overline{\gamma}_b}$
BFSK (coherent detection)	$\frac{1}{2\overline{\gamma}_b}$
BFSK (incoherent detection)	$\frac{1}{\overline{\gamma}_b}$

Each signal transmission scheme (modulation) is described by an approximately linear function now as a result of Rayleigh fading, which in the channel with AWGN gave a graph in the waterfall form.

Table 3 compares the approximate analytical BER expressions for M signals ensembles with phase (Phase Shift Keying, PSK) and quadrature amplitude (Quadrature Amplitude Modulation, QAM) modulation in Rayleigh and Gaussian channels [35].

5.4 Proposals to Increase the Noise Immunity for MWSN with FITP Based on the Developed Mathematical Model for Wireless Communication Channels Description

When transmitting signals in channels with fading, the reliability is usually distinguished between such options as "good", "bad" or "terrible" (Fig. 10).

The leftmost curve ("good") has an exponential shape and corresponds to the dependence $P_B = f(\gamma_b)$ expected behavior, when using any nominal modulation schemes at AWGN, i.e. at small values γ_b it is possible to achieve good transmission reliability.

The middle curve is called the Rayleigh limit and shows a deterioration in transmission reliability due to γ_b decrease, which is amplitude characteristic and slow fading in the channel and direct visibility absence between transmitter and receiver.

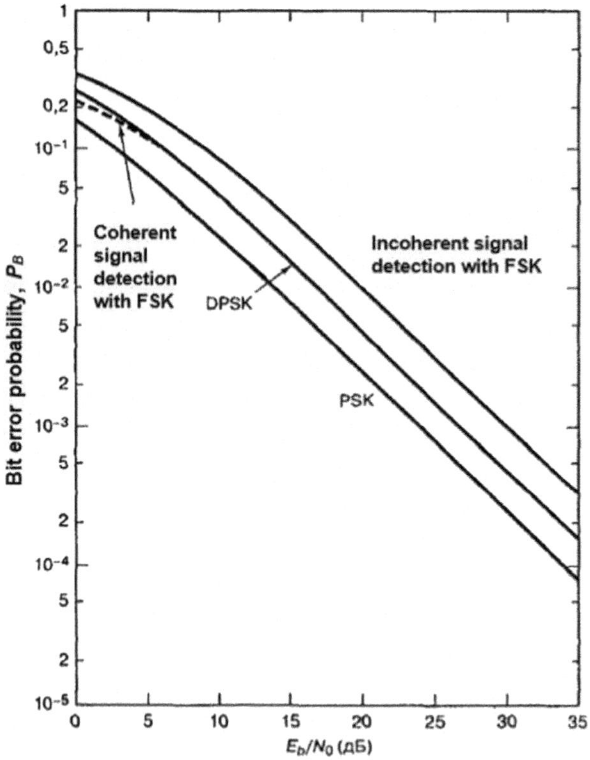

Fig. 9. BER for several binary modulation types in channel with slow Rayleigh fading

Table 3. BER ($P_в$) for some M-PSK/QAM modulation types in Rayleigh and Gaussian channels

Modulation	$\hat{l}_3$ in a channel with AWGN	$\hat{l}_3$ in the Rayleigh channel with AWGN
BPSK, QPSK, 4-QAM, MSK (coherent detection)	$Q\sqrt{2\gamma_b}$	$\frac{1}{2}\left(1-\sqrt{\frac{\overline{\gamma}_b}{\overline{\gamma}_b+1}}\right)$
DBPSK (differential coherent detection)	$\left(\frac{1}{2}\right)\exp(-\gamma_b)$	$\frac{1}{2(\overline{\gamma}_b+1)}$
BFSK (coherent detection)	$\left(\frac{1}{2}\right)\exp(-\gamma_b)$	$\frac{1}{2(\overline{\gamma}_b+1)}$

(*continued*)

Table 3. *(continued)*

Modulation	$\hat{l}_3$ in a channel with AWGN	$\hat{l}_3$ in the Rayleigh channel with AWGN
BFSK (incoherent detection)	$Q\sqrt{\gamma_b}$	$\frac{1}{2}\left(1-\sqrt{\frac{\overline{\gamma}_b}{\overline{\gamma}_b+2}}\right)$
M-PSK, $M > 4$ (Gray code, coherent detection)	$\left(\frac{1}{2}\right)\exp\left(-\left(\frac{1}{2}\right)\gamma_b\right)$	$\frac{1}{\overline{\gamma}_b+2}$
M-QAM, $M > 4$ (Gray code, coherent detection)	$\frac{2}{\log_2 M} Q\left(\sqrt{2\log_2 M\gamma_b}\cdot \sin\left(\frac{\pi}{M}\right)\right)$	$\frac{M-1}{M\log_2 M}\cdot\left(1-\sqrt{\frac{3\overline{\gamma}_b\log_2 M\,(M^2-1)}{3\overline{\gamma}_b\log_2 M\,(M^2-1)+1}}\right)$
M-QAM, $M > 4$ (Gray code, coherent detection)	$\frac{4\left(1-M^{-\frac{1}{2}}\right)}{\log_2 M} Q\left(\sqrt{\frac{3\gamma_b\log_2 M}{M-1}}\right), M =$ $2^k, k = 4, 6, \ldots$	$\frac{M-1}{M\log_2 M}\cdot\left(1-\sqrt{\frac{3\overline{\gamma}_b\log_2 M\,(M^2-1)}{3\overline{\gamma}_b\log_2 M\,(M^2-1)+1}}\right)$

The “terrible” curve is often called the “error floor” when $P_{\mathrm{B}} \approx 0.5$, which corresponds to the effect of severe deterioration due to frequency-selective or rapid fading. In this case, none γ_b increase exists that ensure the required information transfer reliability level.

Reliability improvement methods should be used to eliminate or reduce distortions level. The control method choice depends on the channel fading type and causes. First, the signals distortion for the transition from the “terrible” curve to the Rayleigh limit (“bad” curve) is reduced. Further approximation to the Gaussian channel curve is possible by using signal diversity methods and powerful correction codes.

Combating methods with fading are classified as following [37]:

– DSSS – Direct Sequence Spread Spectrum – a way to combat signal distortion due to frequency-selective interference ISI, that supposes expansion;

– FHSS – Frequency Hopping Spectrum Spreading – combating method with signal distortion due to frequency-selective fading, which consists in expanding the spectrum by the operating frequency pseudo-random (hopping) adjustment method;

– OFDM – Orthogonal Frequency-Division Multiplexing – combating method with signal distortion due to frequency-selective fading, by increasing the symbol transmission duration.

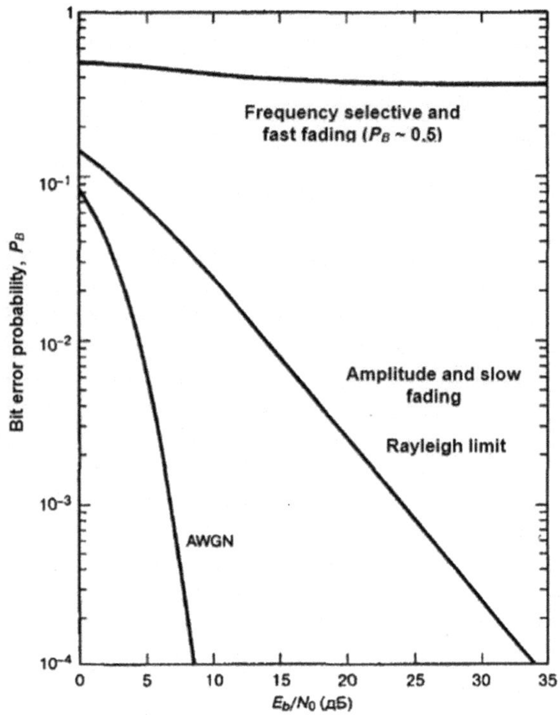

Fig. 10. Signal transmission reliability: good, bad, terrible [35]

After eliminating the signal attenuation causes due to frequency-selective and rapid fading, use diversity methods to move the wireless system operating point from the "bad" transmission curve (Fig. 5) to a curve approaching the AWGN characteristics.

The first and most effective combating method for fading is the signals spatial diversity, which was implemented in 1927 and was based on the several receiving antennas Rx use. It is also possible to organize spaced channels by using several transmitting antennas Tx. With the multi-element antennas MIMO technology development (Multiple Input - Multiple Output) [38] there is an opportunity to significantly increase the noise immunity (energy efficiency, EE) of the wireless systems by simultaneous signals diversity at reception and transmission. According to the leading specialist in the digital telecommunications field R. Calderbank, the spatial resources use on the modern flexible and universal methods basis of space-time signals coding (Space-Time Coding, STC) in multi-antenna MIMO systems can significantly improve the EE exchange capabilities and conditions for spectral efficiency (SE) [39, 40]. MIMO technology has become an integral part of modern wireless standards (LTE, WiMAX IEEE 802.16, Wi-Fi IEEE 802.11) [41] and the basis for next-generation wireless systems. The Orthogonal Space-Time Block Coding (OSTBC) scheme [40], proposed in 1998 by S. Alamouti and named after him, is defined as the basic standard. In [42, 43] the authors proposed improved methods for signals orthogonal space-time block coding.

Figure 11 shows BER/SNR graphs at the input of the receiver with using BPSK for different diversity options on reception and transmission (borrowed from the original article by S. Alamouti [44]). Obviously, Single Input - Single Output (SISO) cannot provide a high reliability level due to lack of diversity ("no diversity"), even in a quasi-static Rayleigh channel. It is also clear that SISO systems cannot meet the ever-increasing users' demands on the information transmission speed, as the increase in wireless systems spectral efficiency (SE) is directly related to the large signal ensembles use and the signal points convergence, which impairs noise immunity.

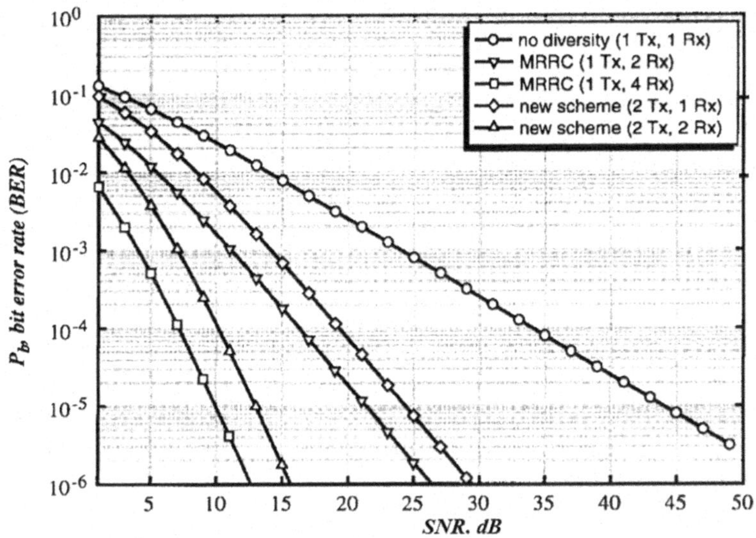

Fig. 11. Spaced transmission and reception at BPSK noise immunity comparison (coherent detection)

Exact expressions to calculate BER probability P_e for spaced reception methods with MRRC, which are equivalent to the noise immunity of spaced OSTBC transmission, in the channel with slow Rayleigh fading for BPSK and QPSK are given in [44].

6 Conclusion

1. New architectural, algorithmic and technical solutions for intelligent control systems construction and high-performance sensor networks based on the use of robotic objects are developed.
2. Algorithmic and technical solutions for semi-natural modeling modernization proposed in the study. These are used for technological solutions effectiveness decisions verification, and to confirm the sensor system with real data the stated tactical and technical characteristics compliance.
3. New packet transmission routes construction and maintenance intelligent methods have been developed for intelligent adaptive flying information and telecommunication robots with the multi-path transmission mode.

4. MWSN with FITP operation analysis shown that they belong to the dynamic, complex, multifunctional and multilevel systems class, that have problems in developing and maintaining a structure with a given connectivity, quality of service, increase network bandwidth. Such complex networks operation requires an effective control system, part of which is the FITP network subsystem management topology (location).
5. MWSN with FITP increasing bandwidth methods analysis shown that the existing methods, which are embedded in the FITP topology (location) management system, solve only partial problems that ensure MWSN disconnected components geometric connectivity, do not take into account channel resources capacity, load distribution and packet service in MWSN nodes. Also, most methods solve only static problems, and do not take into account the MSN nature mobility and FITP maneuverability, and therefore are subject to improvement. Existing MWSN ground planning methods are also not effective, as they have a high complexity and computational time, which does not allow FITP to work out the obtained solutions in real time.
6. The method development for increasing the MWSN with FITP capacity in the rapid and unpredictable MSN movement conditions is relevant and has not only theoretical but also applied practical significance to ensure the effective modern MWSN operation.
7. The MWSN with FITP analysis for noise protection increase methods development is carried out.
8. Ways to reduce the frequency-selective fading effects in the channel are the equalizers use on reception and / or signals generation potentially invariant to frequency-selective distortion on transmission (modulation MFSK and DPSK; expanding the signals range by DSSS or FHSS; OFDM technology).
9. Masures aimed at increasing SNR are error correction codes and signal diversity methods (time or space, frequency or polarization, combined diversity, etc.). The most effective combating fading method is the signals spatial diversity, which can be implemented at the reception and / or transmission.

References

1. Ilchenko, M., Uryvsky, L., Osypchuk, S.: The Main Directions of Improving Infocommunications in the Global Tendencies. Advances in Information and Communication Technologies. Lecture Notes in Networks and Systems. Monograph. Springer, Cham, (2020), pp. 3–22 (Scopus) DOI: https://doi.org/10.1007/978-3-030-58359-0
2. Ilchenko, M., Uryvsky, L., Osypchuk, S.: World trends of modern information and telecommunication technologies development. In: International Conference Radio Electronics & Info Communications (UkrMiCo). - IEEE Xplore Digital Library. - https://ieeexplore.ieee.org/document/9165461/. DOI: https://doi.org/10.1109/UkrMiCo47782.2019.9165461 (Scopus)
3. Uryvsky, L., Budishevsky, A.: FOG-cloud-strategies of dynamic telecommunication networks management. Inf. Telecommun. Sci. **2**, 74–80 (2020)

4. Novikov, V.I., Lysenko, O.I., Valuysky, S.V., Guida, O.G.: Mathematical models, methods and algorithms for optimizing the performance of wireless sensor networks with mobile sensors and telecommunications aircraft platforms. Scientific notes of Tavriya National University named after VI Vernadsky. Series: Technical Sciences, vol. 31 (70), № 3 2020. Part 1, pp. 54–64 Magazine page: www.tech.vernadskyjournals.in.ua. ISSN 2663–5941 (Print). ISSN 2663–595X (Online)
5. Lysenko, O., Romaniuk, V., Tachinina, O., Valuiskyi, S.: Integrated computer technologies in mechanical engineering. Chapter: The problems of control in wireless sensor and mobile ad-hoc networks, pages 385–404. Copyright: © 2020. Publisher Springer International Publishing, DOI 978-3-030-37618-5_33 (Scopus)
6. Romanchenko, I.S., et al.: Models of application of information and telecommunication technologies on the basis of unmanned aviation complexes in emergency situations. VI Novikov. НАУ, 2016, p. 332 (2016)
7. Uryvsky, L., Shmigel, B.: Complex Methodology for Efficiency Evaluation of Discrete Information Transmission Systems. Sciences of Europe (Praha, Czech Republic), vol. 2, 53 (2020), pp. 55–61 (2020). ISSN 3162-2364
8. Basu, P., Redi, J., Shurbanov, V.: Coordinated flocking of UAVs for improved connectivity of mobile ground nodes. In: IEEE MILCOM'04: Military Communications Conference, October 31 - November 3 2004: proceedings. - Monterey, vol. 3, pp. 1628–1634 (2004)
9. Minochkin, A.I., Romanyuk, V.A.: Methodology of operational management of mobile radio networks. Communication **2,** pp. 53–58 (2005)
10. Totsenko, V.G.: Methods and systems of decision support: algorithmic aspect. Naukova dumka, 381 p. (2002)
11. Minochkin, A.I., Romanyuk, V.A.: Tasks of topology management of the network of unmanned aerial vehicles of the mobile component of military communication networks. In: Collection of Scientific Works of VITI NTUU "KPI", vol. 2, 83–90 (2005)
12. Minochkin, A.I., Romanyuk, V.A.: Methods of multiple access in mobile radio networks. Communication **2**, 46–50 (2004)
13. Bunin, S.G., Voiter, A.P.: Computing networks with packet radio communication. Tehnyka, 223 p. (1989)
14. Leung, K.K., Clark, M.V., Mc. Nair, B., Kostic, Z., Cimini, L.J., Winters, J.H.: Outdoor IEEE 802.11 cellular networks: radio and MAC design, and their Performance. In: IEEE Communications (ICC 2002): International Conference, April 28 - May 2 2002: Proceedings, New York, 2002, vol. 1, pp. 512–516 (2002)
15. Chandrashekar, K., Dekhordi, M.R., Baras, J.S.: Providing full connectivity in large ad-hoc networks by dynamic placement of aerial platforms. In: IEEE MILCOM 2004: Military Communications Conference, 31 October–3 November 2004: Proceedings, Monterey, vol. 3, pp. 1429–1436 (2004)
16. Ilchenko, M.E.: Telecommunication systems based on high-altitude air platforms.ME Ilchenko, SA Kravchuk. - K .: Наукова думка, 580 с (2008)
17. Han, Z., Swindlehurst, A.L., Liu, K.J.R.: Smart deployment/movement of unmanned air vehicle to improve connectivity in MANET. In: IEEE Wireless Communications and Networking: conference, 3–6 April 2006: Proceedings. - Las Vegas, 2006, pp. 252–257 (2006)
18. Dousse, O., Thiran, P., Hasler, M.: Connectivity in ad-hoc and hybrid networks. In: IEEE INFOCOM'2002: The 21st Annual Joint Conference, June 23–27 2002: Proceedings, New York, pp. 1079–1088 (2002)
19. Gupta, P., Kumar, P.R.: Critical power for asymptotic connectivity. In: Decision and control: 37th IEEE Conference, December 16–18 1998: proceedings. - Tampa FL, pp. 1106–1110 (1998)

20. Miller, L.E.: Probability of a two-hop connection in a random mobile network. In: Information Sciences and Systems: Conference, March 21–23 2001: Proceedings. - Baltimore MD, pp. 1381–1388 (2001)
21. Santi, P., Blough, D.M.: The critical transmitting range for connectivity in sparse wireless ad hoc networks. IEEE Trans. Mob. Comput. **2**(1), 25–39 (2003)
22. Xue, F., Kumar, P.R.: The number of neighbors needed for connectivity of wireless networks. ACM Wireless Network J. **10**(2), 169–181 (2004)
23. Li, N., Hou, J.C.: Improving connectivity of wireless ad hoc networks . In: Mobile and Ubiquitous Systems (MobiQuitous 2005): 2nd Annual International Conference, July 17–21 2005: proceedings. - San Diego, 2005, pp. 314–324 (2005)
24. Khuller S. Approximation Algorithms for Finding Highly Connected Subgraphs, 364 p. PWS Publishing Co., Boston (1996)
25. Bettstetter, C.: On the minimum node degree and connectivity of a wireless multihop network. In: Mobile Ad Hoc Networking and Computing (Mobihoc 2002): ACM International Symposium, June 9–11 2002: Proceedings. - Lausanne, pp. 80–91 (2002)
26. Zhang, H., Hou, J.C.: On the critical total power for asymptotic k-connectivity in wireless networks. In: IEEE INFOCOM'2005: The 24th Annual Joint Conference, 13–17 March 2005: Proceedings, Miami, pp. 466–476 (2005)
27. Bakhtin, A.A.: Development of methods of management of connectivity and maintenance of quality of service in a mobile episodic network with retransmission: author's ref. dis. at the request of scientists. Ph.D. tech. Science: special. 05.12.13 "Systems, networks and devices of telecommunications", 27 p. (2009)
28. Fokin, G.A.: Management of self-organizing packet radio networks on the basis of radio stations with directional antennas: author's ref. dis. at the request of scientists. Ph.D. tech. Science: special. 05.13.13 "Telecommunication systems and computer networks". GA Fokin, 19 p. (2009)
29. Basu, P., Redi, J., Shurbanov, V.: Coordinated flocking of UAVs for improved connectivity of mobile ground nodes. In: IEEE MILCOM'04: Military Communications Conference, October 31–November 3 2004: Proceedings. - Monterey, vol. 3, pp. 1628–1634 (2004)
30. Reidt, S., Wolthusen, S.: Connectivity augmentation in tactical mobile ad hoc networks. In: IEEE MILCOM'08: Military Communications Conference, November 17–19 2008: Proceedings, San Diego, vol. 3, pp. 1441–1448 (2008)
31. Skulish, M.A., Globa, L.S.: Telephone traffic. Statement of an optimization problem for a network with emulation of the ATM service according to the PWE3 protocol. Modeling Inf. Technol. **54**, 230–236 (2009)
32. Skulysh M.A. Improving the algorithm for managing information flows in the nodes of telecommunications networks. Bull. Kharkiv National Univ. **863,** 236–245 (2009)
33. Skulysh, M.A., Globa, L.S.: Method of traffic management in a multiservice switching center. Telecommun. Sci. **1**(2), 30–40 (2011)
34. Uryvsky, L., Martynova, K.: Complex analytical model of priority requires service on cloud server. In: International Conference Radio Electronics & Info Communications (UkrMiCo). - IEEE Xplore Digital Library. https://ieeexplore.ieee.org/document/9165323/, doi:https://doi.org/10.1109/UkrMiCo47782.2019.9165323 (Scopus)
35. Proakis, J.G., Salehi, M.: Digital Communications, 5th edn., 1150 p. McGraw-Hill, NY (2008)
36. Uryvsky, L., Solianikova, V.: Analysis of spatial-time characteristics of a radio line with multipath within 5G technology. Information and telecommunication sciences, vol. 11, no. 1, pp. 87–91 (2020). https://doi.org/10.20535/2411-2976.12020.87-91
37. Goldsmith, A.: Wireless Communications. Cambridge University Press, Cambridge, 419 p. (2005)
38. Bakulin, M.G., Varukyna, L.A., Kreindelin, V.B.:: MIMO technology: principles and algorithms. Moscow: Hotline - Telecom, 242 p. (2014)

39. Naguib, A., Calderbank, R.: Space-time coding and signal processing for high data rate wireless communications. Wirel. Commun. Mob. Comput. **1**, 13–34 (2001)
40. Calderbank, A.R., Tarokh, V., Jafarkhani, H.: Space-time block coding from orthogonal designs. IEEE Trans. Inf. Theory **45**(5), 1456–1467 (1999)
41. Hanzo, L., Akhtman, Y., Wang, L.: MIMO-OFDM for LTE, WiFi and WiMAX. Coherent Versus Non-Coherent and Cooperative Turbo-Transceivers, UK, 658 p. (2011)
42. Solodovnyk, V.I., Naumenko, M.I., Osypchuk, S.O., Urivsky, L.O.: Method of orthogonal space-time block coding of signals. Patent for utility model № 146345 from 10.02.2021 (Ukraine)
43. Solodovnyk, V.I., Naumenko, M.I., Osypchuk, S.O., Urivsky, L.O.: Method of orthogonal space-time block coding of signals. Patent for utility model № 146346 from 10.02.2021 (Ukraine)
44. Alamouti, S.: A simple transmitter diversity scheme for wireless communications. IEEE J. Sel. Areas Commun. **16**, 1451 – 1458 (1998)

Technologies for Building Intelligent Video Surveillance Systems and Methods for Background Subtraction in Video Sequences

Anatolii Babaryka(✉), Ivan Katerynchuk, and Oksana Komarnytska

Bohdan Khmelnytskyi National Academy of the State Border Guard Service of Ukraine, Khmelnytskyi, Ukraine

Abstract. The research is devoted to the analysis and improvement of video analytics functions in video surveillance systems in order to increase the efficiency of detection of dynamic objects in the sectors of video surveillance. It has been established that video analytics methods using background subtraction and object recognition methods have significant disadvantages, namely: algorithms cannot select an object from the background at low contrast; some moving objects can be recognized as backgrounds; algorithms critically depend on lighting conditions, etc. Thus, the aim of the study is to improve the method of detecting dynamic objects in video sequences, which uses methods of subtraction of the background, based on pixel analysis of frames using elements of the theory of expert systems. The advanced method of detecting dynamic objects in video sequences is based on the ViBe algorithm, and differs from the original in that it uses a color model U * V * W * with double threshold levels and expert systems to eliminate uncertainties in the classification of pixels (Dempster-Schaefer theory) and the dynamic method of updating background pixel models.

Keywords: Algorithm · Method · Video sequence · Background subtraction · Dynamic object · Colour model · Pixel · Background · ViBe

1 Introduction

The use of intelligent video surveillance systems of state border protection bodies and units will facilitate development of automated analysis of non-standard situations, violations of the state border crossing rules, perimeter control, provide opportunities for biometric analysis (automatic identification of persons by facial images), analysis of vehicle license plates, automatic object tracking, automatic detection and classification of objects, search of objects in the database of video archives, etc. A significant role in intelligent video analytics systems belongs to the detection of dynamic objects in the video stream. Existing models and algorithms for background selection and object recognition have significant shortcomings that limit their application in practice, namely: under low contrast algorithms are unable to select an object from the background; some moving objects can be recognized as a background, algorithms are critical to lighting conditions, and so on. That is why the relevance of the study is determined by the need to

M. Klymash et al. (Eds.): *Future Intent-Based Networking*, LNEE 831, pp. 468–480, 2022.
https://doi.org/10.1007/978-3-030-92435-5_26

eliminate the discrepancy between the necessity to ensure high efficiency of operational and service activities of border guards by introducing modern technical means of border protection and imperfection of scientific and methodological apparatus of processing information.

Intelligent Video Surveillance System (IVSS) is a video surveillance system that has the ability to automatically analyse data received from VSS cameras and perform necessary tasks, such as generating alarms or warnings. The use of intelligent video surveillance systems makes it possible to automate such areas of activity as perimeter control analytics, situational analysis (automatic detection of crisis situations associated with the accumulation of large numbers of people), biometric analysis (allows automatic identification of persons by facial images), analysis by several cameras (allows automatic monitoring of the object by several video surveillance cameras), automatic detection and classification of objects, search for objects in the database of the video archive, etc. [1].

2 Review of the Literature

The research of intelligent video surveillance systems is a scope of the works of a wide range scientists, such as Ainsworth T., Bouwmans T, Antoine Vacavant, Sobral Andrews, Zivkovic Z., Tourani Ali, Velastin Sergio, Li Ying and others [2–11].

Information from VSS cameras is stored in video archives and is broadcasted in real time on the operator's monitors. The operator's attention, in this case, is divided proportionally between the channels of the video surveillance system. Therefore, the operator is not physically able to simultaneously monitor the situation in each sector of the surveillance cameras. Thus, there is a possibility that the operator will not detect important information. After 12 min of continuous viewing of information from VSS cameras, the operator may not detect up to 45% of cases of activity on VSS channels, and after 22 min this figure may increase up to 95%. After 20–40 min of continuous surveillance of the situation from VSS cameras, a so-called "video blindness" occurs, when the operator is unable to detect some information on the monitor screen. Therefore, for the effective use of video surveillance systems, it is advisable to automate the processes of detecting "objects of interest" and "suspicious actions".

An important issue, which is insufficiently paid attention to in modern research, is the development of intelligent information technologies for information processing in video surveillance systems.

The process of automated information analysis in video surveillance systems includes the following steps: foreground subtraction; selection and classification of moving objects; tracking the trajectory of the detected objects; recognition and classification of actions of "objects of interest".

The purpose of foreground subtraction is to separate the moving fragments of the image from the still ones (background). In the second stage, the foreground image is segmented (compact areas moving at the same speed are detected, which can be considered as elements of one object). This is followed (if necessary) by tracking the trajectory of moving objects (tracking). The next step is to identify and analyze the behavior of the detected objects.

To solve problems at each of these stages, scientists use a variety of techniques. For example, methods based on background subtraction [2], probabilistic approaches [3], mathematical models such as co-occurrence matrices [4], methods of time difference and optical flux [5], etc. are most often used to construct the foreground. Also developing methods of building a background based on the use of neural networks.

There are also a large number of methods for solving the problem of selection and classification of objects [6]. The video may contain a number of moving objects that have been separated from the background. These can be people, groups of people, vehicles, animals, etc. To classify these objects, it is necessary to perform segmentation, i.e. to separate objects from each other. [7]. The structural and functional scheme of the video surveillance process with the functions of intelligent information processing is shown in Fig. 1.

In modern video surveillance systems with the functions of automated video information processing, one of the important tasks is detection of moving objects, construction of their trajectories and analysis of such trajectories. For example, one of the detectors used in modern intelligent video surveillance systems is a conditional line crossing detector. The logic of this detector is that the operator in the software interface builds a conditional line crossing of which by certain objects (persons, vehicles, or any moving objects) is visually (and / or audibly) signalised. To implement the abovementioned task, it is necessary to solve the following partial tasks: detection of dynamic objects, their localization, tracking from frame to frame and fixing the moment of crossing of a group of pixels belonging to the "object of interest" with pixels belonging to the "conditional line" [8].

An important step in the process of detecting moving objects in the video sequence obtained from stationary (fixed) VSS cameras is background subtraction. The general approach is to select parts in the video frame that are significantly different from the background model, i.e. to create a foreground mask. The simplest background subtraction algorithm is to use a video frame that does not contain any moving objects as a reference.

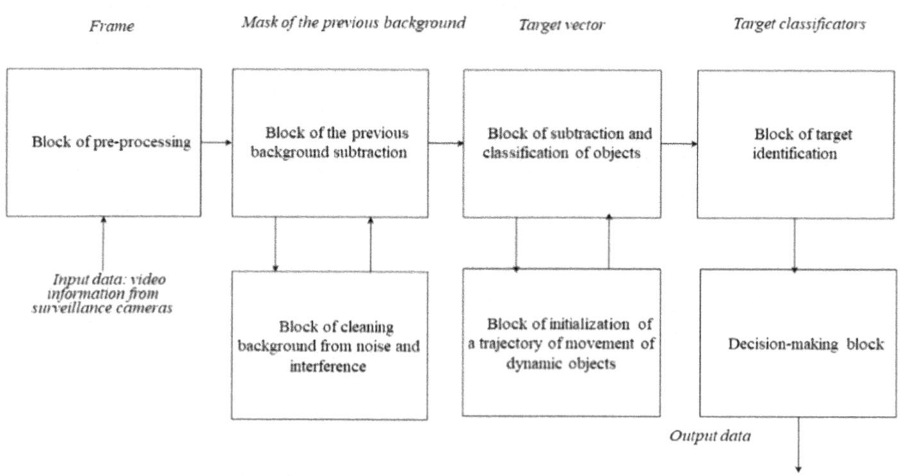

Fig. 1. Structural and functional scheme of the video surveillance process with the functions of intelligent data processing [16]

Then, by subtracting the background from the following video frames, we can detect moving objects. However, in real conditions there are a number of problems that complicate the process of background subtraction [9] – Fig. 2.

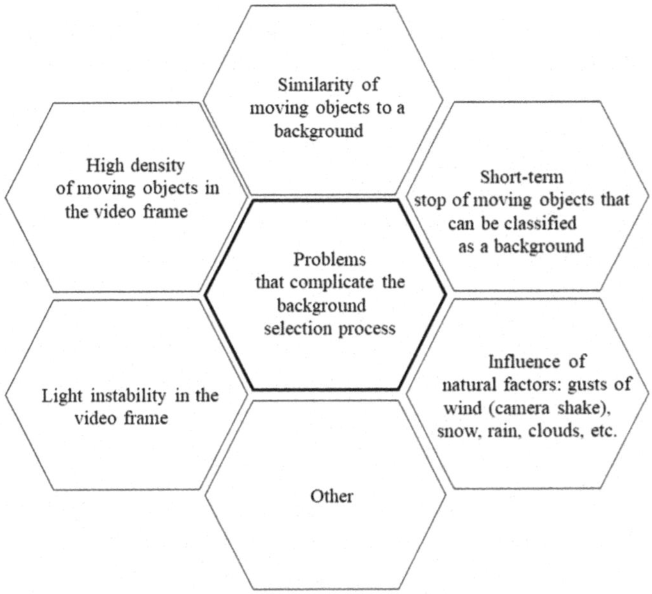

Fig. 2. Problem factors that complicate the background selection process

All the variety of methods and algorithms for background subtraction according to the theories used [9–14] can be divided into categories based on:

1) basic methods and methods that operate on average and variance values;
2) methods of fuzzy logic;
3) Gaussian processes;
4) non-parametric methods;
5) the use of neural networks, etc.

In [15], a classification of methods for detecting moving objects is proposed: methods based on interframe difference, calculation of optical flux and background subtraction.

The most common method of detecting moving objects is the background subtraction, the main idea of which is to subtract the current frame from the pre-formed background mask – Fig. 3 [16]. This method provides the ability to process video streams in real time. Creating a background model is to calculate the absolute difference between the current frame and a predefined still image (Frame Difference) that does not contain moving objects. This method uses only one previous frame, so it is unable to detect pixel motion inside a large object that moves evenly and is sensitive to interference such as camera shake, gusts of wind, treetops, water waves, and so on. In addition, the disadvantage of this method is the high sensitivity to the dynamic background and abrupt changes in the frame (abrupt changes in lighting, weather conditions, camera shake, etc.). In the study [16], typical algorithms and methods for selecting dynamic objects in video sequences were analyzed.

Thus, the authors C. Stauffer and W. E. L. Grimson [17] proposed a method in which the color distribution of each pixel is represented by the sum of normal distributions of pixel radiation intensities and each background pixel is described by a mixture of k Gaussian distributions. Eric Hayman and Jan-Olof Eklundh improved this algorithm [18] and named it Mixture of Gausian (MOG). The MOG algorithm made it possible to highlight the background model in the presence of small fluctuations in lighting. But with abrupt changes in lighting or frame noise, this algorithm erroneously determines the background model. To solve these problems, the MOG algorithm continued to be improved by many scientists. The research resulted in improved algorithms MOG-2, GMM, GMG, TLGMM, STGMM, SKMGM, TAPPMOG and others. For example, the MOG2 background extraction algorithm is based on the principles of a method for restoring the background and detecting moving objects from static cameras using Gaussian mixture models. The improvement of the MOG2 algorithm is that it selects a certain number of Gaussian distributions for each pixel. This approach allowed to achieve better adaptability to such a factor as abrupt changes in lighting. The peculiarity of the GMG algorithm [19] is that it uses the first n frames (according to the authors' recommendations $n = 120$) to model the background. The algorithm combines methods for statistical evaluation of the background model and Bayesian approach to foreground pixel segmentation. Approximation of approaches such as the Kalman filter bank and the Gale-Shapley algorithm is also used to solve the problem of tracking dynamic objects.

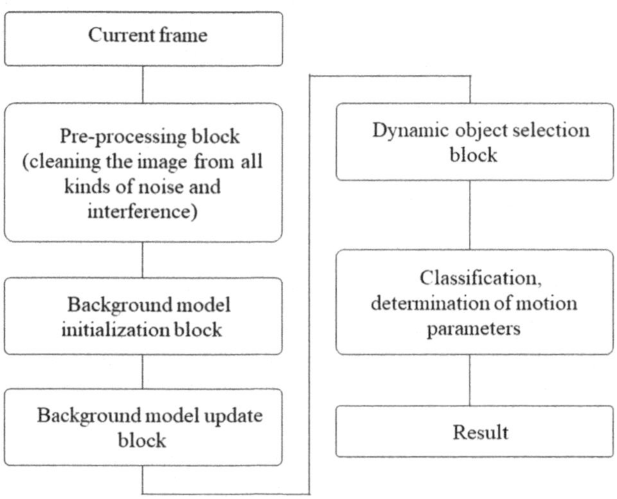

Fig. 3. Block-scheme of a typical algorithm for detecting moving objects based on background subtraction methods [16]

ViBe (Visual Background extractor) – a method proposed in [20] is quite simple in computational terms and fast. ViBe algorithm combines such positive characteristics as high speed, quality of selection of moving objects, low resource consumption in terms of computing. But, despite these advantages, ViBe has a number of disadvantages: sensitivity to sudden changes in lighting, insufficient level of "suppression" of shadows of dynamic objects. The original ViBe algorithm uses the RGB colour scheme. The authors proposed to use images in grayscale in order to increase the speed of the proposed method in the future.

Based on the analysis of the advantages and disadvantages of existing models and algorithms, the authors [16] made assumptions about the possibility of using other color schemes besides RGB, which are characterized by the best photometric invariant features.

3 Improved Method of Detecting Dynamic Objects in Video Sequences

Improved method of detecting dynamic objects in video sequences differs from the original ViBe by using the colour scheme U*V*W*, using double threshold levels and elements of expert systems theory to eliminate uncertainties in the classification of pixels, as well as using a dynamic approach to update the background model with the neighbouring pixels.

According to the analysis of the advantages and disadvantages of colour models XYZ, I1I2I3, HSI, YIQ, Lab, YCrCb, RGB, HSV, C1C2C3, Opp, Nopp, Copp, Luv, xyz, YES, CMY, YUV, HSL, UVW, xyY, etc. [16], the assumption was made about the possibility of using the model U*V*W*, which is characterized by the best photometric invariant features.

The operation of the algorithm can be divided into the following stages: initialization of the background model; foreground detection (dynamic objects); updating the background model [16].

At the stage of initialization of the background model, for each pixel p with coordinates (x, y) a certain number N of its previous values $v(p)$ is allocated. Then, for each pixel of the current frame, you can build a model:

$$M(p) = \{v_1(p), v_2(p), \ldots, v_N(p)\}, \tag{1}$$

where $v(p)$ is the pixel p value; $v(p_i)$ is the pixel p_i value pi; $v_n(p)$ is the value of the n-th pixel.

At the foreground detection stage firstly it's necessary to check to see if the current pixel belongs to the background model. To do this, let's denote the value of the pixel p with coordinates (x, y) in the current frame as $v_n(p)$ and construct a sphere of radius R around it in the colour space U*V*W*. Then let us determine the number of K values of $v(p)$ that fall into this sphere. To do this, we need to determine the distance between two pixels in Euclidean space, and compare it with the value of R:

$$M\,\Delta E(v(p_i), v(p_j)) = \begin{cases} \left|v(p_i) - v(p_j)\right| > R, \\ \left|v(p_i) - v(p_j)\right| \le R. \end{cases}, \tag{2}$$

where $v(p_i)$ is the value of the pixel with the coordinates in the current i-th frame; $v(p_j)$ is the value of the pixel with the coordinates in the previous j-th frame.

The Euclidean distance between $v(p_i)$ and $v(p_j)$ in the colour space U*V*W* is represented as follows:

$$\Delta E\big(\nu(p_i), \nu(p_j)\big) = \sqrt{\begin{array}{l}(U^*(\nu(p_i)) - U^*(\nu(p_j)))^2 + \\ (V^*(\nu(p_i)) - V^*(\nu(p_j)))^2 + \\ (W^*(\nu(p_i)) - W^*(\nu(p_j)))^2.\end{array}}, \quad (3)$$

If the value of the absolute difference between $v(p_i)$ and $v(p_j)$ is greater than a certain threshold value *R*, the pixel is considered a candidate for belonging to the foreground (belonging to a dynamic object), otherwise - to the background.

A fixed threshold value of *R*, when applying the algorithm in difficult conditions (sudden changes in lighting, camera shake, dynamic background, etc.), in our opinion, is not an effective solution. Analysing empirically the results of experimental studies of the original ViBe algorithm, we came to the following conclusions:

- if you manually define a low value of *R*, then the background pixels will be determined only those that have indicators very close to the reference background. At the same time, we will get a certain number of other pixels that really belong to the background and were mistakenly identified as belonging to dynamic objects.
- if you manually define a high value of *R*, the pixels with the indicators "farthest" from the reference samples will be determined belonging to the dynamic objects.

Thus, it is proposed to apply the dynamic value of the threshold level *R*. The essence of the approach is to apply double threshold levels and elements of the theory of expert systems to eliminate uncertainties in the classification of pixels. Let us denote by R_{low} is the relatively low value of the threshold level, R_{high} is the relatively high value of the threshold level (Fig. 4). Then, to make a decision, we will no longer have two cases, but three:

$$\Delta E(\nu(p_i), \nu(p_j)) = \begin{cases} |\nu(p_i) - \nu(p_j)| < R_{low}, \\ R_{low} \le |\nu(p_i) - \nu(p_j)| \le R_{high}, \\ |\nu(p_i) - \nu(p_j)| > R_{high}. \end{cases} \quad (4)$$

The next step is to calculate the number of points belonging to the foreground and the background. Calculations according to formula (4) are performed *N* times and the result of the number of matching pixels is obtained, which is denoted by *K*. Next, the minimum number of *K* elements that are candidates for the background is determined empirically so that the pixel could be classified as background. Otherwise, it is believed that this pixel belongs to the foreground. In the original ViBe algorithm, the authors proposed the following rule:

$$N_i = \begin{cases} 1 < R & K < \#_{min} \\ 0 \ge R & K \ge \#_{min} \end{cases}. \quad (5)$$

If $K < \#\text{min}$, then the pixel belongs to a dynamic object, otherwise - the background.

In our case, we obtained an interval of uncertainty, falling into which pixel can belong to both the dynamic object and the background. To decide on an unambiguous

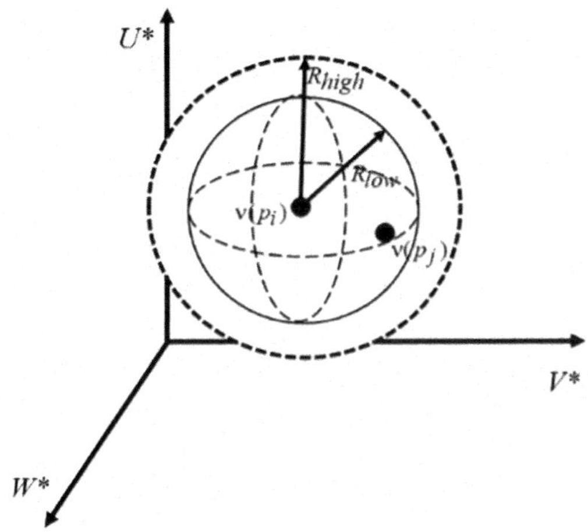

Fig. 4. Visualization of pixel values $\nu(p_i)$ and $\nu(p_j)$ in the color space U*V*W * with threshold levels R_{low} and R_{high}

classification, we use the mathematical apparatus of the Demster-Schaefer theory [21]. According to the main provisions of this theory, the reliability function reflects the sum of all weights of subsets B of the set A (hypothesis A) and has the following form:

$$bel(A) = \sum\nolimits_{B \subseteq A} m(B). \tag{6}$$

where $m(B)$ is the weight function that reflects the distribution of certainty weights.

The likelihood function is the sum of the weights of the sets B that intersect with the set A:

$$pl(A) = \sum\nolimits_{B|B \cap A \neq \varnothing} m(B), \tag{7}$$

where $bel(A) \leq P(A) \leq pl(A)$, and $P(A)$ the exact probability of the hypothesis A.

Let us denote $A^{R_{low}}$ as a hypothesis of the correct classification of the pixel belonging to the background, and $A^{R_{high}}$ as a hypothesis of the correct classification of a pixel belonging to a dynamic object. Then the weight functions of these events can be displayed as follows:

$$m\left(A^{R_{low}}\right) = \frac{K_{low}}{K}, \tag{8}$$

$$m\left(A^{R_{high}}\right) = \frac{K_{high}}{K},. \tag{9}$$

$$m\left(A^{R_{low}} \cup A^{R_{high}}\right) = \frac{K - K_{low} - K_{high}}{K}. \tag{10}$$

To consolidate these weight functions, we will use Transferable Belief Model of Philippe Smets [22]:

$$P_{Bet}(x) = \sum\nolimits_{x \in A \subseteq X} \frac{m(A)}{|A|}. \tag{11}$$

Having applied this model to the problem of consolidation of these weight functions, we obtain two functions of Belief Models $P_{Bet}(A^{R_{low}})$ and $P_{Bet}(A^{R_{high}})$. If $P_{Bet}(A^{R_{low}}) \geq P_{Bet}(A^{R_{high}})$, the current pixel is classified as belonging to the background, otherwise – as belonging to a dynamic object.

After the foreground is detected, the background model is updated. If the pixel p_i in the current frame has been classified as background, then the following two procedures occur:

- first, an element is randomly selected from the set $M(p)$, which is replaced by the value of the pixel in the current *i*-th frame $\nu(p_i)$;
- an element is randomly selected from around the pixel p_i, the value of which will also be replaced by $\nu(p_i)$.

This ensures the spatial consistency of the background model, because the values of the background model of one pixel also fall into the background model of neighbouring pixels. One of the disadvantages of the original ViBe is that in case of the appearance of dynamic objects during the initialization of the background model, the appearance of so-called "phantom objects", which were accidentally classified as a background was noticed.

In order to remove these artefacts, it was proposed to use a dynamic approach in updating the background model by means of neighbouring pixels. The essence of the proposed approach is to build a three-level neighbourhood. It is considered that the neighbourhood of the control pixel consists of three levels: on the first level it is a circle of 3×3, on the second level it is a circle of 5×5, on the third level it is a circle of 7×7.

The implementation of this approach involves the probabilistic selection of the value of the neighbourhood pixel when updating the background model based on empirically obtained coefficients ($K^1_{okol} = 0.83$, $K^2_{okol} = 0.11$, $K^3_{okol} = 0.06$) and the application of an equal law of distribution of random variables within each of the three levels.

4 Results and Discussion

Software-algorithmic implementation of the improved method of dynamic object detection was developed on the basis of C ++ in Visual Studio 2019. The research was conducted on test video sequences from the resource ChangeDetection.NET (CDNET), which are sequences of frames in jpg format with such environmental features as: bad weather, dynamic background, use in normal conditions (baseline). The proposed algorithm was studied in comparison with the original ViBe, the implementation of which was obtained from the materials of O. Barnich and M. Van Droogenbroeck [20].

Evaluation of the efficiency of algorithms was performed on such metrics as "*precision*", "*recall*" and the metric W proposed in [22]. The BGS Library was used for the experiment. Parameters of the original ViBe: $N = 20$, $R = 20$. Parameters of the proposed improved ViBe: $N = 20$, $R_{low} = 8$, $R_{high} = 25$.

According to the indicators obtained during the experimental study, the consolidated results were formed, which are given in Table 1.

Table 2 shows the visual results of the study, the analysis of which shows the improvement of the proposed method in bad weather conditions and the suppression of the dynamic background. The consolidated results of the experimental study (see Table 1) indicate an improvement in the results of the proposed method compared to the original ViBe by an average of 6.7%.

Table 1. Generalized characteristics of the efficiency of algorithms

Algorithm	Metrics		
	Precision	Recall	W
Original ViBe	0.7521	0.6982	0.93321
Proposed method	0,8126	0,7910	0,94185

Visual results of the study (Table 2) allow us to assess the advantages and disadvantages of the proposed method compared to the original ViBe.

Thus, the use of the U*V*W * colour scheme, which is characterized by better photometric invariant features than RGB, made it possible to improve the selection of blocks that have similar colour parameters (gray car on gray asphalt, man in green clothes on grass, etc.).

With the help of a dynamic approach to the selection of the threshold level *R* we managed to partially eliminate the appearance of small artefacts that occurred when such types of dynamic background as a small movement of trees, trembling leaves, wave oscillations on water surface. This approach to the selection of the threshold level *R* also allows to suppress the noise that occurs due to small displacements of the CCTV camera or the appearance of "vibration" caused by a strong wind.

Thus, the use of the U * V * W * color scheme, which is characterized by better photometric invariant features than RGB, made it possible to improve selection of blocks that have similar color parameters (gray car on gray asphalt, man in green clothes on grass, etc.).

Using a dynamic approach to the selection of the threshold level R helped to partially eliminate the appearance of small artifacts that occurred when such types of dynamic background as a small movement of tree crowns, trembling leaves, wave oscillations on the water surface. This approach to the selection of the threshold level R also allows you to suppress the noise that occurs due to small displacements of the VSS camera or the appearance of "vibration" in strong gusts of wind.

During the experiment the values of the probability coefficients of the neighboring level were empirically selected. The optimal values of the studied test sets of video sequences were the values $K^1_{okol} = 0.83$, $K^2_{okol} = 0.11$, $K^3_{okol} = 0.06$. However, fixed values of probability coefficients are optimal only for these sets of video sequences, and in our opinion, for greater versatility of the proposed method, it is advisable to continue research on the implementation of a dynamic approach to the choice of these coefficients. Also, one of the ways to improve the proposed approaches is to study other methods for selecting the threshold levels R_{low} та R_{high} to decide on a unique classification in addition to the mathematical apparatus of the Demster-Schaefer theory. Since, when transforming from the RGB color scheme to U * V * W *, the color components change disproportionately, so it would be advisable to investigate the influence of the angular parameters of the studied pixel on the values of the threshold levels R_{low} and $R_{high.}$.

The application of the above approaches negatively affected the performance of the algorithm. However, the speed of processing frames with a resolution of 320 × 240, 720 × 576 and 720 × 480 is sufficient for real-time operation.

Table 2 Comparative analysis of the operation of algorithms for testing video sequences

The frame under study	Reference frame	The frame obtained using the original ViBe algorithm	The frame obtained using an advanced ViBe algorithm

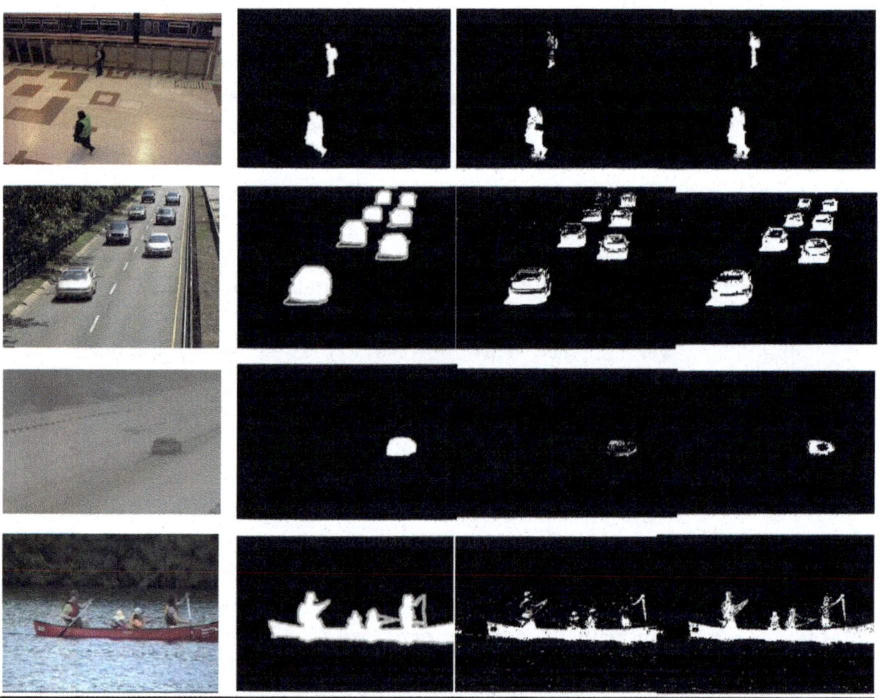

5 Conclusions

The chapter presents an improved method of detecting dynamic objects in video sequences based on the ViBe algorithm.

The scientific novelty of the obtained results is the development of an improved method for detecting dynamic objects in video sequences, which is based on the ViBe algorithm. Improvements were made by using the U*V*W* color scheme, using double threshold levels and elements of expert systems theory to eliminate uncertainties in pixel classification (mathematical apparatus of Demster-Schaefer theory and a transformable confidence model developed by Philippe Smets), and using a dynamic approach in updating background model due to neighboring pixels. In order to implement the proposed solutions and confirm the effectiveness of these approaches, an experimental study of the proposed method in comparison with the original ViBe was carried out. The experiment was performed using test frames from a set of CDNET in various variants of the environment, which is as close as possible to the actual application in video surveillance systems and with different variants of resolution. The consolidated results of the experiment on the metrics "*precision*", "*recall*" and the author's metric *W* proposed in [6] indicate an improvement in the results of the proposed method compared to the original ViBe by an average of 6.7%. The obtained visual results of the study are the best in terms of segmentation of dynamic objects, in bad weather conditions and during suppression of the dynamic background.

The disadvantages of the proposed method include the reduction of speed, which is uncritical and allows it to be used in software systems in real time.

References

1. Katerynchuk, I., Babaryka, A.: Analysis of technologies of functioning of departmental systems of video surveillance and definition of directions of their improvement. Coll. Sci. Works Natl. Acad. State Border Guard Serv. Ukr. Ser. Military Tech. Sci. **3**(77), 246–259 (2018)
2. Cristani, M., Farenzena, M., Bloisi, D., Murino, V.: Background Subtraction for Automated Multisensor Surveillance: a comprehensive Review. EURAS1P J. Adv. Signal Process. **2010**, 1–24 (2010)
3. Napoli, C., Pappalardo, G., Tramontana, E., Nowicki, R.K., Starczewski, J.T., Wozniak, M.: Toward work groups classification based on probabilistic neural network approach. Artif. Intell. Soft Comput. **9119**, 79–89 (2015)
4. Capizzi, G.,et al.: Automatic classification of fruit defects based on co-occurrence matrix and neural networks. In: IEEE Federated Conference on Computer Science and Information Systems (FedCSIS), pp. 861–867 (2015)
5. Yang, M., Tao, J., Shi, L., Mu, K., Che, J.: An outlier rejection scheme for optical flow tracking. In: IEEE International Workshop on Machine Learning for Signal Processing, pp. 18–21 (2011)
6. Viola, P., Jones, M., Snow, D.: Detecting pedestrians using patterns of motion and appearance. Int. J. Comput. Vision **63**(2), 153–161 (2005)
7. Yaryshev, S.: Digital methods of video information processing and video analytics. St. Petersburg (2011)
8. Anatolii, B.: Study of detection and tracking algorithms of moving objects in video sequences from video surveillance cameras. Conc. Sci.-Method. Prin. Realization Policy Field State Bord. Secur. Ukr. **6**, 89–105 (2019)
9. Vacavant, A., Chateau, T., Wilhelm, A., Lequièvre, L.: A benchmark dataset for outdoor foreground/background extraction. In: Park, J.-Il., Kim, J. (eds.) Computer Vision - ACCV 2012 Workshops, pp. 291–300. Springer Berlin Heidelberg, Berlin, Heidelberg (2013). https://doi.org/10.1007/978-3-642-37410-4_25
10. Background Subtraction Website. https://sites.google.com/site/backgroundsubtraction/test-sequences/human-activities

11. Bouwmans, T.: Traditional and recent approaches in background modeling for foreground detection: an overview. Comput. Sci. Rev. **11**, 31–66 (2014)
12. Brutzer, S., Hoferlin, B., Heidemann, G.: Evaluation of background subtraction techniques for video surveillance. In: Proceedings of the 24th IEEE Conference on Computer Vision and Pattern Recognition. Colorado Springs, CO, USA, pp. 1937–1944 (2011)
13. Benezeth, Y., et al.: Comparative study of background subtraction algorithms. J. Electron. Imaging **19**(3), 033003 (2010)
14. Andrews, S., Antoine, V.: A comprehensive review of background subtraction algorithms evaluated with synthetic and real videos. Comput. Vision Image Underst. **122**, 4–21 (2014)
15. Weiming, H.: A survey on visual surveillance of object motion and behaviors. IEEE Trans. Syst. Man Cybern. Part C (Appl. Rev.) **34**, 334–352
16. Katerynchuk, I., Babaryka, A.: Improvement of the algorithm for detecting dynamic objects in video sequences. Radio Electron. Comput. Sci. Manag **3**, 88–98 (2020)
17. Stauffer, C., Grimson, W.: Adaptive background mixture models for real-time tracking. In: Computer Society Conference on Computer Vision and Pattern Recognition, pp. 2246–2252 (1999)
18. Hayman, E., Eklundh, J.: Statistical background subtraction for a mobile observer. In: Proceedings Ninth IEEE International Conference on Computer Vision, pp. 67–74 (2003)
19. Godbehere, A., Matsukawa, A., Goldberg, K.Y.: Visual tracking of human visitors under variable-lighting conditions for a responsive audio art installation. In: American Control Conference (ACC), Montreal, QC, Canada, pp. 4305–4312 (2012)
20. Barnich, O., Van Droogenbroeck, M.: ViBe: a universal background subtraction algorithm for video sequences. IEEE Trans. Image Process. **20**(6), 1709–1724 (2011). https://doi.org/10.1109/TIP.2010.2101613
21. Deng, Y.: Generalized evidence theory. Appl. Intell. **43**, 530–543 (2015)
22. Babaryka, A.: Substantiation of the indicator of selection of the optimal algorithm for background selection in video sequences from video surveillance cameras of departmental video surveillance systems. Mod. Inf. Technol. Field Secur. Defense **3**(36), 97–102 (2019)

An Ontological Approach to Detecting Irrelevant and Unreliable Information on Web-Resources and Social Networks

Mykola Dyvak(✉), Andriy Melnyk, Svitlana Mazepa, and Mykola Stetsko

West Ukrainian National University, Ternopil, Ukraine
mdy@wunu.edu.ua

Abstract. This paper considers an important scientific and applied task of identifying irrelevant and unreliable information on web resources, which is an important area of development and implementation of methods of data mining. The analysis of modern methods and means of estimation of irrelevant and unreliable information from the point of view of estimation of information sources is carried out and the basic problem directions which arise in the course of their functioning are allocated.

A system of indicators for filtering unreliable and irrelevant information, which is obtained on the basis of several sources, is proposed. Based on this system, a method of checking information from web resources for relevance and reliability has been implemented. This approach is based on the possibility of using a predefined resource, the data from which are only reliable.

A method of detecting inaccurate and irrelevant information has been developed, taking into account the peculiarities of its distribution through relevant pages in social networks and the use of multitasking classification of information obtained from various data sources.

The proposed intelligent data processing methods together with other methods of intellectual analysis used to evaluate information obtained from the Internet, will significantly increase the efficiency of the process of establishing irrelevance and inaccuracy of information, and will build an assessment of a particular web resource for publishing and disseminating such information.

Keywords: Irrelevant and unreliable information · Web-resource · Information source · Social networks

1 Introduction

Detection of irrelevant and unreliable information is an important element in the process of information awareness and acceptance. The large amount of information disseminated through the Internet requires its comprehensive evaluation, and this in turn also raises the question of evaluating the relevant information source [1–3].

Unlike printed materials, such as books and articles, web pages do not have to meet standards of quality, accuracy and statistical reliability. This lack of quality control

M. Klymash et al. (Eds.): *Future Intent-Based Networking*, LNEE 831, pp. 481–492, 2022.
https://doi.org/10.1007/978-3-030-92435-5_27

creates particular problems for establishing the appropriate level of trust [4], within which it is necessary to evaluate each web page for accuracy, reliability, relevance and objectivity [5].

On the Internet, each website address has a domain as part of the address, which allows you to identify the owner of the website. A domain or its membership in a trusted zone can be a quick way to evaluate the quality of a web resource before you visit it. Typically,.edu,.gov, and.org are more likely to have better information than.com domains [5–7].

This paper implements the procedure for assessing the reliability and relevance of information from a particular web resource. The use of such a procedure also allows for a general assessment of the relevant information source for reliability and relevance.

2 Analysis of Research and Publications

Most publications on this topic focus on the use of methods of intellectual analysis [2, 3, 5, 8], namely artificial neural networks, decision trees, the use of symbolic rules in constructing the resulting characteristics, rules, approaches based on the methods of the nearest neighbor, the method of reference vectors, Bayesian networks, linear regression, methods of correlation-regression analysis; methods of cluster analysis, in particular hierarchical and non-hierarchical methods of cluster analysis [8], methods of searching for associative rules, apriority method; methods of limited or partial search, genetic algorithms in the framework of evolutionary programming, various methods of data visualization [2]. Most of these methods have been implemented in the framework of artificial intelligence [4–7]. Within the framework of data mining there are several key problems that require a solution based on the specifics of the subject area in which they are formed [7–9]. Such tasks include the following: classification, clustering, forecasting, association, visualization, analysis of detection of deviations, evaluation, analysis of relationships, summarizing [2, 4, 7–11].

At the same time, the specifics of information that accumulates on the Internet, or obtained from it, requires the implementation of additional approaches that can be easily implemented and programmatically interpreted within a given subject area [12–14]. Especially relevant is the direction of development of methods of information analysis for relevance and reliability [15–18].

3 Statement and Solution of the Problem

Checking web resources for out-of-date and inaccurate information manually is time consuming. Such verification of inaccurate information does not scale according to the amount of newly created information, especially when an organization has several resources, including resources in different social networks. The use of tools aimed at automating such verification, or the partial use of automatic verification methods are mainly based on the automation of information retrieval processes, natural language processing methods, machine learning, graph theory [5, 6, 11].

In general, the relevant information can be presented by the following cortege [11, 13, 15, 19, 20]:

$$\mathrm{Kw} = \langle \mathrm{S}, \mathrm{P}, \mathrm{O} \rangle, \tag{1}$$

where *Kw* – relevant information, *S* – subject area, *P* – predicate (relationship of the object to the subject area), *O* – object in subject area.

For example, let's consider the following information resource, which contains news on the official Facebook page of the Ministry of Internal Affairs: «The Ministry of Internal Affairs of Ukraine (MIA) is the central executive body». In this block you can select the following information blocks:

$$Kw = \langle MIA,\ \mathrm{is},\ central\ executive\ body \rangle \tag{2}$$

Most known means of automatic retrieval of information are based on the presented (1) and its corresponding interpretation.

The data is extracted from open sources, which belong to one category or group of relevant media resources of an institution or organization [19–21]. Then there is the search of non-relevant information and construction of the system for its classification. This process often involves the extraction of data or the corresponding relationship between them. Data mining can be classified as data mining from a single source or open source [22–24]. One source of data extraction is mainly based on a relatively reliable resource (for example, the official website of the institution).

This method of obtaining relevant information is relatively effective, often leading to the establishment of incomplete information, as it depends on the level of content or relevance of the posted content.

Searching for current news based on data from several open sources is less effective, but will significantly expand the formation of complete and reliable information blocks.

We form the following sets:

$$\mathrm{At} \in \mathrm{Kw}, \tag{3}$$

$$\mathrm{BaseAt} = \{\mathrm{At}\}, \tag{4}$$

$$\mathrm{Base\ Relation} = \{\mathrm{BaseAt}, \mathrm{RelationAt}\}, \tag{5}$$

where *At* - represents a piece of data that has been verified; *BaseAt* - represents a set of verified data; *Base Relation* - represents a set of connections between entities, which are described by the corresponding relations RelationAt.

In order to form a database with real news, which is obtained on the basis of several sources, they need to be further filtered, taking into account the following features:

Redundancy of the presented data in one context (for example, $Kw = \langle MIA, is, central\ executive\ body \rangle$) and

$$Kw_1 = \left\langle \frac{\text{Ministry of Internal Affairs, } is,}{central\ executive\ body} \right\rangle \tag{6}$$

Kw_1 will be redundant because «MIA» and «Ministry of Internal Affairs» correspond to the same entity;

– Invalidity - an indicator that depends on a specific time interval, for example, $Kw = \langle$Ministry of Internal Affairs, Ministry, Ukraine$\rangle$ irrelevant information and should be updated. One of the options to overcome this problem is to present data with a certain set of time characteristics or to expand the obtained set by establishing additional statements. Contradiction of the provided data (for example $Kw = \langle MIA$, address, Kyiv$\rangle$) and $Kw = \langle MIA$, address, Kharkiv$\rangle$) are conflicting data that require further analysis);

Web-resource (the complexity of identifying the resource to the trusted area within a given subject area). The authenticity of a resource can be established, for example, by the fact that its address belongs to a trusted zone; Completeness - the availability of information obtained from resources does not always allow to establish its reliability.

3.1 The Procedure for Checking Information from Web Resources for Relevance and Reliability

To assess the falsity of information posted on web resources, it is necessary to compare the data obtained and presented using the relationship (1) with the information obtained from reliable sources.

Typically, the data validation procedure for the cortege $Kw = \langle S, P, O\rangle$ is to estimate the possibility that the boundaries denoted by the predicate P are formed taking into account the affiliation of S to the node representing O in the set $BaseAt$. In particular, this process can be described using the following steps, which are presented by used scheme on Fig. 1.

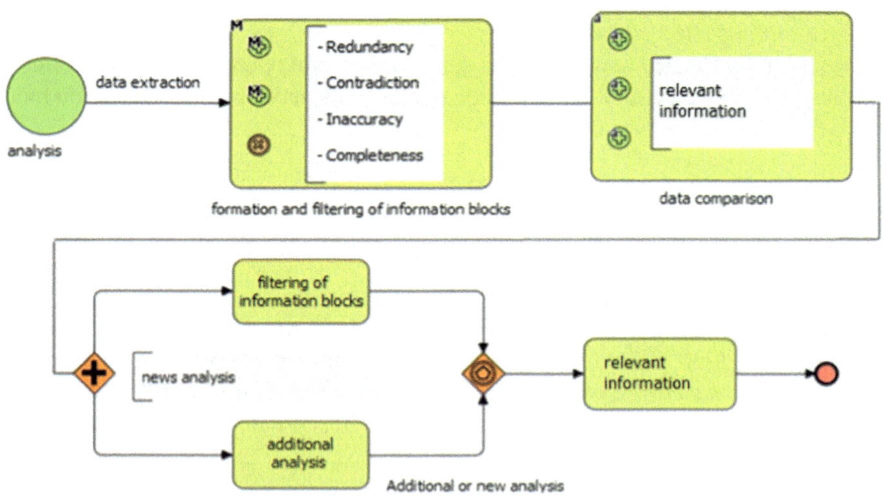

Fig. 1. The scheme of analysis of information from web resources using indicators to assess the reliability of information

1. The location of the essence. Topic S, as well as object O, must first match the element of the set of valid data in $BaseAt$, which represents the same entity.

2. Check the relationship. The information described by (1) is not considered non-relevant information if the boundary is indicated by a predicate from the element representing S to the element representing O, exists in the plural of descriptions of *BaseRelation* relations. Otherwise, the information is considered fake, or in need of further clarification.
3. Establishment of relevant data. When $Kw = \langle S, P, O\rangle$ is not included in the set *BaseRelation*, the probability for the boundaries of the entity, which is denoted by a predicate within a certain subject area can be determined using the semantic proximity index [15, 16, 25–29], which allows a pairwise evaluation of elements from the set ⟨S, P, O⟩, which are in semantic relations (synonyms, hyper-hyponymic relations, associativity), and zero values for all other pairs. Consider the specifics of the described verification process. Assume that the information block verified on the official resource can be represented as a formalized set of statements $Kw = \langle s_i, p_i, o_i\rangle, i = 1, 2, \ldots n$. Set W_{BaseAt} is based on verified data sets $st_j, pt_j, ot_j, j = 1, 2, \ldots m$.

Checking the information blocks that are placed on the adjacent to the studied resource will be reduced to the construction of *Fa* functions with the found authenticity indicators $Af_i \in [0; 1]$ for each block $\langle s_i, p_i, o_i\rangle$, comparing them in pairs with $\langle st_j, pt_j, ot_j\rangle$ taking into account the description of the relationship with *Base Relation* :

$$Af_i = 1, \textit{if the block is verified as authentic};$$

$$Af_i = 0, \textit{if the block is verified as non} - \textit{authentic}.$$

The overall indicator of the authenticity of the news Af will be defined as the aggregate value of all Af_i:

$$Fa : (s_i, p_i, o_i) \overset{W_{BaseAt}}{\rightarrow} Af_i \tag{7}$$

$$Af = O(Af_i), \partial ei = 1, \ldots n, \tag{8}$$

where O – the aggregation function is selected (for example, weighted average).

The information to be checked is relevant data provided $Af = 1$, and id $Af = 0$ the mentioned news are completely unreliable:

$$\begin{aligned} &Fa((s_i, p_i, o_i)W_{BaseAt}) = \\ &= P(\textit{limit } p_i, \textit{which connects } s_i \textit{ and } o_i \textit{ in } W_{BaseAt}), \end{aligned} \tag{9}$$

where P – probability of conformity s_i, o_i to reliable data st_j, ot_j in W_{BaseAt}.

$$s_i = \underset{st_j}{argmin}\left[U\left(s_j, st_j\right)\right] < \theta, \tag{10}$$

$$o_i = \underset{ot_j}{argmin}\left[U\left(o_j, ot_j\right)\right] < \theta, \tag{11}$$

where U– the distance between two elements of the corresponding entities, which can be calculated using the Jakard distance formula [15, 16, 30, 31].

The value of U will reflect the identification of two entities, for example, if $U(o_i, o_{i+1}) = 0$, then o_i and o_{i+1} are the same entities.

As a result, we obtain a set of extracted data and a set of verified reliable data. One of the main problems in constructing a set of reliable data will be the source (sources) from which this data can be obtained. As a suitable source can be used a predefined resource, the data from which are only reliable.

3.2 A Method of Verifying Information from Web Resources Based on the Analysis of Data Obtained from Social Networks

Detection of unreliable information, taking into account the peculiarities of its dissemination through the relevant pages in social networks, can be implemented using methods of analyzing the structure of information in the process of their dissemination or duplication. Detection of irrelevant information based on its dissemination can be formulated as a task of multitasking classification of information obtained from different data sources. The input data of this method of distribution can be an ordered set, the elements of which are information blocks and the corresponding indicators of their identification (time, source, profile).

Such an ordered set can be represented through a tree-like structure, which, taking into account the relevant indicators, fixes the degree of dissemination of certain information through social networks. The root element of such a structure identifies the source from which the information was originally taken, and other nodes represent the users who disseminated the information, given their relationship to the parent source (for example, the organization's official social network page and user pages or just "friends").

The assessment of the reliability of information disseminated through the relevant profiles in social networks can be done by taking into account the power of the set or by assessing them through time characteristics.

Figures 2 and 3 illustrate the assessment of the reliability of information obtained through social networks using the above assessment methods.

The main criteria used when using iterative set estimation are:

- number of iterations - the maximum number of iterations in which the information blocks were distributed;
- width of the iterative step - the number of sources of information within the iterative step;
- dimensionality - the total number of sources of information through which dissemination takes place;

Figure 3 shows the time-dependent process of forming the elements of the sets. The following indicators were used for this:

- period of existence of the information element - the time during which the information was disseminated through the relevant information sources;

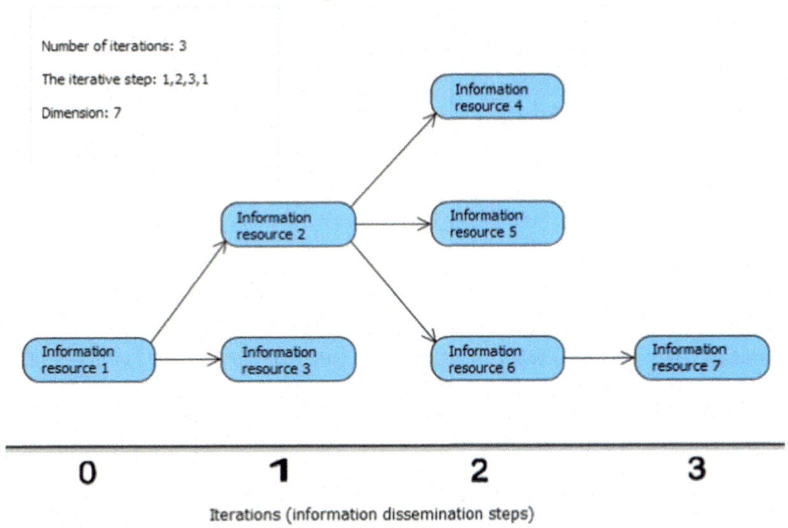

Fig. 2. The scheme of calculation of reliability of the information on the basis of an estimation of processes of formation of elements of sets

- real-time mode - the number of sources of information that disseminated information at a certain time;
- number of sources of information dissemination - the total number of sources of information through which dissemination takes place;

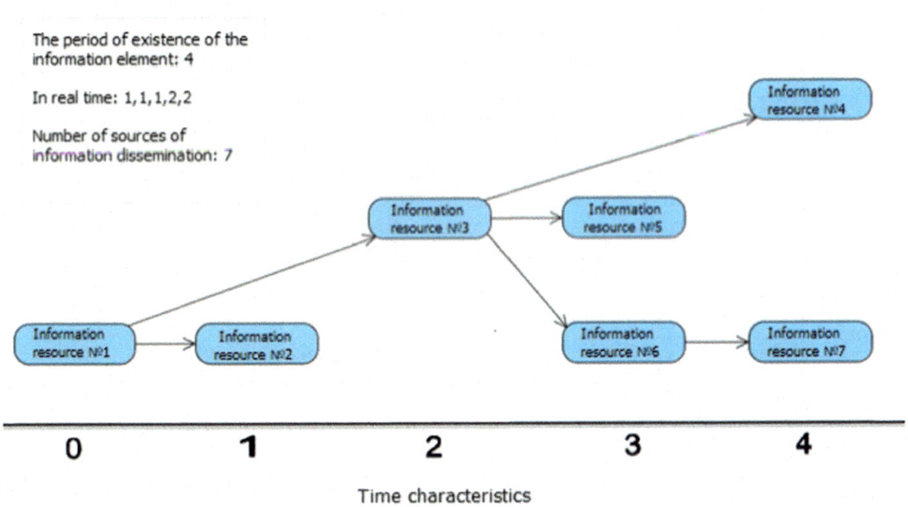

Fig. 3. The scheme of calculating the reliability of information based on the analysis of time characteristics of the formation of elements of sets

The difficulty of applying this method of determining reliable information is the increased number of nodes at each stage of the iteration, for example, due to the large number of active users belonging to the root node. In addition, you often have to filter out additional metrics that can build up at each iteration. In the general case, the reliability of the information will be determined based on its proximity to the root distribution node, taking into account the current time characteristics.

4 Experiments

Based on the proposed approach, a number of experimental studies were conducted. The essence of these studies was to assess the information of the web resource of the Ministry of Internal Affairs of Ukraine (https://mvs.gov.ua/) and the relevant sources of information that are fully linked to it, partially or not. Figure 4 shows a diagram that shows the results of the evaluation of the relevant sources of information, taking into account the parameters at each iterative step.

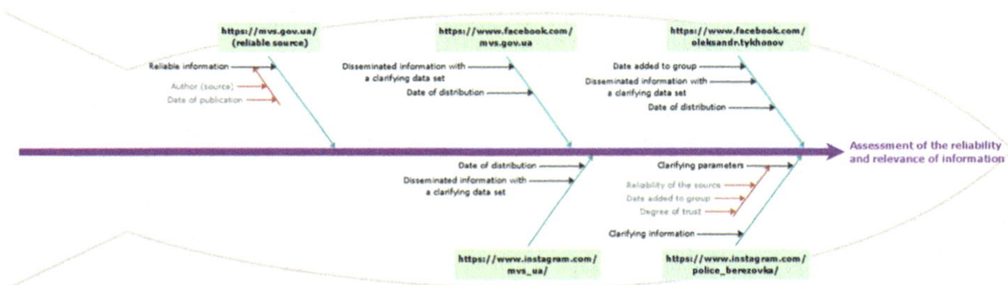

Fig. 4. The scheme of assessing the reliability of information based on iterative and temporal characteristics on the example of data obtained from the resource https://mvs.gov.ua/

As a result of expert assessment with elements of the use of automated tools for content analysis, the reliability of information was assessed on the basis of indicators of context-semantic analysis, and the construction of appropriate weighted average estimates of $O(Af)$. The corresponding results are shown in Table 1.

As a result of the assessment of resources on the basis of the indicators described above, a generalized assessment of the reliability of the source was carried out, in accordance with the proposed assessment scale (Table 2):

$$\begin{cases} if\ 0 \leq O(Af) < 0.5\ then - Invalid\ source(low\ reliable) \\ if\ 0.5 \leq O(Af) < 0.75\ then - Medium\ reliable\ of\ source \\ if\ 0.75 \leq O(Af) \leq 1\ then - High\ reliable\ of\ source \end{cases} \quad (12)$$

Reliability assessment was performed on the basis of an aggregate assessment of each information source. As a result, it is established that on two information resources, the information that is disseminated can be considered reliable, and on two - partially reliable.

Table 1. The results of assessing the authenticity of web resources

Indicators		Web resources				
		https://mvs.gov.ua/ (reliable source)	https://www.facebook.com/mvs.gov.ua	https://www.facebook.com/oleksandr.tykhonov	https://www.instagram.com/mvs_ua/	https://www.instagram.com/police_berezovka/
Indicators for assessing the reliability of information	Redundancy	1.000	0.972	0.515	0.950	0.795
	Invalidity	1.000	0.879	0.625	0.855	0.654
	Contradiction	1.000	0.915	0.500	0.915	0.596
	Inaccuracy	1.000	0.900	0.650	0.900	0.545
	Completeness	1.000	0.875	0.500	0.795	0.750
	The overall indicator of the authenticity of the resource $O(Af)$	1.000	0.908	0.558	0.883	0.668

Table 2. The results of assessing the authenticity of web resources

Evaluation parameters	https://mvs.gov.ua/ (reliable source)	https://www.facebook.com/mvs.gov.ua	https://www.facebook.com/oleksandr.tykhonov	https://www.instagram.com/mvs_ua/	https://www.instagram.com/police_berezovka
Author	Press center	Press center	Tykhonov	Press center	District Police Department
Date of publication (distribution)	30.04.2021	30.04.2021	02.05.2021	30.04.2021	01.05.2021
Reliability of the source	High	High	Medium	High	Medium
Source registration date in the trusted domain	–	20.09.2016	31.10.2020	21.09.2020	30.09.2020
Degree of trust	High	High	Medium	High	Medium
Reliability assessment	–	High	Medium	High	Medium

5 Conclusions

The paper considers the features of the process of evaluating information for relevance and reliability. The lack of appropriate approaches, which are easy to use and implement, prompted the development of an appropriate system of indicators and methods for assessing inaccurate and outdated information. The use of the methods proposed in this work allowed to increase the degree of recognition of the source of information for the placement of irrelevant and inaccurate information.

The created methods are software implemented as an add-on to existing solutions within individual software modules and can be effectively used to assess the reliability of information. The effectiveness of the proposed methods was confirmed experimentally on the example of evaluation of web resources.

References

1. Vo, N., Lee, K.: The Rise of Guardians: Fact-checking URL Recommendation to Combat Fake News. In: The 41st International ACM SIGIR Conference on Research & Development in Information Retrieval (SIGIR '18), pp. 275–284. Association for Computing Machinery, New York (2018). https://doi.org/10.1145/3209978.3210037
2. Ye, J., Skiena, S.: MediaRank: Computational Ranking of Online News Sources, pp. 2469–2477 (2019). https://doi.org/10.1145/3292500.3330709.
3. Yu, B., Zhang, Z., Liu, T., Wang, B., Li, S., Li, Q.: Beyond word attention: using segment attention in neural relation extraction. IJCAI (2019)
4. Wang, C.: Relation extraction. In: Proceedings of the Twenty-Eighth International Joint Conference on Artifcial Intelligence. International Joint Conferences on Artifcial Intelligence Organization, pp. 5401–5407. https://doi.org/10.24963/ijcai.2019/750
5. Zhou, X., Zafarani, R.: A survey of fake news: fundamental theories, detection methods, and opportunities. ACM Comput. Surv. **53**(5), 1–40, Article 109 (2020). https://doi.org/10.1145/3395046
6. Ray, O., Qian, J., Wang, W.Y.: A survey on natural language processing for fake news detection (2020). ArXiv abs/1811.00770
7. Dyvak, M., Papa, O., Melnyk, A., Pukas, A., Porplytsya, N., Rot, A.: Interval model of the efficiency of the functioning of information web resources for services on ecological expertise. Mathematics **8**(12), 2116 (2020). https://doi.org/10.3390/math8122116
8. Nørregaard, J., Horne, B.D., Adalı, S.: NELA-GT-2018: a large multi-labelled news dataset for the study of misinformation in news articles. In: Proceedings of the International AAAI Conference on Web and Social Media, vol. 13, pp. 630–638 (2018)
9. Trivedi, R., Sisman, B., Dong, X., Faloutsos, C., Ma, J., Zha, H.: LinkNBed: multi-graph representation learning with entity linkage, pp. 252–262 (2018). https://doi.org/10.18653/v1/P18-1024
10. Vilone, G., Longo, L.: Explainable artificial intelligence: a systematic review (2020)
11. Kovbasistyi, A., Melnyk, A., Dyvak, M., Brych, V., Spivak, I.: Method for detection of non-relevant and wrong information based on content analysis of web resources. In: 2017 XIIIth International Conference on Per-spective Technologies and Methods in MEMS Design (MEMSTECH), Lviv, 2017, pp. 154–156. https://doi.org/10.1109/MEMSTECH.2017.7937555
12. Dyvak, M.P., Kovbasistyi, A.V., Melnyk, A.M., Turchyn, L.Y., Martsenyuk Y.O.: System for web resources content structuring and recognizing with the machine learning elements. Radio Electron. Comput. Sci. Control (3) (2018). https://doi.org/10.15588/1607-3274-2018-3-14

13. Dyvak, A., Melnyk, A., Shevchuk, R., Kovbasistyi, A., Huhul, O., Tymchyshyn, V.: Mathematical modeling of the estimation process of functioning efficiency level of information web-resources. In: Proceedings of the 2020 10th International Conference "Advanced Computer Information Technologies" – Deggendorf, Germany, 16–18 September 2020, pp. 492–496 (2020)
14. Bian, T., et al.: Rumor Detection on Social Media with Bi-Directional Graph Convolutional Networks (2020). arXiv preprint arXiv:2001.06362
15. Sample, C., McAlaney, J., Bakdash, J.Z., Thackray, H.: A cultural exploration of social media manipulators. In: Proceedings of the 17th European Conference on Cyber Warfare and Security, Oslo, Norway, pp. 342–341 (2018)
16. Sharma, K., Qian, F., Jiang, H., Ruchansky, N., Zhang, M., Liu, Y.: Combating fake news: a survey on identification and mitigation techniques. ACM Trans. Intell. Syst. Technol. **10**, 1–42 (2019)
17. Hashimoto, T., Shepard, D.L., Kuboyama, T., Shin, K., Kobayashi, R., Uno, T.: Analyzing temporal patterns of topic diversity using graph clustering. J. Supercomput. **77**(5), 4375–4388 (2020). https://doi.org/10.1007/s11227-020-03433-5
18. Ruchansky, N., Seo, S., Liu, Y.: CSI: a hybrid deep model for fake news detection. In: Proceedings of the 2017 ACM on Conference on Information and Knowledge Management, CIKM 2017, pp. 797–806 (2017)
19. Dutta, H.S., Dutta, V.R., Adhikary, A., Chakraborty, T.: HawkesEye: detecting fake retweeters using Hawkes process and topic modeling. IEEE Trans. Inf. Forensics Secur. **15**, 2667–2678 (2020)
20. Gontier, C., Pfister, J.P.: Identifiability of a binomial synapse. Front. Comput. Neurosci. **14**, 86 (2020). https://doi.org/10.3389/fncom.2020.558477. PMID: 33117139
21. Gao, S., Ma, J., Chen, Z.: Modeling and predicting retweeting dynamics on microblogging platforms. In: Proceedings of the Eighth ACM International Conference on Web Search and Data Mining, WSDM 2015, pp. 107–116 (2015)
22. Lukasik, M., Srijith, P.K., Vu, D., Bontcheva, K., Zubiaga, A., Cohn, T.: Hawkes processes for continuous time sequence classification: an application to rumour stance classification in twitter. In: Proceedings of the 54th Annual Meeting of the Association for Computational Linguistics, ACL 2016, pp. 393–398 (2016)
23. Shevchuk, R., Melnyk, A., Opalko, O., Shevchuk, H.: Software for automatic estimating security settings of social media accounts. In: 2020 10th International Conference on Advanced Computer Information Technologies (ACIT), pp. 769–773 (2020). https://doi.org/10.1109/ACIT49673.2020.9208879
24. Ganji, M.D.; Rahmanzadeh, A.: Chapter 6—Mathematical modeling and simulation. In: Nguyen-Tri, P., Do, T.-O., Nguyen, T.A. (eds.) Smart Nanocontainers. Micro and Nano Technologies, pp. 89–102. Elsevier, Amsterdam (2020). ISBN 978-0-12-816770-0
25. Brainard, J., Hunter, P., Hall, I.: An agent-based model about the effects of fake news on a norovirus outbreak. Rev. D'épidémiologie Santé Publique **68**, 99–107 (2020)
26. Wahid-Ul-Ashraf, A., Budka, M., Musial, K.: Simulation and Augmentation of Social Networks for Building Deep Learning Models (2019). arXiv arXiv:1905.09087
27. Zhou, X., Zafarani, R.: A survey of fake news: fundamental theories, detection methods, and opportunities. ACM Comput. Surv. **53**, 1–40 (2020)
28. Shrivastava, G., Kumar, P., Ojha, R.P., Srivastava, P.K., Mohan, S., Srivastava, G.: Defensive modeling of fake news through online social networks. IEEE Trans. Comput. Social Syst. **7**, 1159–1167 (2020)
29. Varol, O., Ferrara, E., Davis, C.A., Menczer, F., Flammini, A.: The online human-bot interactions: detection, estimation, and characterization. In: Proceedings of the International AAAI Conference on Web and Social Media, Palo Alto, CA, USA, 25–28 June 2018 (2018)

30. Yang, K., Varol, O., Davis, C.A., Ferrara, E., Flammini, A., Menczer, F.: Arming the public with artificial intelligence to counter social bots. Hum. Behav. Emerg. Technol. **1**, 4861 (2019)
31. Saura, J.R., Ribeiro-Soriano, D., Palacios-Marqués, D.: From user-generated data to data-driven innovation: a research agenda to understand user privacy in digital markets. Int. J. Inf. Manag. 102331 (2021)

Application Peculiarities of Deep Learning Methods in the Problem of Big Datasets Classification

Bohdan Rusyn(✉), Oleksiy Lutsyk, Rostyslav Kosarevych, and Yuriy Obukh

Karpenko Physico-Mechanical Institute of the NAS of Ukraine, Naukova Street 5, Lviv 79060, Ukraine
yobukh@ipm.lviv.ua

Abstract. The chapter proposes a new approach to estimate the quality of training datasets for convolution neural networks with deep learning. It is shown that the accuracy of image classification depends on the following parameters such as completeness, imbalance, uniformity of coverage of feature vectors, data compactness and particular class accuracy. A comparative assessment of the dependence of the training time of the model on the sample size of the proposed approach is carried out in comparison with VGG net, Alex net, CNN ensemble, Google net. It is established that the minimization of the Kulbak-Leibler distance allows forming the parameters of the training sample with much lower computational costs and a more compact representation of the feature space. This allows creating highly efficient information systems that carry out the process of classification of big data with high accuracy.

Keywords: Deep learning · Quality estimation · Big data · Training sample · Computational costs

1 Introduction

In recent years, there has been an exponential growth in the amount of digital information. As of 2020, about 40–44 ZB of information were generated. However, it is known that not all accumulated big data contains useful information. This stimulates the development of modern information technology for processing, analyzing and interpreting large amounts of data in real time. Data processing allows to quickly structure the information, their analysis opens up the possibility of identifying random patterns, which cannot always be effectively classified or predicted. All this together opens up a wide range of opportunities for the use of big data, from modern info communication systems of the new generation 5G, to the problems of protection and preservation of the environment.

Machine learning methods based on the collection and analysis of big data have been widely used in various fields of human activity. Currently, there is a steady trend that artificial intelligence systems are increasingly displacing traditional approaches that are based on heuristic decision-making methods.

M. Klymash et al. (Eds.): *Future Intent-Based Networking*, LNEE 831, pp. 493–506, 2022.
https://doi.org/10.1007/978-3-030-92435-5_28

This is due, the fairly rapid development of computing tools, video controllers on the one hand and on the other hand the availability of a large amount of data for machine learning. Obviously, there are important circumstances that complicate or prevent the effective use of such intelligent systems. The main reason, in our opinion, that prevents this is a difficulty in obtaining a representative training sample in which data processing and classification will be carried out with high accuracy and in real time.

In machine learning tasks, there is a relationship between the amount of data used to train the model and the subsequent accuracy of its work on both test and validation data. This is most often appears in the problem of lack of data to create a quality training sample [1]. A representative training sample is largely responsible for the correct training of the model in the classification.

There is no universal approach that would give an unambiguous answer to the question of how much data is needed and what size is needed to train a particular model with predictable accuracy.

As a rule, the data in the training sample form a vector of features of a given length. Using statistical methods and elements of cluster analysis, it is possible to estimate the quality of the training sample [2]. There are a number of works on data evaluation in the case of using classical models for their classification [4, 5]. However, these approaches are used to work in the feature space [6]. To a large extent, the result of assessing the quality of the sample primarily depends on the process of finding features, which is often heuristic in nature and not amenable to strict justification.

There are a large number of applications where the objects of analysis are images, such as remote sensing, robotics, biometric authentication and others [7–10]. The information in the form of images is more difficult to generalize and requires intermediate stages of processing such as their improvement by pre-processing, as well as the construction of invariant system of features to affine transformations. Without these intermediate stages of data processing, it is almost impossible to assess the coverage of classes in the feature space and as a result to make assumptions about the adequacy and quality of the training sample for testing a particular model. One of such effective methods of solving this problem is the use of convolutional layers of the neural network, which generalize the choice of features in the image and reduces them to the learning process [11, 12]. It is known that the main tool for establishing the representativeness of the training sample in deep learning for a long time used the so-called learning curves. With their help, it is possible to estimate the quality of the training sample not directly, but through its impact on a particular model under training [13]. However, the most accurate approach to the evaluation of the training sample is based on methods that combine the information of the trained model with a posteriori data [14]. This approach is highly reliable and in many works is taken as a reference. Its disadvantage is the high computational complexity, which imposes limitations on the efficiency of evaluation. For example, the quality of images strongly influences the informativeness of the training sample, which is formed on the basis of these images [15].

Remote sensing of the Earth's surface involves the acquisition and analysis of large amounts of data generated by high-resolution optical and infrared scanners placed on spacecraft or drones. These data are usually presented in the form of large image matrices of 10000×10000 pixels in grayscale. Applied remote sensing tasks involve building of automated methods for classifying objects of interest, which leads to the emergence of large databases that can be used for deep learning [16, 17]. Training neural networks is a computationally complex process that does not always give the desired result. According to research, the process of correct training of the neural network strongly depends on the properties of the training sample, namely how well it represents a set of discriminatory features.

Experiments on training samples show that it does not make sense to have large amounts of data if their quality is poor. Under the quality of the data we will understand the complex characteristics that describe the properties of the data that perform the task, namely: compactness of presentation, imbalance, class consistency, class deviation within the sample and accuracy.

Based on this, there is a need to carry out a preliminary assessment of the training sample, and this leads the need to develop approaches for estimating the quality of the training sample [18, 19].

The essence of the proposed approach involves a preliminary assessment of the training sample quality based on the above-mentioned components. We will always have only two cases when preliminary estimation of the sample quality will have an effect:

- If the training sample is not very informative, then training the model will not give the expected result. Therefore, such a sample will be supplemented in such a way as to correspond to the complex quality characteristics;
- If the training sample is redundant, the training process of the model is computationally complex, and reducing its redundancy will not significantly affect the accuracy of training, but can significantly reduce training time.

This approach allows making a preliminary assessment of the properties of the training sample before training the neural network in this sample. As a result, we optimize the total time spent on the creation of the training sample, its building and evaluation, as well as make it possible to automate the process of creating a quality training sample for arbitrary classification models.

2 Reduction of the Data Dimension

The procedure of reducing the dimensionality of the data is illustrated by the example of processing a database containing images obtained by remote sensing. If the input database is denoted as $X = \{x_1, x_2, ..., x_n\}$, where $x_1, x_2, ..., x_n$ - is the vector of features of individual images with a length of 4096 samples, the database of significantly reduced dimension will be written as $M = \{m_1, m_2, ..., m_k\}$, where $M << X$.

In the proposed approach to determine the quality of the image database in the role of a feature generator is proposed to use a multilayer convolutional network, which train to generate image features. It is obvious that the architecture and the number of layers of the network will depend on the type and number of images. Therefore, we will use the VGG architecture, which is shown in Fig. 1.

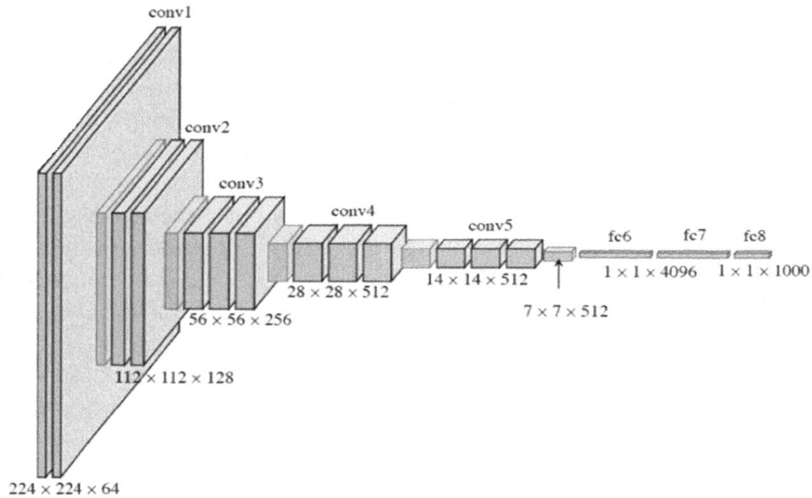

Fig. 1. VGG convolution network architecture

It has a simple uniform structure of sequentially arranged convolutional and combining layers; the depth of the architecture is 16 layers. The operation of the network in the mode of the feature generator is balanced, and the filters effectively capture useful features [22]. Another positive advantage of such a structure is the ability to perform training only once for a large enough sample to train convolutional filters and apply the calculated coefficients without the need for retraining.

The convolutional layers output the data as a three-dimensional array of values, where the slice in the third coordinate corresponds to the filter used by the input of the next layer. The information represented by fully connected layers at the network output is a combination of characteristics that were generated by the previous layer. Based on this convolution can be written as:

$$conv(a^{l-1}, Z^n)_{x,y} = \psi^l(\sum_{i=1}^{n_H^{l-1}} \sum_{j=1}^{n_W^{l-1}} \sum_{k=1}^{n_C^{l-1}} Z^n_{i,j,k} a^{l-1}_{x+i-1,y+j-1,k} + b^l_n), \quad (1)$$

where, n_H, n_W are the size of the image in height and width; n_C is the number of image channels; Z is the filter, which in this case has a square shape; a is the size of a specific network layer; b is the initial threshold; l is the dimension.

Pre-selection of image features is needed to optimize a large amount of data that are in the input image database. Then, with the help of combination characteristics, it is possible to increase the discriminatory properties of features in the feature space.

The goal of dimensionality reduction is to preserve a high-dimensional data structure in a reduced-dimensional space. The classical dimensionality reduction approaches include principal component analysis (PCA), which is a representative of linear methods and is aimed at obtaining low-proportional representations of data that differ from each other. It was found that classical approaches cannot fully represent the local and global data structure. Therefore, it was proposed to use probabilistic approaches to data reduction based on the Kullback-Leibler distance (KL) [23].

As the similarity between the vectors of features responsible for individual images x_i and x_j, we take the conditional probability $p_{i|j}$, written in the form:

$$p_{i|j} = \frac{\exp(\frac{-\|x_j - x_i\|^2}{2\sigma_j^2})}{\sum_n \exp(\frac{-\|x_j - x_n\|^2}{2\sigma_j^2})}, \tag{2}$$

where σ_j^2 is the variance is localized in near x_j.

Obviously, a similar conditional probability expression can be written for data m_i and m_j in a reduced dimension:

$$q_{i|j} = \frac{\exp(-\|m_j - m_i\|^2)}{\sum_k \exp(-\|m_j - m_k\|^2)}. \tag{3}$$

If in the reduced dimension M correctly reflects X, then the conditional probabilities will be proportional. In this case, the task is to find a low-dimensional representation of the data while minimizing the differences between $p_{i|j}$ and $q_{i|j}$. The measure of KL is such a measure that indicates how different one distribution is from another. In our case, the probability distributions P and Q are discrete and definite in the same probability space, then the objective function will take the form:

$$\sum_j D_{KL}(P_i||Q_i) = \sum_j \sum_i p_{i|j} \log p_{i|j} - p_{i|j} \log q_{i|j}. \tag{4}$$

We will minimize this expression using numerical gradient methods that need to be initialized with initial values. In this case, random values from near origin are selected.

With the described approach it is possible to reduce the dimensionality of the data based on the established conditional probability of pairwise metric distances. This allows moving to the building a quality features of the database with much lower computational costs and a compact feature space representation.

According to the proposed method the quality estimation of the training sample is carried out on the basis of all sample data, and then thinning and revaluation. Thinning of the sample is realized randomly, and in the presence of additional information it is possible to extract both redundant data and those that make noise in the learning process. The reduction of the training sample is justified only to the moment when its representativeness begins to decline (Fig. 2).

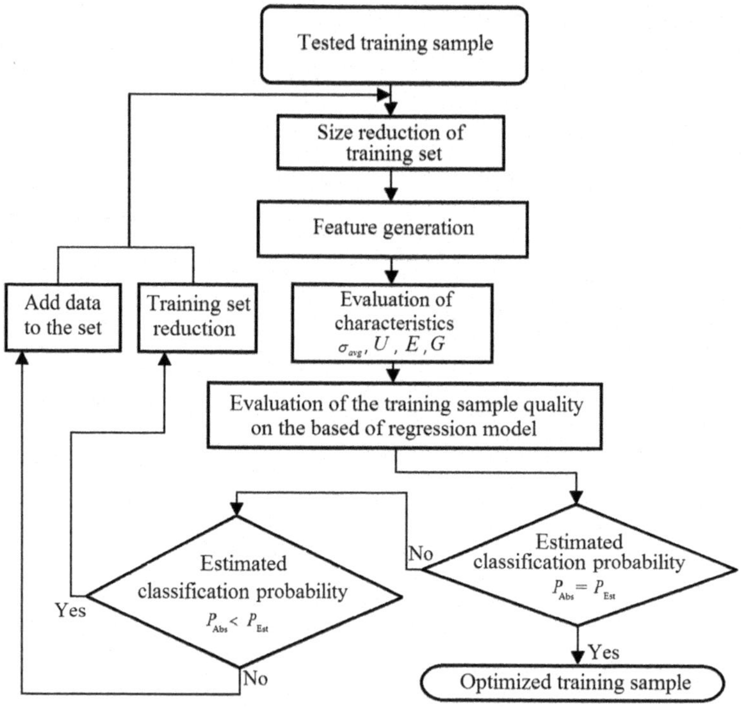

Fig. 2. Block diagram of the training sample reduction method

3 Database Estimation

A good quality training sample for deep learning will be considered a set of data that in the process of training the network provides the desired level of interpretation or classification in real time. Quantitative assessments of the training sample are embedded in the data structure itself. The peculiarities of the data structure primarily include their clustering properties, overlap and imbalance of classes [22, 24], sparse data [25], compactness of representation and representativeness.

We will quantify this data. One of the informative parameters is completeness, which is characterized by the percentage of data that includes one or more values. A characteristic feature for completeness will be the average deviation of the class within the sample, which will be calculated according to the following expression:

$$\sigma_{avg} = \exp\left(\sum\nolimits_{c}^{N} \left(\frac{1}{N} - \frac{K_c}{K}\right)^2\right), \tag{5}$$

where N is the total number of classes, K_c is the number of components belonging to a particular class, c is the serial number of a particular class. K is the total number of components.

An important requirement is that, first of all, the most informative data should be checked first, since the quantitative assessment of completeness practically does not depend on unimportant data.

The second most important assessment is the unevenness or imbalance of the training sample by classes, which will be determined by the following relation:

$$U = \frac{1}{(N - K_c)^2} \sum\nolimits_c^N \left(K_c - \frac{K}{N} \right). \tag{6}$$

It should be noted that if the unevenness indicator show a large imbalance, then this indicates that a particular class is represented by a significantly smaller number of feature vectors and in the process of the neural network training, it is ignored or treated as noise.

Intrinsic features of the training set include the uniformity of coverage of feature vectors in the feature hyperspace. Then the third quantitative estimate takes the following form:

$$E = \frac{1}{K^2} \sum\nolimits_i^K \sum\nolimits_{d=1}^L \sum\nolimits_{c=1}^N \exp\left(\left(x_d - \frac{1}{2N}(2r - 1)(\max(x_{i,c}) - \min(x_{i,c}) \right)^2 \right). \tag{7}$$

where r is the coverage ratio.

In the course of a series of experiments, it was found that the uniformity of feature vector coverage in the feature hyperspace is inversely related to the clustering properties of the training sample.

Since the clustering analysis of the training sample is a computationally complex and time-consuming process it is proposed to replace it with a property of data compactness representation in the feature space. This quantitative feature is directly proportional to the clustering properties of the training sample and quite accurate marker that indicates the simplicity of constructing a classifier in the process of training a neural network:

$$G = 1 - \frac{\sum_{c=1}^N \sum_{d=1}^N \sum_{i=1}^K (x_{i,c} - x_{i,d})}{(K_c^2 - K_c) \sum_{i=1}^N (\max_c(x_{i,c}) - \min_c(x_{i,c}))^2}. \tag{8}$$

The next parameters for evaluating data are accuracy, which reflects how well the data describes a particular class, what it identifies, and consistency, which shows how well the data fits with the object it describes.

In addition, during the verification of the input data, the presence of duplicated data on the learning process was found to have a negative impact, which subsequently leads to an imbalance. This includes the cases of the presence in the training set a several records of the same feature vector, as well as cases when the feature vectors are very close to each other. Therefore, when such a case occurs during the preliminary assessment of the training sample, it is advisable to localize and remove this data.

The building of features that are invariant to a specific training set will allow using a set of these features for a preliminary assessment of the quality of the training set without performing the computationally complex process of training a neural nework. The proposed approach will make it possible to correct the already existing training sample by significantly reducing it, which in turn will reduce the network training time. The simplest option for reducing the training sample is to randomly extract data after each step of checking it for representativeness. A more difficult but accurate option for reducing the training sample is to remove redundant data that introduce noise into the training process.

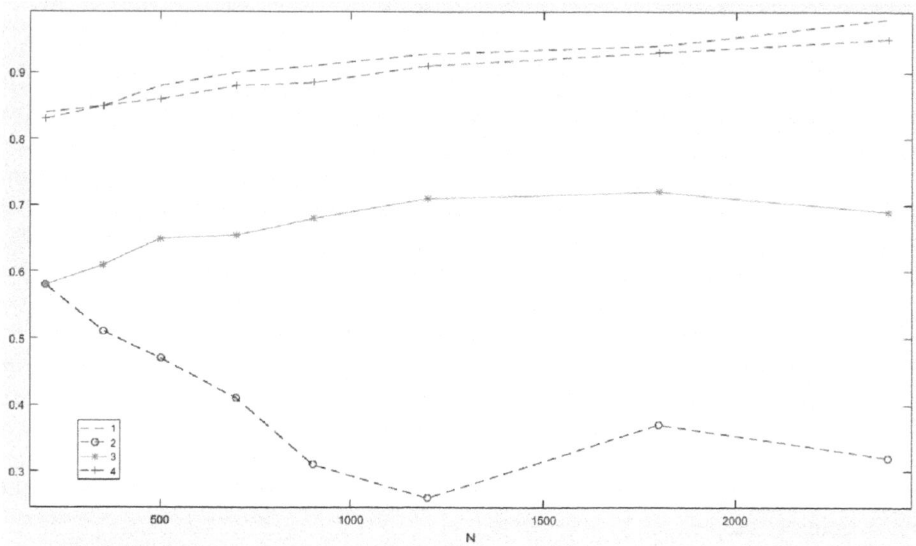

Fig. 3. Dependence of the proposed quantitative estimates on the training sample size

As a result of the application of the main quantitative estimates to determine the quality of the training sample, the plotted dependences are shown in Fig. 3. The training sample was created on the basis of the synoptic base for the study of cloudiness of images. Curve 1 - displays the dependence of the characteristics of the class deviation within the sample, 2 - the dependence of the imbalance characteristics, 3 - the dependence of the characteristics of the compactness of the representation of classes, 4 - the dependence of the characteristics of the consistency of classes depending on the size of the training sample. As it can be seen, three of the four curves increase with an increase in the number of elements of the training sample. This property allows setting a threshold at which the training efficiency of the neural network remains at the desired level.

4 Estimation of the Training Sample Quality

As a result of numerous experimental studies, a relationship was established between the set of characteristics obtained from the training sample described in the previous section and the classification probability of a deep learning model trained on the same training sample. Based on this, an assumption was made about the correspondence of the set of characteristics of the training sample to the possible recognition probability, which can be achieved by the deep learning model as a result of training on the same sample.

One of the options for establishing the relationship between the characteristics of the training sample and the probability of training is to use linear regression, in this case, multiple linear regression in a multidimensional space, since we are working with a set of a number of characteristics. This will make it possible to obtain an integral estimate, which, with some probability, will describe the quality of a particular sample.

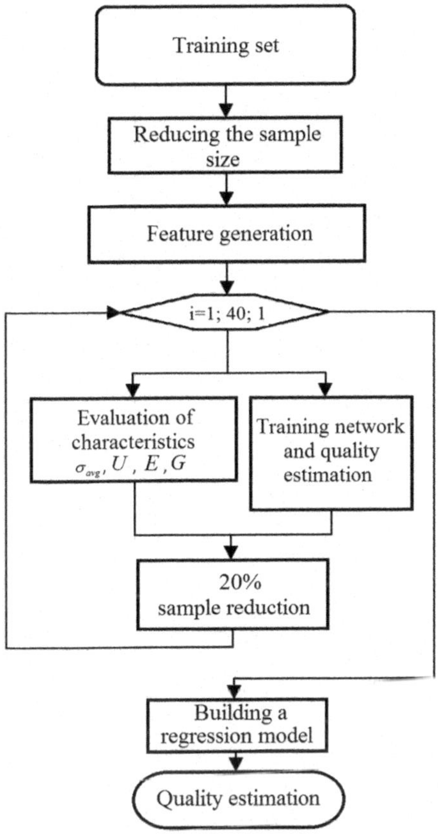

Fig. 4. Block diagram of training set quality estimation

The characteristics of the training sample are well described by a regression model of the linear type, which makes it possible to use this approach to pre-estimate the quality of an arbitrary training sample.

According to the proposed approach, the evaluation of the training sample is carried out on the basis of all sample data. After that, the training sample is gradually thinned and re-evaluated. Thinning of the sample, in the simplest case, is implemented randomly, and in the presence of additional information, can remove redundant data that make noise in the training process. And this limit is set by the proposed approach, which is presented in the form of block diagrams in Fig. 4.

To test the proposed approach, a training base for equipment error research with a total size of 95,000 data tapes was used. The test consisted of the simultaneous application of a regression model to assess the predicted result of the classification accuracy and the full cycle with the training of the deep learning model and the assessment of the actual accuracy. The test results are shown in Table 1.

Table 1. The probability of correct classification of the model trained on the basis of training samples and with the help of evaluation based on the proposed approach

The size of the training sample	Evaluation using the proposed approach	Probability of correct classification %
95000	96	94
90000	95	94
85000	96	94
80000	94	94
75000	93	93
70000	91	94
65000	89	91
60000	87	90
55000	81	83
50000	78	81
45000	77	74
40000	71	73
35000	67	65
30000	61	58
25000	55	51
20000	52	49
15000	50	46
10000	43	38
5000	37	33

As can be seen from Table 1, the evaluation using the proposed approach shows the results comparable with those corresponding to the results of the real neural network trained in the test training sample. Some difference in the results is due to the fact that the regression linear model used for evaluation was created based on the results of the classification probabilities of another model of deep learning with a different structure and number of internal parameters (Fig. 5).

Individual deep learning models give different results for the probabilities of correct classification when trained on the same training set. In turn, the proposed approach is based on the results of the model chosen as the standard. Table 2 shows the results of

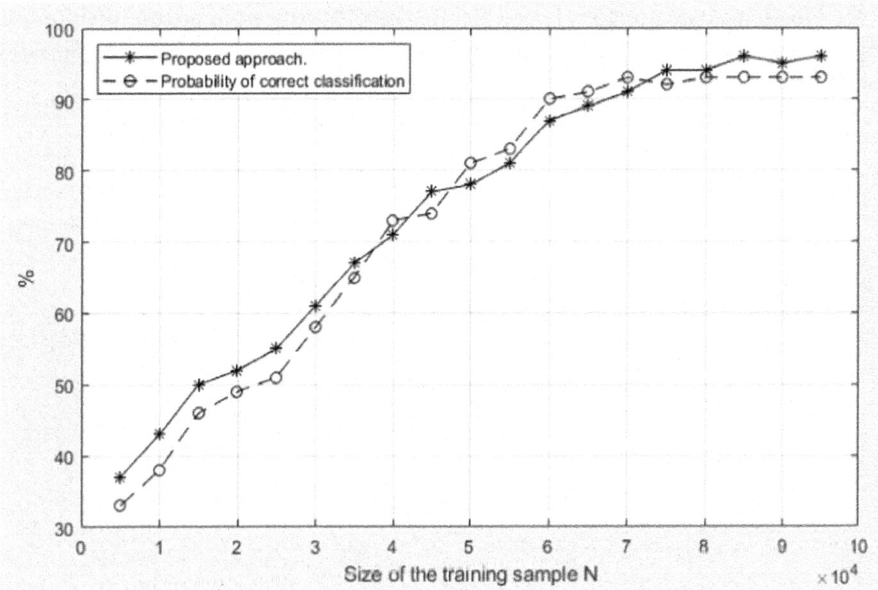

Fig. 5. Estimating the probability of correct classification depending on the size of the training sample for the proposed and standard approaches

comparing the probabilities of correct classification of such deep learning models as VGG16, Google network, Alexnet, proposed by the approach. All models studied on the basis of Cifar-10 images with different sizes of training subsamples. The results confirm the effectiveness of the proposed approach, which makes it possible to make a preliminary assessment of the training sample, and detects when its further increase does not lead to an increase in the results of the correct classification.

Table 2. The probability of correct classification of models trained on the basis of training samples Sifar-10 and by evaluating the proposed approach

The size of the training sample	Evaluation using the proposed approach	VGG16%	Alexnet%	Google network%
40000	93	92	84	98
35000	91	92	83	98
30000	88	92	83	98
25000	84	91	81	97
20000	79	90	78	97
15000	75	84	76	95
10000	70	81	63	93
5000	64	72	61	85

An important characteristic of deep learning models is the training time in the training sample. The dependences of the learning time of deep learning models on the size of the training sample are shown in Fig. 6.

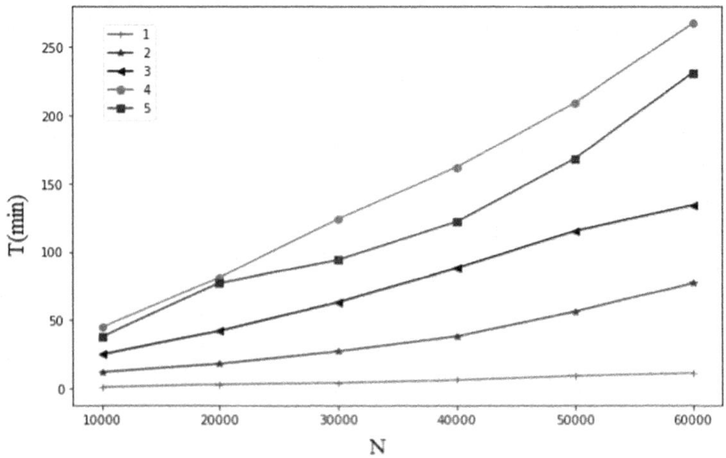

Fig. 6. Dependences of training time of deep learning models and the number of training sample images (1- proposed approach, 2- VGG net, 3- Alex net, CNN ensemble, Google net)

When evaluating the quality of the training sample, the computational complexity and estimation time are important. From Fig. 5. it is seen that the time of evaluation of the training sample by the proposed approach is more faster than direct evaluation after training, because the learning process of the deep learning model is computationally complex, and the proposed approach requires only the selection of features. The experiment used a core i5 processor, 16 GB of RAM and Nvidia GTX 1060 graphics accelerator.

5 Conclusions

A new method for estimating the quality of the training sample of large databases has been developed. The method is based on the assumption that the quality of the training sample can be represented by a set of a finite number of characteristics, each of which describes certain properties of the data. Establishing a relationship between the characteristics of the training sample and the accuracy of the classifier trained on the basis of this sample is carried out using a linear regression model.

It is shown that the quality of the training sample in the classification of large databases can be effectively determined by calculating, analyzing and comparing the following sample parameters: the average class deviation within the sample σ, its imbalance U, uniformity of coverage of the vector E, compactness of data representation in space G and particular class accuracy.

To reduce the computational complexity, it was proposed to use the method of the dimensionality reduction of the data without losing the data structure, based on minimizing the Kulbak-Leibler distance. This allowed us to move to the building of the characteristics of the training sample with much lower computational costs and a compact representation of the feature space.

Experiments obtained on different test training samples show that the method gives results commensurate with those obtained as a result of neural network training. At the same time, the time of evaluation of the training sample by the proposed method increases the speed of obtaining the result by an order of magnitude. This allows to effectively using it to pre-evaluate the training sample, making it possible to adjust its size before learning the network on large databases.

References

1. Pang, B., Nijkamp, E., Wu, Y.: Deep learning with tensor flow: a review. J. Educ. Behav. Stat. **45**, 227–248 (2019). https://doi.org/10.3102/1076998619872761
2. Rusyn, B.P., Lutsyk, O.A., Tayanov, V.A.: Upper-bound estimates for classifiers based on a dissimilarity function. Cybern. Syst. Anal. **48**(4), 592–600 (2012). https://doi.org/10.1007/s10559-012-9439-2
3. Azzopardi, G., Petkov, N.: Trainable cosfire filters for keypoint detection and pattern recognition. IEEE Trans. Pattern Anal. Intell. **35**, 490–503 (2013). https://doi.org/10.1109/TPAMI.2012.106
4. Vapnik, V.: The Nature of Statistical Learning Theory. Springer, New York (2000). https://doi.org/10.1007/978-1-4757-2440-0
5. Bishop, C.M.: Pattern Recognition and Machine Learning (Information Science and Statistics). Springer, London (2006)
6. Chen, Y., Lin, Z., Zhao, X., Wang, G., Yanfeng, G.: Deep learning-based classification of hyperspectral data. IEEE J. Sel. Topics Appl. Earth Obs. Remote Sens. **7**(6), 2094–2107 (2014). https://doi.org/10.1109/JSTARS.2014.2329330
7. Kosarevych, R., Lutsyk, O., Kapshii, O., Rusyn, B.: Random point patterns and bags of visual words for remote sensing imagery. J. Appl. Remote Sens. **13**(3) (2019). https://doi.org/10.1117/1.JRS.13.034521
8. Li, Y., Zhang, H., Xue, X., Jiang, Y., Shen, Q.: Deep learning for remote sensing image classification: a survey. WIREs Data Min. Knowl. Discovery (2018). https://doi.org/10.1002/widm.1264
9. Cheng, G., Xie, X., Han, J., Guo, L., Xia, G.: Remote sensing image scene classification meets deep learning: challenges, methods, benchmarks, and opportunities. IEEE J. Sel. Topics Appl. Earth Obs. Remote Sens. **13**, 3735–3756 (2020). https://doi.org/10.1109/JSTARS.2020.3005403
10. Hoque, M., Burks, R., Kwan, C., Li, J.: Deep learning for remote sensing image super-resolution. In: IEEE 10th Annual Ubiquitous Computing, Electronics & Mobile Communication Conference (UEMCON), pp. 286–292 (2019). https://doi.org/10.1109/UEMCON47517.2019.8993047.
11. Van Niel, T.G., McVicar, T.R., Datt, B.: On the relationship between training sample size and data dimensionality: Monte Carlo analysis of broadband multi-temporal classification. Remote Sens. Environ. **98**(4), 468–480 (2005). https://doi.org/10.1016/j.rse.2005.08.011
12. Emmert-Streib, F., Yang, Z., Feng, H., Tripathi, S., Dehmer, M.: An introductory review of deep learning for prediction models with big data. Front. Artif. Intell. **3**, 1–23 (2020). https://doi.org/10.3389/frai.2020.00004

13. Hinterstoisser, S., Lepetit, V., Wohlhart, P., Konolige, K.: On pre-trained image features and synthetic images for deep learning. In: Leal-Taixé, L., Roth, S. (eds.) ECCV 2018. LNCS, vol. 11129, pp. 682–697. Springer, Cham (2019). https://doi.org/10.1007/978-3-030-11009-3_42
14. Shorten, C., Khoshgoftaar, T.M., Furht, B.: Deep Learning applications for COVID-19. J. Big Data **8**(1), 1–54 (2021). https://doi.org/10.1186/s40537-020-00392-9
15. Dodge, S., Karam, L.: Understanding how image quality affects deep neural networks. In: IEEE 2016 Eighth International Conference on Quality of Multimedia Experience (2016). https://doi.org/10.1109/QoMEX.2016.7498955.
16. Choi, R., Coyner, A., Kalpathy-Cramer, J., Chiang, M., Campbell, J.: Introduction to machine learning, neural networks, and deep learning. Transl. Vision Sci. Technol. **9**, 14 (2020). https://doi.org/10.1167/tvst.9.2.14
17. Purushotham, S., Meng, C., Che, Z., Liu, Y.: Benchmarking deep learning models on large healthcare datasets. J. Biomed. Inform. **83**, 112–134 (2018). https://doi.org/10.1016/j.jbi.2018.04.007
18. Ma, X., Geng, J., Wang, H.: Hyperspectral image classification via contextual deep learning. EURASIP J. Image Video Process. **2015**(1), 1–12 (2015). https://doi.org/10.1186/s13640-015-0071-8
19. Subbotin, S.A.: The training set quality measures for neural network learning. Opt. Memory Neural Netw. **19**(2), 126–139 (2010). https://doi.org/10.3103/S1060992X10020037
20. Forsati, R., Moayedikia, A., Safarkhani, B.: Heuristic approach to solve feature selection problem. In: Cherifi, H., Zain, J.M., El-Qawasmeh, E. (eds.) DICTAP 2011. CCIS, vol. 167, pp. 707–717. Springer, Heidelberg (2011). https://doi.org/10.1007/978-3-642-22027-2_59
21. Huang, K., Aviyente, S.: Wavelet feature selection for image classification. IEEE Trans. Image Process. **17**(9), 1709–1720 (2008). https://doi.org/10.1109/TIP.2008.2001050
22. Muschelli, J.: ROC and AUC with a binary predictor: a potentially misleading metric. J. Classif. **37**(3), 696–708 (2019). https://doi.org/10.1007/s00357-019-09345-1
23. Belov, D., Armstrong, R.: Distributions of the Kullback-Leibler divergence with applications. Br. J. Math. Stat. Psychol. **64**(2), 291–309 (2011). https://doi.org/10.1348/000711010X522227
24. Prati, R.C., Batista, G.E.A.P.A., Monard, M.C.: Class imbalances versus class overlapping: an analysis of a learning system behavior. In: Monroy, R., Arroyo-Figueroa, G., Sucar, L.E., Sossa, H. (eds.) MICAI 2004. LNCS (LNAI), vol. 2972, pp. 312–321. Springer, Heidelberg (2004). https://doi.org/10.1007/978-3-540-24694-7_32
25. Shepperd, M., Cartwright, M.: Predicting with sparse data. In: 7th IEEE International Software Metrics Symposium, pp. 28–39 (2001). https://doi.org/10.1109/METRIC.2001.915513

Author Index

M. Klymash et al. (Eds.): *Future Intent-Based Networking*, LNEE 831, pp. 507–508, 2022.
https://doi.org/10.1007/978-3-030-92435-5

Printed by Printforce, the Netherlands